AF891051

Petroleum Geology

Hua Liu · Youlu Jiang

Petroleum Geology

中国石油大学出版社
CHINA UNIVERSITY OF PETROLEUM PRESS

Springer

Hua Liu
Department of Petroleum Geology
School of Geosciences
China University of Petroleum (East China)
Qingdao, China

Youlu Jiang
Department of Petroleum Geology
School of Geosciences
China University of Petroleum (East China)
Qingdao, China

ISBN 978-981-96-7085-7 ISBN 978-981-96-7086-4 (eBook)
https://doi.org/10.1007/978-981-96-7086-4

Jointly published with China University of Petroleum Press.
The print edition is not for sale in China (Mainland). Customers from China (Mainland) please order the print book from: China University of Petroleum Press.

© China University of Petroleum Press 2025

This work is subject to copyright. All rights are solely and exclusively licensed by the Publisher, whether the whole or part of the material is concerned, specifically the rights of reprinting, reuse of illustrations, recitation, broadcasting, reproduction on microfilms or in any other physical way, and transmission or information storage and retrieval, electronic adaptation, computer software, or by similar or dissimilar methodology now known or hereafter developed.
The use of general descriptive names, registered names, trademarks, service marks, etc. in this publication does not imply, even in the absence of a specific statement, that such names are exempt from the relevant protective laws and regulations and therefore free for general use.
The publishers, the authors, and the editors are safe to assume that the advice and information in this book are believed to be true and accurate at the date of publication. Neither the publishers nor the authors or the editors give a warranty, express or implied, with respect to the material contained herein or for any errors or omissions that may have been made. The publishers remain neutral with regard to jurisdictional claims in published maps and institutional affiliations.

This Springer imprint is published by the registered company Springer Nature Singapore Pte Ltd.
The registered company address is: 152 Beach Road, #21-01/04 Gateway East, Singapore 189721, Singapore

If disposing of this product, please recycle the paper.

Preface

Oil and natural gas geology is a branch of economic geology and is an emerging discipline summarized from the extensive practice of petroleum and natural gas exploration and exploitation. Oil and natural gas geology mainly includes four parts: the generation, migration, accumulation, and distribution of oil and gas, with the core content being the principles of oil and gas reservoir formation.

The formation of oil and gas reservoirs is an event that occurs in geological history. To economically and efficiently find these deeply buried oil and gas deposits, we need the theory of petroleum geology as a guide to understand the formation process and distribution patterns of oil and gas in the Earth's crust. How are oil and gas reservoirs formed and distributed in the Earth's crust? This is the main question that this book aims to answer.

This textbook mainly elaborates on the basic principles and distribution patterns of oil and gas generation, migration, accumulation, and reservoir formation in the Earth's crust. The content system follows the "positive sequence" of oil and gas reservoir formation, that is, it discusses around the oil and gas generation, migration, accumulation, and distribution, in the order of occurrence and development of things.

The main content of the textbook is divided into seven chapters and establishes the following system: First, it starts with understanding the basic characteristics of oil, gas, and oilfield water, then it discusses the genesis of oil and gas, as well as the source rocks, reservoirs, and cap rocks that generate, store, and seal oil and gas. Next, it focuses on the basic principles of oil and gas migration, accumulation, and reservoir formation, after that, it introduces the basic unit of oil and gas accumulation—the oil and gas reservoir and its characteristics, as well as oil and gas fields, oil and gas accumulation belts, petroliferous basins, and the distribution patterns and controlling factors of oil and gas in the Earth's crust.

Petroleum Geology (English Edition) is based on the national "Twelfth Five-Year Plan" textbook Petroleum and Natural Gas Geology and Exploration (Third Edition). The structure of the textbook has been adjusted by referring to relevant books at home and abroad, removing the part on oil and gas exploration and retaining only the part on petroleum geology theory.

This textbook is edited by Prof. Youlu Jiang and Prof. Hua Liu from China University of Petroleum (East China). Professor Youlu Jiang is responsible for the application, outline formulation, content arrangement, and organization of the textbook. The specific writing assignments are as follows: The Preface is written by Hua Liu, Chap. 1 by Jiangxiu Qu, Chap. 2 by Shengyu Yang and Xiujian Ding, Chap. 3 by Liqiang Zhang, Jingdong Liu, and Shang Xu, Chap. 4 by Hua Liu, Chaps. 5 and 6 by Youlu Jiang, Chap. 7 by Hua Liu and Jingdong Liu. The entire book is translated and typeset by Changjiang Liu, and reviewed and finalized by Hua Liu.

It is sincerely hoped that readers can put forward valuable opinions on the problems existing in this textbook, so that they can be revised in the next edition.

Qingdao, China
January 2025

Hua Liu
Youlu Jiang

Contents

Chapter 1
Basic Characteristics of Oil, Natural Gas, and Oilfield Water

Abstract Oil, natural gas, and oilfield water are all stored in the pores of underground rocks. Influenced by their sources and geological factors, their chemical compositions and physical properties vary from one region to another. Understanding the basic properties of oil, gas, and water is of great significance for clarifying the formation, migration, and accumulation patterns of oil and gas, and for developing rational exploitation and processing plans. This chapter mainly introduces the chemical composition, physical properties, and classification of petroleum, natural gas, and oilfield water. It also briefly discusses the stable isotopes in oil and gas, particularly the compositional characteristics of stable carbon isotopes.

1.1 Oil

Oil is a flammable organic mineral that is naturally generated in the pores of underground rocks, with liquid hydrocarbons as the main chemical components. The composition of oil is highly complex, with thousands of organic compounds identified, primarily hydrocarbons, and also containing varying amounts of non-hydrocarbon compounds and multiple trace elements. It is predominantly in a liquid state, often dissolving varying amounts of hydrocarbon gases, non-hydrocarbon gases, solid hydrocarbons, and non-hydrocarbon substances. Therefore, oil is essentially a mixture of various organic compounds, without a definite chemical composition or physical constants. The composition and properties of oil vary by regions. Studying the chemical composition and physical properties of oil is crucial for understanding the generation, migration, accumulation, and distribution patterns of oil, formulating development and processing plans, and evaluating the quality of oil products.

© China University of Petroleum Press 2025

H. Liu and Y. Jiang, *Petroleum Geology*,
https://doi.org/10.1007/978-981-96-7086-4_1

1.1.1 Chemical Composition of Oil

1.1.1.1 Elemental Composition of Oil

The elemental composition of oil from different regions and eras is relatively close, but there are certain differences (Table 1.1). The chemical elements that make up oil mainly include carbon, hydrogen, sulfur, nitrogen, and oxygen, with carbon and hydrogen being dominant.

(1) Carbon and Hydrogen

In terms of mass fraction, carbon generally accounts for 83–87%, and hydrogen for 10–14%. These two elements together make up over 95%, mainly in the form of hydrocarbons, constituting the main parts of oil. Due to the significant variation in the content of heteroatoms in oil, it is not easy to compare oil only based on its carbon or hydrogen content. The hydrogen-to-carbon atomic ratio [n(H)/n(C)] is often used to represent their proportion, which can be an important parameter reflecting the chemical composition of oil.

(2) Sulfur, Nitrogen, and Oxygen

Table 1.1 Elemental composition of some domestic and foreign oils (according to Zhang et al., 1999)

Production area		Elemental composition (%)				
		C	H	S	N	O
China	Daqing Oilfield (Sartu mixed oil)	85.74	13.31	0.11	0.15	0.69
	Shengli Oilfield (101 mixed oil)	86.26	12.20	0.80	0.41	
	Gudao Area of Shengli Oilfield	84.24	11.74	2.20	0.47	
	Dagang Oilfield	85.67	13.40	0.12	0.23	
	Jianghan Oilfield (mixed oil)	83.00	12.81	2.09	0.47	1.63
	Karamay Oilfield (mixed oil)	86.13	13.30	0.04	0.25	0.28
Former Soviet Union	Yarek Sue	80.61	10.36	1.05		8.97
	Ukraine	84.60	14.00	0.14	1.25	1.25
	Old Grolzny	86.42	12.62	0.32		0.68
	Kara-Brak	87.77	12.37			0.46
America	Ventura (California)	84.00	12.7	0.4	1.70	1.20
	Colinga (California)	86.40	11.7	0.60		
	Beaumont (Texas)	85.70	11.00	0.70	2.61	
	Kansas	84.20	13.00	1.60	0.45	0.45

Sulfur, nitrogen, and oxygen in oil mainly exist in the form of compounds. The total content of these three elements and trace elements is generally only 1–4%, but it can reach 3–7% due to increased sulfur content.

a. Sulfur Content

The sulfur content in oil varies greatly. For example, the Renqiu Oilfield in China averages 0.33–0.43%, while the Karamay Oilfield averages 0.05%. However, the sulfur content in some oilfields can be as high as 4–5%, such as Mexican oil with a sulfur content of 3.6–5.3%. Based on the sulfur content, oil extracted to the surface is typically divided into high sulfur (greater than 1% sulfur content) and low sulfur (less than 1% sulfur content) categories. Some also use a three-tier method, classifying oil into high sulfur oil (greater than 2% sulfur content), sulfur-containing oil (0.5–2% sulfur content), and low sulfur oil (less than 0.5% sulfur content).

The sulfur content in oil has environmental implications. Oil from marine, nearshore lacustrine, and salt lake facies in brackish to saline sedimentary strata generally has a higher sulfur content, typically above 1%. In contrast, oil from inland freshwater lake sedimentary strata generally has a lower sulfur content, usually less than 1%.

Sulfur in oil is a harmful impurity that can form compounds such as hydrogen sulfide (H_2S), iron sulfide (FeS), sulfurous acid (H_2SO_3), mercaptans $[RS]_2Fe$, and even sulfuric acid (H_2SO_4), which are highly corrosive to metal equipment such as machinery, pipelines, oil tanks, and distillation towers. Therefore, the sulfur content is an important indicator for evaluating the quality of oil.

b. Nitrogen and Oxygen Content

The nitrogen content in oil is very low, generally less than 0.2%, with most being only a few parts per ten thousand to a few parts per thousand. However, some samples may have a nitrogen content of over 0.5%, with the highest reaching 1.7% (oil from the Ventura Basin in the United States). A nitrogen content of 0.25% is often used as the boundary between low-nitrogen and high-nitrogen crude oils.

The oxygen content in oil is relatively low and is not directly measured. It is commonly estimated by the difference method.

(3) Trace Elements

In addition to the 5 major elements, more than 50 trace elements have been identified through the analysis of oil ash, with their content varying from a few parts per ten thousand to a few parts per hundred thousand. The following 33 trace elements are listed in order of their abundance and commonality: Fe, Ca, Mg, Si, Al, V, Ni, Cu, Sb, Mn, Sr, Ba, B, Co, Zn, Mo, Pb, Sn, Na, K, P, Li, Cl, Bi, Be, Ge, Ag, As, Gd, Au, Ti, Cr, and Cd.

The elemental composition of oil is very similar to that of organic matter in the natural world, which is one of the evidence for the organic origin of oil. In particular, vanadium (V) and nickel (Ni) are two trace elements that are widely distributed and have genetic significance. Analyzing the content and ratio (V/Ni) of vanadium and nickel in the ash of oil can help distinguish oil of marine or terrestrial origin.

1.1.1.2 Compound Composition of Oil

The compound composition of oil is mainly hydrocarbons and non-hydrocarbons.

(1) Hydrocarbon Compounds

Based on structural characteristics, hydrocarbon compounds in oil can be divided into three categories: alkanes, cycloalkanes, and aromatic hydrocarbons. Alkanes and cycloalkanes are saturated hydrocarbons, while aromatic hydrocarbons are unsaturated hydrocarbons.

a. Alkanes

Also known as aliphatic hydrocarbons or paraffins, they have the general formula C_nH_{2n+2} and are chemically less reactive. In fact, the term "paraffin" itself implies "little affinity". The structural characteristic of alkanes is that carbon atoms are connected by single bonds (C–C), arranged in a straight chain. Those without branches are called normal alkanes; those with branches are called isoalkanes. At room temperature and atmospheric pressure, alkanes with 1–4 carbon atoms (C_1–C_4) are gaseous; straight-chain alkanes with 5–16 carbon atoms (C_5–C_{16}) are liquid; and high molecular weight alkanes with more than 17 carbon atoms (C_{17}) are solid. The density, melting point, and boiling point of alkanes all increase with increasing molecular weight. The density of all alkanes is less than 1 and they are almost insoluble in water (except for gaseous hydrocarbons).

Normal Alkanes

Normal alkanes from C_1 to C_{45} have been identified in oil, with most having fewer than 35 carbon atoms. However, a few normal alkanes with more than 200 carbon atoms have also been found in the oil from the Altamont and Bluebell oil fields in the Uinta Basin, Utah, USA. Normal alkanes generally account for 15–25% of the mass of oil, over 30% in light crude oil, and less than 15% in heavy crude oil.

In oil, normal alkanes with different carbon atom numbers are continuously distributed within a certain range. The distribution curves of normal alkanes in different types of oil have different characteristics (Fig. 1.1), where the carbon number corresponding to the maximum value on each curve is the main peak carbon of that curve.

The characteristics of the normal alkane curve are closely related to the type of original organic matter, oil formation environment, and the organic matter maturity. Generally, oil formed from terrestrial organic matter has a higher content of high carbon number (above C_{22}) normal alkanes, while oil formed from marine lower planktonic organisms (bacteria, algae) has a higher content of low carbon number (below C_{22}) normal alkanes. Oil with a higher degree of thermal evolution, older age, and greater burial depth tends to have a higher content of low carbon number normal alkanes; conversely, oil with a lower degree of thermal evolution has a higher carbon number for normal alkanes. In addition, oil that has been strongly degraded by microorganisms often has selectively degraded normal alkanes, generally with lower content and fewer low carbon number normal alkanes (Fig. 1.2).

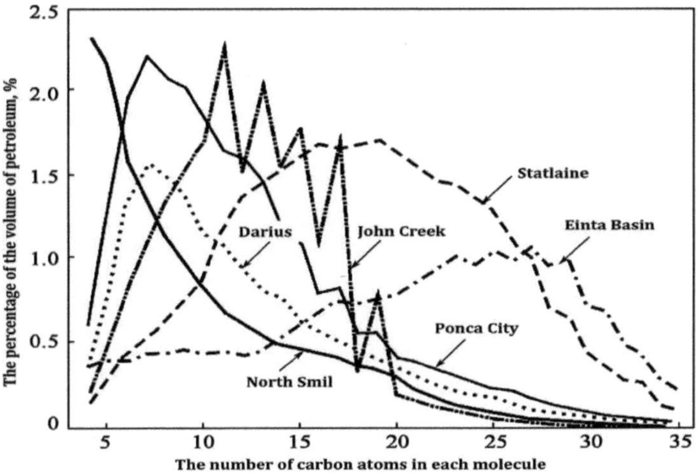

Fig. 1.1 Distribution curves of n-alkanes in different types of petroleum (according to Martin,1963)

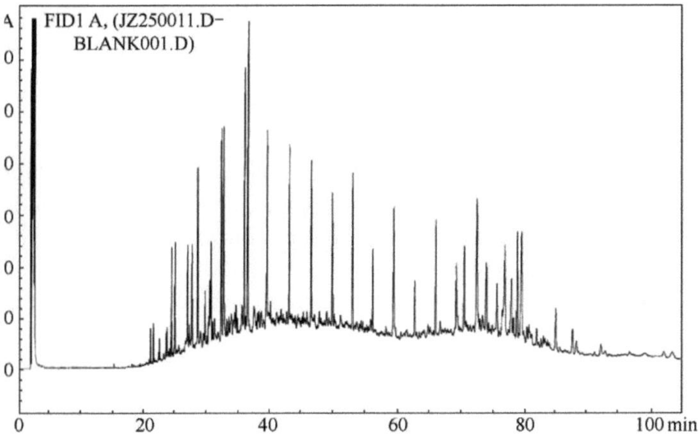

Fig. 1.2 Total hydrocarbon chromatogram of biodegradable oil

Isoalkanes

Isoalkanes in oil are mainly less than C_{10}, and the most important isoalkanes in the medium relative molecular mass range are the homologs of isoprenoids (C_9–C_{25}). The structural characteristic of regular isoprenoids is that there is a methyl branch on every four linear carbon atoms (Fig. 1.3). They are the second most common compounds in oil after normal alkanes.

Because the types and amounts of isoprene-like hydrocarbons contained in homologous oil are similar and come directly from biological organisms, they can be used for oil source comparison and are therefore referred to as biomarker

2,6,10,14-Tetramethylhexadecane (Phytane) :

2,6,10,14-Tetramethylpentadecane (Pristane) :

2,6,10-Trimethylpentadecane (Nor-pristane) :

2,6,10-Trimethyltridecane (Isohexadecane) :

2,6,10-Trimethyldodecane (Farnesane) :

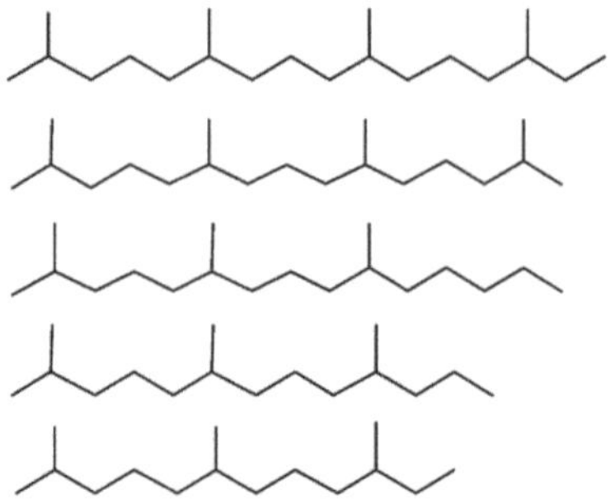

Fig. 1.3 Schematic diagram of common isoprene-like hydrocarbon structures

compounds. Biomarker compounds refer to organic compounds that originate from biological bodies, essentially retain the original molecular structure, and record the special molecular structural information of the original oil source material. These compounds are also known as "molecular fossils," "geochemical fossils," and "fingerprint compounds." Their structural stability plays an important role in identifying the source of organic matter, determining the original depositional environment, oil-source correlation, and tracing the direction of oil and gas migration.

Among the isoprenoids, pristane (Pr) and phytane (Ph) are the most commonly biomarker compounds, mainly derived from the phytyl side chains of chlorophyll. The relative abundance of pristane and phytane can not only reflect the redox conditions of the original depositional environment but is also related to the pH of the aqueous medium. Generally, under strong reducing conditions, phytane is predominantly formed, while under weak oxidizing conditions, pristane is predominantly formed; an acidic aqueous medium environment favors the formation of pristane, while a slightly alkaline aqueous medium environment favors the formation of phytane (Table 1.2).

b. Cycloalkanes

Cycloalkanes are saturated hydrocarbons containing cyclic structures, composed of multiple methylene groups forming rings. Depending on the number of carbon atoms forming the ring, they can be referred as three-membered, four-membered, five-membered rings, etc. Cycloalkanes can be classified according to the number

Table 1.2 Variations in Pr/Ph of Oils formed in different sedimentary facies (according to Mei and Liu, 1980)

Sedimentary facies	Aqueous condition	Pr/Ph	Type of oil
Saline deep lacustrine facies	Strong reduction	0.2–0.8	Predominance of phytane
Freshwater-brackish deep lacustrine facies	Reduction	0.8–2.8	Phytane equilibrium
Limnetic facies	Weak oxidation-weak reduction	2.8–4.0	Predominance of pristane

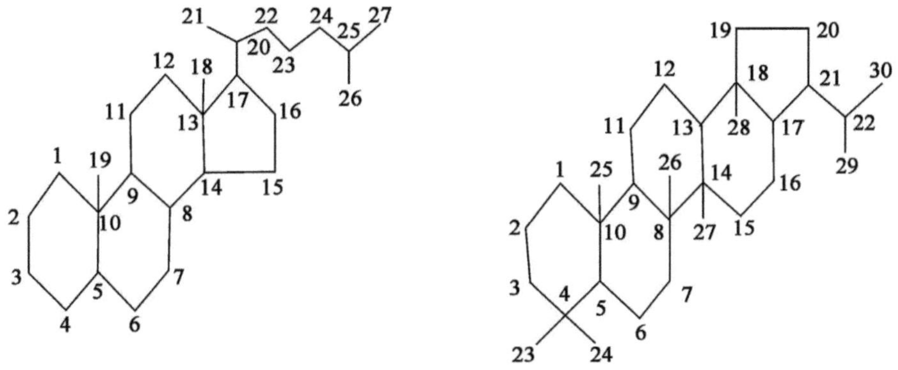

Fig. 1.4 Structure diagram of polycyclic cycloalkanes in oil

of carbon rings in the molecule, such as monocyclic, bicyclic, tricyclic, and polycyclic cycloalkanes. The cycloalkanes in oil are mostly five-membered or six-membered rings and their derivatives, with monocyclic and bicyclic cycloalkanes being predominant. Among the polycyclic cycloalkanes, four-membered and five-membered cycloalkanes are the most important, and their structures are directly related to the tetracyclic steroid compounds and pentacyclic triterpenes in biological organisms. They are important markers for the organic origin of oil and are widely used in the analysis of hydrocarbon source rock maturity and oil-source correlation (Fig. 1.4).

The content of polycyclic cycloalkanes in oil decreases significantly with increasing maturity, with 1–2 ring cycloalkanes being predominant in highly mature oil. Since all the valences of carbon atoms are saturated, cycloalkanes are as stable as alkanes. The density, melting point, and boiling point of cycloalkanes are higher than those of alkanes with the same number of carbon atoms, but the relative density is still less than 1.

c. Aromatic Hydrocarbons

Aromatic hydrocarbons (shortened as aromatics) refer to compounds containing benzene ring structures. Based on structure, aromatic hydrocarbons can be divided into monocyclic aromatics, polycyclic aromatics, and polynuclear aromatics (Fig. 1.5). The most abundant aromatics in oil are benzene, naphthalene, and phenanthrene.

Naphthenoaromatic hydrocarbons are special aromatic hydrocarbons that often contain one or several condensed aromatic rings combined with saturated rings and aliphatic chains. Among them, the four-ring and five-ring naphthenoaromatic hydrocarbons are the most important ones. They are mostly related to the structures of steroid compounds and terpenes, and are biomarker compounds.

With the increase of oil maturity, the series of aromatic hydrocarbons evolve towards lower ring structures. Monocyclic aromatics are insoluble in water but soluble in organic solvents such as gasoline, ether, and ethanol. They are toxic matter

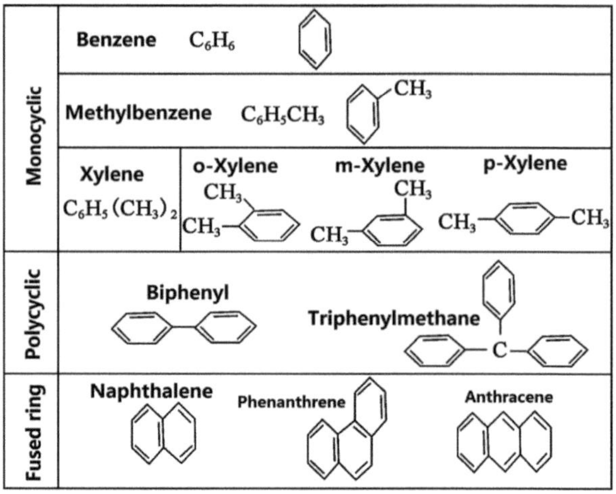

Fig. 1.5 The basic structures of aromatic hydrocarbons

with special odor, and generally have a relative density of 0.86–0.9, which is lighter than water.

(2) Non-hydrocarbon Compounds

Non-hydrocarbon compounds in oil refer to the compounds that contain heteroatoms such as sulfur, nitrogen, and oxygen in addition to carbon and hydrogen atoms in their molecular structure. These non-hydrocarbon compounds are mainly concentrated in the heavy fractions with high boiling points, with a small total amount but a wide variety of types.

a. Sulphocompounds

Organic sulfur compounds in oil, apart from elemental sulfur, are mainly divided into three major categories: thiols, sulfides, and thiophenes.

Thiol compounds refer to sulphocompounds with sulfur radical (–SH–), formed by the substitution of one hydrogen atom in hydrogen sulfide with an alkyl or cycloalkyl group. Thiols are not abundant in oil and are mainly distributed in the lower fractions, with their contents decrease as the boiling point of the fraction increase.

Sulfide compounds refer to sulphocompounds with a sulfur ether bond (–S–), formed by the substitution of two hydrogen atoms in a sulfur compound with alkyl, cycloalkyl, or aromatic groups. Sulfides have higher thermal stability, and their content increases with the boiling point of the fraction, being relatively abundant in oil and generally concentrated in the middle fractions.

Thiophene compounds are unsaturated five-membered ring compounds containing one sulfur atom and four carbon atoms, with good thermal stability. Thiophenes themselves are rare, but benzothiophene, dibenzothiophene, and benzonaphthothiophene are important components of high sulfur oil. In addition, in oil with higher content of resins and asphaltenes, thiophenic derivatives are particularly abundant.

b. Nitrogenous Compounds

The content of nitrogen in oil is generally low and mainly exists in the form of aromatic heterocycles. Nitrogenous compounds in oil can be divided into basic and neutral (non-basic) categories. Basic nitrogenous compounds mainly include homologues of pyridine, quinoline, isoquinoline, and acridine, as well as porphyrins. Neutral nitrogenous compounds include pyrroles, indoles, carbazole homologues, and amides. Most nitrogenous compounds are found in high fractions.

Porphyrins in oil exist in the form of metal complexes with vanadium and nickel (Fig. 1.6), and their content varies greatly. A significant portion of oil does not contain or only has trace amounts of porphyrin compounds. Taking the Dongying Depression as an example, samples from the edge of the depression almost do not contain porphyrins, while the porphyrin content in the central part of the depression is nearly 2,000 mg/l. This is related to the sedimentary environment and burial depth. Generally, oil formed in the Mesozoic and Cenozoic has more porphyrins, while oil formed in the Paleozoic has very low or no porphyrins. This may be related to the poor stability of porphyrins, which can undergo ring-breaking lytic reactions and be destroyed and decomposed under high temperatures (>250 °C) or oxidative conditions. Therefore, the presence of porphyrins proves that the formation temperature and temperature resistance of oil are not too high.

Porphyrins, as the first biomarker compound isolated from oil, have a structure very similar to that of animal hemoglobin and plant chlorophyll, showing a clear kinship relationship, and provide strong evidence for the organic origin theory of oil.

c. Oxy-Compound

Oxy-compounds can be divided into acidic and neutral categories. The former includes cycloalkane carboxylic acids, fatty acids, and phenols, collectively known as petroleum acids; the latter includes aldehydes, ketones, etc., and are present in very

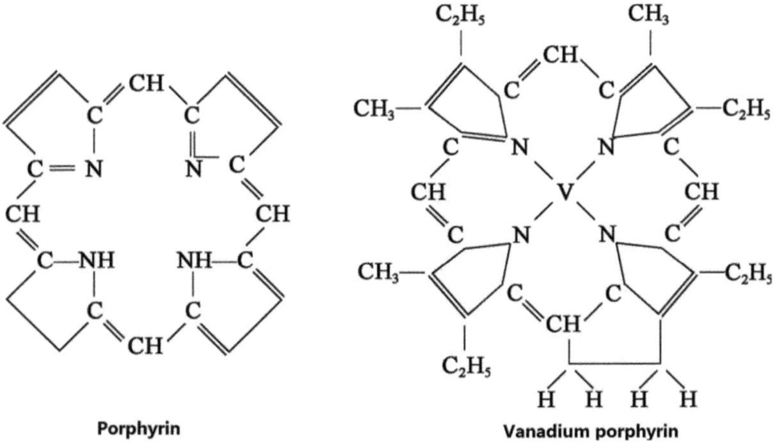

Fig. 1.6 The structural formulas of porphyrin and vanadyl porphyrin

small amounts. Among the petroleum acids, cycloalkane carboxylic acid is the most important one, accounting for about 90% of the petroleum acids, and its content in oil is mostly below 1%. The solubility of cycloalkane carboxylic acid in water is very low. High molecular weight cycloalkane carboxylic acids are essentially insoluble in water but are easily soluble in petroleum hydrocarbons. Cycloalkane carboxylic acid can easily form various salts, among which the alkali metal salts of cycloalkane carboxylic acid can dissolve well in water, and are often found in groundwater in contact with oil, which can be used as a sign for oil exploration.

1.1.1.3 Fractional Composition of Oil

Since oil is a mixture of various organic compounds, its boiling point varies largely, from normal temperature to over 500 °C. Therefore, it is necessary to distill oil when studying or utilizing it. Distillation refers to the separation of oil into several parts with different boiling point ranges (i.e., fractions) according to the different boiling points of the compounds in the oil through heating distillation. The fractions can be unequally or equally spaced. Practically, the weight or volume percentage of the fraction distilled within a certain temperature range (fraction) is often used to represent the composition of oil, known as the fractional composition of oil.

The distillation process of petroleum is shown in Fig. 1.7. The fractions are often named after oil products such as gasoline, kerosene, and diesel, but they are not actual oil products and require further processing after distillation.

The composition of different fractions in oil is extremely different. Generally, the low-boiling light fractions are mainly composed of low molecular weight alkanes and cycloalkanes; the middle fractions are dominated by alkanes and cycloalkanes

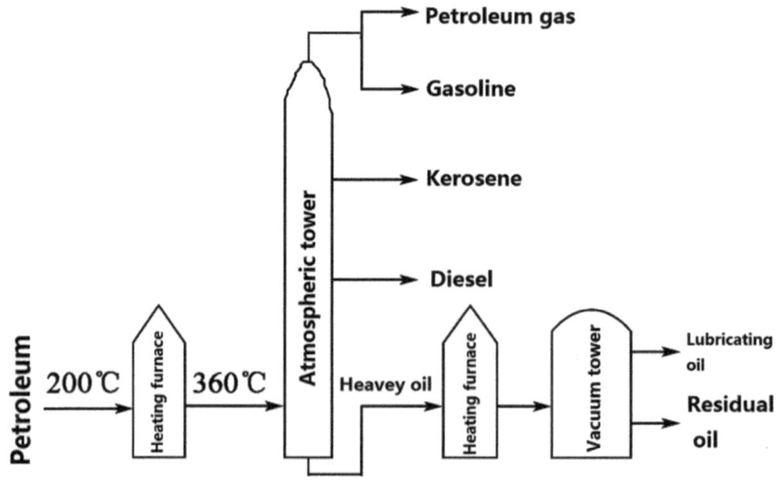

Fig. 1.7 Schematic diagram of the oil distillation process

Table 1.3 Fractional composition of oil

Fraction name		Boiling point (°C)	Carbon atom number	Compounds
Light fractions	Oil gas	<35	1–4	Alkanes, cycloalkanes
	Gasoline	50–200	5–12	
Middle fractions	Kerosene	130–250	12–14	Alkanes, cycloalkanes, with aromatic hydrocarbons and S, N, O compounds
	Diesel	180–350	14–18	
Heavy fractions	Lubricating Oil	350–500	18–20	High carbon number, high molecular weight cycloalkanes, aromatic hydrocarbons, and S, N, O compounds
	Residue	>500	>20	

with medium molecular weight and higher carbon numbers, and contain a certain amount of aromatic hydrocarbons and naphthenoaromatic hydrocarbons, as well as a small amount of sulfur, nitrogen, and oxygen compounds; while the heavy fractions are composed of high molecular weight cycloalkanes, aromatic hydrocarbons, naphthenoaromatic hydrocarbons, and sulfur, nitrogen, and oxygen compounds (Table 1.3).

Comparing the fractional composition of crude oils from different regions at home and abroad (Table 1.4), the crude oils from the main oil areas in our country generally have a heavier fraction, especially more residue content (>500 °C) and less gasoline fraction (<200 °C).

It should be noted that the boiling point ranges (distillation ranges) used by refineries in different regions are not completely consistent, resulting in different hydrocarbon compositions in the obtained fractions. For example, the hydrocarbon composition of kerosene with a distillation range of 150–300 °C mainly consists of hydrocarbons with 10–15 carbon numbers, while the hydrocarbon composition of kerosene with a distillation range of 190–260 °C mainly consists of hydrocarbons with 12–14 carbon numbers. In addition, the original properties of oil from different oil fields vary greatly, leading to significant differences in the composition of the separated fractions and the hydrocarbon composition within the same fraction.

1.1.1.4 Group Composition of Oil

Oil contains many individual compounds, but the representation is too complicated, so in practical applications, we often use group composition. The so-called "group" refers to a class of compounds with similar chemical structures. The mixtures obtained by selectively separating different types of compounds in oil using different organic solvents, which have similar physical and chemical properties, are the group composition of oil.

Considering the strong volatility of the light fractions and the significant human errors due to different storage and transportation conditions, the light fractions with

Table 1.4 Fractional composition of some domestic and foreign crude oils (according to Xu and Yang, 2009)

Production area	Fractional composition (mass fraction) (%)			
	Initial distillation point ~200 °C	200–350 °C	350–500 °C	>500 °C
Daqing	11.5	19.7	26.0	42.8
Shengli	7.6	17.5	27.5	47.4
Gudao	6.1	14.9	27.2	51.8
Liaohe	9.4	21.5	29.2	39.9
Huabei	6.1	19.9	34.9	39.1
Zhongyuan	19.4	25.1	23.2	32.3
Xinjiang (pipeline oil)	15.4	26.0	29.9	29.7
Xinjiang (Kuerle)	19.6	31.1	26.1	23.2
Xinjiang (Jiuqu)	2.3	18.9	28.9	49.9
Shanjiasi	1.2	12.2	18.3	68.3
Saudi (light)	23.3	26.3	25.1	25.3
Saudi (light-heavy mixture)	20.7	24.5	23.2	31.6
UAE (Murban)	31.5	30.6	23.2	14.7
UK (North Sea)	29.0	27.6	25.4	18.0
Indonesia (Minas)	11.9	30.2	24.8	33.1

boiling points below 210 °C should be removed before component separation, and only the fractions with boiling points above 210 °C (i.e., topped crude oil) should be used for component separation (Fig. 1.8). After this process, oil can be divided into four components: saturated hydrocarbons, aromatic hydrocarbons, resins (also known as non-hydrocarbons in experimental analysis reports), and asphaltenes.

It should be noted that the selective dissolution of organic solvents is not absolute and there is always a certain degree of mixed dissolution effect. When flushing with n-heptane or n-hexane on alumina or silica gel columns, besides saturated hydrocarbons, a small amount of unsaturated hydrocarbons is also eluted; and when flushing with benzene, besides aromatic hydrocarbons, resins are also eluted.

1.1.2 Geochemical Classification of Oil

Considering different research purposes, people classify oil from various perspectives. For example, based on the source environment, oil is divided into marine oil and terrestrial oil; according to the maturity of organic matter, oil is divided into immature oil, mature oil, and over-mature oil. Petroleum geologists pay more attention to the relationship between the composition of oil and its source rocks and

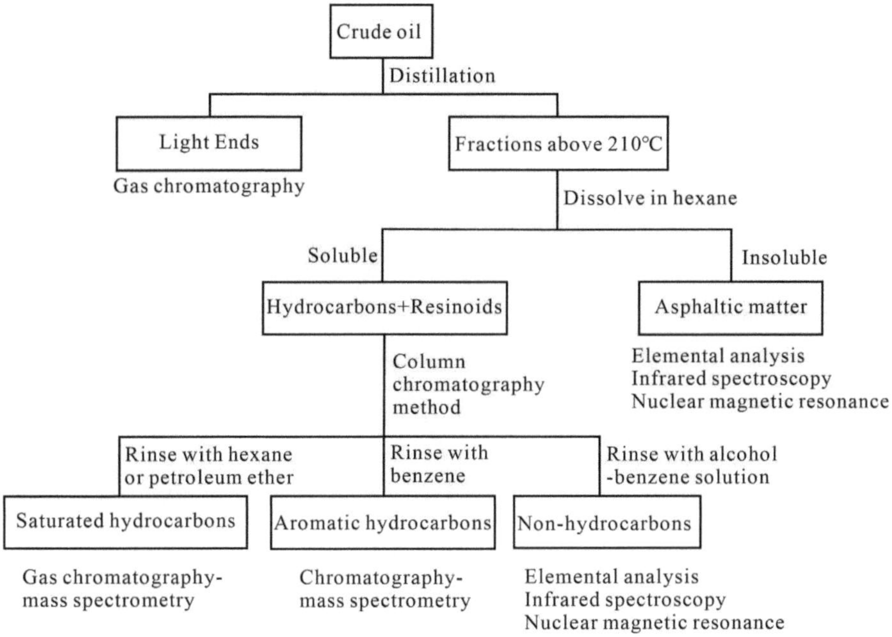

Fig. 1.8 Schematic diagram of crude oil component separation process (according to Chen, 1994, with modifications)

their evolutionary effects. The representative classification scheme is the geochemical classification proposed by Tissot and Welte (1978). This classification uses a triangular diagram with the relative content of alkanes (paraffins), cycloalkanes, and aromatic hydrocarbons + sulfur, nitrogen, oxygen compounds as the three end members of the triangular diagram, and also refers to the sulfur content in oil; all data refer to the analysis data of the oil fraction with a boiling point greater than 210 °C under atmospheric pressure. The classification scheme and results are shown in Fig. 1.9 and Table 1.5.

Paraffin Type consists of light oils and a certain amount of high wax, high boiling point oil, rich in high molecular weight normal alkanes, with resin and asphaltene content below 10%. The relative density is generally less than 0.85, and the viscosity is also lower. Paraffin-Cycloalkane Type generally has a relative content of 5–15% for resins and asphaltenes, 25–40% for aromatic hydrocarbons, and the viscosity and density are generally higher than those of paraffin type oil. Cycloalkane Type is less common and is an immature oil, or the product of biodegradation of paraffin type and paraffin-cycloalkane type.

Aromatic-Intermediate Type has a relative content of 10–30% for resins and asphaltenes, 40–70% for aromatic hydrocarbons, and the relative density is generally higher than 0.85. Aromatic-Cycloalkane Type and Aromatic-Asphalt Type are both secondary altered oil, with heavy and viscous oil quality, and the relative content of resins and asphaltenes can be as high as more than 25%.

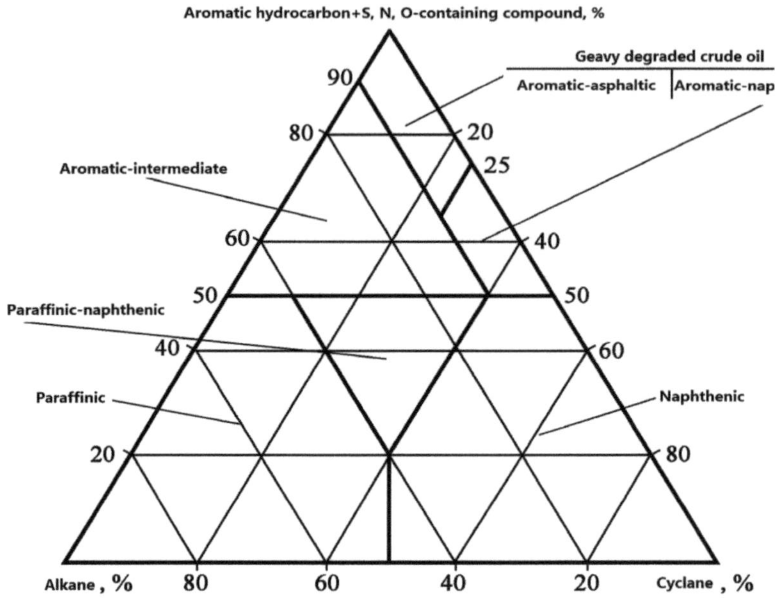

Fig. 1.9 Triangular diagram representing six types of oil (according to Tissot & Welte, 1978)

Table 1.5 Tissot and Welte's geochemical classification scheme for oil

Hydrocarbon component content		Sulfur content (%)	Type of oil
S > 50% AA > 50%	P > 40%, and P > N	<1	Paraffin type
	P ≤ 40%, and N ≤ 40%		Paraffin-cycloalkane type
	P < N, and N > 40%		Cycloalkane type
S ≤ 50% AA ≥ 50%	P > 10%	>1	Aromatic-intermediate type
	P ≤ 10% and N > 25%	<1	Aromatic-cycloalkane type
	P ≤ 10%且N ≤ 25%	>1	Aromatic-asphalt type

Note S = saturated hydrocarbons; P = paraffins (paraffins); N = cycloalkanes, then S = P + N; Let AA = aromatics + N, S, O compounds (resins, asphaltenes)

In terms of frequency of occurrence, Paraffin-Cycloalkane Type, Aromatic-Intermediate Type, and Paraffin Type are the three most common types of oil.

The distribution of marine and terrestrial oil on the oil classification triangular diagram is shown in Fig. 1.10. Their chemical compositions have obvious differences, which are specifically as follows:

(1) Marine oil is mainly Aromatic-Intermediate Type and Paraffin-Cycloalkane Type, with saturated hydrocarbons accounting for 25–70% of the oil, and aromatic hydrocarbons accounting for 25–60% of the total hydrocarbons. Terrestrial oil is mainly Paraffin Type, partly Paraffin-Cycloalkane Type,

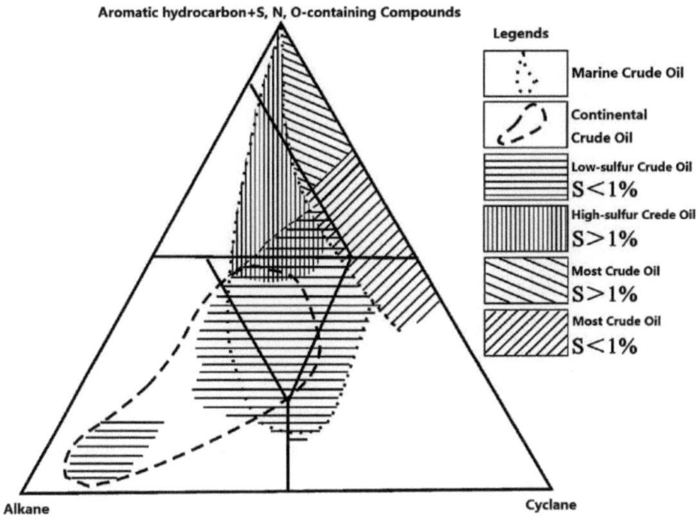

Fig. 1.10 Distribution of marine and terrestrial oil on the oil classification triangular diagram (according to Tissot & Welte, 1978, with modifications)

with saturated hydrocarbons accounting for 60–90% of the oil, and aromatic hydrocarbons accounting for 10–20% of the total hydrocarbons.

(2) High wax content is one of the basic characteristics of terrestrial oil. According to the analysis data of terrestrial oil in China, the wax content is generally greater than 5%, usually 10–30%, and some can reach more than 40%. However, the wax content in marine oil is less than 5%, generally only 0.5–3%.

(3) Marine oil is generally high sulfur oil, especially oil in marine carbonate and evaporite rock series, with higher sulfur content. Terrestrial oil is generally low sulfur oil, but some oil in salt lakes or evaporite rock series can also be high sulfur oil.

(4) The content and ratio of trace elements vanadium and nickel are one of the important signs to distinguish marine and terrestrial oil. Marine oil has high content of vanadium and nickel, and V/Ni is greater than 1; while terrestrial oil has lower content of vanadium and nickel, and V/Ni is less than 1.

(5) There is also a significant difference in the carbon stable isotope composition between marine and terrestrial oil. In terms of C_{13}/C_{12} ratio, marine oil is generally higher than terrestrial oil.

1.1.3 Physical Properties of Oil

The physical properties of oil depend on its chemical composition. Oil from different regions, different strata, and even the same stratum in different structural positions may have significant differences in physical properties.

(1) Color

The color of oil varies greatly, ranging from colorless, light yellow, yellow–brown, dark brown, dark green to black. For example, some wells in the Sichuan Huangguashan Oilfield and Dagang Banqiao Oilfield produce light yellow oil, while the oil produced by the Karamay Oilfield is brown to black, and the oil from the Daqing, Shengli, and Yumen Oilfields is mainly black.

The color of oil is related to the content of resins and asphaltenes in its components. The higher the content of resins and asphaltenes, the darker the color. Most of the oil presents different shades of dark color.

(2) Density and Relative Density

The density of oil refers to the mass of oil per unit volume, with units of g/cm^3 or kg/m^3. Since the density of oil decreases with increasing temperature, the temperature should be indicated when measuring the density. In our country, the density of oil at 20 °C is usually defined as the standard density, denoted as ρ_{20}.

The relative density of oil extracted to the surface is generally referred to as the weight ratio of unit volume of crude oil at 1 atm and 20 °C to unit volume of pure water at 4 °C, denoted as $d_4{}^{20}$. The relative density of oil varies greatly, generally between 0.75 and 1.00. Oil with a relative density greater than 1 or less than 0.75 has also been found in nature. For example, Iran has produced oil with a relative density of 1.016, while the former Soviet Union Surakhan produced oil with a relative density of 0.71. Table 1.6 shows the physical property parameters of crude oil from some domestic oilfields.

In the United States, the API degree (American Petroleum Institute) is commonly used to represent the relative density of oil, while in Western Europe, the Baumé degree is generally used to represent the relative density of oil.

$$\text{API} = \frac{141.5}{d_4^{60F}} - 131.5, \text{Baumé degree} = \frac{140}{d_4^{60F}} - 130, \ (60\,°\text{F} = 15.5\,°\text{C})$$

d_4^{60F} is the mass ratio of unit volume of crude oil to unit volume of pure water at 1 atmosphere and 60 °F. Therefore, the API degree and Baumé degree are exactly the opposite in numerical value.

The density of oil mainly depends on its chemical composition. Oil with high molecular components or high content of resins and asphaltenes has a larger density. Generally, oil with a lower density and lighter color is often of paraffinic nature, containing more oil, and can produce more gasoline and lubricating oil after processing. Oil with a higher density and darker color is rich in high molecular weight asphaltenes.

In addition, the density of oil is also related to the amount of dissolved gas, temperature, pressure, and other factors. When other conditions remain unchanged, if there is no gas cap in the oil reservoir, the density of oil decreases with increasing temperature and increases with increasing pressure. When there is a gas cap in the oil reservoir, as the pressure increases, the amount of dissolved gas in the oil gradually

Table 1.6 Physical property parameters of crude oil from some domestic oilfields

Crude oil source	Stratum	Density (kg/m^3, 20 °C)	Viscosity (mm^2/s, 50 °C)	Freezing point (°C)	Wax content (%)	Asphaltenes (%)	Asphaltenes (wt%)
Chenjiazhuang Oilfield	Ngx	946–1,084.4	761–107,178	−13 to 35	1.1–31.5	3.92–19.41	20.36–59.2
Shengli Zhengjia Oilfield	$Es_{1\text{-}3}$	930.1–1,046.8	101–89,107	−11 to 52	1.12–12.77	9.1–18.82	33.2–41.15
BohaiJZ25-1S	Es	756.4–896.6	1.04–12.24	−35 to 18	0.4–12.0	0.16–5.65	1.11–5.84
BohaiLD5-2N	N	1,003.8–1,011.4	33,595–74,462	22–33	0.88–2.23	13.39–27.8	14.04–23.3
Liaohe Yulou Oilfield	Es	804–883.3	2.39–30	18–39	3.93–29.72		4.71–19.8
Liaohe Yuehai Oilfield	Es	858.5–1,037.8	2.15–23,660	−24 to 28	2.28–25.05	6.48–41.82	
Xinjiangshi West Bulge	P_{1j}	791.8–813.7	2.07–3.34	0–10	2.54–15.19		

increases, causing the volume of oil to increase and the density to decrease, until it reaches the saturation pressure. After that, the amount of dissolved gas can no longer increase, and if the pressure continues to increase, the volume of oil begins to shrink, and the density also increases.

(3) Viscosity

When oil is subjected to force and flows, there is an internal friction force between its molecules that hinders the relative movement of the molecules. This property is called the viscosity of oil, and its magnitude is measured by viscosity (μ). The higher the viscosity of oil, the more difficult it is to flow. The viscosity of oil is a very important physical property that directly affects the extraction, storage, transportation, and refining.

The viscosity of oil can be expressed in three ways: dynamic viscosity, kinematic viscosity, and conditional viscosity.

Dynamic viscosity, also known as absolute viscosity, is defined as the condition where a shear force of 1 Pascal (Pa) is applied to a liquid, causing it to flow at a constant relative velocity between two layers of the liquid that are 1 m apart and have an area of 1 m^2. If the flow velocity is exactly 1 m/s, then the viscosity of the liquid is 1 Pa s. The dynamic viscosity of oil varies greatly. For example, the viscosity of Cretaceous oil from Daqing Oilfield is 190–220 mPa s, the viscosity of the middle to upper Proterozoic oil from Renqiu Oilfield is 530–840 mPa s, and the viscosity of the Triassic oil from Karamay Oilfield is 500 mPa s.

The magnitude of the dynamic viscosity of oil depends on its chemical composition and the external temperature and pressure conditions. Oil with a high content of small molecules such as alkanes and cycloalkanes has low dynamic viscosity, while oil with a high content of paraffin, resins, and asphaltenes has high viscosity. As the temperature increases, the dynamic viscosity decreases, so oil flows more easily underground than on the surface. At a depth of 1500–1700 m, the dynamic viscosity value of oil is usually only half of that at the surface. An increase in pressure also increases the dynamic viscosity, while an increase in the amount of dissolved gas in oil will reduce the dynamic viscosity.

The kinematic viscosity of oil is the ratio of its dynamic viscosity to the density of the oil at the same temperature and pressure, with the unit of mm^2/s. Conditional viscosity refers to the viscosity expressed in conditional units measured by different viscometers. It is determined by the flow time (in seconds) of a certain volume of oil from a specific instrument at a certain temperature, or the ratio of this flow time to the flow time of the same volume of water. It can be further divided into Engler viscosity, Saybolt viscosity, Redwood viscosity, etc.

(4) Solubility

Since hydrocarbons are difficult to dissolve in water, the solubility of oil in water is very low. When comparing molecules with the same number of carbon atoms, the order of solubility in water from high to low is: non-hydrocarbons → aromatics → cycloalkanes → alkanes. Except for methane, the solubility of hydrocarbons in water decreases with increasing molecular weight. The solubility of oil in water increases with an increase in temperature or an increase in the CO_2 content in water. If the salinity of the water increases, the solubility of hydrocarbons will decrease. Understanding the solubility of oil in water helps to understand the phase state of the initial migration of oil.

Although oil is difficult to dissolve in water, it is easily soluble in many organic solvents, such as benzene, chloroform, carbon disulfide, carbon tetrachloride, and ether. The solubility of oil in organic solvents can be used to identify the content and properties of oil in rocks.

(5) Fluorescence

Most oil and its products, except for light gasoline and paraffin, can emit visible light when exposed to ultraviolet light, regardless of whether they are in their original state or dissolved in organic solvents. The unsaturated hydrocarbons and their derivatives in the oil can absorb electrons with shorter wavelengths and higher energy from the ultraviolet light, and then emit low-energy visible light, which is called fluorescence.

The fluorescence of oil depends on the composition of the compounds. The polycyclic aromatic hydrocarbons and non-hydrocarbons in oil cause the light emission, while the saturated hydrocarbons do not emit light at all. The fluorescence color of light oil is lighter (light blue), the fluorescence color of heavy oil is darker (yellow), and the fluorescence color of asphalt is the darkest (brown). The color of the emitted light varies with the properties of the oil or asphalt substances and is not affected by the properties of the solvent. The intensity of the light emission is related to the concentration of the oil or asphalt substances. In the low concentration range, the intensity of the light emission is directly proportional to the concentration of the oil substances, but after the concentration exceeds a certain critical value, the intensity of the light emission decreases instead. This is called concentration quenching. Concentration quenching is reversible, and the light emission intensity increases when diluted with a solvent.

The luminescence phenomenon of oil is very sensitive, and it can emit light as long as the content of oil or asphalt substances in solution is one hundred-thousandth. Therefore, in oil and gas exploration work, people often use fluorescence analysis to identify the oil content of rock samples and can roughly determine their components and content. This method is simple, fast, economical, and practical.

(6) Optical Activity

When polarized light passes through crude oil, the plane of polarization will rotate a certain angle, which is called the optical rotation angle. This property of oil is called optical activity. The optical rotation angle of oil is generally a few tenths of a degree to a few degrees, and most of them rotate to the right (clockwise). The reason why oil has optical activity is that some nitrogen-containing compounds, steranes, and triterpenes, and other biomarker compounds in oil often have chiral carbon atoms, which give oil optical activity. Organic compounds with chiral carbon atoms can form two stereoisomers (Fig. 1.11), one molecular isomer rotating the plane of polarization to the left, and the other molecule rotating the plane of polarization to the right. When the two isomers are present in equal amounts, the rotation is canceled out, and when the two isomers are not present in equal amounts, optical activity is produced.

(7) Pour Point and Wax Content

The highest temperature at which oil loses its flow ability is called the pour point. The pour point of oil does not have a fixed value, and oil rich in asphaltenes gradually thickens as the temperature decreases without a clear solidification phenomenon.

The wax in oil is mainly composed of long-chain normal alkanes and isoalkanes, with a melting point of 30–35 °C. Under reservoir conditions, wax is in a dissolved state in oil. When oil is extracted from the reservoir to the ground, with the decrease of pressure and temperature, wax gradually precipitates out of the oil and adheres to the walls of the oil pipe, which is the wax deposition phenomenon of the oil well.

The pour point of oil is positively correlated with the wax content and the carbon atom number of alkanes. Oil rich in wax stops flowing when the temperature drops to the wax precipitation point, accompanied by the crystallization and precipitation of wax. Low pour point oil is high-quality oil, and oil with a high pour point is prone to wax deposition at the bottom of the well, which brings trouble to oil production. Therefore, the pour point is also an important parameter that must be referred to in oil transportation.

(2) Heating Value

The heating value, also known as the calorific value, refers to the amount of heat released when oil is burned. When 1 kg of oil is burned, it can produce 10,000–11,000 kcal of heat, making it a high-quality fuel. The main elements in oil that release heat when burned are carbon and hydrogen. When 1 kg of carbon is completely

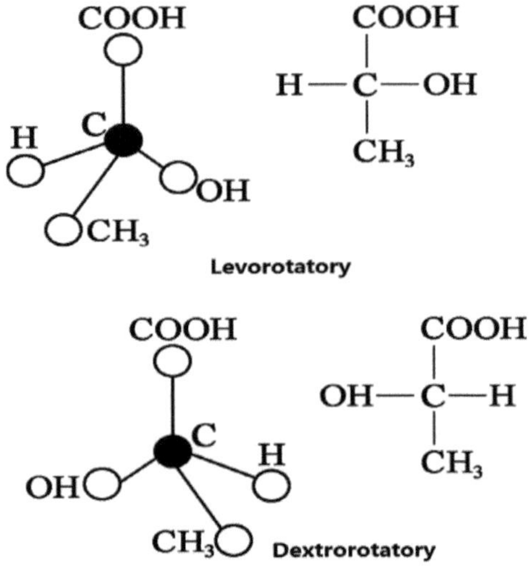

Fig. 1.11 Two stereoisomers of lactic acid molecule

burned, it can release 8,000 kcal of heat, and the heat released by 1 kg of hydrogen reaches 34,200 kcal, which is more than four times the heat released by carbon. The hydrogen content in oil is higher than that in coal and oil shale, so the heating value is also higher, 40–60% higher than that of bituminous coal.

1.1.4 Heavy Oil and Bitumen

Heavy oil and bitumen (also known as natural bitumen) generally refer to crude oil with high viscosity, high density, and difficult or impossible to flow under reservoir conditions. Different countries have different definition standards. This book adopts the definition of heavy oil and bitumen by Zou et al. (2014): Oil with a viscosity greater than 10,000 mPa s and a relative density greater than 1.0 (10° API) under reservoir temperature conditions is considered bitumen. Oil with a viscosity of 50–10,000 mPa s and a relative density of 0.934–1.0 (10°–20° API) under reservoir temperature conditions is considered heavy oil, also known as heavy oil or viscous oil. Oil sand, also known as bitumen sand or tar sand, is a mixture of bitumen or heavy oil with sand, clay, minerals, and water, mainly due to the thickening effects of biodegradation, light hydrocarbon volatilization, water washing, and free oxidation, resulting in the enrichment of polar heteroatom components—resins and asphaltenes.

Heavy oil and bitumen resources hold an important position in the world's oil and gas resources and are widely distributed around the world. Some asphalt sands formed on the surface or in the shallow layer can become asphalt sand oilfields with mining value, such as the Athabasca, Cold Lake, Wabasca, and Peace River oilfields in the Western Canada Basin, with a total geological reserve of asphalt sand exceeding 0.177×10^{12} t. The Orinoco heavy oil belt in the Eastern Venezuela Basin, as one of the world's largest heavy oil accumulation areas, has a geological resource volume of about 0.197×10^{12} t. China's heavy oil and bitumen resources are also very rich, and a large number of heavy oil reservoirs have been discovered in many large and medium-sized oil and gas basins and regions. It is preliminarily estimated that the heavy oil resource volume alone reaches 250×10^8 t, mainly produced from the clastic rock reservoirs of the Mesozoic and Cenozoic, with a small part from the Paleozoic marine strata.

1.1.4.1 Chemical Composition of Heavy Oil and Bitumen

Like conventional crude oil, heavy oil and bitumen are mainly composed of carbon, hydrogen, sulfur, nitrogen, and oxygen—the five elements, with carbon and hydrogen being the main components, and also containing trace amounts of metallic elements such as nickel, vanadium, iron, and copper. However, the hydrogen content is relatively low, and the hydrogen-to-carbon atomic ratio is relatively small, mostly below 1.7.

Table 1.7 Elemental composition of Chinese heavy oil (according to Zou et al., 2014)

Source	Carbon (%)	Hydrogen (%)	Sulfur (%)	Nitrogen(%)	Ni (μg/g)	V (μg/g)
Huanxiling, Liaohe	87.2	11.8	0.27	0.37	40.0	0.5
Shuguang, Liaohe	86.6	12.3	0.35	0.40	60.0	0.9
Gaosheng, Liaohe	85.8	11.5	0.56	1.06	122.5	3.1
Gudao, Shengli	85.1	11.6	2.07	0.33	21.1	2.0
Shanjiasi, Shengli	84.7	11.2	0.45	0.69	42.3	3.4
Bamianhe, Shengli	84.9	12.2	1.87	0.53	32.5	
Caoqiao, Shengli	85.5	12.2	1.54	0.44	47.6	
Jiuqu, Xinjiang			0.15	0.35	13.9	0.2
Wuerhe, Xinjiang			0.32	0.78	43.8	1.1
Hongqianse			0.09	0.21	12.2	<0.1
Jinglou, Henan	85.8	12.5	0.32	0.74	21.8	1.4
Gucheng, Henan			0.36	0.73	35.4	1.1
Yangsanmu			0.33	0.34	25.0	0.9
Chengbei, Bohai			0.36	0.52	22.5	0.8
Suizhong 36-1			0.31	0.37	42.5	1.0

The chemical composition of heavy oil and bitumen in China has significant differences from that of foreign countries. Except for a few areas, the sulfur content is generally low, less than 1%, while the nitrogen content is relatively high. It also has the characteristics of high nickel content and low vanadium content, which is exactly the opposite of heavy oil and bitumen from abroad, indicating that China's heavy oil and bitumen come from terrestrial organic matter (Tables 1.7 and 1.8).

1.1.4.2 Physical Properties of Heavy Oil and Bitumen

Compared with conventional oil, heavy oil and bitumen have a high content of high molecular weight hydrocarbons and heteroatom compounds, characterized by high density, high viscosity, and a heavy fraction composition (Table 1.9). As can be seen

Table 1.8 Chemical composition of foreign heavy oil and Bitumen (according to Zou et al., 2014)

Source	Sulfur (%)	Nitrogen (%)	Trace elements (μg/g)		Resin content (%)
			Ni	V	
Cerro Negro (Venezuela)	4.0	0.75	108.6	430	
Jobo (Venezuela)	3.9	0.52	94.4	390	
Bachaquero (Venezuela)	2.9	0.38		470	
Boscan (Venezuela)	5.7	0.44	147	1,220	
Tia (Venezuela)	2.5	0.30		397	
Orinoco (Venezuela)	4.0				35.4
Cold Lake (Canada)	4.4				28.7
Lloydminster (Canada)	4.0	0.32	52.7	10.5	38.4
Athabasca (Canada)	4.9	0.43	68.1	144	34.1
Qayarah (Iraq)	8.4				36.1

from Table 1.9, the content of fractions in heavy oil and bitumen below 200 °C is very low, generally less than 5%, while the content of atmospheric residue above 350 °C basically accounts for more than 80%, and can even reach up to 90%.

1.1.4.3 Formation and Distribution of Heavy Oil and Bitumen

Heavy oil and bitumen are both secondary formed from low-mature or mature oil through thickening processes such as water washing, biodegradation, and free oxidation, which reduce light components and increase heavy components.

The formation and distribution of heavy oil and bitumen are closely related to the tectonic movements of the Mesozoic and Cenozoic, controlled by the distribution of the global Mesozoic and Cenozoic orogenic fold belts. The Mesozoic and Cenozoic tectonic movements led to the destruction of ancient oil reservoirs, causing conventional oil to migrate to shallow layers or even the surface, where it underwent thickening processes to form heavy oil and bitumen. At present, the vast majority of heavy oil and bitumen are found in Cretaceous, Paleogene, and Neogene strata. Geographically, they are mainly distributed along the edge slopes, uplifts, or edges of basins (or depressions), as well as in the shallow parts of fault zones.

1.2 Natural Gas

In a broad sense, natural gas refers to all naturally occurring gases in nature, including gases of different compositions, origins, and production states. Соколов (1971) classified natural gas into eight categories based on the environment in which they exist:

Table 1.9 Physical properties of Chinese heavy oil (according to Zou et al., 2014)

Source	Density (20 °C) (g/cm^3)	Dynamic Viscosity (50 °C) (mPa s)	Kinematic viscosity (m^2/s)			Pour point (°C)	Fractional composition (%)			
			50 °C	80 °C	100 °C		<200 °C	200–350 °C	350–500 °C	>500 °C
Huanxiling, Liaohe	0.9469	268.5	287.9	33.85	17.33	−20	4.3	20.3	35.3	39.1
Shuguang, Liaohe	0.9123	142.3	159.0	48.5		20	6.0	16.9	28.2	46.1
Gaosheng, Liaohe	0.9443	225.8	243.5			13	4.5	12.4	23.6	59.5
Gudao, Shengli	0.9333	201.5	219.9	56.3		13	6.1	14.9	27.2	51.8
Shanjiasi, Shengli	0.9719	6,355.4	6,656.3			14	1.2	12.2	18.3	67.7
Baminahe, Shengli	0.9302	556.5	609.5	136.6		4	4.0	17.7	25.0	52.9
Caoqiao, Shengli	0.9268	265.8	292.2	68.6		4	2.8	15.8	23.6	57.4
Jiuqu, Xinjiang	0.9284	272.5	299.0		66.3	−22	2.4	19.4	27.1	51.2
Wuerhe, Xinjiang	0.9622			1,896.6	542.9	9	0.1	10.9	27.8	59.4
Hongqianqu, Xinjiang	0.9233			580.8		−19	0.1	9.8	27.8	59.4
Jinglou, Henan	0.9489	1,436.8	1,542.0	229.9	86.3	10	0.9	8.3	32.9	58.0
Gucheng, Henan	0.9437	1,344.0	1,442.4		81.1	10	0.6	9.0	28.5	61.0
Yangsanmu, Dagang	0.9492	594.6	637.9	172.9 (70 °C)		−2	0.8	15.0	33.3	50.8

(continued)

Table 1.9 (continued)

Source	Density (20 °C) (g/cm^3)	Dynamic Viscosity (50 °C) (mPa s)	Kinematic viscosity (m^2/s)			Pour point (°C)	Fractional composition (%)			
			50 °C	80 °C	100 °C		<200 °C	200–350 °C	350–500 °C	>500 °C
Chengbei, Bohai	0.9537	767.3	819.2	129.7		10	2.23	12.0	21.9	63.4
Suizhong36-1, Bohai	0.9677	743.2	781.8	119.0		−6	3.14	17.4	21.3	58.1

atmospheric gas, gas in surface sediments, gas in sedimentary rocks, gas in the ocean, gas in metamorphic rocks, gas in magmatic rocks, mantle exhaust gas, and cosmic gas.

The natural gas studied in petroleum geology (narrow sense) refers to the flammable gases generated from the evolution of sedimentary organic matter, that is, gases dominated by gaseous hydrocarbons in sediments or sedimentary rocks.

1.2.1 Chemical Composition of Natural Gas

The natural gas in oil (gas) fields mainly consists of gaseous hydrocarbons, along with varying amounts of various non-hydrocarbon gases.

Hydrocarbon gases are mainly alkanes with 1–4 carbon atoms, generally dominated by methane (CH_4), followed by heavier hydrocarbon gases (hydrocarbon gases with more than 2 carbon atoms, commonly represented as C_{2+}). Among them, ethane (C_2H_6) and propane (C_3H_8) are the most common, while butane (C_4H_{10}) and higher homologues are less common.

Natural gas can be divided into wet gas and dry gas based on the content of heavier hydrocarbon gases, that is, hydrocarbon gases with $C_{2+} < 5\%$ are collectively called dry gas, also known as lean gas; hydrocarbon gases with $C_{2+} \geq 5\%$ are collectively called wet gas, also known as rich gas. The chemical composition of natural gas in different regions varies greatly (Table 1.10).

Under reservoir conditions, the total amount of non-hydrocarbon gases is not much, but there are many types, including nitrogen, carbon dioxide, carbon monoxide, hydrogen sulfide, hydrogen, and other gases. Non-hydrocarbon gases also contain trace amounts of inert gases such as helium, argon, and neon, with contents ranging from a few thousandths to a few percent. Inert gases, due to their extremely low content, cannot form gas reservoirs on their own and are usually coexisting with other gases in gas reservoirs, with a small portion dissolved in oil and formation water.

1.2.2 Production States of Natural Gas

Natural gas in the Earth's crust can be divided into dispersed and aggregated types based on its distribution characteristics, into free state, dissolved state, adsorbed state, and solid state gas hydrate based on its phase state, and into associated gas and non-associated gas based on its relationship with oil production.

Dispersed natural gas belongs to unconventional natural gas, mainly including oil-dissolved gas, water-dissolved gas, shale gas, coalbed methane, and solid state gas hydrates. For more details about these types of natural gas, please refer to the following chapters.

Table 1.10 Chemical composition of some oil (gas) field gases (percentage content) (according to Zhang et al., 1999)

Country	Oil (gas) field	Strata	CH_4	Heavy hydrocarbon	CO_2	N_2	H_2S	H_2	O_2	H_e
China	Daqing Oilfield	K_1	83.82	13.0	0.11	2.58				
	Dagang Oilfield	ES_3	75.21	23.22						
	Shengdengshan Gasfield	P_1	94.57	0.99	0.24	2.43		0.02		
	Shiyougou Gasfield	Tc	97.80	0.40	0.20	1.10	0.1			
	Yanhu Gasfield	Q	95.50	0.50		3.5				
America	Mottled-Dome	J			12.2	79.7			0.92	7.18
	August (Kansas)	C_2	10.5	1.6	0.1	85.6				2.13
	Heller (Utah)	J	5.1	2.3	1.1	84.4				7.16
	Bend Bulge	P	0.1		0.8	89.9				8.6
Former Soviet Union	Grozny	R	47.0	51.3	1.7					
	Ishimba	R	42.9	47.3	0.3	4.8	4.6			0.03
	Duymaz	D	61.4	25.4	0.2	14.0				
	Krasnokamsk		19.4	48.6	0.4	21.2	0.4			

Aggregated natural gas is free gas, that is, gas that has migrated and aggregated separately, including gas reservoir gas and gas cap gas. This is the basic form of natural gas existing in conventional gas reservoirs or oil and gas reservoirs, and only large-scale free gas aggregation can be effectively developed and utilized.

1.2.2.1 Gas Reservoir Gas

Gas reservoir gas refers to a certain scale of separate natural gas aggregation in the trap, that is, the gas in the pure gas reservoir, which is basically not associated with oil. It can be hydrocarbon gas or non-hydrocarbon gas. Some gas reservoir gases in oil and gas fields maintain a certain connection with the underlying or lateral distributed oil (gas) reservoirs or oil reservoirs in the vertical and horizontal directions, which is the result of oil and gas migration under specific geological conditions.

When non-hydrocarbon gases are relatively enriched, they can become non-hydrocarbon gas reservoirs. For example, the high-pressure hydrogen sulfide gas reservoir in the Paleogene of the Zhaolanzhuang structure in the Jizhou Depression has a hydrogen sulfide content as high as 92%. The carbon dioxide content in the natural gas produced by the Fangpingwang oil and gas field in the Paleogene of the Jiyang Depression is as high as 63–66%. The carbon dioxide gas field in Shatouwei, Sanshui Basin, Guangdong, has a carbon dioxide content as high as 99.53%.

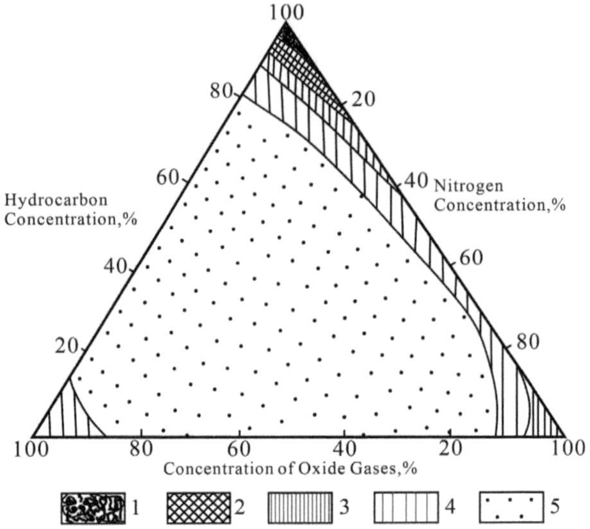

Fig. 1.12 World gas reservoir gas composition diagram (according to Воронов & Тихомировдидр, 1976). Frequency (%): 1. >50, 2. 10–50, 3. 1–10, 4. 0.1–1, 5. <0.1

Воронов and Тихомировдидр (1976) summarized the distribution characteristics of the chemical composition of gas reservoir gas based on the analysis of about 2,000 gas reservoirs and 15,000 gas samples from gas and oil-bearing basins around the world (Fig. 1.12): the vast majority of gas reservoir gases are mainly hydrocarbon gases, accounting for more than 80% of the total number of gas reservoirs; gas reservoirs dominated by nitrogen are less than 10%. Gas reservoirs dominated by acidic gases such as carbon dioxide or hydrogen sulfide are even fewer, far less than 1%.

1.2.2.2 Gas Cap Gas

Gas cap gas refers to the natural gas that coexists with oil in oil and gas reservoirs in a free gas cap state. It is closely related to petroleum in terms of genesis and distribution, with heavy hydrocarbon gas content reaching several percent to tens of percent, second only to methane, and belongs to wet gas (rich gas). With the increase or decrease of formation pressure, gas cap gas can dissolve in oil or precipitate. The size of the gas cap volume in oil and gas reservoirs is related to the chemical composition of the gas and the formation pressure.

1.2.3 Physical Properties of Natural Gas

Natural gas is generally colorless, with a smell of gasoline or hydrogen sulfide, and is combustible. Under normal temperature and pressure, the hydrocarbons existing in the gaseous state include methane, ethane, propane, butane, and isobutane, while the non-hydrocarbons include hydrogen, nitrogen, carbon dioxide, hydrogen sulfide, and inert gases. Under higher temperature and pressure conditions underground, C_4–C_7 alkanes and some cycloalkanes, aromatic hydrocarbons, and organic sulfides can also exist in the gaseous state.

(1) Density and Relative Density

The density of natural gas is defined as the mass of the gas per unit volume. Since wet gas contains more heavy hydrocarbon gases, the density of wet gas is greater than that of dry gas.

Under standard conditions (10^5 Pa, 15.55 °C), the density of the components of surface natural gas is 0.6773 kg/m^3 (methane) to 3.0454 kg/m^3 (pentane). The density of natural gas mixtures is generally between 0.7 and 0.75 kg/m^3, while the density of associated oil gas, especially oil-dissolved gas, can reach up to 1.5 kg/m^3 or even higher. The density of natural gas under reservoir temperature and pressure conditions can generally reach 150–250 kg/m^3, which is much greater than the density of natural gas under surface temperature and pressure conditions; the density of condensate gas can reach up to 225–450 kg/m^3.

The relative density of natural gas generally refers to the ratio of the density of natural gas to the density of air under the same temperature and pressure conditions (such as 10^5 Pa, 15.55 °C, 10^5 Pa, 20 °C). The relative densities of the hydrocarbon components in natural gas are 0.5539 (methane) to 2.4911 (pentane), and the relative density of hydrocarbon gas mixtures is generally between 0.56 and 1.0.

The density and relative density of natural gas increase with the increase of heavy hydrocarbon content as well as the content of carbon dioxide and hydrogen sulfide.

(2) Viscosity

The viscosity of natural gas reflects the magnitude of the internal friction force between its molecules and is a parameter for measuring the resistance of natural gas to flow, commonly referred as dynamic viscosity (i.e., absolute viscosity). The viscosity of natural gas is very small, generally less than 0.01 mPa s under surface conditions, much lower than that of oil.

The viscosity of natural gas is related to pressure, temperature, and gas composition. Under near-atmospheric pressure conditions, viscosity is independent of pressure, increases with temperature, and decreases with increasing molecular weight. Under higher pressures, viscosity increases with pressure, decreases with increasing temperature, and increases with increasing molecular weight. In addition, the viscosity of natural gas also increases with the content of non-hydrocarbon gases.

(3) Solubility

Natural gas can dissolve to varying degrees in oil and water. The mutual solubility of natural gas and water is poor, while it has a strong solubility with oil.

The solubility of natural gas in oil refers to the amount of gas that is dissolved in oil under reservoir temperature and pressure conditions and can be precipitated under surface temperature and pressure conditions (temperature and pressure both decrease) (m^3 gas/m^3 oil). When natural gas is dissolved in oil, it can reduce the density, viscosity, and surface tension of the oil.

Under the same temperature and pressure conditions, the solubility of hydrocarbon gases in oil is much higher than that in water. For example, the solubility of methane in oil is about ten times than that in water. The main factors affecting the solubility of natural gas in oil include formation pressure, gas composition, and the content of light components in oil. When the content of heavy hydrocarbons in natural gas or the content of light fractions in oil increases, the solubility of natural gas in oil increases. Lowering the temperature or increasing the pressure can also achieve the same effect until the liquid is saturated with gas at the bubble point pressure.

(4) Heating Value

The heating value of natural gas varies greatly. Hydrocarbon gases have a high hydrogen content, thus they have a high heating value and burn more completely, making them clean and high-quality fuels. The higher the carbon number in hydrocarbon gases, the higher the heating value. For example, the heating value of methane is 37,112 kJ/m^3, and that of butane is 125,876 kJ/m^3. The heating value of wet gas in natural gas can be as high as 83,680 kJ/m^3, while the heating values of coal and oil are 16,736 kJ/m^3 and 41,840 kJ/m^3, respectively.

1.3 Oilfield Water

In a broad sense, oilfield water refers to the groundwater related to the generation, migration, and accumulation of oil and gas within the oil and gas field area, including water in oil layers and non-oil layers. The focus of petroleum geology research is on the groundwater that has a direct relationship with oil and gas reservoirs, exists in oil and gas reservoirs or oil and gas traps, and is directly connected with oil and gas layers, that is, reservoir water, which is also the narrow sense of oilfield water.

Oilfield water has a physical property that is quite different from pure water due to the dissolution of various components. The relative density of oilfield water is generally greater than 1, and the viscosity is significantly higher than that of pure water. The higher the salinity, the greater the relative density and viscosity. Generally, the transparency is poor, appears turbid, and often has a certain color, pale green when containing H_2S, and light red, brown, or pale yellow when containing iron colloids. When oilfield water contains a small amount of oil, it often has a smell of gasoline or kerosene, a rotten egg smell when containing H_2S, and a bitter taste when containing $MgSO_4$. Oilfield water contains more ionic components, which can conduct electricity, and the higher the ion content, the stronger the conductivity.

1.3.1 Source of Oilfield Water

It is generally believed that there are four main sources of oilfield water: sedimentary water, infiltrated water, deep-formed water, and transformed water.

Sedimentary water (also known as buried water) refers to the water preserved during the accumulation of sediments, which is essentially ancient seawater or ancient surface water (such as ancient lake water, ancient river water, etc.). Therefore, the salinity and chemical composition of sedimentary water are closely related to ancient seawater (lake) water, and there are obvious differences in mineralization of oilfield water formed in different environments.

Infiltrated water refers to the water that seeps into underground pores and permeable rock layers through condensation water, atmospheric precipitation, and surface water. Its mineralization degree is low and can dilute high-mineralization groundwater. This dilution effect is particularly evident in oilfield water near the unconformity interface.

Deep-formed water (also known as primary water) refers to the native water and metamorphic water originating from the upper mantle and deep crust, which is released from magma. It is a high-temperature, high-mineralization, gas-saturated groundwater, which plays an important role in the formation of metal ore deposits.

Transformed water (also known as regenerated water) refers to the water released from the interlayer water of clay minerals during sedimentary rock formation and the water produced by the decomposition of organic matter during the transformation into hydrocarbons. This transformation is mainly influenced by temperature and pressure, accompanied by ion exchange reactions, etc.

In fact, oilfield water can be considered as a mixture of sedimentary water, infiltrated water, deep-formed water, and transformed water in different proportions, after a series of complex physical and chemical actions, and is associated with reservoir water that coexists with oil and gas.

1.3.2 Mineralization of Oilfield Water

Mineralization refers to the total amount of solid substances in a unit volume of water in a dissolved state, that is, the total content of various ions, elements, and compounds in a unit volume of water, expressed in g/L, mg/L, ppm.

The mineralization of oilfield water is generally higher than that of general groundwater, mainly related to the deep high-temperature evaporation and concentration effect experienced by the original sedimentary water in a relatively closed environment. The mineralization of oilfield water in a relatively closed, slow water exchange, reducing environment is high, and the closer to the basin center, the higher the mineralization. In areas with good water exchange replenished by infiltrated water, the mineralization is significantly reduced. Vertically, the mineralization generally increases with increasing burial depth.

Due to differences in sources and formation processes, there are significant differences in the mineralization of oilfield water in various regions. The mineralization of terrestrial oilfield water is generally lower than that of marine oilfield water, and the range of variation is large. For example, the mineralization of oilfield water in the Cretaceous sandstone of the Burqan Oilfield in Kuwait is 154 g/L. The mineralization of oilfield water in a certain oilfield in the Jiuquan Basin, China, is 30–80 g/L. The mineralization of terrestrial oilfield water in China is generally below 50 g/L, with those below 10 g/L being predominant.

1.3.3 Chemical Composition of Oilfield Water

The chemical composition of oilfield water actually refers to the chemical composition of the dissolved solutes in the oilfield water. In the process of oilfield water formation, due to the interaction between water and oil and gas, the oilfield water not only contains inorganic components and dissolved gases but also has organic components that are not commonly found in general groundwater.

(1) Inorganic Components

The inorganic composition of oilfield water is mainly composed of ions such as Na^+ (including K^+), Ca^{2+}, Mg^{2+}, and Cl^-, $SO_4{}^{2-}$, HCO^{3-} (including $CO_3{}^{2-}$), among which Cl^- and Na^+ are dominant. In addition, it also contains dozens of trace elements, such as I, Br, B, Sr, Ba, etc., whose enrichment is often related to the formation and evolution of oil, and can indicate the source of oilfield water, hydrogeochemical environment, oil and gas content, etc.

(2) Organic Components

Common organic components in oilfield water include benzene, phenol and their homologues, two-ring and polycyclic aromatic hydrocarbons, ammonia, cycloalkane carboxylic acids and their salts. Benzene, phenol and their homologues are monocyclic aromatic hydrocarbon compounds in oil, which are soluble in water and are considered direct indicators for exploring oil and gas fields due to their high mobility, thermodynamic stability, and close connection with the genesis of oil and gas. The content of benzene, phenol and their homologues in oilfield water is relatively high, and their composition is significantly different from that in non-oilfield water (Fig. 1.13).

Two-ring and polycyclic aromatic hydrocarbons have high solubility, thus they are also present in high amounts in oilfield water and can provide more reliable information about oil and gas. High content of ammonia is also a good indicator of oil and gas.

(3) Gas Components

Common gas components in oilfield water include two major categories: hydrocarbon gases and non-hydrocarbon gases. Hydrocarbon gases are mainly methane, and there

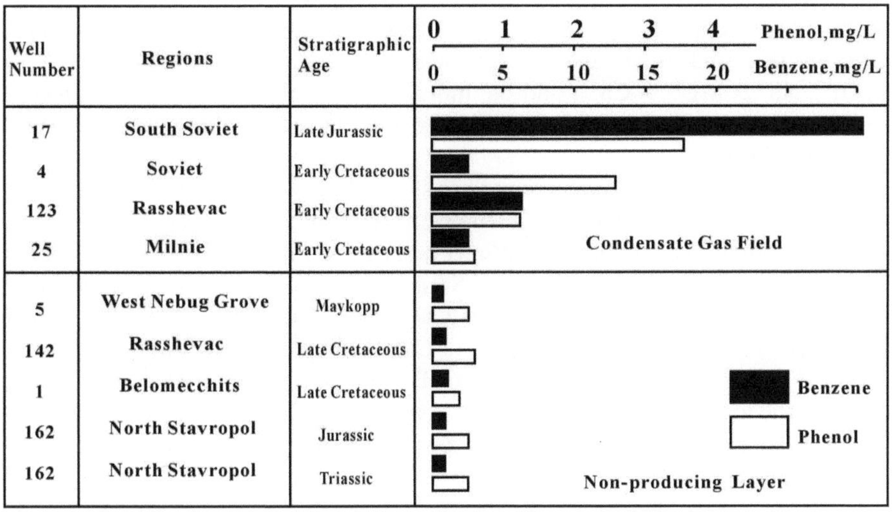

Fig. 1.13 Distribution of benzene and phenol content in oil-bearing and non-oilfield waters of a certain condensate gas field in the former Soviet Union (from Zhang et al., 1999)

are also heavier hydrocarbon gases such as ethane. Non-hydrocarbon gases include a variety of gases, including carbon dioxide, hydrogen sulfide, nitrogen, helium, argon, etc. Oilfield water generally does not contain oxygen. The presence of heavy hydrocarbon gases is a major characteristic of oilfield water and can be used as one of the important signs for exploring oil and gas fields. Hydrogen sulfide and carbon dioxide gases are usually the products of the biogeochemical action of sulfate minerals and organic matter in a closed reducing environment, so they can also be used as one of the signs of oilfield water.

1.3.4 Types of Oilfield Water

Since the first classification of oilfield water was proposed by Palmel in the United States in 1911, there have been various classification schemes for oilfield water, most of which are based on the content and combination relationships of dissolved Na^+ (including K^+), Ca^{2+}, Mg^{2+}, and Cl^-, $SO_4{}^{2-}$, $HCO_3{}^-$ (including $CO_3{}^{2-}$) in the water. In these classification schemes, the Shchukin classification proposed by the Soviet scholar Shchukin (Щулин, 1946) is more concise and widely used.

Shchukin believed that the chemical composition of water mainly depends on the environment in which it is located, and different environments can form waters with different properties and chemical compositions. Natural water, in terms of its formation environment, is mainly divided into two major categories: continental water and seawater. Continental water has a low salinity (generally less than 0.5 g/L), and its chemical composition is characterized by $HCO_3{}^- > SO_4{}^{2-} > Cl^-$, $Ca^{2+} > Na^+ >$

Mg^{2+}, and the molar ratio $r(Na^+)/r(Cl^-) > 1$. Seawater has a higher salinity (generally about 35 g/L), and its chemical composition is characterized by $Cl^- > SO_4^{2-} > HCO_3^-$, $Na^+ > Mg^{2+} > Ca^{2+}$, and the molar ratio $r(Na^+)/r(Cl^-) < 1$. Continental freshwater is dominated by $NaHCO_3$ and contains Na_2SO_4; while seawater does not contain Na_2SO_4.

Through analysis and comparison, Shchukin believed that the main ions in water form salts according to the order of their chemical affinity. The order of combination of ions according to their chemical affinity is shown in Fig. 1.14. If $r(Na^+) > r(Cl^-)$, then the excess Na^+ will combine with SO_4^{2-} or HCO_3^-, which may result in the formation of Na_2SO_4 water type or $NaHCO_3$ water type. If $r(Na^+) < r(Cl^-)$, then the excess Cl^- will combine with Mg^{2+} or Ca^{2+}, which may result in the formation of $MgCl_2$ water type or $CaCl_2$ water type. Therefore, combining Mg^{2+} and SO_4^{2-}, and using the ion ratios $[r(Na^+) - r(Cl^-)]/r(SO_4^{2-})$ and $[r(Cl^-) - r(Na^+)]/r(Mg^{2+})$, natural water can be divided into four types: Na_2SO_4 type, $NaHCO_3$ type, $MgCl_2$ type, and $CaCl_2$ type (Table 1.11, Fig. 1.15).

$CaCl_2$ type water is formed in a well-sealed, stagnant water body in the deep crust, which is conducive to the preservation of oil and gas reservoirs. Oilfield water

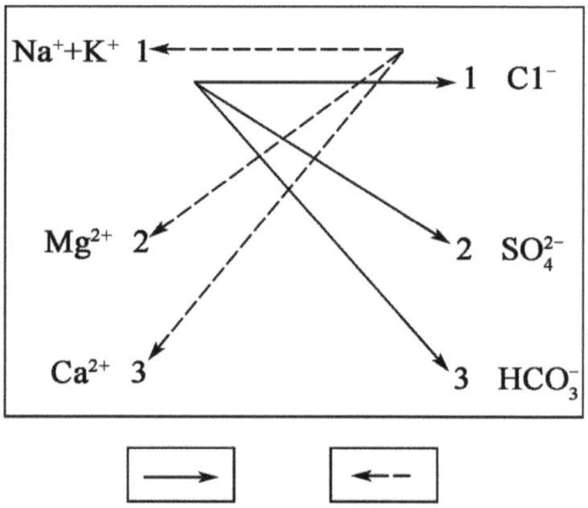

Fig. 1.14 Schematic diagram of ion combination order (according to Liu & Yan, 1991). 1—When $r(Na^+) > r(Cl^-)$, the order of ion combination. 2—When $r(Na^+) < r(Cl^-)$, the order of ion combination

Table 1.11 Shchukin's classification table of the genesis of natural water (quoted from Zhang et al., 1999)

Type of water	Genesis coefficient	$r(Na^+) > r(Cl^-)$	$[r(Na^+) - r(Cl^-)]/r(SO_4^{2-})$	$[r(Cl^-) - r(Na^+)]/r(Mg^{2+})$
Continental water	Na_2SO_4 type	>1	<1	<0
	$NaHCO_3$ type	>1	>1	<0
Marine water	$MgCl_2$ type	<1	<0	<1
Deep-formed water	$CaCl_2$ type	<1	<0	>1

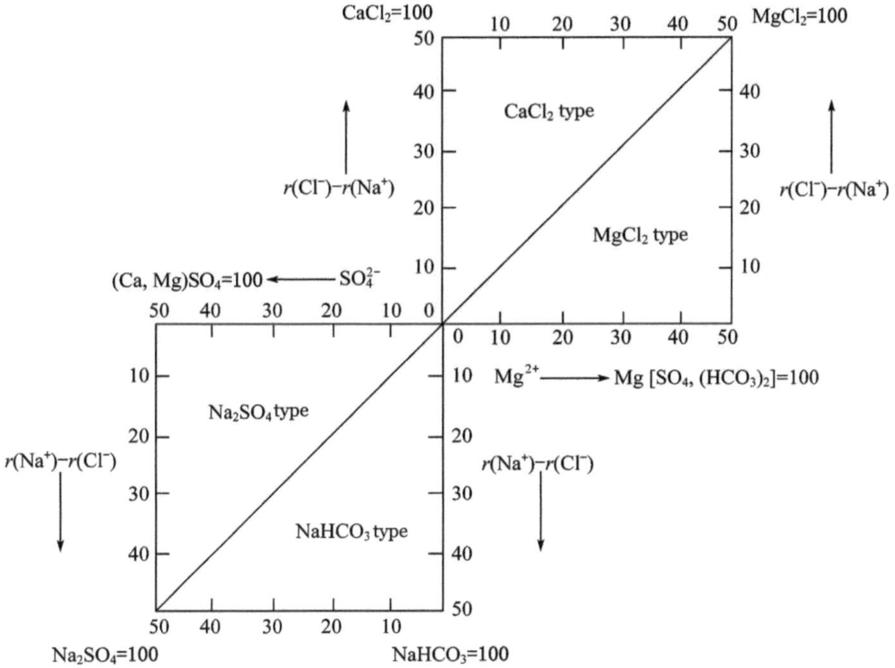

Fig. 1.15 Classification of the origins of natural water (according to Shchukin, 1946, cited from Zhang et al., 1999)

is often a high-mineralization $CaCl_2$ type water, but it has no genetic relationship with oil and gas substances.

$NaHCO_3$ type water is generally divided into high-mineralization and low-mineralization types based on mineralization. High-mineralization $NaHCO_3$ type water is a product of the reducing environment where oil and gas substances exist, and it is related to oil and gas fields in terms of genesis. It is one of the basic water types of oilfield water.

$MgCl_2$ type water is mainly formed by the evaporation and concentration of seawater in lagoons. It can also be formed by the dissolution of salts retained in marine sedimentary rocks by continental freshwater, or by the mixing of deep $CaCl_2$ type water with upper $NaHCO_3$ type or Na_2SO_4 type low-mineralization water. The environment of $MgCl_2$ type water generally has no or few oil and gas fields.

Na_2SO_4 type water is the most widely distributed type of water in surface or shallow groundwater active areas, usually indicating poor hydrogeological sealing of the crust, which is not conducive to the preservation of oil and gas reservoirs. Therefore, its distribution area generally has no oil and gas reservoirs. Of course, some oilfields also have Na_2SO_4 type water, but this is when the oil and gas reservoirs are at the stage of destruction.

The above four types of natural water, due to their different formation conditions, have a certain regularity in their distribution in the vertical section of the crust.

Generally, with increasing burial depth, the mineralization of groundwater increases, and the water type gradually changes. In oil and gas-bearing areas, as the burial depth increases, the vertical section of the oilfield shows Na_2SO_4 type or $MgCl_2$ type, $NaHCO_3$ type, and $CaCl_2$ type from top to bottom in sequence. In some areas, there is also a reversal of water types, that is, high-mineralization water in the upper part, and low-mineralization water in the lower part, such as in some folded areas.

1.3.5 Application of Oilfield Water Chemistry in Oil and Gas Exploration

(1) Use of Oilfield Water Chemical Characteristics to Find Favorable Exploration Areas

Due to the filtration and diffusion effects of oil and gas, the chemical composition of shallow groundwater or surface water changes, such as the increase or decrease of certain components, or the formation of special components or component combinations. Especially in areas with more developed fractures, such as structural high points and inflection points, these chemical anomalies can be used to roughly delineate the range of oil and gas reservoirs.

The chemical components in the oilfield water whose content changes include both organic components directly from oil and gas or related to oil and gas, such as direct indicators including soluble gaseous hydrocarbons, benzene, phenol, cycloalkane acids, etc., and special components or component combinations that appear in the water due to the mixing of formation water with shallow water or environmental changes. Although they are not directly derived from the components of oil and gas, they have a certain genetic connection with oil and gas and are indirect indicators of the presence of oil and gas underground (Liu & Yan, 1991).

(2) Use of Oilfield Water Chemical Characteristics to Infer Oil and Gas Preservation Conditions

Although the mineralization, water type, trace components, hydrogen and oxygen isotope composition, and some ion ratios of formation water cannot directly indicate oil and gas accumulation areas, they can reflect the geochemical environment and the degree of hydrogeological sealing of the formation water. This can be used to infer the preservation conditions of oil and gas.

Oilfield water is generally in a deep burial, well-sealed, and reducing environment conducive to the generation and preservation of oil and gas. In this environment, the high temperature, strong evaporation and concentration, and extensive dissolution of salt minerals lead to high mineralization, mainly with $CaCl_2$ type water. The enrichment of trace elements in the formation water is related to the original sedimentary environment and the source of organic matter, and often indicates the existence of a strong evaporation and concentration effect with strong sealing. In addition, the sodium chloride coefficient $r(Na^+)/r(Cl^-)$, the metamorphic coefficient $[r(Cl^-)–r(Na^+)]/r(Mg^{2+})$, the desulfurization coefficient $100 \times r(SO_4{}^{2-})/[r(Cl^-) + r(SO_4{}^{2-})]$, and the carbonate equilibrium coefficient $[r(HCO_3{}^-) + r(CO_3{}^{2-})]/r(Ca^{2+})]$ are also commonly used to study the migration, changes, and storage state of the formation water, and further analyze the migration, aggregation, and preservation environment conditions of oil and gas.

1.4 Stable Isotopes in Oil and Gas

Isotopes can be divided into stable isotopes and radioactive isotopes according to their properties. Stable isotopes refer to isotopes whose atomic nuclei do not change spontaneously. There are 274 stable isotopes in nature. Stable isotopes have the same proton number and atomic structure, similar chemical properties, and relative stability, but they have different neutron numbers and certain differences in atomic mass. Therefore, there is a fractionation effect when they participate in biological, chemical, and physical processes. Radioactive isotopes refer to isotopes with unstable atomic nuclei that can spontaneously undergo radioactive decay or nuclear fission, forming new atoms with different proton numbers, such as the decay of ^{238}U to ^{206}Pb.

Studies have shown that the isotopic composition of oil and gas is closely related to the source of the oil and gas and the geological and geochemical changes experienced by the parent material and the products. Therefore, understanding the isotopic composition and evolutionary characteristics of oil and gas helps us to identify the source of oil and gas, perform oil (gas) source correlation, trace the secondary migration path of oil and gas, and analyze the secondary changes of oil and gas.

Table 1.12 shows the isotopic characteristics of the main elements in oil and gas, and the relative content of isotopes is mostly analyzed by a mass spectrometer.

1.4.1 Carbon Isotopes in Oil and Gas

Carbon has three isotopes, namely ^{12}C, ^{13}C, and ^{14}C, among which ^{12}C and ^{13}C are stable isotopes, and ^{14}C is a radioactive isotope.

Table 1.12 Isotopic characteristics of the main elements in oil and gas

Element symbol	Element	Proton number	Neutron number	Atomic mass	Relative abundance (%)
H	Hydrogen	1	0	1	99.9844
			1	2	0.0156
			2	3	–
He	Helium	0	1	3	1.3×10^{-4}
			2	4	99.9999
C	Carbon	6	6	12	98.892
			7	13	1.108
			8	14	–
N	Nitrogen	7	7	14	99.635
			8	15	0.365
O	Oxygen	8	8	16	99.759
			9	17	0.0374
			10	18	0.2039
S	Sulfur	16	16	32	95.1
			17	33	0.74
			18	34	4.2
			20	36	0.016

1.4.1.1 Representation of Stable Carbon Isotopes

The ratio of stable carbon isotopes can be expressed as the relative abundance ratio ($^{13}C/^{12}C$). In practical work, the relative measurement method is used, that is, the δ value expressed by comparing the sample to be measured with the standard sample:

$$\delta^{13}C = \frac{(^{13}C/^{12}C)_{sample} - (^{13}C/^{12}C)_{standard}}{(^{13}C/^{12}C)_{standard}} \times 1000/00$$

where: $(^{13}C/^{12}C)_{sample}$—$^{13}C/^{12}C$ value of the samples to be tested,

$(^{13}C/^{12}C)_{standard}$—$^{13}C/^{12}C$ value of the standard sample.

To facilitate comparison, internationally there is a tendency to use a unified standard, that is, the carbon isotope of the Cretaceous belemnite from South Carolina, USA, referred to as the PDB standard, with $^{13}C/^{12}C = 1{,}123.7 \times 10^{-5}$. In China, the Ordovician limestone from Zhoukoudian, Beijing, is commonly used as the standard, with $^{13}C/^{12}C = 1{,}123.6 \times 10^{-5}$, which is close to the PDB standard.

1.4.1.2 Compositional Characteristics of Stable Carbon Isotopes in Oil and Gas

The composition of stable carbon isotopes in oil and gas is mainly influenced by the original material that generated the oil and gas, the physical and chemical conditions of the oil and gas generation process, and secondary changes. The smaller the $\delta^{13}C$ value of the original material, the smaller the $\delta^{13}C$ value of the generated oil and gas. The stable carbon isotope composition in oil and gas mainly has the following characteristics:

(1) The $\delta^{13}C$ of oil is generally between −22 and −33‰, with marine oil having higher $\delta^{13}C$ values, generally between −22 and −27‰. Terrestrial oil has lower $\delta^{13}C$ values, generally between −29 and −33‰.
(2) There are differences in the stable carbon isotope composition among different homologous components in oil. If the oil has not undergone secondary changes such as biodegradation and water washing, the homologous components range from saturated hydrocarbons to aromatics, non-hydrocarbons, and asphaltenes, with the $\delta^{13}C$ gradually increasing.
(3) The $\delta^{13}C$ of natural gas components varies greatly, with methane ranging from −3.2 to −100‰; ethane from −18.5 to −44‰; propane from −18 to −38‰; and butane from −23 to −34‰.
(4) The $\delta^{13}C$ increases with the maturity of natural gas. Generally, natural gas formed at low temperatures and shallow depths has lower $\delta^{13}C$ values; while natural gas formed at high temperatures and deep depths has higher $\delta^{13}C$ values.

1.4.2 Hydrogen Isotopes in Oil and Gas

Hydrogen has three isotopes, namely ^1H (protium, H), ^2H (deuterium, D), and ^3H (tritium, T). Among them, ^3H is a radioactive isotope with a half-life of 12.46 years. During radioactive decay, ^3H emits a beta particle, forming the stable isotope ^3He.

The relative abundance of the two stable hydrogen isotopes in a substance is usually represented by the $^2H/^1H$ ratio (i.e., $D/^1H$) or δD, which can be calculated by the following formula:

$$\delta D = \frac{(D/^1H)_{sample} - (D/^1H)_{standard}}{(D/^1H)_{standard}} \times 1000^0/_{00}$$

where: $(D/^1H)_{sample}$—$D/^1H$ value of the samples to be tested;

$(D/^1H)_{standard}$—$D/^1H$ value of the standard sample.

The standard value for D/H is taken from standard seawater, abbreviated as SMOW. A positive δD indicates that the sample is enriched in D compared to the standard, while a negative value indicates that the sample is enriched in ^1H.

The δD of oil is generally between −80 and −160‰, and that of natural gas is generally between −105 and −270‰. There is a certain positive correlation between the δD of natural gas and $\delta^{13}C$, generally, the higher the $\delta^{13}C$, the higher the δD. The δD of oil and natural gas is related to the depositional environment in which they are formed, with oil and gas formed in marine environments generally having higher δD values than those formed in terrestrial environments.

In addition, sulfur, nitrogen, and oxygen stable isotopes are also widely used in the geological field, and their representation methods are like those of the aforementioned isotopes. They can also be represented by $^{34}S/^{32}S$ (or $\delta^{34}S$), $^{15}N/^{14}N$ (or $\delta^{15}N$), and $^{18}O/^{16}O$ (or $\delta^{18}O$), but their research in the field of petroleum geology is relatively less.

Questions for Discussion

1. What are the characteristics of the elemental composition and compound composition of oil?
2. What is the normal alkane distribution curve? What factors are related to the shape of the curve?
3. What are biomarker compounds? What are the common biomarker compounds?
4. What are the basis and scheme for the geochemical classification of oil? How can marine and terrestrial oil be distinguished?
5. What are the main factors affecting the relative density, viscosity, and solubility of oil?
6. What is the fluorescence and optical activity of oil? What is the reason for the fluorescence and optical activity of oil?
7. According to the state of production, what types of natural gas can be divided?
8. What is oilfield water? What are the characteristics of its chemical composition and physical properties? Briefly describe Shchukin's water type classification scheme.
9. How are the carbon and hydrogen isotopes in oil and gas represented? What are their characteristics?

References

Chen, R. (1994). *Petroleum and natural gas geology*. China University of Geosciences Press.

Соколов, В. А. (1971). Геохимия природных газов, Москва, Недра.

Liu, F., & Yan, W. (1991). *Principles of hydrogeology in oil and gas fields*. Petroleum Industry Press.

Mei, B., & Liu, X. (1980). Distribution of isoprenoid hydrocarbons in Chinese crude oils and its relationship to the geological environment. *Oil & Gas Geology, 1*(2), 106.

Tissot, B. P., & Welte, D. H. (1978). *Petroleum formation and occurrence*. Springer.

Xu, C., & Yang, Z. (2009). *Petroleum refining engineering*. Beijing: Petroleum Industry Press.
Zhang, H., Fang, C., Gao, X., et al. (1999). *Petroleum geology* (3rd ed.). Petroleum Industry Press.
Zou, C., Tao, S., Hou, L., et al. (2014). *Unconventional petroleum geology*. Geological Publishing House.

Chapter 2
Formation of Oil and Natural Gas

Abstract The formation mechanism and evolution process of petroleum are one of the core contents of petroleum geology. This chapter focuses on the organic origin of petroleum, introducing the stages and characteristics of the formation of petroleum from precursors such as proteins, carbohydrates, lipids and lignin under the action of biological and thermal effects. The formation mechanism and identification technology of hydrocarbon and non-hydrocarbon natural gas will also be introduced. At the same time, source rock evaluation and oil source comparison will also be discussed in a targeted manner.

2.1 An Overview of the Genesis of Oil and Gas

The genesis of oil and natural gas is an important issue in natural science and one of the fundamental questions in petroleum geology. Only after the formation of oil and gas can there be a series of geological phenomena such as migration and accumulation. Studying the genesis of oil and gas is not only theoretically significant but also essential for understanding the formation and distribution patterns of oil and gas reservoirs to guide oil and gas exploration.

However, there has long been debate over the genesis of oil and gas, particularly in terms of the original substances, environment, and transformation conditions. The focus of this debate has been constantly changing over time. The first reason for this is that oil and gas are fluid minerals that easily flow underground, and the places where oil and gas are produced are generally not where they are generated. Moreover, the composition of oil and gas will continue to change during the underground migration process. Furthermore, oil and gas, especially oil, are chemically complex organic mixtures that are very sensitive to the changes in external conditions, with different components possibly having different generation and evolution processes. At the same time, solving the problem of the genesis of oil and gas involves a wide range of geology, chemistry, and other disciplines. These factors have brought many difficulties to the study of the genesis of oil and gas but have still attracted many geologists, geochemists, and astronomers to explore it.

© China University of Petroleum Press 2025

H. Liu and Y. Jiang, *Petroleum Geology*,
https://doi.org/10.1007/978-981-96-7086-4_2

Since the 1870s, the debate on the genesis of oil and gas has mainly been between two schools of thought: inorganic origin and organic origin, with the fundamental differences lying in the understanding of the sources of hydrocarbons. The former believes that oil and natural gas are derived from inorganic substances, while the latter believes that oil and gas originate from organic matter, gradually formed by the biological organic matter preserved in sedimentary rocks during the geological history after the origin of life on Earth.

2.1.1 Inorganic Genesis of Oil and Gas

2.1.1.1 Inorganic Genesis of Oil

As early as in 1876, Russian chemist Mendeleev (Д.И. Менелеевд) proposed the carbon theory, suggesting that there are heavy metal carbonides (such as iron carbide) deep underground. These heavy metal carbonides react with water to produce hydrocarbons. Subsequently, Russian scholar Sokolov (В.Д. Соколов) first proposed the cosmic theory at the annual meeting of the Moscow Natural Science Researchers Association, suggesting that "hydrocarbons are inherent in the universe." In the mid-twentieth century, a group of former Soviet scholars led by the famous petroleum geologist Kudriavtsev (Кудрявцев) actively advocated the "magma theory" of oil and gas genesis, suggesting that the formation of oil is related to the synthesis of hydrocarbons when basic magma cools. At present, the most influential theory of the inorganic genesis of oil is related to the Fischer–Tropsch geological synthesis theory.

Former President of the Royal Society of London and chemist Robinson (1963, 1966) noticed that the distribution of normal alkanes in crude oil is the same as that in "hydrogenation reforming" oil of Fischer–Tropsch synthesis. Based on this, he proposed that the original oil on Earth may have been generated through the following Fischer–Tropsch reaction 2 billion years ago.

$$CO + H_2 \xrightarrow[300-400\,^{\circ}C]{\overset{\text{(Catalyst)}}{\text{Fe, Co, Ni, V}}} C_nH_m + H_2O + Q\ \text{(Heat)}$$

Due to the various conditions required for the Fischer–Tropsch synthesis reaction, some scholars have studied the possibility of this reaction occurring under geological conditions. Studies have shown that there is a release of H_2 during the serpentinization process of common ultramafic rocks, so as long as there is serpentinization of ultramafic rocks in the crust, a large amount of H_2 can be produced. Mid-ocean ridges, subduction zones, and rifts are all favorable places to produce H_2 through the alteration of ultramafic rocks. Serpentinite, komatiite, and other ultramafic rocks often coexist with carbonate minerals, and these carbonate minerals may release CO_2 during the serpentinization process. Geological environments such as subduction, magmatic intrusion, and rifts are suitable for the emission of CO_2.

In the Fischer–Tropsch synthesis reaction, not only does metallic iron have catalytic activity, but ionized (oxidized) iron also has the same catalytic activity. During the synthesis process, due to the dissociation of CO_2, surface magnetite will continuously oxidize into hematite. At the same time, under the presence of H_2, it is reduced back to magnetite. Studies have also shown that at the temperature of 500 °C, iron oxide can exchange cations with its carrier (silica or alumina), that is, iron ions enter a more stable position in the carrier lattice, thus obtaining good catalytic activity. It is speculated that iron silicate may also be an active catalyst for the Fischer–Tropsch synthesis reaction, and magnetite, hematite, and iron silicate are all common minerals in the crust, which can fully meet the needs of the Fischer–Tropsch synthesis reaction.

In the continental lithosphere, if there is serpentinization to produce H_2, decarburization to generate CO_2, and a temperature below 500 °C required for the Fischer–Tropsch reaction, the Fischer–Tropsch synthesis reaction can occur. It seems that the most suitable parts are the contact zones of subduction zones, the ophiolite thrust sheets, and the crust thinned by rift action. The hydrocarbons synthesized by the Fischer–Tropsch reaction rise along with fractures and magmatic activity and are transported to the reservoir to form oil and gas reservoirs. H.O. Klemme found through statistical research that more than half of the world's oil and gas are related to plate subduction and the various fractures associated with it, and the probability of discovering large oil and gas fields near these fractures is higher. For example, in the offshore of Canada, the eastern, central, and western parts of North America, the North Sea, and Saudi Arabia, the basement of oil and gas basins is all related to deep large fractures associated with plate subduction. Szatmarı (1989) proposed a general model for the geological synthesis of oil by Fischer–Tropsch based on this. It should be pointed out that although the Fischer–Tropsch reaction is objectively existent and humans can synthesize petroleum-like substances through the Fischer–Tropsch reaction. However, there has been no direct evidence of the existence of the Fischer–Tropsch reaction found in nature so far, and there is no direct evidence that oil and gas are produced by the natural Fischer–Tropsch reaction.

2.1.1.2 Inorganic Genesis of Natural Gas

In the theory of inorganic genesis, the mantle degassing theory has a significant impact. Based on the model of the formation and evolution of the solar system and the Earth, Gold (1993) believes that there is a large amount of methane and other non-hydrocarbon gases deep in the Earth. During the evolution of the Earth, these gases are heated from the deep Earth and released through deep large fractures or volcanic activities, and some are transported upward, forming natural gas reservoirs in the crust and upper mantle; some escape into the atmosphere. At present, more and more volcanic rock gas reservoirs are being discovered, and volcanic rock areas such as the East Pacific Rise, the Red Sea, Iceland, China's Wudalianchi, and Tengchong in Yunnan all have such gas reservoirs with this genesis.

The Geological Research Institute of the former Soviet Union, based on theoretical analysis and experimental simulation of this deep source gas, proposed that under the unique temperature and pressure conditions of the upper mantle, the liquid–gas phase is a huge gas storage of H_2 and hydrocarbons. During the migration process of deep source gas to the surface of the crust, the degassing action of H_2 and methane is controlled by structures. Thermodynamic calculations and simulations at a pressure of 65×10^6 Pa and a temperature of 1700 °C show that high pressure can suppress the thermal decomposition of hydrocarbons on the one hand, and promote the cyclization, polymerization, and condensation of hydrocarbons on the other hand. Thus evolving towards high molecular weight hydrocarbons can provide an experimental basis for the synthesis of oil from deep source gas.

In fact, this deep source gas theory has been used to guide the exploration of ultra-deep wells. The former Soviet Union had formulated an ultra-deep well plan for 11 areas to verify the deep source gas of the Earth. Among them, the SG-3 well on the Kola Peninsula of the Baltic Shield, which started in May 1975 and by December 1983, had drilled to 12,066 m, becoming the deepest well in the world. The well found bitumen inclusions and high concentrations of H_2, CH_4, He, N_2, and brine in the Archean gneiss and amphibolite of the Kola Group at a depth of 7,000 m, proving that there are non-biological methane and other gases deep in the crust.

In addition, inorganic methane gas with a mantle source has been seen in the places where new oceanic crust is formed in the Pacific and Atlantic ridges, in the Zaotang River of Tengchong County, Yunnan Province, and in the Wudalianchi volcanic group in Heilongjiang Province.

2.1.2 Organic Genesis of Oil and Gas

As early as in the mid-eighteenth century, the famous Russian scientist Lomonosov proposed the coal distillation theory, suggesting that oil is a product of coal undergoing high-temperature distillation underground. This is the earliest scientific hypothesis of oil genesis and the earliest organic genesis theory of oil and gas. The organic genesis theory of oil and gas that now holds the dominant position has been generated, deepened, and continuously improved in the large amount of production practice and scientific research in oil and gas exploration and development, and in turn, has effectively guided the world's oil and gas exploration practice.

2.1.2.1 Main Basis of Organic Genesis Theory of Oil and Gas

The organic genesis theory of oil and gas can be established not only because of the rationality of the theory itself but also because the distribution and composition characteristics of global oil and gas also support this theory.

(1) More than 90% of the oil fields discovered in the world are produced in sedimentary rocks, and there is no oil production in large areas of magmatic and metamorphic rocks. There are no industrial oil and gas accumulations found in areas with no relation to sedimentary rocks, such as shields and large crystalline basement uplifts.
(2) Oil has been found in sedimentary rock layers from the Precambrian to the Holocene of the Quaternary, but the distribution of oil and natural gas in the strata of each geological era is uneven. Most of the oil and gas are distributed in the strata since the Mesozoic, where the average content of dispersed organic matter is relatively high. Moreover, the oil from different places and periods has both similarities and differences. The temporal distribution of combustible organic minerals such as coal and oil shale also have this characteristic. This indicates that there is a genetic correlation between oil and coal, oil shale, and dispersed organic matter in sedimentary rocks.
(3) Analysis shows that the ash content of petroleum is rich in elements such as vanadium (2000 times), nickel (1000 times), copper (50 times), and cobalt (30 times) compared to the Earth's crust, and it also contains lead, tin, zinc, barium, silver, and other elements, with enrichment coefficients of more than 10, while the basic elements of sedimentary rocks (oxygen, silicon, aluminum, calcium, magnesium, sodium, potassium) have an enrichment coefficient in the ash content of petroleum not exceeding 5. The ash content of coal and petroleum has similarities in trace element composition. In living organisms, trace elements also have the similar distribution. These phenomena may indicate that coal and petroleum have a genetic correlation with organisms.
(4) Many oilfield test results show that the temperature of oil layers rarely exceeds 100 °C. In the light aromatic hydrocarbons of petroleum, the content of xylene > toluene > benzene, and petroleum contains porphyrins and other organic compounds that only exist stably at low temperatures. Obviously, petroleum may be generated under relatively low-temperature conditions.
(5) In addition to porphyrins, many biomarker compounds such as isoprenoid alkanes, terpenes, and steroids, which are unique to biological organic matter, have been found in petroleum. The similarity of the carbon isotope composition of petroleum to that of biological organic matter also reflects their genetic correlation.
(6) All types of hydrocarbons contained in petroleum have been detected in modern sediments and ancient sedimentary rocks. Many scholars have also shown that there is indeed an oil and gas generation process in modern sediments, and the amount of oil and gas generated is considerable. These all provide favorable scientific evidence for the organic genesis theory of oil and gas.

2.1.2.2 Development of Organic Genesis Theory of Petroleum

The establishment of the organic genesis theory of oil and gas is a long process, and many researchers have explored it and achieved effective results. In the early

1920s, the founder of geochemistry and Soviet mineralogist Vernadsky systematically studied the geological process of organic matter. In his main works "Introduction to Geochemistry" and "Biosphere," he detailed the organic composition of petroleum and the main basis for the organic genesis theory of petroleum, and proposed a carbon cycle model.

In the 1930s, Treibs (1933) first discovered and confirmed that porphyrin compounds are widely present in oil and asphalt of different ages and origins, and believed that these porphyrin compounds originate from plant chlorophyll. These chlorophylls can be transformed into deoxophyllin (DPEP) through biochemical and geochemical actions, which provided a very important basis for the organic genesis theory of petroleum. In addition, it was also found that petroleum has rotary polarization, which is also a strong support for the organic genesis theory, thus ushering in the prevalence of the organic theory, and various organic theories emerged one after another. Such as "plant theory", "animal and plant theory", "mixed theory", "fat theory", "carbohydrate theory", "protein theory", etc. In 1932, Gubkin (И.М. Иван Михайлович Губкин) proposed the concept of source rock and "mixed theory", believing that the mud containing various types of dispersed organic matter is the source rock for generating hydrocarbons. These dispersed organic matters can come from marine biological remains or from the decomposition products of terrestrial organisms. The source rock mainly produces dispersed petroleum (micro-oil) due to bacterial action in the early stage of rock formation, and then enters the porous rock layer with water during the compaction process, forming oil and gas reservoirs through further transportation and accumulation. Gubkin's views are considered the prototype of modern organic oil theory.

In the 1940s, Whitmore and others found hydrocarbons in algae and bacteria. It is estimated that the annual formation of hydrocarbons from modern marine plants alone can reach 600 million barrels (1 ton $\approx$ 7 barrels). Therefore, he believes that the oil generation process is simply the separation and aggregation of hydrocarbons in the organism. In 1947, Smith and others formed a research group and analyzed samples from the modern sediments of the Gulf of Mexico (0–7 m) and found liquid hydrocarbons similar to the components of petroleum hydrocarbons. By the early 1950s (1952–1954), Smith and Soviet scholars such as Vereshchagin successfully detected the major types of liquid hydrocarbons (paraffin, cyclane, aromatic hydrocarbon) that make up petroleum in various modern water body sediments (salty, brackish, and fresh), making the view that petroleum directly originates from the organic matter of modern sediments widely spread, and the early oil theory emerged. This theory believes that the original organic matter contained in sediment gradually transforms into oil and natural gas during the rock-forming process and is transported to the adjacent reservoir.

Since the mid-1950s, with the promotion and application of chromatography technology and the accumulation and comprehensive analysis of many organic geochemical data, the discussion on the micro-soluble organic components in the geological body and the evolution of sedimentary organic matter has become deeper. Research has found that the carbon number distribution of n-alkanes in modern sediments and organisms has odd–even predominance, and the carbon number distribution of normal fatty acids has even–odd predominance, while ancient sedimentary rocks and crude oil (or oilfield water) do not have this advantage (Bray et al., 1961). This discovery has strongly criticized the theory of direct oil formation from sedimentary organic matter, and more importantly, it has opened the prelude to the study of the evolution mechanism of organic matter and its relationship with oil formation.

Abelson (1963) proposed that oil is produced from the insoluble part (kerogen) that constitutes 70–90% of the organic matter in sedimentary rocks through a certain degree of burial evolution, and generated by pyrolysis during the late stages of diagenesis. Philippi (1965) pointed out that the transformation of a large amount of sedimentary organic matter into hydrocarbons requires a certain burial depth and temperature. Vassoevich (1969) first proposed that there is a main stage and phase in the generation of oil. Pusey (1973) found that liquid hydrocarbons are mainly distributed in the geothermal range of 65.6–148.9 °C; above this temperature, most are condensate gas fields, gas fields, and dry gas fields, while below this temperature, it is biogenic gas. This indicates that the distribution of oil has a certain temperature and depth range and is not in the shallow surface layer, thus supporting the theory of late-stage oil formation. Pusey called the temperature range in which liquid hydrocarbons (oil) exist in the crust the "liquid window", which defines the main range of oil accumulation. In areas with a high geothermal gradient, oil is shallowly distributed, and the liquid window is very narrow, such as the Minas oil field in central Sumatra, where the liquid window is only 911 m; conversely, in areas with a low geothermal gradient, oil is deeply distributed, and the liquid window is very wide. The burial depth of continental oil fields in the eastern part of China is generally between 1000 to over 3000 m.

In the mid-1970s, Albrecht and Durand (1976) studied a complete organic matter evolution profile from immature to mature and overmature in the Cretaceous Douala Basin in Cameroon. This is one of the best examples of organic matter evolution research.

The above research achievements have played an important role in the establishment and improvement of the theory system of the late oil formation theory. In the 1970s, the famous French geochemist Tissot established the kerogen thermal degradation hydrocarbon evolution model based on summarizing and generalizing the research results of predecessors. He proposed and perfected the late kerogen hydrocarbon theory, and summarized the laws of oil and gas formation, evolution, and distribution. So far, the modern theory of oil and gas formation has been basically established. It not only conforms to geological facts and has gradually been accepted by many petroleum geologists but also plays an important role in oil and gas exploration.

The origin of oil and natural gas is a very complex theoretical issue. Although the organic origin theory of oil and gas is increasingly perfected and plays an important role in oil and gas exploration practice, it does not deny the scientific value of the inorganic origin theory of oil and gas. In the past two decades, the rise of cosmic chemistry and new theories of Earth formation, the development and application of plate tectonics theory, and the deepening of isotope geochemical research have provided new theoretical basis for the inorganic origin theory of oil and gas. What's more, more and more researchers have noticed the important role of materials from the deep Earth in the transformation of sedimentary organic matter into oil and gas (hydrogenation and catalysis), which can be regarded as the integration of organic and inorganic origin theories of oil and gas.

In short, whether it is the organic origin theory or the inorganic origin hypothesis, there are still many issues to be further studied, such as the origin and distribution of hydrocarbons in the deep Earth and cosmic space, the detailed mechanism of the transformation of various original substances (including organic and inorganic substances) into oil and gas, what are the characteristics of the oil or natural gas generated by different original substances, and how to quantitatively determine the source rock layer and its hydrocarbon generation and expulsion efficiency. It is believed that with the development of modern science and technology and experimental means, the scientific research of the theory of the origin of oil and gas will become more in-depth.

2.2 Primary Materials for the Formation of Oil and Gas

The modern organic theory of oil and gas formation indicates that oil and gas are formed from the biological organic matter that has been preserved in sediments through processes of sedimentation and burial, undergoing certain biochemical and physicochemical changes. Moreover, oil and gas are merely one of the many forms of existence that this preserved organic matter takes during the process of burial and evolution.

2.2.1 Biological Organic Matter and Its Chemical Composition

Apart from water, organisms primarily consist of four biochemical components: proteins, carbohydrates, lipids, and lignin, all of which have relatively stable chemical compositions and structures (Fig. 2.1). Additionally, there are substances present in smaller amounts but with significant roles, such as nucleic acids, vitamins, and enzymes. Compared to petroleum, biological organic matter contains higher amounts of sulfur, oxygen, and nitrogen, yet carbon and hydrogen remain the main components, indicating the similarity between petroleum and these organic components. These components can be transformed into petroleum by simply removing the impurities like sulfur, oxygen, and nitrogen (Table 2.1).

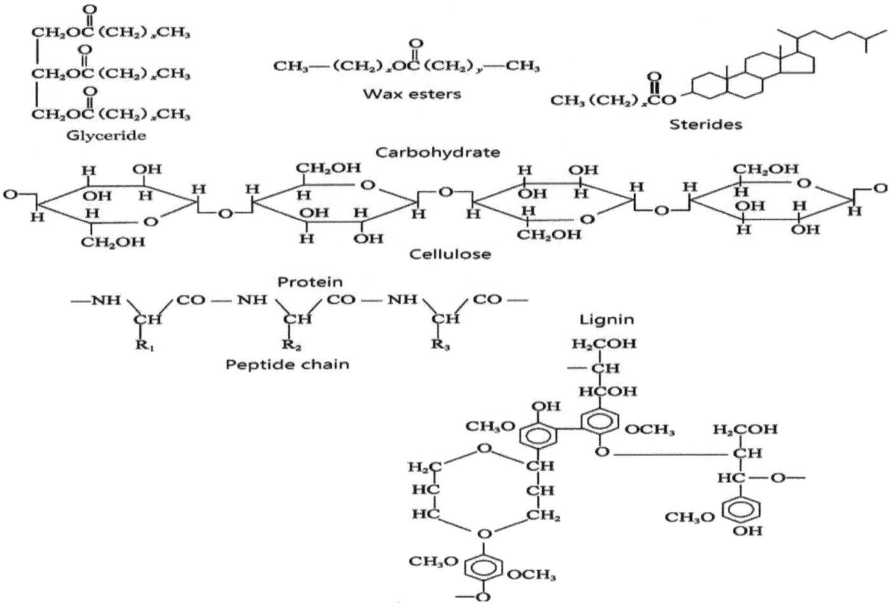

Fig. 2.1 Schematic diagram of the structure of several biochemical polymers (according to Huc, 1980; quoted from Houfu, 1999)

Table 2.1 Comparative elemental composition of biological organic matter, kerogen, and petroleum

Organic matter type	Elemental composition (%)				
	C	H	O	S	N
Lipids	76	12	12	/	/
Proteins	53	7	22	1	17
Carbohydrates	44	6	50	/	/
Lignin	63	5	31.6	0.1	0.3
Kerogen	79	6	8	5	2
Petroleum	84.5	13	0.5	1.5	0.5

2.2.1.1 Lipids

Also known as lipid compounds, they are one of the indispensable substances for an organism to maintain its vital activities, including fats, organic acids, steroid compounds, waxes, pigments, etc., with carbon and hydrogen being the main elements. They are generally composed of aliphatic chains or rings with a small number of oxygen-containing functional groups such as ester groups, hydroxyl groups, ether groups, and carboxyl groups. Lipids primarily originate from lower aquatic organisms with high reproductive rates, large biomass, and rich in lipids, such as fungi and algae.

Lipid compounds are insoluble in water but readily soluble in low polarity organic solvents such as diethyl ether, chloroform, and benzene. These compounds have the highest relative hydrogen content among organic matter in biological organisms and are easily hydrolyzed into simpler organic compounds in geological bodies. They often exist in the form of acids, esters, and alkenes. Therefore, lipid compounds found in geological bodies are typically various forms of organic acids and alcohols.

Lipid compounds have stable chemical properties and strong resistance to corrosion. They can be appropriately transformed under geological conditions and preserved in sediments. Moreover, they are the biological organic matter most like petroleum in chemical composition and structure. They can be transformed into petroleum by undergoing simple chemical changes to remove a small amount of oxygen and have historically been considered the most important source of petroleum.

2.2.1.2 Proteins

Proteins are highly ordered polymers composed of various amino acids and are the fundamental components of all tissues in organisms, serving as the material basis for the survival of organisms. Proteins are the primary carriers of nitrogen in biological organisms, with nitrogen accounting for about 16% of the mass of proteins. Nitrogen-containing compounds in petroleum may have a genetic connection with proteins in biological organisms.

In the elemental composition of proteins, carbon, oxygen, and nitrogen are the most abundant, and their chemical properties are unstable. Once they leave the biological organism and enter the water body, soil, and sediment, they are prone to break down into amino acids under the action of acids, bases, or enzymes. Amino acids are relatively stable in nature. Through processes such as hydrolysis and low-temperature pyrolysis, they can be transformed into hydrocarbons with low carbon numbers and nitrogen-containing compounds by decarboxylation and deamination. They can also form more complex geological polymers through condensation reactions.

It is generally believed that proteins are more conducive to oil generation and are the main source of low carbon number hydrocarbons and nitrogen-containing compounds in petroleum.

2.2.1.3 Carbohydrates

Carbohydrates, also known as sugars, include glucose, maltose, sucrose, starch, cellulose, etc. Their elemental composition is primarily carbon, hydrogen, and oxygen. They are widely distributed organic substances in nature and are most abundant in plants. As a source of hydrocarbons, carbohydrates attract attention due to their abundance.

Carbohydrates can be reduced by hydrogen to form hydrocarbons. However, most of them are easily consumed by aerobic bacteria or hydrolyzed into water-soluble substances, and they cannot stably exist in water bodies, soil, and sediments. The possibility of direct conversion into hydrocarbons is very small, and they may not be the main source of oil. But because they are easily decomposed and utilized by various microorganisms and transformed into microbial organisms, or directly converted into methane gas by microorganisms, they participate in the formation of oil and gas. Cellulose, which has relatively good stability, if in stagnant swamp environments of water bodies, can be piled up and evolved into coal, and is one of the sources of aromatic hydrocarbons and natural gas.

2.2.1.4 Lignin and Tannins

Both lignin and tannin have aromatic structural characteristics and are derived from higher plants, with the composition marked by having the structure of phenols. Among them, lignin is a high-molecular-weight polyphenolic compound and the basic constituent of the woody part of higher plants. It is an aromatic cluster polymer compound with strong anti-corrosion ability (>cellulose) and is very stable in nature and not easily hydrolyzed, but can be oxidized to aromatic acids and fatty acids. In anoxic water bodies, it can be decomposed under the action of water and microorganisms and generate humic substances with other compounds. It is the main substance for coal formation and can also generate natural gas and aromatic hydrocarbons. The tissue and characteristics of tannin are between lignin and cellulose and mainly appears in higher plants. In addition, there is a series of phenols and aromatic acids and their derivatives widely distributed in plants. They are the main source of aromatic structure in sedimentary organic matter and an important organic component for coal formation.

Different organisms contain different biochemical components, hence the compositions and characteristics of organic matter provided by different organisms in different geological periods and sedimentary environments must also be different. Plants are rich in carbohydrates while animals are rich in proteins, and the content of lipids varies greatly in organisms. It is usually higher in animals, lower plants, and certain tissues of higher plants. Lignin is a characteristic component of higher plants. During the evolutionary history of Earth's life, the status of different types of organisms in providing oil-prone organic matter is different. Planktonic plants, due to their rapid reproduction, vast living space, and long evolutionary process, are considered the main providers of organic matter in geological history. Bacteria, due to their physiologically great variability and adaptability to the environment, are almost ubiquitous, and their contribution to organic matter is second only to planktonic plants. Higher plants appeared in the Silurian period, much later than planktonic plants and bacteria, and their contribution to organic matter is third. Planktonic animals are also considered to be the main providers of organic matter in geological history. As their survival depends on the development of planktonic plants, so planktonic animals have significant importance in the contribution to organic matter in areas of high planktonic plant productivity. The contribution of other large swimming animals and terrestrial animals to organic matter can be neglected.

2.2.2 *Sedimentary Organic Matter*

In a broad sense, sedimentary organic matter refers to all organic substances that are preserved in sediments or sedimentary rocks. They are the general term for all organic matters that are directly preserved or further evolved after the deposition of biological remains and biological secretions and excretions along with inorganic particles.

Considering the source of biological matter, there are three sources of sedimentary organic matter: primarily from the autochthonous organic matter of the basin itself, allochthonous organic matter carried into the basin by rivers, wind, etc., from the land, and redeposited organic matter formed by the re-deposition of previously deposited organic matter due to factors such as rock weathering.

The biological organic matter that enters sediment undergoes varying degrees of decomposition under different redox conditions. A portion of the decomposition products is utilized by microorganisms as an energy source, thereby participating in the recycling of organic carbon in the biosphere. Another part of the decomposition products is transformed into simple molecules through physical–chemical actions, such as CO_2, H_2O, etc. Only an extremely small part of the organic matter does not undergo complete recycling and physical–chemical decomposition, thus forming sedimentary organic matter. The organic matter formed in this way includes some molecular products obtained through selective decomposition of biological precursors, partially or entirely inheriting their original structures, such as amino acids, peptides, monosaccharides, polysaccharides, lipids, phenols, lignins, etc. Some are preserved directly in the form of biological tissues without significant alteration. These organic matters with different chemical components in the sediment can be recombined into some complex components, such as humic acid, kerogen, etc., as the burial depth of the sediment increases, and evolve into various types of products and by-products in different stages of diagenesis, thus forming a series of sedimentary organic matter on the sedimentary profile (Fig. 2.2).

2.2.3 *Kerogen*

2.2.3.1 The Definition, Distribution and Formation Process of Kerogen

The term "kerogen" comes from the Greek word keros, which means a substance capable of generating oil or wax-like materials. In 1912, A. G. Brown first introduced this term to represent the organic matter in Scottish oil shales, which can produce substances like petroleum when pyrolyzed. Subsequently, this term was often used

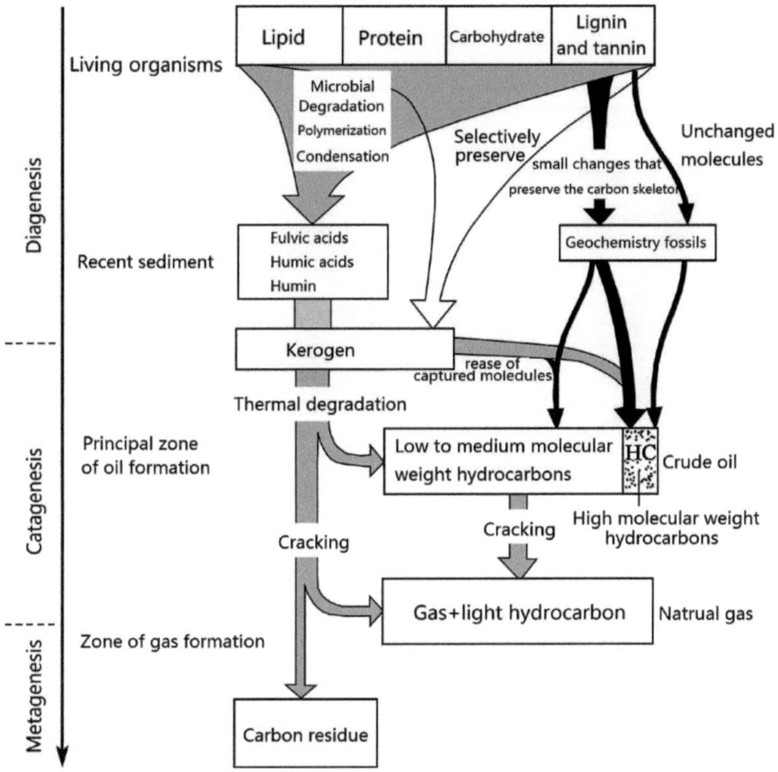

Fig. 2.2 The transformation of organic matter in nature (slightly modified from Tissot & Welte, 1984)

to represent the organic matter in oil shales and algal coals. It was not until the 1950s that it was officially defined to represent the insoluble organic matter in sedimentary rocks. In 1979, Hunter defined kerogen as all dispersed organic matter in sedimentary rocks that is insoluble in non-oxidative acids, bases, and common organic solvents. Correspondingly, the part of the rock that is soluble in organic solvents is called bitumen.

Kerogen is finely dispersed in sediments and sedimentary rocks, appearing as a brown to black powder, and under a microscope, it exhibits spherical, rod-like, and amorphous shapes. It is mainly found in clay rocks and micritic carbonate rocks, and it is the most important form of organic carbon in the Earth's crust. It contains 1000 times more organic carbon than coal and petroleum in reservoirs, and 50 times more than the total amount of dispersed soluble organic carbon in non-reservoirs. Statistically, in terms of the quantity of organic matter, kerogen is the most widely distributed and significant type of sedimentary organic matter, accounting for approximately 95% of the total organic matter in geological bodies (Fig. 2.3). It is estimated

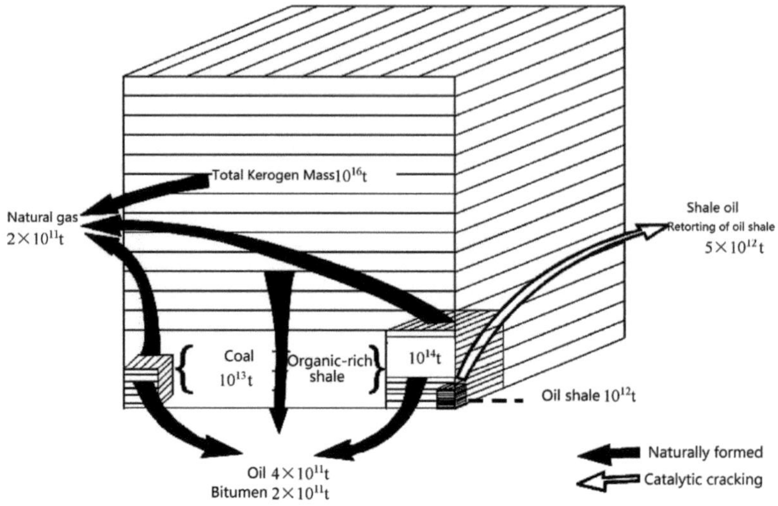

Fig. 2.3 Comparison of the quantity of kerogen with the maximum resources of fossil fuels (according to Durand, 1980)

that the average content of kerogen in rocks is 0.3%, and the total amount of kerogen in the Earth's crust is about 10^{16} tons.

The formation of kerogen is a complex issue. The traditional view is that after the deposition of sediment and during the early stages of diagenesis, various biological polymers from the aqueous environment will undergo a series of transformations. Large molecular biological polymers (lignin, carbohydrates, proteins, lipids) are degraded into soluble monomeric organic matter (phenols, monosaccharides, amino acids, fatty acids) through hydrolysis and the action of microbial enzymes. Among these monomeric organic matters: some are decomposed, oxidized and consumed by microorganisms. Some are converted into biogas CH_4 under the action of bacteria. Some of the components mainly with fatty acids are converted into sapropelic substances after sapropelization, including a small amount of petroleum hydrocarbons, non-hydrocarbon compounds containing sulfur, nitrogen and oxygen and other insoluble lipid polymers, and a small amount of soluble sapropelic substances can also be synthesized through pure chemical decomposition of original lipids. Most of the rest are converted into humic substances (fulvic acid, humic acid, humin) through humification. The insoluble humin and other components go through re-condensation, and some oxygen-containing groups and nitrogen-containing groups are removed in the forms of H_2O, CO_2 and NH_3, and sulfur in the environment will also join into the organic structure under reducing conditions, thereby forming a small amount of soluble bitumen (hydrocarbons, non-hydrocarbons) and many neutral organic polymer kerogens that are insoluble in acids, alkalis and organic solvents.

However, an increasing number of studies indicate that the above genesis only applies to the amorphous components of kerogen, and cannot explain the components with obvious biological structures that are abundant in kerogen. Therefore, researchers proposed the theory of selective preservation. This theory suggests that there are many stable tissues in organisms, such as spores, pollen, cuticles, cork layers, and algal bodies, etc., which have relatively strong resistance to weathering and microbial degradation (Nip et al., 1986). Although they are present in small amounts in organisms, during the diagenesis process, they become relatively enriched as the more abundant and rapidly hydrolyzable biological polymers (such as proteins and carbohydrates) continue to degrade, thus becoming a part of the kerogen.

2.2.3.2 The Macerals of Kerogen

Observing kerogen under various microscopes reveals that it is composed of Macerals with different colors, shapes, and structures (Table 2.2). The sapropelic group mainly originates from algae and other aquatic organisms as well as bacteria. The exinite is derived from the spores, pollen, cuticles, resins, waxes, and cork layers of terrestrial plants; the vitrinite comes from the structured and unstructured woody fibers of plants; the inertinite originates from the carbonized wood fiber parts. Although the exinite has a relatively large number of sub-microscopic components, it only accounts for 2–10% of the total amount of kerogen.

Alginite is the algal remains with a certain structure, having a relatively complete form with a clear outline. The outer edge of colonial algae is irregular, and the surface shows a honeycomb-like or sponge-like structure. Under transmitted light, it appears yellow, yellowish-brown, or light greenish-yellow. Under reflected light, it appears dark gray with slight protrusions. It has strong fluorescence.

Amorphous refers to organic components that do not have a fixed shape and structure when observed under a microscope, often appearing as irregular lumps, flocculent, or cloud-like structures. In terms of weight, it is the most important type of Macerals. Under transmitted light, it appears bright yellow, brown-yellow, or brown, and is transparent to opaque. Under electron microscopy, it appears as granular (or particulate), overlapping or piled up. The origin of amorphous material is relatively complex, and it is generally considered to be the product of complete decomposition of aquatic organisms (such as algae).

Sporo-pollinite includes spores and pollen from herbaceous, woody, aquatic, and terrestrial plants. They often appear as single units in shapes such as round, oval, triangular, or polygonal, and sometimes as aggregates. The surface may have various decorations or protrusions, and the color ranges from yellow-green to brown.

Gutinite originates from the epidermal tissue of plants, typically consisting of a single layer of cells that encase leaves, herbaceous stems, buds, and young roots. Under the microscope, it often appears as a slender ribbon, with a smooth outer edge and a serrated or wavy inner edge.

Resinite comes in many shapes, often oval or spindle-shaped, with a clear structureless outline. Under the microscope, it often appears lemon-colored.

Table 2.2 Classification of kerogen Macerals based on transmitted light

Micro-component	Sub-component	Original organic matter	Oil generation potential	Reflectance
Sapropelic group	Alginite, amorphous	Algae and other lower aquatic organisms and bacteria. Products of sapropelization, relatively hydrogen-rich	↓ Oil	↓ Reflectance
Exinite	Sporo-pollinite, cutinite, resinite, suberinite	Terrestrial plant spores, pollen, cuticle, resin, wax, and cork layers, relatively hydrogen-rich	Generation	
Vitrinite	Telinite	Plant structural and non-structural woody fibers, from higher plants	Potential	Increases ↓
Inertinite	Fusinite	Carbonized wood fibers, from forest fires, redeposited organic matter, relatively oxygen-rich	Decreases ↓	

Suberinite has distinct cell wall and cell cavity structures. The cells resemble slabs or large mesh patterns, arranged regularly, with no gaps between cells. The outlines are generally straight. The color is yellow or brown-yellow.

Telinite has a relatively clear woody structure, and even after intense decomposition, the gelified components can still be distinguished by color to identify cell traces.

Colinite is a general term for gelified components that have undergone intense decomposition and have completely lost their cellular structure. Under transmitted light, it often appears as uniform elongated slabs, small segments, irregular or regular bands, mostly in colors ranging from orange-red to brown–red, transparent to translucent. Because this Maceral is a typical humic substance with a uniform structure, it is commonly used to determine reflectance values, and its reflectance (R_o) can reflect the degree of organic matter evolution and maturity.

Fusinite is formed from the woody parts of higher plants through intense carbonization. It has shapes such as broken blocks, fragments, bands, or ovals. Under transmitted light, it is black and opaque.

2.2.3.3 The Elemental Composition and Chemical Structure of Kerogen

(1) The Elemental Composition of Kerogen

Kerogen is primarily composed of carbon, hydrogen, oxygen, and trace amounts of sulfur and nitrogen. The average weight percentages are approximately 76.4% for carbon, 6.3% for hydrogen, 11.1% for oxygen, 3.65% for sulfur, and 2.02% for nitrogen, which means that the content of carbon and hydrogen is lower than that in petroleum, while the content of oxygen, sulfur, and nitrogen is much higher than that in petroleum (Table 2.3). Looking at the relative ratios of the constituent elements in kerogen, for every 1000 carbon atoms, there are approximately 500–1800 hydrogen atoms, 25–300 oxygen atoms, 5–30 sulfur atoms, and 10–35 nitrogen atoms. Table 2.3 shows the elemental composition of some representative kerogens from both domestic and international sources.

Both the Macerals and elemental composition of kerogen are not fixed and unchangeable. It not only changes with the original parent material of kerogen and the change of the generation environment, but also changes with the geological and geochemical actions suffered by kerogen. Therefore, the lab-measured Macerals and elemental composition usually are only the composition characteristics at a certain specific geological time–space point where kerogen is located.

(2) The Chemical Structure of Kerogen

Studies have shown that kerogen is a high molecular weight polymer, without a fixed chemical composition, nor a fixed molecular formula and structural model. Therefore, the structure of kerogen refers to a structural model that is artificially constructed through a reasonable synthesis of internal composition information obtained by various analytical methods.

Tissot and Espitalie (1975), based on information provided by various analytical techniques, proposed a general structural model for amorphous kerogen. The general amorphous kerogen is a large molecule that forms a three-dimensional network system, which is likely composed of nuclei cross-linked by bridges, and both the nuclei and bridges can have functional groups. In addition, lipid molecules can be trapped within the kerogen matrix, acting similarly to a molecular sieve (Fig. 2.4).

a. Nuclei

Each nucleus is composed of 2–4 basic building blocks stacked to varying degrees of parallelism, with each basic building block typically containing two layers of aromatic plate-like bodies. Each aromatic cluster plate or laminar body contains a smaller number (less than 10) of fused aromatic cluster cyclic compounds and a small amount of sulfur-, nitrogen-, and oxygen-containing heterocyclic compounds, with the diameter of the plate-like body being less than 10×10^{-10} m. The number of layers in each stack is often two, with the interlayer distance of the two plate-like bodies being around 3.4×10^{-10} m. This distance is wider in the shallow layers and narrower in the deeper layers, with the maximum value being around 3.7×10^{-10} m. The stack is the basic structural unit of kerogen.

Table 2.3 Elemental composition of kerogen in some continental oil source layers at home and abroad (Difan, 1991)

Types of kerogen	Basin or areas	Layers	Weight percentage					Atomic ratio	
			C	H	O	N	S	H/C	O/C
I_1	Biyang	HTY Fm. (O)	63.14	8.68	13.54	1.36		1.68	0.16
	Nanyang	HTY Fm. (O)	62.96	8.71	11.33	0.86		1.66	0.14
	Daqing	QSK Fm. (LK)	79.95784	9.21	3.61			1.39	0.03
	Einta Basin	QSK Fm. Shale		9.6	8.6	0.7	2.7	1.44	0.12
I_2	Biyang	HTY Fm. (O)	76.55	8.85	12.91	2.03		1.37	0.13
	Nanyang	HTY Fm. (O)	77.85	8.49	7.86	1.54		1.32	0.08
	Daqing	QSK Fm. (LK)	72.40	8.06	7.65			1.34	0.08
	Fushun	Oil shale (Pa)	75.65	9.04	11.56	2.40	1.35	1.43	0.12
II	Nanyang	HTY Fm. (O)	74.47	7.78	19.86	1.89		1.25	0.20
	Fushun	Oil shale (Pa)	71.82	7.94	15.95	2.73	1.56	1.33	0.17
	Maoming	Oil shale (Pa)	73.74	8.22	13.30	2.69	2.05	1.34	0.14
	Paris Basin	Lower Toarcian	72.7	7.9	12.5	2.1	4.8	1.30	0.13
III_1	Ordos	YC system (T)	68.28	5.47	13.01			0.96	0.14
	Ordos	YC system (T)	79.47	5.55	9.23			0.84	0.09
III_2	Ordos	YA system (T)	77.56	5.08	15.04			0.79	0.15
	Fushun	SC (Pa)	69.62	5.39	21.03	3.43	0.53	0.93	0.23
	Maoming	Lignite (Pa)	70.54	5.45	20.21	2.83	1.01	0.92	0.22
	Douala Basin	Upper Cretaceous	72.8	6.01	18.9	2.28		0.99	0.19

Note HTY Fm.—Hetaoyuan Formation; QSK Fm.—Qingshankou Formation; SC—Subbituminous Coal; O—Oligocene; LK—Lower Cretaceous; Pa—Paleogene; T—Triassic

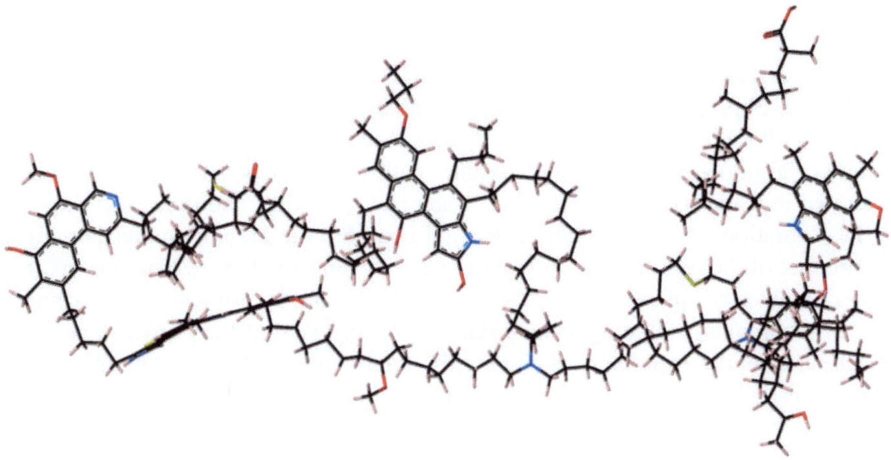

Fig. 2.4 Schematic diagram of immature Type I kerogen structure, with black representing carbon, gray representing hydrogen, red representing oxygen, blue representing nitrogen, and yellow representing sulfur (Ungerer et al., 2014)

b. Bridging Bonds Linked with Nuclei

The nuclei are connected by various bridging bonds, which are typically aliphatic chains, functional groups containing sulfur or oxygen, and so on. For example: Ketone—C(=O)—, Ester—C(=O)—O—, Ether—O—, Sulfide—S—or disulfide—S—S—, aliphatic ester—C(=O)—O—R。

c. Surface Functional Groups Located on the Nuclei or Bridging Bonds

Nuclei or bridging bonds can have various functional groups, such as hydroxyl groups, Carboxyl group, Methoxy group. For example: hydroxyl—OH, carboxyl—C(=O)—O—H, methoxyl—O—CH_3.

d. Kerogen Has the Characteristics of a Molecular Sieve

Similar to what is observed in coal. After thorough extraction, the kerogen separated by acid treatment can still release hydrocarbon molecules upon further extraction.

(3) Classification of Kerogen

Due to the fact that the composition of kerogen is controlled by the environment in which it forms and the source of the original organic matter, there are different types of kerogen. The issue of kerogen types is a question of the quality of the source material for oil generation. It not only controls the direction of kerogen evolution but also the rate and quantity of hydrocarbon generation.

a. Classification Based on Original Organisms

Based on original organisms, organic matter is divided into two types: sapropelic and humic. Sapropelic organic matter is primarily the product of the decomposition and polymerization of aliphatic organic matter under anaerobic conditions. It mainly originating from aquatic plankton, benthic organisms, and aquatic plants. It is formed in water basins with stagnant water and reducing conditions, and can lead to the formation of oil, oil shale, and algal coal. Humic organic matter is mainly the product of the peatification of plant material, primarily derived from the organic matter of terrestrial plants (especially higher plants), and is formed in oxic sedimentary environments, including swamp lakes or related sedimentary environments. It is rich in aromatic structures such as lignin, tannin, and cellulose. It mainly forms natural gas and humic coal, and under certain conditions, it can also generate liquid petroleum.

This classification method is overly simplistic and far from meeting the needs of oil and gas exploration. With the advancement of analytical techniques, researchers from different disciplines have proposed classification schemes for kerogen from various perspectives. Currently, the classifications that are widely adopted are primarily based on Macerals and elemental composition.

b. Classification Based on the Relative Content of Macerals

Kerogen is divided into several types based on the relative content of its Macerals (Table 2.4). Two main methods are used: one is to calculate the ratio of the sum of sapropelic and exinite to the vitrinite; the other is to calculate the type index (T value) based on the relative percentage content of each Maceral in the kerogen, and then classify the types of kerogen according to the T value. The classification criteria can be found in Table 2.4.

$$\begin{aligned} T = {} & (\text{sapropelite content} \times 100 + \text{exinite content} \times 50 + \text{vitrinite content} \times (-75) \\ & + \text{inertinite content} \times (-100))/100 \end{aligned}$$

Table 2.4 Microscopic identification and classification criteria of kerogen

Type	Index		
	Relative content method		T value method
	Sapropelic + exinite (%)	Vitrinite (%)	(%)
I	>90	<10	>80
II_1	65–90	10–35	40–80
II_2	25–65	35–75	0–40
III	<25	>75	<0

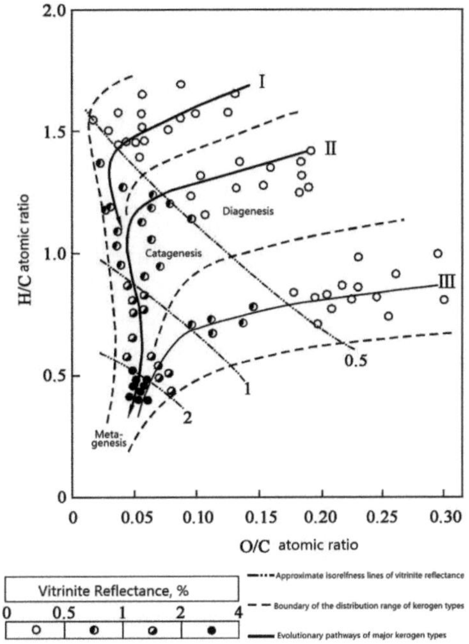

Fig. 2.5 Van Krevelen diagram of kerogen types (simplified after Tissot & Welte, 1984)

c. Classification Based on Elemental Composition

The French Institute of Petroleum, based on the carbon, hydrogen, and oxygen elemental analysis results of the kerogen samples, uses the Van Krevelen diagram to classify the kerogen into three main types (Fig. 2.5).

Type I Kerogen

This kind of kerogen has a high original H/C atomic ratio (1.25–1.75) and a low original O/C atomic ratio (0.026–0.12). There are many chain structures, rich in lipids and decomposition products of proteins, with low content of aromatic structures and heteroatomic bonds. It is mainly derived from the remains of aquatic lower organisms such as algae and bacteria, and is rich in ^{12}C. The main Maceral is sapropel. It is usually formed in quiet and oxygen-poor shallow water rich in organic matter silt. Pyrolysis analysis shows that this kind of kerogen has very high oil generation potential and is the most important oil-generating parent material.

Compared with other types, type I kerogen appears less frequently in nature. Typical examples are the kerogen in the Eocene Green River oil shale in Colorado, Utah, and Wyoming in the United States, and the kerogen in typical lacustrine sediments such as the first member of the Qingshankou Formation and the first member of the Nenjiang Formation in the Lower Cretaceous of the Songliao Basin in China, and the Hetaoyuan Formation of the Paleogene in the Biyang Basin.

Type II Kerogen

Type II kerogen is one of the most common kerogens, with the original H/C atomic ratio of 0.65–1.25 and the O/C atomic ratio of 0.04–0.13. It contains a large amount of medium-length straight-chain alkanes and cycloalkanes, as well as many polycyclic aomatic hydrocarbons and heteroatomic functional groups. It is a mixed organic matter mainly from marine plankton, plants and microorganisms. When pyrolyzed, the hydrocarbon types produced are less than those of Type I kerogen, and the oil generation potential is medium. The kerogens in the Lower Toarcian of the Jurassic in the Paris Basin of France, the Silurian in North Africa, the Cretaceous in the Middle East, the Devonian in Western Canada, and the third member of the Paleogene Shahejie Formation in the Dongying Sag of China all belong to this type.

Type III Kerogen

Type III kerogen has a relatively low original H/C atomic ratio (0.46–0.93) and a relatively high original O/C atomic ratio (0.05–0.30). There are many aromatic structures and oxygen-containing functional groups. There are very few saturated hydrocarbons and only a small number of aliphatic structures with mainly methyl and short chains. It is mainly derived from the organic matter of terrestrial higher plants rich in lignin, cellulose and tannin. Compared with Type I and Type II kerogens, the pyrolysis products are very few and the oil generation potential is poor. But at the high-maturity stage, a considerable amount of methane gas can be formed, and the main mineralization products are coal, aromatics, and natural gas. The kerogens in the Upper Cretaceous of the Douala Basin in Cameroon and the Yan'an Formation of the Lower Jurassic in the Shaanxi-Gansu-Ningxia Basin in China belong to this type.

In addition to the above three types of kerogens, people define the organic matter originated from the residual organic matter after forest fires and the organic matter caused by redeposition as Type IV kerogen (residual type). The original H/C atomic ratio of this type of kerogen is extremely low (0.5–0.6), and the O/C atomic ratio is as high as 0.25–0.3. It contains a large number of aromatic nuclei and oxygen-containing groups, and the main maceral is the inertinite. Type IV kerogen hardly produces oil, but can generate a small amount of gas.

Combined with the characteristics of continental hydrocarbon source rocks in China, Hu Jianyi and Huang Difan (1991) put forward a five-classification scheme that is widely accepted in the petroleum geology community in China. Kerogens are divided into three categories and five types: standard sapropelic type (I_1), sapropelic type with humus (I_2), intermediate type (II), humic type with sapropel (III_1) and standard humic type (III_2). The specific classification criteria are shown in Table 2.3, Figs. 2.6 and 2.7.

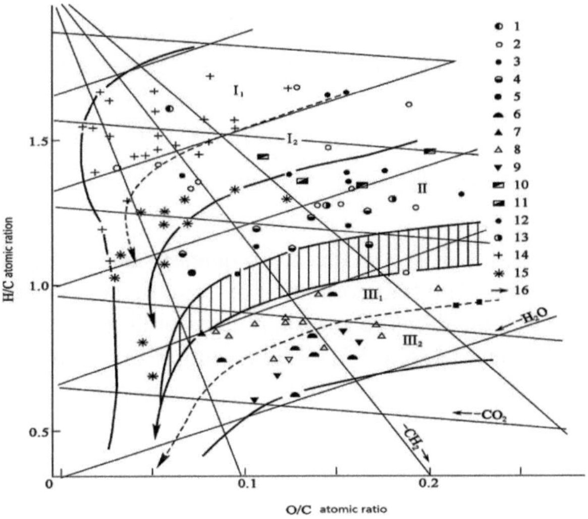

Fig. 2.6 Evolution diagram of continental kerogen types (Difan, 1991). 1—Daqing Oilfield; 2—Nanyang Oilfield; 3—Biyang Basin; 4—Langgu Sag; 5—Liaohe Depression; 6—Qaidam Basin; 7—Sichuan Basin; 8—Ordos Basin (T); 9—Ordos Basin (J); 10—Fushun Oil Shale; 11—Maoming Oil Shale; 12—Fushun and Maoming Coal; 13—Dongying Sag; 14—Green River Shale and Algae; 15—Lower Toarcian Shale; 16—Evolutionary Path

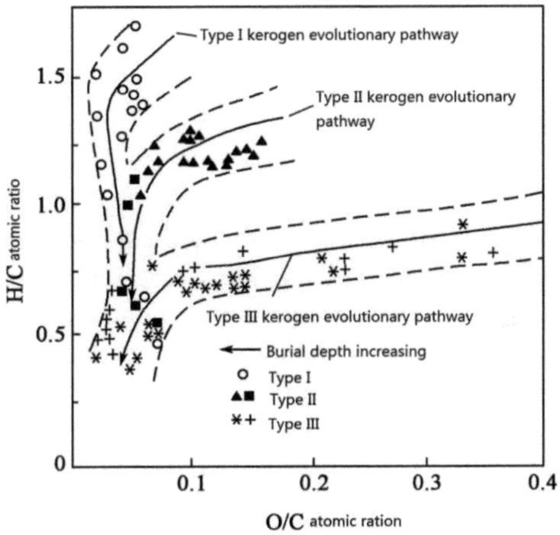

Fig. 2.7 Elemental analysis diagram of kerogens from different sources. Type I: ○ Green River Shale in the Uinta Basin of the United States (Tissot & Welte, 1978). Type II: ▲ Lower Toarcian Shale in the Paris Basin of France (Durand et al., 1972), ■ Posidonia Shale in the Rias period in Germany (B. Durand). Type III: * Logbaba Shale in the Douala Basin of Cameroon (Durand et al., 1976) + humic coal (Durand et al., 1977)

2.3 Factors Influencing Organic Matter Evolution and Hydrocarbon Generation and Their Patters

2.3.1 *Factors Affecting the Evolution of Organic Matter into Hydrocarbons*

The Earth's crust contains a vast quantity and variety of organic matter with complex structures. The transformation of these organic matters found in sedimentary rocks into hydrocarbons requires a process of deoxygenation, hydrogenation, and carbon enrichment (Table 2.5). The evolution of sedimentary organic matter, including kerogen, into oil and natural gas is essentially a series of complex chemical reactions. The realization of these chemical reactions is the result of a combination of various environmental factors. Therefore, the generation of oil and gas must also be realized under specific environmental conditions. Oil and gas exploration practices and theoretical research worldwide have shown that temperature and time are the main factors affecting oil and gas generation, while others such as bacteria, catalysts, radioactivity, and pressure also have certain impacts.

2.3.1.1 Temperature and Time

Research has shown that the transformation of sedimentary organic matter into petroluem is a thermal degradation process, and an increase in temperature is conducive to accelerating the speed of chemical reactions. French scholar Connan (1974) proposed that the transformation of sedimentary organic matter into oil conforms to a first-order reaction in chemical kinetics. At any given moment, the reaction rate is only related to the concentration of the reactant.

$$-dC_A/dt = KC_A \tag{2.1}$$

Table 2.5 Elemental composition of organic matter and oil at different depths (according to C.E. ZoBell, 1947)

Depth	Type of material	Carbon (%)	Hydrogen (%)	Oxygen (%)	Nitrogen (%)	Sulfur (%)
Shallow ↓ Deep	Marine humic mud	52	6	30	11	0.8
	Recent sediments	58	7	24	9	0.6
	Ancient sediments	73	9	14	0.3	0.3
	Oil	85	13	0.5	0.4	0.1

where: C_A—the concentration of the reactant (kerogen) at time t; t—reaction time; K—reaction rate constant (degradation rate); the negative sign—the concentration of the reactant decreases as the reaction proceeds and the product increases.

The value of K can be obtained from the Arrhenius equation (Arrhenius, 1889):

$$K = A \cdot e^{-\frac{E}{RT}} \tag{2.2}$$

where: A—the pre-exponential factor (frequency factor): the number of collisions of particles per unit time per unit volume (times/S cm^3);

R—the gas constant (8.315 J/K mol); E—activation energy of the reactant;

T—absolute temperature (K) (273.15 + t°C); e—the base of natural logarithm (2.71828).

According to the Arrhenius equation, the rate of oil generation from the thermal cracking of organic matter increases exponentially with an increase in temperature. Increasing the temperature or reducing the activation energy can speed up the reaction. When the temperature is too low, K is small. As the temperature rises, K increases and the rate of hydrocarbon formation accelerates.

By establishing a connection between Eqs. (2.1) and (2.2) and after some calculations, the time–temperature quantitative relationship equation can be obtained:

$$lnt = \frac{E}{R} \cdot \frac{1}{T} - Constant \tag{2.3}$$

That is, the logarithm of the reaction time ln(t) is inversely proportional to the reaction temperature T, indicating that there is a compensatory relationship between time and temperature during the oil formation process.

The degree of thermal evolution of organic matter is equivalent to the extent of the reaction. For the reaction of a specific type of sedimentary organic matter to evolve into oil and gas, the geothermal temperature is an important factor in determining the reaction rate. To achieve the same degree of reaction, when the reaction rate is fast, the required time is short, and when the reaction rate is slow, the required reaction time is long. Therefore, temperature and time jointly determine the degree of evolution of organic matter, and temperature and time can compensate for each other. The artificial simulation experiments conducted by the French Petroleum Institute have proven that the results and laws obtained from high-temperature rapid simulation in the laboratory are consistent with the low-temperature slow evolution in nature.

When the temperature rises to a certain value, organic matter begins to be transformed into oil and natural gas in large quantities. This temperature limit is called the maturity temperature or threshold temperature of organic matter, and the corresponding depth is called the threshold depth, which is the hydrocarbon generation threshold. The age of the source rock, the type of kerogen, and the geothermal gradient all have an impact on the threshold temperature for hydrocarbon generation. Connan (1974) synthesized an analysis of the threshold temperatures of different types of hydrocarbon source rocks of different ages in several oil and gas basins around the world and found that the older the hydrocarbon source rocks, the lower the threshold temperature for hydrocarbon generation, and the younger the hydrocarbon source rocks, the higher the threshold temperature for hydrocarbon generation. Difan (1991)

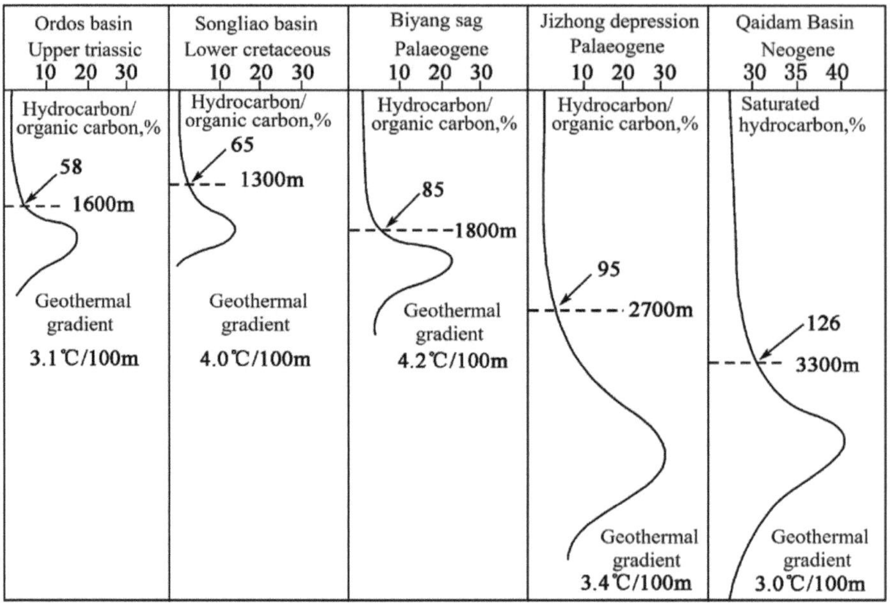

Fig. 2.8 The relationship between the burial depth of hydrocarbon source rocks and oil and gas generation in different basins in China (Difan, 1991)

summarized the relationship between the burial depth of hydrocarbon source rocks of different ages in China's main oil and gas basins and the generation of oil and gas (Fig. 2.8), also indicating that the older the hydrocarbon source rocks, the lower the threshold temperature for hydrocarbon generation. These studies have confirmed the compensatory effect of time in the process of oil and gas generation.

It should be emphasized that among the two factors of temperature and time, the impact of temperature on oil and gas generation is decisive, while the role of time is secondary. When the temperature is low enough, the rate of oil and gas generation will be very slow, and the time required for the organic matter to evolve to maturity is very long, which is meaningless for oil and gas generation. In fact, there will be no commercially significant oil and gas generation. This indicates that for the chemical reactions of oil and gas generation, the compensation of time only makes sense when the temperature reaches a certain level. The basins that are conducive to the generation and preservation of oil and gas should be young thermal basins (with high geothermal gradients) and ancient cold basins (with low geothermal gradients).

2.3.1.2 Other Influencing Factors

(1) Bacterial Activity

Bacteria are the most widely distributed and rapidly reproducing organisms on Earth, found in most natural water systems and in shallowly buried (less than 1000 m) sediments. They play an important role in the transformation of organic matter, the generation, and degradation of oil and gas. The essence of the role of bacteria in the process of oil and gas generation is to separate elements such as oxygen, sulfur, nitrogen, and phosphorus from organic matter, enriching carbon and especially hydrogen, and the longer the bacterial action time, the more thorough this element separation and enrichment is carried out.

For oil and gas generation, the most significant are anaerobic bacteria. Under reducing conditions lacking free oxygen, on the one hand, by consuming carbohydrates in the original sedimentary organic matter and continuously adding bacterial remains, the oxygen content in organic matter is reduced and the hydrogen content is increased, making it more conducive to oil generation On the other hand, bacterial activity directly participates in the process of oil and gas generation. Organic matter is decomposed by anaerobic bacteria to produce methane, hydrogen, carbon dioxide, organic acids, and other hydrocarbons. Bacteria can also selectively decompose plants, separating many hydrocarbons originally synthesized and directly burying them in sediments.

(2) Catalytic Action

In the 1960s, geologists found that the enrichment of oil has a certain positive correlation with the content of montmorillonite in the adjacent strata. Experimental simulations have shown that clay minerals have varying degrees of catalytic action during the thermal cracking process of kerogen, with montmorillonite having the strongest catalytic ability, followed by illite. Under normal circumstances, the presence of clay minerals will increase the total yield of oil and gas and the relative proportion of gaseous hydrocarbons and aromatic compounds. In recent years, research has shown that the catalytic action of clay minerals is related to the experimental heating rate, that is, rapid heating will amplify the catalytic action of clay. Therefore, the catalytic action of clay under geological temperature conditions is more moderate than under thermal simulation conditions.

In addition, Mango (1992) pointed out that transition metals are key catalysts for the conversion of normal alkanes (hydrogenation) into light hydrocarbons and natural gas. Heavy metal elements have obvious catalytic effects on the hydrogenation of olefins, acetylenes, and aromatics, hydrocarbon dehydrogenation cyclization, aromatization, isomerization, and hydrogenolysis of C–C bonds. Leventhal (1986) found through thermal simulation experiments that the presence of uranium elements would promote the transformation of kerogen into oil and gas at lower temperature conditions, and the gas-oil ratio of the pyrolysis products without adding uranium elements was higher.

(3) Pressure Effect

With the development of organic geochemistry and the breakthrough in deep oil and gas exploration, the role of pressure in the process of organic matter generating hydrocarbons has attracted more and more attention. From the perspective of chemical theory, the increase in pressure can inhibit the progress of chemical reactions that increase in volume. The hydrocarbon generation reaction of kerogen, especially the generation of gaseous hydrocarbons at higher temperatures, is obviously a reaction that increases in volume. Therefore, theoretically, high pressure will hinder or reduce the process of oil and gas generation. However, many simulation experiments and example observations conducted by scholars from various countries have found that some simulation experiments show that pressure can inhibit the thermal evolution of organic matter, while others show that pressure has no obvious effect on thermal evolution parameters such as the reflectivity of vitrinite. In many basins, it has been proven that overpressure has an inhibitory effect on the thermal evolution of organic matter, and in many basins, it has been proven that overpressure has no recognizable effect on thermal evolution parameters such as the reflectivity of vitrinite (Hao, et al. 2004). This has led to three contradictory views: pressure has no obvious effect on the thermal evolution of organic matter and hydrocarbon generation (Tissot & Welte, 1984); the increase in pressure promotes the thermal evolution of organic matter, especially the thermal cracking of hydrocarbons (Braun, 1990); the increase in pressure inhibits the thermal evolution of organic matter and the action of hydrocarbons (Blanc, 1992).

In the static water pressure environment, the thermal evolution of organic matter is controlled by time and temperature, and the increase in pressure will not inhibit the thermal evolution of organic matter. In the overpressure environment, overpressure can become an important factor affecting the thermal evolution of organic matter, that is, it has an inhibitory effect on the hydrocarbon generation process (Hao et al., 2003). Different pressure increase mechanisms, as well as the same mechanism at different stages, have varying impacts on the formation of hydrocarbons from organic matter. When the increase in pressure is manifested as an increase in the effective stress received by the sedimentary organic matter, it can promote the generation of hydrocarbons from organic matter, such as the early stage of compaction and the high pressure formed by tectonic stress. When the increase in pressure is manifested as an increase in pore fluid pressure, resulting in abnormal pressure, the process of hydrocarbon generation from organic matter will be inhibited, such as entering the poorly-compactness stage and the overpressure caused by the volume expansion of organic matter cracking (Wang et al., 2006). Therefore, in the early stage of the evolution of organic matter, since the increase in pressure mainly increases the effective stress of the strata, it can promote the cracking of large molecular hydrocarbons and increase the catalytic activity of clay minerals, thereby promoting the generation of hydrocarbons from organic matter. In the gas generation stage of the evolution of organic matter, the generation of abnormal pressure will often inhibit the cracking of liquid hydrocarbons and reduce the conversion of liquid hydrocarbons to gaseous hydrocarbons.

Therefore, the main view currently believes that abnormal high pressure mainly shows an inhibitory effect on the process of organic matter generating hydrocarbons. It has also been found in nature that when there is abnormal high pressure, even if the strata temperature exceeds 200 °C, there can still be liquid hydrocarbons, while under normal pressure, it is all gaseous hydrocarbons. For example, in the Washington Lake Oilfield (6,540 m), the Bar Lake Oilfield (6,060 m), the strata temperature exceeds

200 °C, and it is still an oil field. This may be due to the abnormal high pressure preventing the cracking of liquid hydrocarbons.

(4) Radioactive Action

In shale, there is a significant positive correlation between the content of organic matter and the content of radioactive uranium elements, mainly due to the similar sedimentary enrichment processes of the two. In addition, there are different amounts of radioactive elements such as thorium and potassium in the hydrocarbon source rock. Radioactive elements have a direct or indirect impact on the generation and evolution of oil and gas. Alpha particles released during the decay process of elements such as uranium can change the structure of kerogen or petroleum molecules, thereby forming oil or further cracking oil into gas (Yang et al., 2018), which is the direct action mechanism of radiation on hydrocarbon generation. Radioactive elements release heat during the decay process, and radiation heat is one of the driving forces for the thermal evolution of oil and gas, which is the indirect action mechanism of radioactive action on hydrocarbon generation.

2.3.2 Stages and General Patterns of Organic Matter Evolution and Hydrocarbon Generation

As the burial depth gradually increases and the geothermal temperature rises continuously, the sedimentary organic matter will undergo a regular evolution, gradually and continuously transforming into oil and natural gas. The generation of oil and gas is an integral part of this evolution process, or rather, oil and gas are the forms of existence of sedimentary organic matter under certain environmental conditions. Since the physical and chemical conditions of sedimentary organic matter at different depths vary, the transformation reactions and main products of organic matter also have obvious differences. That is to say, the transformation process of primitive organic matter into oil and natural gas has distinct stages. Therefore, many scholars at home and abroad have proposed various schemes for the division of stages of organic matter generation (Table 2.6), but their basic connotations are roughly the same. This book mainly draws on Houfu's (1999) four-stage division model, dividing the oil and gas formation process into four gradually transitional stages: the biochemical gas generation stage, the thermal catalytic oil and gas generation stage, the thermal cracking wet gas generation stage, and the deep high-temperature gas generation stage, corresponding to the immature stage, mature stage, high maturity stage, and over-mature stage of the evolution of sedimentary organic matter (Fig. 2.9).

Table 2.6 Stages of organic matter evolution and hydrocarbon formation (according to Zhang et al., 1999)

Depth,km (T,°C)	Coal Rank	Vitrinite Reflectance (R_o,%)	Sporopollenin Color(C.Gutjahr,1966)	Kerogen Color (K.Peters, 1977)	Н.Б.Вассоевич (1970)		Pu Shi (1973)	Jiamo FU (1975)		Houfu ZHANG (1981)	Tissot, Welt (1984)		Zhongxiang PAN et al. (1986)	Difan HUANG et al. (1991)	
<1.5 10~60	Peat		Yellow	Yellow Light Yellow Brown	Early Stage	Early diagenesis Methane		Oil and gas formation stage	Initial methane generationsStage	Biochemical gas generation stage	Diagenesis	Biomethane	Biochemical methane stage	Immature stage	
	Lignite	0.5													
1.5~4.0 (60~180)	Long flame coal		Yellow Dark brown	Dark brown Deep dark brwon						Thermal catalytic oil and gas generation stage			Heavy-light oil stage	Mature stage	Low maturity
	Gas coal				Main stage	Associated gas	Liquid window		Low maturity		Catagenesis	Oil			Middle maturi
	Fat coal	1.0													
4.0~7.0 (180~250)	Coking coal			Deep dark brown	Final stage	Condensate gas, Wet gas		Oil and gas maturity stage	High maturity crude oil stage	Thermal cracking condensation gas stage		Wet gas	Condensate gas-wet gas stage		High maturity
	Lean coal	1.5													
	Meager coal	2.0													
7.0~10.0 (250~375)	Semi-anthracite	2.5	Black	Deep dark brown Black		Thermogenic methane			Final gas stage	Deep high-temperature gas generation stage	Metagenesis	Methane	Dry gas stage	Over-mature stage	
	Anthracite	3.0													

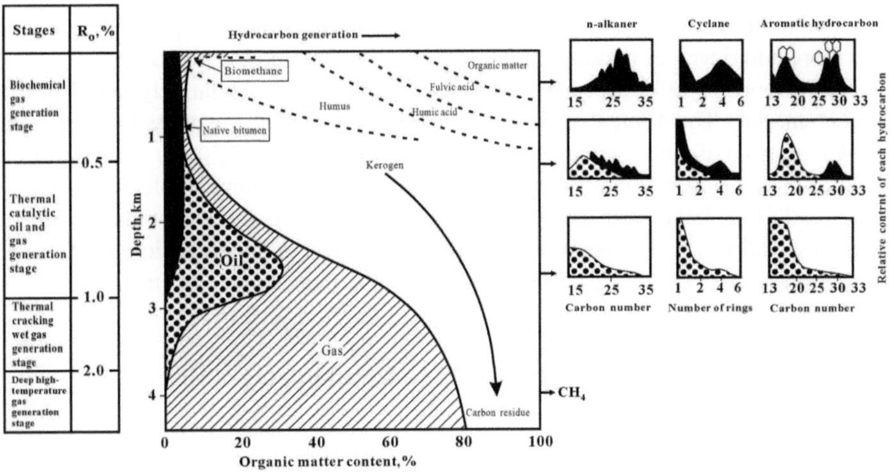

Fig. 2.9 General model of the relationship between oil and gas formation and the burial depth of source rocks (after Houfu, 1999, modified)

2.3.2.1 Biochemical Gas Generation Stage (Immature Stage)

The sedimentary organic matter has been in the biochemical gas generation stage since its formation. The depth range of this stage starts from the top of the sediment to 1500–2500 m, with temperatures between 10–60 °C, which is basically consistent with the early sedimentary diagenesis stage. In the shallow layer, bacterial biochemical action is predominant, and in the deeper layer, chemical action is predominant.

At the end of sedimentation to the early diagenesis, the burial depth is relatively shallow, and the temperature and pressure are low, suitable for the survival of various bacteria. The original sedimentary organic matter is degraded into soluble biochemical monomer organic matter (fatty acids, amino acids, monosaccharides, phenols) through hydrolysis and microbial action. Some of these monomer organic matters are consumed as nutrients by microorganisms, some are converted into simple molecules such as CO_2, CH_4 (biogenic gas), NH_3, H_2S, and H_2O under the biochemical action of bacteria (methanogenic bacteria). Most of the monomer organic matters are transformed into kerogen through polymerization and condensation. Since the organic matter is immature, the kerogen is yellow to light brown, and the type II kerogen Ro is less than 0.5%. A small amount of monomer organic matter (especially lipid compounds) forms special hydrocarbons that retain some of the original biochemical structure through simple reactions, known as biomarker compounds. In the later period of this stage, as the burial depth increases and the temperature approaches 60 °C, a small amount of liquid oil begins to be generated.

The biogas generated in this stage, also known as biochemical gas, has a methane content of over 95%, which belongs to dry gas. The stable carbon isotope $\delta^{13}C_1$ value is exceptionally low, ranging from −55 to −85‰. When the preservation conditions are suitable, it can form large gas reservoirs. The shallow burial depth makes it easy to explore and develop, making it a high economic benefit research object.

The proportion of hydrocarbons formed in this stage is relatively small in the organic matter. The typical products are mainly methane, usually lacking light (C_4–C_8) normal alkanes and aromatic hydrocarbons. However, some hydrogen-rich (such as resins, cutans, and botryococcus), sulfur-rich or large molecular organic compounds that have been modified by bacteria, due to their unstable chemical properties and lower hydrocarbon activation energy, can generate liquid oil in the depth and temperature range corresponding to this stage, known as immature oil.

2.3.2.2 Thermal Catalytic Oil and Gas Generation Stage (Mature Stage)

As the burial depth of the sediment gradually increases and the geothermal temperature rises, the evolution of organic matter enters the thermal catalytic oil and gas generation stage. The depth of this stage exceeds 1,500–2,500 m, up to 4,000–4,500 m. The geothermal temperature experienced by organic matter rises to 60–180 °C. In this stage, the main factor promoting oil and gas generation is thermal catalysis, and the main catalyst is the clay minerals in the rocks. Under the action of heat and catalysts, kerogen evolution reaches maturity (enters the oil generation threshold). Many chemical bonds begin to break, and heteroatom (O, N, S) bonds break to produce volatile substances such as carbon dioxide, water, nitrogen, and hydrogen sulfide. A large number of medium and low molecular weight liquid hydrocarbons (carbon number range C_{15}–C_{30}) and a certain amount of gaseous hydrocarbons are obtained, entering the main oil generation period, which is commonly referred to as the "oil window" abroad. With the increase in burial depth, the amount of soluble organic matter rapidly decreases after reaching the maximum value. When the burial depth reaches below 3000 m, it drops to a very low value, which is due to the thermal cracking action generating light hydrocarbons. The remaining kerogen becomes more compact, richer in C and less in H, O, and other heteroatoms.

The liquid hydrocarbons generated at this stage are different from those produced in the early stage, with R_o generally being 0.5–1.3%. The hydrocarbons produced have a decreasing number of carbon atoms and molecular weight in the normal alkanes, mainly composed of medium and low molecular weight molecules, and the odd-numbered carbon preference disappears The number of cyclic and aromatic hydrocarbons also decreases, characterized by low rings and low carbon atom number molecules being dominant. The polycyclic and polyaromatic compounds are significantly reduced.

2.3.2.3 Thermal Cracking Wet Gas Generation Stage (High Maturation Stage)

When the burial depth of the sediment exceeds 3500–4000 m and the geothermal temperature reaches 180–250 °C, the organic matter enters the thermal cracking wet gas generation stage. At this stage, the geothermal temperature exceeds the critical temperature of liquid hydrocarbon substances. In addition to the residual kerogen continuing to break the heteroatom functional groups and side chains, generating a small amount of water, carbon dioxide, nitrogen, and low molecular weight hydrocarbons, the main reaction is the massive breaking of C–C chains, including the opening and rupture of cycloalkanes, a sharp decrease in liquid hydrocarbons, and the content of high molecular weight normal alkanes above C25 tends to zero, with only a small amount of low carbon atom number cycloalkanes and aromatic hydrocarbons. The low molecular weight normal alkanes increase dramatically. The reactions in this stage include two different mechanisms. One is the further cracking of kerogen under the action of high temperature, forming some short-chain hydrocarbons. The other is the formation of low molecular weight gaseous hydrocarbons from the C–C bond breaking of the formed liquid hydrocarbons. Regardless of the mechanism, the main products of this stage are methane and its gaseous homologs, among which ethane and heavy hydrocarbon gases account for a larger proportion, which is wet gas. The kerogen residue in this period is more compact, dark brown, and the kerogen R_o is generally 1.3–2.0%. Some studies have shown that in the early stage of this stage ($R_o < 1.6\%$), kerogen gas generation is important. In the later stage ($R_o > 1.6\%$), the cracking of crude oil into gas is dominant.

2.3.2.4 Deep High-Temperature Gas Generation Stage (Overmature Stage)

When the depth exceeds 6000–7000 m, the sediment has entered the metamorphic stage, reaching the end of organic matter transformation. It is the overmature stage of organic matter evolution, with type II kerogen R_o greater than 2.0%. The temperature exceeds 250 °C, characterized by high temperature and high pressure. The cracking reaction of kerogen continues. Due to the removal of hydrogen in the form of methane, kerogen further condenses and the H/C atomic ratio drops to a very low level. The potential for hydrocarbon generation gradually dries up. The formed liquid hydrocarbons and heavy hydrocarbon gases will also be cracked into the most stable methane thermodynamically. The final kerogen will form pitch or graphite. In recent years, the research team of Academician Horsfield from the GFZ German Research Center for Geosciences found that when R_o is around 3.0%, shale still has additional hydrocarbon generation potential (late gas), and the mechanism is that short-chain alkyl groups with ultra-high activation energy on kerogen break under high-temperature conditions. The main product is methane (Horsfield et al., 2021).

These phenomena have been confirmed in the laboratory, field observations, and deep well drilling results. The Institute of Geochemistry, Chinese Academy of Sciences, conducted high-temperature and high-pressure experiments on oil and found that when the pressure remains constant, oil clearly differentiates into two poles with the increase of temperature, finally forming gas and solid asphalt. The evolution process is oil → oil + gas → oil + gas + solid asphalt + liquid asphalt → gas + solid asphalt (Fig. 2.10). This experimental result matches the field observation phenomena very well. For example, in the dolomite of Weiyuan uplift in Sichuan

Fig. 2.10 Photos of asphalt veins in the field

Basin, the final products of oil thermal evolution are methane and solid asphalt. The latter is irregularly stained or granularly distributed in the cracks or caves of dolomite, with high maturity, usually pitch and tar.

The four stages of organic matter evolution and oil and gas generation are a summary by scientists represented by Tissot, based on the study of the actual situation of organic matter evolution in the basin and many simulations test results. It reflects the general law of organic matter evolution and oil and gas generation and is called the kerogen thermal degradation hydrocarbon generation model. This model clarifies the stages of organic matter evolution and oil and gas generation. It emphasizes the concept of the oil generation threshold and points out that the oil generation threshold is the starting point for the large-scale generation of oil from organic matter and is also the turning point from immature to mature organic matter. It is worth pointing out that for different sedimentary basins, due to differences in their subsidence process, geothermal history, and types of original organic matter, the process of transformation of sedimentary organic matter into oil and gas may correspond to different depths, temperatures, and vitrinite reflectance limits. For example, in recent years, China's oil and gas exploration has continuously made breakthroughs in the ultra-deep areas, especially in the Tarim Basin. Liquid oil was found in the high-temperature strata more than 8000 m deep, opening a new exploration field. This is mainly due to the burial history of long-term shallow burial in the early stage and rapid deep burial in the later stage, low geothermal field during the deep burial period, and the inhibitory effect of overpressure on the generation and cracking of organic matter. In addition, for sedimentary basins with a more complex geological history, they may have experienced several lifting movements. The organic matter in the source rock may not have matured or only reached a lower maturity stage due to shallow burial, and the organic matter did not generate oil or did not completely generate oil. If the organic matter is not destroyed during the uplift, and it is buried again to a considerable depth, the temperature rises again and reaches the critical thermodynamic conditions required for the reactivation of organic matter, after reaching the oil generation temperature, the organic matter can still generate oil and gas, which is called “secondary oil generation.”

2.4 Genesis Types and Characteristics of Natural Gas

Broadly speaking, natural gas includes all gases that naturally exist in the natural world. In petroleum geology, natural gas typically refers to gaseous hydrocarbons, primarily methane, that are stored in underground porous media and may contain small amounts of associated non-hydrocarbon gases. With breakthroughs in the exploration of various types of natural gas and the advancement of research on the mechanisms of natural gas formation, it has become increasingly recognized that natural gas has a wider range of formation conditions than oil. Natural gas can not only be produced alongside the formation of oil but also be formed in large quantities in many conditions and environments that are not suitable for oil generation.

2.4.1 Genesis Types of Natural Gas

The sources of natural gas are diverse. Focusing on the basic substances that form natural gas, it can be divided into two major categories: inorganic and organic (Table 2.7).

Table 2.7 Classification of natural gas genesis types (modified from Dai et al., 1997)

<table>
<tr><td>Inorganic genesis gas</td><td colspan="6">Mantle-derived gas, magmatic gas, radiogenic gas, metamorphic gas, inorganic salt decomposition gas</td></tr>
<tr><td rowspan="5">Organic genesis gas</td><td>Kerogen genesis type</td><td>Maturity gas</td><td>Immature stage</td><td>Mature stage</td><td>High maturity stage</td><td>Over-mature stage</td></tr>
<tr><td rowspan="2">I–II_A</td><td rowspan="2">Humic gas</td><td rowspan="4">Biogenic gas</td><td colspan="3">Oil–gas type</td></tr>
<tr><td>Crude oil-associated pyrolytic gas</td><td>Pyrolytic condensate gas (wet gas)</td><td>Pyrolytic dry gas</td></tr>
<tr><td rowspan="2">II_B–III</td><td rowspan="2">Sapropelic gas</td><td colspan="3">Coal-related gas</td></tr>
<tr><td colspan="2">Pyrolytic gas (condensate gas)</td><td>Pyrolytic dry gas</td></tr>
<tr><td>Mixed genesis gas</td><td colspan="6">Heterogeneous and multi-source mixed gas, homogeneous and multi-stage mixed gas</td></tr>
</table>

2.4.1.1 Inorganic Genesis Natural Gas

Inorganic genesis natural gas refers to natural gas formed by all actions and processes that do not involve organic matter reactions, that is, non-biogenic natural gas. It mainly includes gases produced by deep crustal magmatic activity, metamorphism, inorganic mineral decomposition, radioactivity, and gases from cosmic space, which can be both hydrocarbon and non-hydrocarbon gases.

Inorganic genesis natural gas is divided into two subcategories based on its source and gas formation characteristics: mantle-derived gas and rock chemical reaction gas. Mantle-derived gas, also known as deep-source gas, refers to gases that were captured during the early formation of the Earth and rose from the mantle to the sedimentary sphere through tectonic magmatic activity, containing CH_4 and non-hydrocarbon gases. Rock chemical reaction gas refers to gases formed during the evolution of inorganic minerals, such as gases formed under high-temperature conditions from inorganic salts and minerals and rare gases of crustal origin.

In the case of inorganic genesis hydrocarbon gases, methane is predominant with very little C_2^+, and the carbon isotope abundance of methane $\delta^{13}C_1$ is $\geq -20‰$. There are few gas reservoirs dominated by inorganic genesis hydrocarbons. In gas reservoirs dominated by hydrocarbons or containing hydrocarbon gases, the hydrocarbon gas is mainly CH_4, that is, dry gas.

Inorganic genesis gas often contains a higher amount of non-hydrocarbon gases, mainly CO_2, CO, N_2, H_2S, H_2, etc., with CO_2 being the most common. There are also rare gases and inert gases such as He, Ar, Hg, Ne, Kr, Xe, Rn, etc. Non-hydrocarbon gases are not only important resources but also important indicators for the formation, evolution, and identification of the genesis types of natural gas. Gases from the mantle source have a helium isotope abundance of $^3He/^4He$ equivalent to $8R_A$ (R_A is the $^3He/^4He$ ratio in the air, about 1.4×10^{-6}). Currently, $\delta^{13}C \geq -20‰$, $^3He/^4He \geq 8R_A$ are generally used as indicators of inorganic genesis natural gas. Natural gas dominated by carbon dioxide is often related to the thermal decomposition of carbonate rock inorganic salts or magmatic genesis, and the carbon isotope of inorganic genesis CO_2 generally ranges from −8 to 0‰, with the highest reaching +27‰.

The distribution of inorganic genesis natural gas is usually related to the activity of deep faults, often migrating to shallow layers along deep faults, or entering the sedimentary rocks adjacent to the crystalline rocks along the unconformity between the crystalline rocks and sedimentary rocks, where they accumulate to form reservoirs. Tectonic activity units, especially ancient strata, are more likely to distribute inorganic genesis natural gas. Although inorganic genesis natural gas is difficult to form independent commercial gas reservoirs, there are many gas reservoirs worldwide that have the characteristics of a mixture of inorganic and organic genesis gases. Inorganic genesis natural gas has important theoretical research value in enriching the mechanisms of natural gas formation and explaining the sources of atypical gas reservoirs.

2.4.1.2 Organic Genesis Natural Gas

Organic genesis natural gas mainly emphasizes that the original parent material of the gas comes from organic matter. The form of these organic matters in sedimentary strata can be dispersed organic matter or combustible organic minerals. There are some differences among different scholars in China on the detailed division scheme of organic genesis gas, but basically, the classification method of parent material typing and evolution stage naming is adopted. First, it is divided into two major types according to the type of parent material, that is, the sapropelic gas generated by sapropelic organic matter and the humic gas formed by humic organic matter. Based on the two major subcategories, the genesis types of natural gas are further subdivided according to the stage of organic matter evolution or the characteristics of the main external forces converting organic matter into natural gas.

The formation mechanism of organic genesis natural gas mainly includes biochemical action, thermal pyrolysis, and radiation action. Biochemical action mainly refers to the process of methane-producing bacteria using carbon dioxide, hydrogen, formate, acetate, methanol, etc., to form methane. Thermal pyrolysis refers to the process of organic matter decomposing into hydrocarbons and large molecule hydrocarbons pyrolyzing into smaller molecule hydrocarbons under the action of temperature. Radiation action refers to the process of kerogen being attacked by alpha and beta ions released during the decay process of radioactive elements (such as uranium, thorium, potassium, etc.), breaking and generating light hydrocarbons. Organic genesis natural gas is generally divided into three categories according to the stage of thermal evolution: biochemical gas, thermal degradation gas, and thermal pyrolysis gas. Some scholars have also divided the transitional types, such as the bio-thermal catalytic transition gas.

2.4.2 Biogas

Biogas refers to methane-rich gas produced by the biochemical degradation of sedimentary organic matter by microorganisms (anaerobic bacteria) under low-temperature (<75 °C) reducing conditions during the early stage of diagenesis or the evolution of organic matter in the shallow part of the Earth's crust. Also known as biochemical gas, biogenic gas, etc. Both sapropelic and humic organic matter can be biodegraded to form biogas, but the organic matter that generates biogas is mainly of mixed type (Type II) and humic type (Type III).

Since the 1960s, a series of large gas fields have been discovered in the Cretaceous sandstone of northern West Siberia, Russia. Methane carbon isotope identification has confirmed them as biogas, forming the largest biogas production area in the world today. Later, large biogas fields were also discovered in Italy, Canada, the United States, and Japan, among which the Antrim shale in the Michigan Basin of the United States is rich in typical biogenic shale gas. Several biogas fields have also been discovered in the Quaternary sandstone reservoirs of the Sanhu area in the eastern Qaidam Basin of China. The discovered biogas is mostly from the Cretaceous, followed by the Tertiary and Quaternary. More than 80% of biogas reserves are concentrated in the West Siberian region. These gas reservoirs are shallowly buried, generally within 1500 m.

2.4.2.1 Formation of Biogas

According to the different types of biological metabolism, microorganisms can be divided into aerobic, anaerobic and facultative microorganisms. Rice & Claypool (1981), after studying the biochemical environment of microbial metabolic activities in open sea sediments rich in organic matter, believed that the water–sediment profile can be divided into two kinds of biological metabolic environments, aerobic and anaerobic, and four biochemical action zones, namely: the photic zone, the aerobic zone, the sulfate reducing zone and the carbonate reducing zone. The microbial species, metabolic types, solutes and biochemical properties in different biochemical action zones are different (Fig. 2.11).

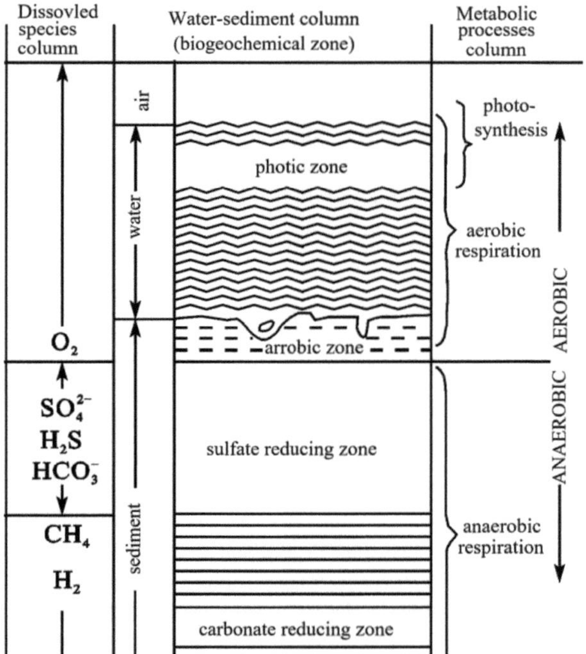

Fig. 2.11 Biochemical environmental profile of microbial metabolism in open sea sediments rich in organic matter (according to Rice & Claypool, 1981)

In the aerobic environment, organic matter is destroyed by aerobic bacteria through aerobic respiration and converted into simple inorganic molecules such as CO_2 and H_2O. When free oxygen is completely consumed, it enters the anaerobic environment. Sulfate-reducing bacteria first reduce sulfate to elemental sulfur or H_2S. Meanwhile, other anaerobic microorganisms can convert organic matter into organic acids, alcohols, CO_2 and H_2 through fermentation. When sulfate is almost completely reduced, it enters the carbonate reduction zone. Then methanogens begin to develop, which can directly decompose organic acids and alcohols into CH_4, and can also use CO_2 and H_2 to synthesize CH_4. Therefore, bacterial methane gas can only be formed in a strictly reduced environment where there is no free oxygen and no sulfate.

To generate a significant amount of biogas, it is necessary to have an abundance of original organic matter, physicochemical conditions conducive to the development of methanogenic bacteria, and an appropriate sedimentary environment.

(1) Abundant Original Organic Matter

Having an abundance of original organic matter, especially humic and mixed types, is the material basis to produce a large amount of methane gas. The nutritional sources for methanogenic bacteria are primarily carbohydrates such as cellulose, hemicellulose, sugars, starch, and fruit acids, which are most abundant in herbaceous plants. This determines that the matrix for biogas is mainly a mixed-type organic matter dominated by herbaceous humus.

(2) Physicochemical Conditions Suitable for the Development of Methanogenic Bacteria

Reductive environments and neutral aqueous media conditions are conducive to the formation of biogas. The toxicity of H_2S formed by sulfate reduction has a significant inhibitory effect on the proliferation of methanogenic bacteria. A strictly anoxic and sulfate-reducing environment is a necessary condition for the proliferation of methanogenic bacteria. The geothermal range for the survival of methanogenic bacteria is 0–75 °C, mainly limited to 4–45 °C, with the most suitable values being 35–42 °C (Zeikus, 1977). The optimal water medium pH for the growth of methane-producing bacteria is 6.5–7.5.

(3) Appropriate Sedimentary Environment

Freshwater lake sediments have low salt content, lack SO_4^{2-}, and have a neutral pH, making it easy for humic and mixed organic matter to decompose into H_2 and CO_2, which is conducive to the proliferation of methanogenic bacteria. Therefore, methanogenic bacteria appear earlier and begin to produce CH_4 in large quantities in the near-surface shallow layer. However, due to shallow burial and poor sealing conditions, most of the biogas is lost or oxidized, making it difficult to form large-scale gas reservoirs. If there is a good sand layer under the organic-rich clay layer, it is possible to form small-scale gas reservoirs.

Semi-saline or saline lakes, especially alkaline saline lakes, are the most favorable environments for the formation and accumulation of biogas. This sedimentary environment can inhibit the premature excessive proliferation of methanogens and is also conducive to the preservation of organic matter. When buried to a certain depth and the decomposition of organic matter reduces the pH of the medium to 6.5–7.5, methanogens can multiply in large numbers, and the CH_4 generated at this stage is easily preserved and can be accumulated into industrial gas reservoirs under certain conditions. The formation of the Quaternary biogas reservoirs in the Qaidam Basin benefits from the semi-saline sedimentary environment, resulting in a relatively deep burial depth of a large amount of gas generation.

2.4.2.2 Main Features of Biogas

The composition of biogas is mainly methane, generally with a content of over 98%, and the content of heavy hydrocarbon gases (C_{2+}) is very low, with a dry coefficient (C_1/C_{2+}) of several hundred, belonging to dry gas. Sometimes it may contain trace amounts of unsaturated hydrocarbons as well as a small amount of CO_2 and N_2.

Methane in biogas is characterized by an enrichment of the light carbon isotope ^{12}C. The range of methane carbon isotope $\delta^{13}C_1$ is −55 to −100‰, with most in the range of −60 to −80‰ (Table 2.8). The isotope can become heavier with the mixing of pyrolytic gas and anaerobic oxidation.

There is less reported data on the hydrogen isotope of biogas, but Schoell (1983) believes that the δD of biogas also exhibits low values, with the lowest for humic-type biogas ranging from −210 to −280‰, and sapropelic-type biogas about −150 to −210‰.

2.4.3 *Oil-Related Gas*

Oil-related gas refers to the natural gas formed when sapropelic organic matter (mainly Types I and II) reaches the mature, high-mature, and over-mature stages during thermal evolution. This includes the wet gas formed by the thermal degradation of kerogen during the oil generation process, as well as the condensate gas and thermal pyrolysis dry gas formed by the thermal pyrolysis of kerogen and liquid hydrocarbons during the high-mature and over-mature stages.

Oil-related gas is widely distributed, and as long as oil reservoirs are found in oil and gas basins, varying amounts of oil-related gas can be found, which can exist in different states.

2.4.3.1 Formation of Oil-Related Gas

The formation process of oil-related gas includes two evolutionary pathways. One is the direct generation of gaseous hydrocarbons from the thermal pyrolysis of kerogen. The other is the thermal degradation of kerogen into oil, and under the condition of

Table 2.8 Composition of biogas in some regions of the world (according to Botz, 1984)

Region or gas field	Reservoir age	Depth (m)	C_1 (%)	C_{2+} (%)	CO_2 (%)	N_2 (%)	$\delta^{13}C$ (‰)
China's Yangtze Delta	Quaternary	8–35.5	90.62–94.61	0.11–0.89	1.85–4.04	1.47–3.35	−73.6
Qaidam Sebei, Qinghai	Quaternary	79.4–1,141	98.94	0.09	/	0.97	−66.4
Urengoy, Russia	Cretaceous	1,117–1,128	98.50	0.10	0.21	1.10	−59.0
Medvezh'ya, Russia	Cretaceous	1,122–1,132	98.60	0.36	0.22	0.73	−58.3
Kenai, USA	Miocene-Pliocene	1,128	99.70	0.18	/	/	−57.0
Cook Inlet North, USA	Miocene-Pliocene	1,280	98.70	0.23	0.134	0.9	−60.7

continuous increase of geothermal temperature, oil is further pyrolyzed into gaseous hydrocarbons.

The thermal evolution process of kerogen can be regarded as a disproportionation reaction, which simultaneously includes two processes in opposite directions. One is the condensation reaction that forms hydrogen-poor polycyclic aromatic structures, and the other is the pyrolysis reaction that forms hydrogen-rich low molecular weight alkanes (Fig. 2.12). The former gradually condenses from low molecular weight phenols to dense polycyclic aromatic hydrocarbons, until graphite, releasing a large amount of hydrogen, while the residual kerogen also becomes hydrogen-poor. The latter is the thermal pyrolysis of oil to generate higher hydrogen content methane and its gaseous homologs and other light hydrocarbons, with the final product being methane.

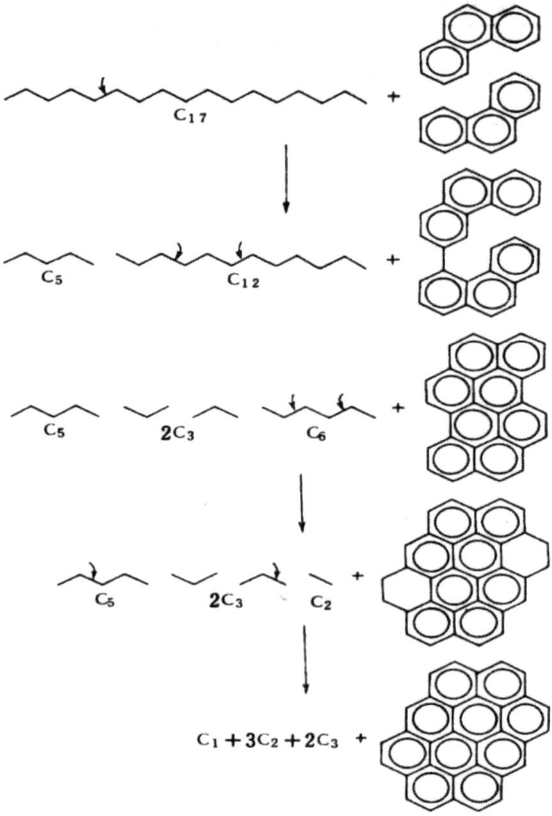

Fig. 2.12 The disproportionation reaction of petroleum thermal evolution

2.4.3.2 Main Features of Oil-Related Gas

Various types of oil-related gas are products of different thermal evolution stages of kerogen, and their chemical compositions differ. The common features of oil-associated gas and condensate-associated gas are the high content of heavy hydrocarbon gases, generally exceeding 5%, sometimes reaching 20–50%, among which, the iC_4/nC_4 ratio is significantly less than 1, and it is about 0.7–0.8 during the thermal catalytic oil and gas stage (according to Héroux et al., 1979). The carbon isotope content of methane $\delta^{13}C_1$ is between −55 and −40‰, oil-associated gas is lighter, $\delta_{13}C^1$ is about −55 to −45‰, condensate-associated gas is heavier, $\delta^{13}C_1$ is about −50 to −40‰. Over-mature pyrolytic dry gas, mainly methane, with very little heavy hydrocarbon gas, less than 1–2%, and the carbon isotope of methane is greater than or equal to −35 to −40‰. The compositional characteristics of some oil-related gases in China are shown in Table 2.9.

2.4.4 *Coal-Related Gas*

Coal-related gas refers to the natural gas formed by various output states of humic organic matter during the thermal evolution process (i.e., the coalification process). The original organic matter for coal-related gas formation is mainly terrestrial higher plants, and the organic composition is mainly carbohydrates and lignin.

In 1959, the Groningen gas field was discovered in the northern part of the Netherlands. Subsequently, it was determined that the huge natural gas accumulation in the Upper Permian aeolian sandstone of the Zechstein Group came from the Carboniferous coal series strata. Coal-related gas began to be valued. Later, more than a dozen coal-related gas fields were discovered in the southern part of the North Sea Basin,

Table 2.9 Compositional characteristics of some oil-related gases in China (Rongshu, 1989)

Oil field or region	Main parameters of natural gas composition analysis				$\delta^{13}C_1$ ‰ (PDB)
	CH_4	Heavy hydrocarbon	C_1/C_{2+}	$C_1/\sum C$	
Daqing Oilfield (oil-associated gas)	53.9–95.61	2.64–38.51	1.40–36.22	0.58–0.975	−37.72 to −49.97
Dongpu Depression (condensate oil-associated gas)	71.04–87.43	10.63–26.91	3.21–20.3	0.75–0.96	−38.9 to −45.1
Panqiao Condensate Gas Field	82.88	15.29	5.42	0.844	
Chuandong Xiangguosi Gas Field (thermal pyrolysis dry gas)	98.15	0.89	110.3	0.991	−33.55

with proven total reserves exceeding 4.5×10^{12} m^3, becoming the world's second-largest gas-producing area. Since then, many countries such as Russia, the United States, and Australia have paid widespread attention to finding coal-related gas reservoirs in coal-bearing basins. By the end of 2017, 13 large coal-related gas fields had been discovered worldwide, with a total original recoverable reserve of 49.995×10^{12} m^3, accounting for 25.8% of the world's total remaining recoverable reserves that year (Jinxing, 2019). China has abundant coal resources, and coal-related gas is one of the important targets for natural gas exploration. In recent years, significant breakthroughs in coal-related gas exploration have been made in the Ordos Basin, Qaidam Basin, and South China Sea.

2.4.4.1 Formation of Coal-Related Gas

The original organic matter of coal-related gas mainly comes from the remains of various types of plants. The types of plants involved in the coal-forming process are different in different eras. Before the Silurian, algae and fungi plants were dominant, forming only sapropelic coal. From the Silurian, terrestrial plants began to appear. Since the Carboniferous, terrestrial higher plants have become the main source of the original organic matter for coal formation. These organic matters are mainly carbohydrates and lignin. If the remains of these plants are accumulated in large quantities in swamps, inland shallow lake basins, and the edges of ocean basins, with almost no minerals participating, under the condition of limited oxygen entry, as the burial depth increases, they can evolve into coal of different ranks through peatification and coalification processes. If these plant remains are deposited in a dispersed state along with minerals, as the burial depth increases, they form humic-type (Type III) kerogen through diagenesis.

The process of coal-derived hydrocarbons is similar to the previously described oil and gas generation, with the main difference being that coal seams or humic-type kerogen have a low H/C ratio (less than 1.0) and a high O/C ratio in their chemical composition. Structurally, they are primarily composed of condensed polynuclear aromatic structures with many short alkyl side chains and oxygen-containing functional groups, hence the gaseous hydrocarbons are predominant during the thermal evolution process. The gas generation from humic-type organic matter can be roughly divided into four stages (Fig. 2.13).

The first stage is from the early stage of peat to lignite (Zone 1, peat-O_1 coal rank), with $R_o < 0.4\%$ and geothermal temperatures generally below 75 °C, corresponding to the biochemical gas generation stage. In the early stage of this phase, dehydration and decarboxylation reactions are predominant, mainly producing CO_2 with a very low rate of CH_4 production. In the later stage, the rate of CH_4 production increases significantly. The gas generated in this stage is biogas.

The second stage is from the middle stage of lignite to long-flame coal (Zone 2, O_2-O_3-I coal rank), with R_o between 0.4 and 0.6%, and geothermal temperatures generally between 70 and 90 °C. In this stage, the total amount of natural gas generated is the greatest. The composition is still mainly CO_2, with CH_4 content gradually

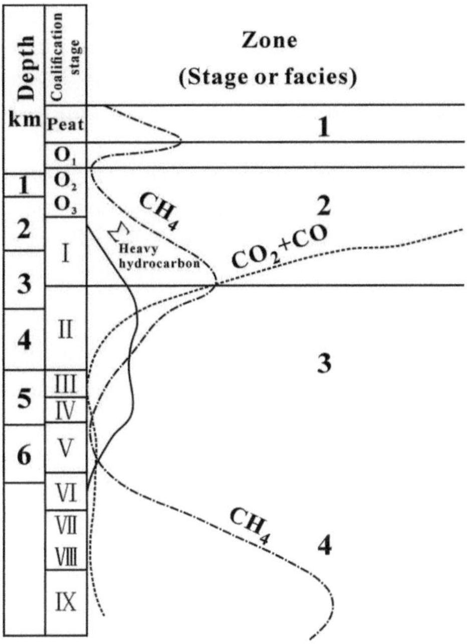

Fig. 2.13 Stages and gas formation patterns in the coalification process of humic organic matter (Vysotsky, 1979)

increasing as geothermal temperatures rise. The content of heavy hydrocarbon gases also increases, with the formation of thermal pyrolysis gas beginning in the later period of this stage.

The third stage is the bituminous coal-lean coal stage (Zone 3, II–V coal rank), with R_o between 0.6 and 1.7%, and geothermal temperatures generally between 90 and 190 °C. In this stage, the composition of natural gas is mainly hydrocarbon gases, which is the main period for the formation of coal-related wet gas and coal-related oil.

The fourth stage is the poor coal-anthracite stage (Zone 4, VI–IX coal rank), with R_o greater than 1.7% and geothermal temperatures exceeding 190 °C, forming coal-related dry gas mainly composed of methane.

2.4.4.2 Main Features of Coal-Related Gas

Coal-related gas is primarily distributed in coal-bearing basins and basins where coal-bearing strata are very developed. The original parent materials are mainly coal and Type III kerogen in the coal seam. The products are mainly hydrocarbon gases, primarily methane, with the content of heavy hydrocarbons rarely exceeding 20%. There is generally a certain amount of non-hydrocarbon gases, such as N_2, CO_2, etc. (Table 2.10). However, the composition of the products formed varies at different stages of coalification. In addition, due to the organic matter parent material, the condensate oil formed together with coal-related gas often contains higher amounts of benzene, toluene, methylcyclohexane, and dimethylcyclopentane.

The isotopic values of methane in coal-related gas vary significantly among different researchers, with $\delta^{13}C_1$ generally ranging from −24 to −52‰, mainly distributed between −32 and −38‰. As the maturity of organic matter increases, the $\delta^{13}C_1$ value increases. Dai et al. (1985) showed that the carbon isotopic values of methane, ethane, and propane in Chinese coal-related gas vary in the ranges of −24.9 to −41.8‰, −23.81 to −27.09‰, and −19.16 to −25.72‰, respectively. The hydrogen isotopic values of methane in coal-related gas vary in the range of −161.4 to −171‰ (Xu, 1985).

Table 2.10 Composition of some domestic and foreign coal-related gases (according to Rongshu, 1989)

Gas field		Reservoir age	Gas source layer age	Natural gas composition (%)				$\delta^{13}C$ (‰, PDB)	Data source
				C_1	C_2^+	N_2	CO_2		
Groningen		P_1	C_2	81.2	3.48			−36.6	According to Stahl (1977)
Lazer		P_1	C_2	89.9	6.10	14.4	0.87	−29.2	
Dalum		P_1	C_2	86.06	0.44			−25.4 to −22.0	
San Juan		K	K					−42.0	Cited from Stahl (1983)
Cooper Basin (Aus)	Mumba 9 well	P_1	P_1	66.02	0.67		33.27	−28.8	According to Rigby (1981)
	Tullawong 9 well	P_1	P_1	71.76	11.62		14.40	−36.3	
Wenliu 22 well, Dongpu		E_2	C-P	96.35	2.35			−27.9	According to Jiawei et al., (1983)
Liuqing 1 well, Shanganning		P_{1x}	C-P	95.0	0.64	4.13			According to Shaochang (1983)
Ren 4 well, Shanganning		P_{1x}		92.52	6.97	0.49	0.01	−30.47	
Zhongba 4 Well, Sichuan		T_{3x}		90.8	8.20	0.17	0.40	−34.8	* Data is from a nearby well of Zhongba 7, Chen (1982)
Zhongba 7 well		T_{3x}		87.33	12.23	0.41	0.03	−36.0 to −35.9*	

Mercury vapor content in coal-related gas is relatively high, generally exceeding 1 μg/m^3. The mercury content in coal-related gas from the Central European Basin can reach 180–400 μg/m^3, and the mercury content in typical coal-related gas reservoirs in China's Dongpu Depression is 1.1–51 μg/m^3 (Yongchang, 1994). The mercury vapor in coal-related gas mainly comes from coal-bearing strata, and due to the strong adsorption capacity of humic-type organic matter for mercury, coal-bearing strata have a higher original mercury abundance. During the thermal evolution process, the adsorbed mercury is transformed into vapor and migrates and accumulates together with hydrocarbon gases, thereby increasing the mercury content in coal-related gas.

2.4.5 Genesis of Non-hydrocarbon Gases

(1) Helium

Helium, as a critical rare gas resource related to national security and the development of high-tech industries, has extensive and irreplaceable applications in fields such as liquid fuel rocket launches, deep diving, and chip manufacturing. The global distribution of natural helium resources and production is highly imbalanced, mainly found in countries like the United States, Qatar, and Algeria.

Helium is formed during the decay process of radioactive elements such as uranium and thorium. High-quality helium source rocks are the material basis for the formation of helium-rich natural gas reservoirs. Most of the natural helium discovered to date is of crustal origin, with a minority being of mantle origin. It is primarily found in three types of tectonic environments: plate edge uplifts, intracontinental rifts, and deep large strike-slip fault zones. Ancient, uranium-rich shales can form large scale helium gas during geological history and are associated with shale oil and gas, forming shale-type helium gas accumulations (Nie et al., 2023).

(2) Hydrogen

Hydrogen, after combustion, only produces water and releases a tremendous amount of energy, making it one of the most promising sustainable energy carriers. Additionally, hydrogen has a wide range of applications in fossil fuel processing, ammonia production, metal smelting, and aerospace. Under the "Dual Carbon Target" background, hydrogen is receiving high attention from governments around the world. Natural hydrogen widely exists in the strata, especially in regions with active tectonics. The Republic of Mali in Africa is one of the few countries that have achieved commercial mining of hydrogen. In 1987, about 50 km north of the capital, a natural hydrogen reservoir with a purity of 98% was discovered and has been mined to this day.

Natural hydrogen mainly includes two major sources: inorganic (mantle degassing, water–rock reaction, and water radiolysis) and organic (biological action, thermal action, and radiation decomposition). In recent years, research has shown that shale organic matter under conditions of high maturity and over-maturity ($R_o > 3.0\%$) can pyrolyze to produce hydrogen, which may form natural hydrogen reservoirs under suitable geological conditions.

(3) Carbon Dioxide

As an important greenhouse gas, carbon dioxide is not highly concentrated in the atmosphere. In geological bodies, occasional discoveries of rich carbon dioxide gas reservoirs are mostly concentrated in areas with frequent volcanic activity or well-developed basement faults. The formation of geological carbon dioxide mainly includes inorganic and organic origins.

Inorganic origins include two mechanisms: chemical and magmatic. The former refers to the generation of CO_2 from minerals such as carbonates through high-temperature thermolysis, low-temperature hydrolysis, or acidic dissolution by groundwater. The latter is the large amount of CO_2 released during the ascent of magma due to the decrease in temperature and pressure. Organic origins mainly include the following three situations: a large amount of CO_2 generated from the biochemical degradation of organic matter under the action of anaerobic bacteria, CO_2 formed from the thermal degradation and pyrolysis of kerogen, especially Type III kerogen, and the oxidation of hydrocarbons can also form CO_2.

(4) Nitrogen

Nitrogen is more commonly found than carbon dioxide in natural gas, with content usually not exceeding 10% (often 2–3%). During the exploration process of shale gas in southern China, multiple nitrogen-rich gas reservoirs have been drilled (with nitrogen content up to 95%).

The sources of nitrogen in natural gas are generally considered to include biological, atmospheric, magmatic, and metamorphic sources. Research has found that the nitrogen in natural gas is mainly generated from the transformation of nitrogen-containing compounds in sedimentary organic matter or petroleum during biochemical or thermal catalytic processes, which is the biological source. Generally, biogas is relatively richer in N_2. Atmospheric nitrogen can be brought into the gas reservoir through the circulation of surface water and groundwater and is often enriched in shallow strata.

(5) Hydrogen Sulfide

Most natural gas reservoirs do not contain or only contain very small amounts of H_2S gas, and gas reservoirs dominated by H_2S are relatively rare. H_2S is mainly concentrated in carbonate and sulfate rock reservoirs. Gas fields with H_2S content greater than 10% are almost all in carbonate rocks and evaporite rocks, such as the Zhaolanzhuang H_2S gas reservoir in Jizhong (with an H_2S content of up to 92%). The current understanding is that the genesis of H_2S gas is mainly due to biological, thermochemical, and magmatic origins.

There are two different pathways for the biological formation of H_2S in nature. One is the formation of sulfur-containing organic compounds through microbial assimilation reduction and plant absorption, which then decompose under certain conditions to produce H_2S. This process is the process of decay, and the H_2S formed in this way is limited to shallowly buried strata, with poor preservation conditions. A large amount of H_2S will dissipate, so generally, the scale and content of H_2S gas reservoirs formed by this genesis will not be very large. The other way for biological action to generate H_2S is through sulfate reduction to directly form H_2S. The prerequisite for the formation of this type of H_2S is the presence of sulfates and sulfate-reducing bacteria. Sulfate-reducing bacteria use the combined oxygen in sulfates for anaerobic respiration, reducing sulfates to generate H_2S.

Thermochemically derived H_2S can be divided into two types based on its formation mechanism: thermolytic and thermoreductive. Thermolytic H_2S is formed from the cleavage of sulfur-containing heterocyclic compounds in organic sulfur compounds under the influence of heat. During this formation process, sulfur-containing organic matter is first transformed into sulfur-containing hydrocarbons and sulfur-containing kerogen. When the temperature increases to a certain extent (about 80 °C), the heterocyclic atoms in the kerogen gradually break, generating a certain amount of gas, including H_2S, but the concentration is relatively low. When the temperature continues to rise to the deep thermal pyrolysis stage (130 °C), the decomposition of sulfur-containing organic compounds begins, producing a large amount of H_2S. Therefore, H_2S of this origin is often found in dry gas and is of thermolytic origin. Thermoreductive H_2S is generated from the reduction of sulfates by organic matter or H_2 under high temperatures to produce H_2S. Most gas reservoirs with high hydrogen sulfide content are distributed in areas with greater burial depth, which reflects those thermochemical origins may have greater significance for the formation of hydrogen sulfide.

H_2S gas can also be released during the ascent of magma. Many scholars at home and abroad have confirmed the existence of this origin of H_2S through studies of volcanic gases. For example, the gas emitted from the Nasu Teine and Kuchinoshima volcanoes, after the removal of water vapor, contains H_2S accounting for 37.5% and 61% of the remaining gas, respectively (Dai et al., 1989).

2.4.6 Identification of Natural Gases of Different Genesis Types

The genetic information of natural gas is mainly contained in the compositional characteristics of its gas components. Therefore, the identification of natural gas genesis types is mainly carried out by analyzing various compositional characteristics and carbon-hydrogen isotope characteristics of its gas components.

2.4.6.1 Identification of Inorganic and Organic Genesis Methane and Its Homologues

The most important criterion for the division and identification of the two major types of natural gas, inorganic and organic, is the carbon isotopic composition of methane.

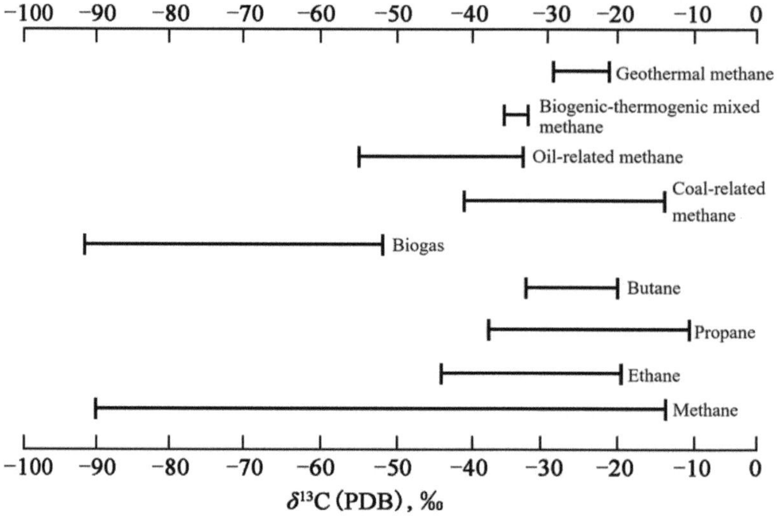

Fig. 2.14 Distribution characteristics of $\delta^{13}C$ values of alkane gases in Chinese natural gas (according to Dai, 1992)

Generally, $\delta^{13}C_1 > -20‰$ is considered as one of the indicators of inorganic methane. Dai et al. (1992) compiled a distribution map of $\delta^{13}C$ for alkane gas in Chinese natural gas (Fig. 2.14) based on the measured results of $\delta^{13}C$ values of alkane gas in China. They proposed that in the absence of coal-bearing strata, this value can be extended to $\delta^{13}C_1 > -30‰$. However, when the thermal evolution degree of coal-bearing organic matter reaches the over-mature stage, the $\delta^{13}C$ value of methane can be greater than $-20‰$.

The relationship between carbon isotopes of methane homologs is also an important basis for the division of the two major types of alkane gases. Organically derived gases typically exhibit the characteristic of $\delta^{13}C_1 < \delta^{13}C_2 < \delta^{13}C_3$, whereas inorganically derived gases exhibit an inverse sequence, that is, $\delta^{13}C_1 > \delta^{13}C_2 > \delta^{13}C_3$, meaning that $\delta^{13}C$ values decrease with the increase in the carbon number of alkane gas molecules.

2.4.6.2 Differentiation Between Coal-Related and Oil-Related Gases

(1) Differentiation of Coal-Related and Oil-Related Gases Using Alkane Carbon Isotope Composition Characteristics

Due to the isotopic inheritance effect of their respective gas-forming parent materials, the isotopic composition characteristics of the two major types of alkane gases are different. Dai (1992) compiled the isotopic characteristics of various natural gases at home and abroad, and summarized a template that can efficiently distinguish oil-related gas, coal-related gas, and mixed gas through the carbon isotopes of methane and ethane (with the assistance of propane carbon isotopes) (Fig. 2.15). This template has shown significant effects in the identification research of natural gas in the Ordos Basin, Bohai Bay Basin, Sichuan Basin, and Tarim Basin in China.

(2) Application of Light Hydrocarbon Compounds to Distinguish Coal-Related and Oil-Related Gases

Natural gas often contains varying amounts of C_1–C_7 light hydrocarbon compounds, which hold significant geochemical information. Light hydrocarbon components derived from sapropelic parent materials are rich in n-alkanes, while those from humic parent materials are abundant in iso-alkanes and aromatic hydrocarbons. Condensates rich in cycloalkanes are an important characteristic of terrestrial parent materials. These characteristics of light hydrocarbons generated from different parent materials can be utilized to identify oil-related and coal-related gases.

Among C_7 light hydrocarbon compounds, methylcyclohexane primarily originates from lignin, cellulose, and sugars in higher plants and has relatively stable thermodynamic properties, making it a good parameter for reflecting the type of terrestrial parent material. Its significant presence is a characteristic of light hydrocarbons in coal-related gas. Various structural isomers of dimethylcyclopentane mainly come from the lipid compounds of aquatic organisms and are affected by maturity. Their significant occurrence is a characteristic of light hydrocarbons in oil-related gas. N-heptane primarily comes from algae and bacteria and is very sensitive to maturation effects, making it a good maturity indicator. This information can also be used to differentiate between coal-related and oil-related gases.

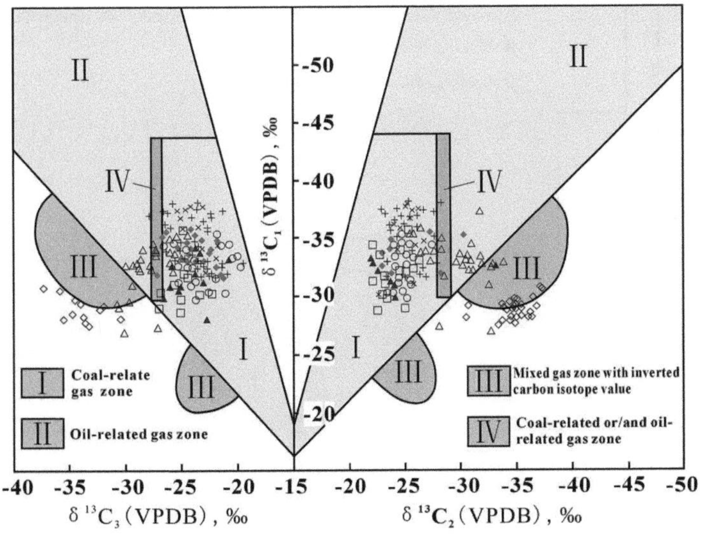

Fig. 2.15 Identification chart for different types of natural gas (according to Dai, 1992)

Additionally, hydrogen-rich kerogen (the parent material of oil-related gas) generates C_{2-7} alkanes that are several orders of magnitude larger than those from hydrogen-poor kerogen (the parent material of coal-related gas), while the content of aromatic hydrocarbons is very low. Therefore, aromatic hydrocarbons and branched alkanes can also be used to distinguish oil-related gas from coal-related gas.

(3) Differentiation of Organic and Inorganic Carbon Dioxide

The primary distinguishing feature between organic and inorganic CO_2 is the stable carbon isotopic composition. In China, the $\delta^{13}C$ range for organic CO_2 is from −8 to −39‰, with the main frequency range being −12 to −17‰. The $\delta^{13}C$ range for inorganic CO_2 is generally between +7 and −10‰, with the main frequency range being −3 to −6‰. Moreover, organic CO_2 in natural gas reservoirs rarely exceeds 20%. Therefore, CO_2 in hydrocarbon gas reservoirs and CO_2 gas reservoirs with a content greater than 20% is almost entirely of inorganic origin. Dai et al. (1989) created Fig. 2.16 based on the $\delta^{13}C$ values and percentage content of CO_2 from over 300 gas reservoirs of different origins worldwide, which can serve as an identification chart for organic and inorganic carbon dioxide (Fig. 2.16).

It should be noted that when discussing the source and genesis types of natural gas, geological and geochemical indicators should be applied scientifically and comprehensively, in conjunction with the specific geological and geochemical context of the gas reservoir for analysis.

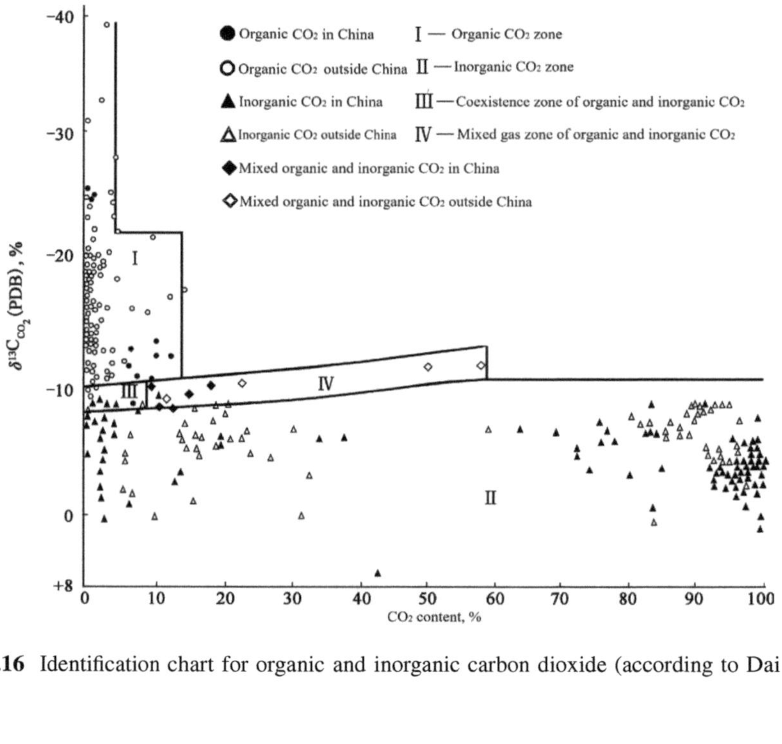

Fig. 2.16 Identification chart for organic and inorganic carbon dioxide (according to Dai et al., 1989)

2.5 Evaluation of Source Rocks

2.5.1 Conception of Source Rocks

Source rocks are also referred to as hydrocarbon source rock. Different researchers provide slightly different definitions. Tissot & Welte, 1984 define it as "rocks that have generated oil, can generate oil, or have the capability to generate oil". Hunt (1979) defines it as "fine-grained sediments that have generated and expelled enough hydrocarbons to form industrial oil and gas accumulations". In China, researchers use different terms such as "oil (gas) source rock", "oil and gas source rock", and "hydrocarbon source rock", and there are also differences in connotation. Houfu's (1987) call "rocks that can generate oil and gas" oil and gas source rocks (or oil and gas parent rocks, hydrocarbon source rocks). Qijun & Jianyu, (1987) refer to "rocks that have the conditions to generate oil and gas, have generated and can expel industrially valuable oil and gas as oil (gas) source rocks".

For the name of the concept itself, all the above terms exist in Chinese literature, but the frequency of use of "hydrocarbon source rock" has been significantly higher in recent years. Thus, we use the term "hydrocarbon source rock" in this textbook. Regarding the connotation of the concept, from the perspective of oil and gas genesis theory research, Tissot's definition has its rationality because if there is organic matter in the rock, hydrocarbon substances can be formed under certain conditions. Our actual research objects of source rocks also include "all rocks with potential hydrocarbon-generating capabilities". However, from the perspective of oil and gas geological exploration practice, the purpose of our research on source rocks is to determine whether there are rocks in the sedimentary basin that can form industrial oil and gas accumulations, and the spatiotemporal distribution of these rocks, to provide guidance for exploration. It is precisely based on this that many researchers have recently proposed the concept of "effective hydrocarbon source rock", specifically referring to rocks that can generate and expel oil and gas and provide commercial oil and gas accumulations. Hunt's definition reduces the connotation of the concept, focusing on the relationship between source rocks and oil and gas formation, which is closely related to oil and gas exploration practice. For this reason, we believe that it is more reasonable to understand source rocks according to Hunt's definition and define source rocks as: rocks that have generated and expelled enough hydrocarbons to form commercial oil and gas accumulations. In the early stages of research, when it has not yet been determined whether the research object has a large amount of hydrocarbon generation and expulsion, it can be called a potential source rock.

Strata composed of source rocks are called source rock formations. Within a certain geological period, several source rock formations with the same lithological-lithofacial characteristics and the non-source rock formations between them are collectively referred to as source rock series. Source rocks are essential conditions for the formation of oil and gas accumulations in sedimentary basins. Therefore, the study of source rock formations is of theoretical significance for exploring the genesis of oil and gas, and it is also one of the main bases for guiding oil and gas exploration practice. The main goal of source rock formation evaluation is to determine the source rock formations from the profile in a sedimentary basin (or depression) based on many geological and geochemical analysis results, to delineate favorable hydrocarbon-generating areas in space, to make a quantitative evaluation of hydrocarbon generation, to analyze the oil and gas prospects of the basin, and to provide a scientific basis for oil and gas exploration.

2.5.2 *Geological Characteristics of Source Rocks*

2.5.2.1 Lithological Characteristics of Source Rocks

Lithological characteristics are the most intuitive indicators for studying source rocks (Fig. 2.17). Although lithology is not the essential factor determining whether a stratum can generate oil and gas, it is related to the basic conditions for oil and gas generation, that is, the presence of organic matter and a reducing environment. Source rocks are generally fine-grained, dark in color, rich in organic matter and microfossils, often contain native dispersed pyrite, and occasionally exhibit native oil seepages. The common source rocks include two major categories: siliciclastic rocks and carbonate rocks.

(1) Siliciclastic Source Rocks

Siliciclastic source rocks mainly include mudstone and shale, they are formed in stable water bodies at a certain depth, requiring a quiet and anoxic sedimentary environment. Under such environmental conditions, various organic matters provided by organisms can be deposited and preserved along with clay minerals, providing a material guarantee for oil and gas generation. Due to the high content of organic matter and low-valence iron compounds in these siliciclastic rocks, they often exhibit dark

Fig. 2.17 Source rocks of different lithologies (left: dark mudstone, right: limestone)

colors. In China's major continental basins such as Songliao, Bohai Bay, Junggar, and Qaidam, which contain oil and gas, the main source rock layers are gray-black, dark gray, gray, and gray-green mudstones, and shales; similar types are also the most common in foreign source rocks.

(2) Carbonate Source Rocks

Carbonate source rocks mainly consist of ordinary limestone, biogenic limestone, and marl that are rich in organic matter and formed in low-energy environments. The rocks often contain argillaceous components and are mostly gray-black, dark gray, brown gray, and gray in color. They have a microcrystalline to fine-grained structure with few grains, dominated by micrite; mostly thick-bedded to massive, with well-developed horizontal or wavy bedding; contain pyrite and fossils, occasionally exhibit native oil seepages, and sometimes an asphaltic odor can be detected when struck with a hammer. Some of the abundant natural gas resources in the Sichuan Basin of China are related to the limestone of the Permian and Triassic systems. The widespread Paleozoic carbonate rocks in South China, the Tarim Basin, and many carbonate rocks in the Middle and Upper Proterozoic, as well as the Lower Paleozoic of the North China Platform, all have excellent hydrocarbon-generating conditions. The Upper Jurassic Arab Formation and the Cretaceous Asmari Limestone in the Persian Gulf Basin are important carbonate source rock formations.

2.5.2.2 Organic Matter Enrichment and Source Rock Formation

The formation of source rocks is accompanied by the enrichment of organic matter. Organic matter is a common component in sediments, but its content varies greatly, ranging from organic ore layers that are almost homogeneously deposited to sedimentary layers that are almost devoid of organic matter. Most of the sedimentary organic matter exists in sedimentary rocks in varying concentrations.

(1) Paleo-tectonic, Paleo-climatic, and Paleo-geographical Settings for Organic Matter Enrichment

The active zones at the edges of tectonic plates, as well as the rifts, depressions, and foreland basins of orogenic belts within the plates, are the main tectonic belts that are conducive to the formation of source rocks in geological history. Areas that have experienced long-term continuous subsidence are beneficial for the formation of source rocks. In regions where subsidence is stable and long-lasting, if the rate of subsidence is similar to or slightly greater than the rate of sedimentation, it can provide favorable water depths for the extensive reproduction of organisms and the preservation of organic matter from oxidation over the long term, thus ensuring the rich deposition and preservation of original organic matter (Fig. 2.18).

In addition, within large sedimentary basins, the unevenness of the basin caused by fault segmentation or differences in subsidence rates leads to the emergence of many secondary highs and lows. This allows organic matter to be deposited nearby without the need for long-distance transportation, thus avoiding oxidation during

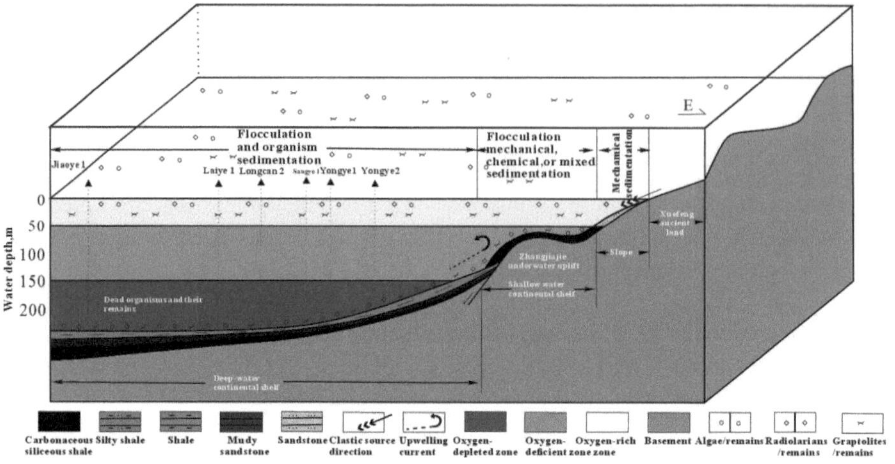

Fig. 2.18 Schematic diagram of the organic-rich shale deposition of the Wufeng Formation—Longmaxi Formation in the northwestern part of Hunan (Mingyang et al., 2019)

the process. Therefore, the segmentation of sedimentary basins is beneficial to the accumulation and preservation of organic matter.

Abundant biological organic matter is the material basis for the formation of organic-rich deposits. The amount of biological organic matter mainly depends on the breeding conditions of organisms. Suitable temperatures, sufficient sunlight, and a humid climate are favorable climates for biological development. Before entering the sediment, most of the biological organic matter is distributed in the water body above the sediment. There are two main pathways for organic matter to enter the sediment: particles and denser organic debris settle directly to the bottom of the water through free settling, and dispersed small particles of organic matter settle by adsorbing and combining with clay minerals to form larger particles. Regardless of the settling way, it is affected by the depth of the water body and the hydrodynamic conditions. The larger the organic matter particles, the shallower the water body, and the quieter the water body, the more conducive it is to settling. Organic matter can only be stably preserved in a reducing environment, so the existence of an oxygen-deficient environment at the bottom of the water body and timely burial are necessary conditions for the preservation of organic matter.

Research on the lithofacies characteristics of source rocks in known oil and gas areas has shown that the most favorable geographical environment for the enrichment of organic matter and the formation of source rocks is the shallow sea, the fore-delta, and deep to semi-deep lake environments.

The carbonate and argillaceous sediments of the shallow sea can both become source rocks. These rocks are generally formed in the limited seas of the continental shelf and the subtidal zone, which are in a persistent low-energy environment. Due to the long-term stable subsidence of the basin bottom, the warm and humid climate, the flourishing of life, the tranquility of the water body, the weak alkalinity of the water medium, and the long-term maintenance of a reducing environment, abundant organic matter can be smoothly accumulated and preserved.

Beyond the coastline, the prodelta zone is an area of rapid, long-term subsidence, characterized mainly by the deposition of dark mudstone and shale rich in organic matter. These are deposited from fine-grained clayey suspended and colloidal materials transported by rivers, containing both marine fossils and terrestrial organic matter, all of which are rapidly buried and preserved. The Niger Delta in Africa, of the Tertiary period, is a world-renowned oil and gas region, where the prodelta dark mudstone and shale are very favorable hydrocarbon source rock layers.

In the deep to semi-deep lake areas, there is an abundant source of organic matter. The water flow is weak, with small waves and low energy, and the bottom of the water body is a reducing environment, making it a favorable setting for the development of terrestrial hydrocarbon source rock series. China's terrestrial hydrocarbon source rock series are characterized by various combinations of shale profiles, especially the shale profiles deposited in deep lake environments are most favorable. For example, the Lower Cretaceous of the Songliao Basin and the Shahejie Formation of the Cenozoic in the Bohai Bay Basin both belong to this deep lake shale-type hydrocarbon source rock layer, and both have formed abundant oil and gas.

In summary, the enrichment of organic matter is a complex process, and any factors affecting the production, preservation, and burial of organic matter will impact the formation of organic-rich sediments. For instance, climate and temperature have a significant impact on the growth of planktonic organisms, affecting organic matter production. The degree of oxidation–reduction and the level of water turbulence affect organic matter preservation, and the rate of organic matter burial also has a certain impact on organic matter enrichment. The enrichment of organic matter requires a certain tectonic background, suitable climate, and appropriate paleo-geographical environment to be realized.

(2) Main Controlling Factors and Models of Source Rock Formation

Previous studies have indicated that paleo-productivity, organic matter preservation conditions, and the dilution effect of organic matter are the main factors controlling the development of source rocks (Katz, 2005).

At the "Marine Source Rocks" (1983) and "Lake Source Rocks" (1985) international symposia organized by the Geological Society of London, debates took place regarding whether the development of source rocks depends more on the high productivity of surface waters or the organic matter preservation conditions of bottom waters. The conclusion was that the former is a more fundamental issue because as long as productivity is high, some organic matter will not have time to oxidize and will accumulate to form source rocks even in oxygen-containing bottom waters.

Statistics show that most organic matter is degraded during the sedimentation process globally, and only a small part of the organic matter is preserved. The proportion of preserved organic matter to the original organic matter is less than 0.5%. Previous studies have analyzed the distribution of global oceanic primary productivity and the total organic carbon content of the surface sediment on the seabed, finding that the correlation between productivity and total organic carbon content in some areas is not obvious and is significantly affected by the redox degree of the water body. This indicates that in addition to paleo-productivity, the preservation conditions of organic matter are also an important factor affecting the development of source rocks.

In response to the two controlling factors of productivity and preservation conditions, the "productivity" model and the "preservation" model have been established as two typical models for the development of source rocks. Under the "productivity" model, the formation of source rocks is mainly controlled by paleo-productivity. The main controlling factors in the development of Mesozoic black shales in the Atlantic and global Cretaceous black shales are primarily productivity. The most typical example of the "productivity" model is the rich nutrient input brought by upwelling currents in modern oceans, which significantly increases productivity. Organic-rich sediments are mainly located in areas of high productivity, such as the Peruvian coastal upwelling current and the East Pacific Equatorial upwelling current, both of which can form high-productivity zones with significant organic matter enrichment. In addition, other factors that significantly increase productivity may also promote the formation of source rocks. For instance, recent studies have shown that volcanic ash entering the water body can bring rich nutrients, promote algal blooms, significantly increase productivity, and facilitate the formation of source rocks.

The "preservation" model for the development of source rocks emphasizes that source rocks are mainly formed in strongly reducing environments where organic matter is well preserved. Good development of source rocks in reducing environments has been observed in places such as Norwegian fjords, British Columbia, and the Black Sea. A typical example of the "preservation" model is the Black Sea, where high salinity water causes stratification, forming a stable strong reducing environment at the bottom of the water, promoting the preservation of organic matter and the formation of source rocks. High salinity water tends to cause water stratification, promoting the formation of strong reducing environments, which is also one of the main reasons why saline lakes commonly develop high-quality source rocks.

In addition to paleo-productivity and organic matter preservation conditions, the dilution effect of organic matter is also an important factor affecting the development of source rocks. The dilution effect of organic matter refers to the process where, with the input of source materials, a large amount of non-organic matter input dilutes the original organic matter abundance. It is generally believed that the greater the sedimentation rate, the greater the dilution effect, and the smaller the sedimentation rate, the smaller the dilution effect. When the dilution effect is significant, a large amount of source debris dilutes the organic matter abundance in the source rock, leading to a reduction in organic carbon content, which is not conducive to the development of source rocks.

In summary, the formation of source rocks is the result of the influence of various factors, with high productivity, static reducing environment, and moderate sedimentation rate being most conducive to the formation of source rocks.

2.5.3 *Geochemical Characteristics of Source Rocks*

To evaluate the hydrocarbon generation potential of source rocks in a sedimentary basin, it is not sufficient to only conduct geological studies of the source rock layers. A systematic study of the quantity, type, and thermal evolution stages of the organic matter contained within the source rocks is also necessary.

2.5.3.1 The Abundance of Organic Matter

The abundance of organic matter in source rocks is the material basis for the formation of oil and gas, and its content in the rock is the main factor that determines the rock's ability to generate hydrocarbons. The relative content of organic matter in the rock is called the abundance of organic matter. The commonly used indicators of abundance currently include total organic carbon (TOC) and the hydrocarbon potential of rock pyrolysis.

(1) Total Organic Carbon (TOC)

Total organic carbon refers to the percentage of the total weight of the rock that is made up of carbon elements contained in all organic matter. The determination of total organic carbon involves first removing inorganic carbon from the sample with hydrochloric acid, then burning in a high-temperature oxygen flow to convert the total organic carbon into carbon dioxide, and finally quantifying the organic carbon using an infrared detector. There is a certain proportional relationship between the content of organic carbon in the sample and the content of organic matter, that is, the content of organic matter = total organic carbon content × K, where K is the conversion factor, mainly depending on the ratio of hydrogen to carbon atoms in the hydrocarbons. Tissot & Welte, (1978) believes that the K value is different for different types of kerogens at different stages of evolution (as shown in Table 2.11).

Modern hydrocarbon migration theories suggest that hydrocarbons formed in source rocks must be effectively expelled only after the rock's own adsorption capacity is satisfied. Therefore, there is a lower limit to the abundance of organic matter in source rocks. This value refers to the number of hydrocarbons generated by organic matter per unit weight of source rock, which is the amount of organic matter required when the rock's adsorption capacity for hydrocarbons is saturated. Thus, the greater the adsorption capacity of hydrocarbons in the source rock, the higher the

Table 2.11 Conversion coefficients (K) for calculating the abundance of organic matter from total organic carbon

Evolution stage	Types of kerogen			Coal
	Type I	Type II	Type III	
Diagenesis stage	1.25	1.34	1.48	1.57
Late diagenetic stage	1.20	1.19	1.18	1.12

required lower limit of organic matter content. Conversely, when the type of organic matter is good and the degree of thermal evolution is high, the corresponding lower limit of organic matter content is lower.

In the oil and gas exploration industry (SY/T 5735-2019), taking the most common shale and carbonate source rocks as examples, a TOC content below 0.5% is considered an ineffective source rock, while a content above 0.5% is considered an effective source rock. Specifically, a TOC between 0.5 and 1.0% is classified as a general source rock, a TOC between 1.0 and 2.0% is considered a good source rock, and a TOC greater than 2.0% is classified as a high-quality source rock. It is important to note that the level of total organic carbon does not directly reflect the hydrocarbon generation potential of the source rock. The proportion of "active" carbon in the organic carbon that can be converted into oil and gas must also be considered.

(2) Rock Pyrolysis Hydrocarbon Potential

Pyrolysis is an important method for evaluating source rocks using a rock evaluation instrument (Rock–Eval) and is one of the routine analysis methods in modern oil and gas exploration. The principle of the analysis is that source rocks contain residual hydrocarbons that have not been expelled and active and inert organic matter (mainly kerogen) that has not generated hydrocarbons. Therefore, by placing the crushed source rock samples in an inert gas environment in a specific heating furnace and conducting programmed temperature increase, different substances will be released at different temperature segments. Recording the quantity of these substances allows obtaining the pyrolysis chromatogram of the rock sample (as shown in Fig. 2.19). S_1 represents the oil and gas components released by the rock at a lower heating temperature (<300 °C) that have been generated and residual in the sample under geological conditions. S_2 represents the hydrocarbon substances generated by the artificial pyrolysis of organic matter in the source rock at a higher pyrolysis temperature (650 °C), representing the hydrocarbon generation potential of kerogen. The temperature corresponding to the S_2 peak of is called T_{max}, which can reflect the ease with which kerogen is converted into oil and gas. S_3 represents the content of CO_2 generated by the pyrolysis of oxygen-containing groups in kerogen. S_4 is the content of inert carbon (carbon that is not easy to generate oil and gas) in kerogen obtained by continuous high-temperature heating of the sample (up to 850 °C). By adding the various types of carbon in S_1, S_2, S_3, and S_4, the total organic carbon content in shale can be calculated.

(3) Bitumen Content

Bitumen refers to the organic matter in the source rock that can be extracted with conventional organic solvents, as opposed to the insoluble kerogen. In the last century, chloroform was commonly used as an extraction reagent, hence the extracted material was called "chloroform bitumen A". Nowadays, more environmentally friendly solvents such as dichloromethane are often used for extraction. The bitumen content reflects the proportion of soluble organic matter in the sample, not the total organic carbon content. It is an important indicator for evaluating oil content.

Fig. 2.19 Pyrolysis analysis cycle and spectrum

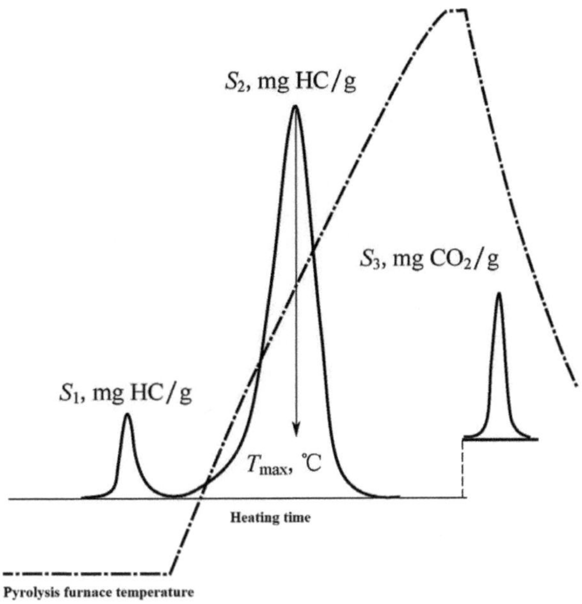

In China's terrestrial freshwater to brackish water sediments, the content of chloroform bitumen "A" in the main source rocks is above 0.1%, with an average value of 0.1–0.3%. Figure 2.20 shows the distribution frequency of chloroform bitumen "A" content in 386 source rock samples from major oil and gas basins in China. Most values are around 0.1%, with higher values reaching up to 1%, and non-source rocks have a content below 0.01%. It should be noted that the bitumen content is greatly affected by the type and maturity of the organic matter in the source rock. In over-mature source rocks, due to severe secondary cracking, the bitumen content is relatively low.

Fig. 2.20 Main Mesozoic and Cenozoic Oil and Gas Basins in China chloroform asphalt "A" content frequency diagram of source rocks (Shang et al., 1982)

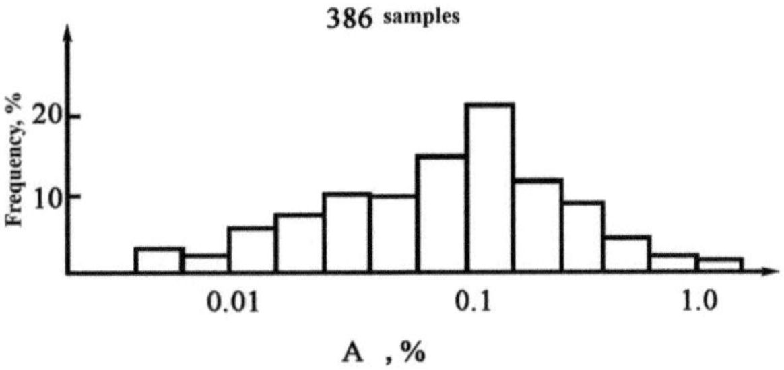

2.5.3.2 Types of Organic Matter

The type of organic matter in source rocks is a key indicator of its hydrocarbon-generating quality and an important parameter for evaluating the hydrocarbon-generating capacity of source rocks. Different types of organic matter possess varying potentials for generating oil and gas and form products of different compositions. Therefore, accurately distinguishing the types of organic matter is another critical issue in source rock research. Current research on the types of organic matter in source rocks generally falls into two aspects: the study of kerogen types and the study of soluble organic matter types.

Determining the type of kerogen is the main body of organic matter type research, with primary methods including elemental analysis, maceral analysis, and rock pyrolysis, among others. Classification schemes for elemental analysis and maceral analysis have been introduced in this chapter. Using pyrolysis analysis to obtain S_2 and S_3, in conjunction with the organic carbon content of the sample, the following two parameters for analyzing organic matter types can be calculated: Hydrogen Index (HI) and Oxygen Index (OI).

$$\mathrm{HI} = \mathrm{S_2/TOC};\ \mathrm{OI} = \mathrm{S_3/TOC}.$$

Research has shown that there is a good correlation between HI and the H/C atomic ratio, and between OI and the O/C atomic ratio (Espitalié et al., 1977). Therefore, in the van Krevelen diagram for kerogen type division, HI can replace the H/C atomic ratio, and OI can replace the O/C atomic ratio (Fig. 2.21), and similar methods can be used for interpretation.

The relative content of soluble organic matter components in source rocks (saturated hydrocarbons, aromatic hydrocarbons, non-hydrocarbons, and asphaltenes) reflects the properties and evolutionary history of the organic matter. Thus, the characteristics of the soluble extractable components in source rocks are also of reference significance for the classification of organic matter types, especially for low-mature source rocks, where the application effect is better. More commonly used information includes: the compositional characteristics of chloroform extracts from source rocks, saturated hydrocarbon gas chromatographic characteristics, and chromatography-mass spectrometry analysis data (Table 2.12). Biomarker characteristics in source rocks can also reflect the properties of the organic matter.

In addition to the above methods, organic matter types can also be determined by assisting with saturated hydrocarbon gas chromatography, biomarker compounds, stable carbon isotope $\delta^{13}C$ (‰) analysis of monohydrocarbons, thermal weight loss characteristics of kerogen, infrared spectral (functional group) characteristics, evolutionary history of organisms, and analysis of the source of organic matter.

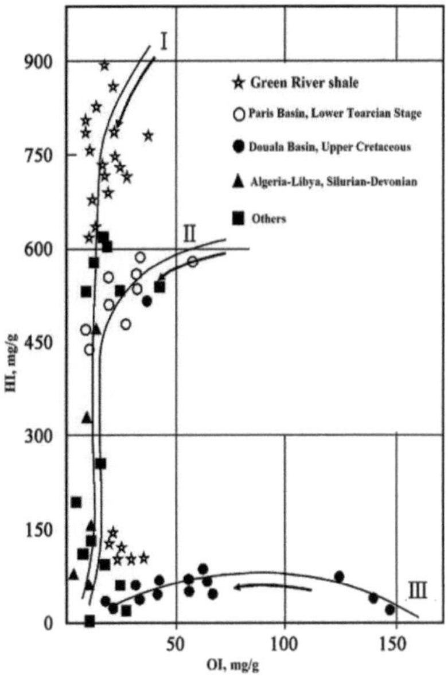

Fig. 2.21 Classification of organic matter types in source rocks using hydrogen index and oxygen index (Espitalié et al., 1977)

2.5.3.3 Organic Matter Maturity

The maturity of organic matter refers to the geological thermal evolution degree of organic matter in source rocks. Since sedimentary organic matter generates different types of hydrocarbons at different stages of evolution, exploration practices have proven that there is a higher success rate in oil and gas exploration in areas where mature source rocks are distributed. Therefore, maturity evaluation is another major content of source rock research.

During the diagenesis and epigenetic evolution process of sedimentary rocks, many physical and chemical properties of organic matter in source rocks undergo corresponding changes, and this process is irreversible. Therefore, the characteristics of changes in certain physical properties and chemical composition of organic matter can be used to determine the degree of thermal evolution of organic matter and divide the stages of organic matter evolution. The conventional geochemical methods currently used to evaluate the maturity of source rocks mainly include the vitrinite reflectance method and the rock pyrolysis Tmax method, among others.

(1) Vitrinite Reflectance (R_o)

Vitrinite is a group of oxygen-rich maceral, chemically composed of aromatic rings as the core with different alkyl side chains. During the thermal evolution process, the alkyl side chains of vitrinite are thermally cracked, the aromatic rings condense, microlamination structures appear, the spacing between aromatic sheets gradually narrows, causing the reflectance to increase, transmittance to decrease, and the color to darken (Fig. 2.22), and this trend is irreversible.

Table 2.12 Summary of organic matter type division schemes for oil (gas) source rocks in China's Mesozoic and Cenozoic (according to Difan, 1991; Huaixian, et al., 2001; Jianzhong, 2005)

Contents		Evaluation parameters	Organic matter types			
			Type I	Type II1	Type II2	Type III
Chloroform asphalt "*A*"		Saturated hydrocarbons (%)	40–60	20–40	20–30	5–17
		Aromatic hydrocarbons (%)	15–25	5.0–15	5.0–15	10–22
		Saturated hydrocarbons/ aromatic hydrocarbons (%)	>3.0	1.0–3.0	1.0–1.6	0.5–0.8
		Non-hydrocarbons + asphaltenes (%)	20–40	40–50	50–60	60–80
		(Non-hydrocarbons + asphaltenes)/total hydrocarbons	0.3–1.0	1.0–3.0	1.0–3.0	3.0–4.5
Saturated hydrocarbon chromatographic characteristics		Peak characteristics	Single high front peak	Double high front peak	Double high rear peak	Single high rear peak
		Main peak carbon number	C_{17}, C_{19}			
		$(nC_{21} + C_{22})/(nC_{28} + C_{29})$	>2.0	1.5–2.0	1.0–1.5	1.2–1.5
		OEP	1.2–0.9	>1.2		>2.0
Molecular hydrocarbon carbon isotopes		$\delta^{13}C$ (‰)	>−29.5	−28 to −26.5	−28 to −25	>−25
Kerogen	Carbon isotopes	$\delta^{13}C$ (‰)	<−28	−29.5 to −2	−26.5 to −25	>−25
	Elemental analysis	Initial (H/C) atomic ratio	>1.5	1.5–1.2	1.2–0.7	<0.7
		Initial (O/C) atomic ratio	<0.1	0.1–0.2	0.2–0.3	>0.25
	Maceral	Exinite (%)	70–90	70–50	50–10	<10
		Vitrinite (%)	<10	10–20	20–70	70–90
		Typing index	>80	80–40	40–0	<0
	Infrared spectroscopy analysis	2,900 cm^{-1}/1,600 cm^{-1}	>1.5	0.9–1.5	0.4–0.9	<0.4
		1,460 cm^{-1}/1,600 cm^{-1}	1.2–0.45		0.45–0.25	<0.25
	Pyrolysis analysis	Hydrogen index (mg/gTOC)	>600	600–250	250–130	<130
		S_2/S_3	>20	20–5.0	5.0–2.5	<2.5

(continued)

Table 2.12 (continued)

Contents	Evaluation parameters	Organic matter types			
		Type I	Type II1	Type II2	Type III
Biomarker compounds	$C_{27}\alpha\alpha\alpha R$ cholestane (%)	>55	55–35	35–20	<20
	$C_{28}\alpha\alpha\alpha R$ cholestane (%)	<15	15–35	35–45	>45
	$C_{29}\alpha\alpha\alpha R$ cholestane (%)	<25	25–35	35–45	>45–55
	$C_{27}\alpha\alpha\alpha R$ cholestane/ $C_{29}\alpha\alpha\alpha R$ cholestane	>2.0	2.0–1.2	1.2–0.8	<0.8

Note Applicable range: $Ro < 0.8\%$

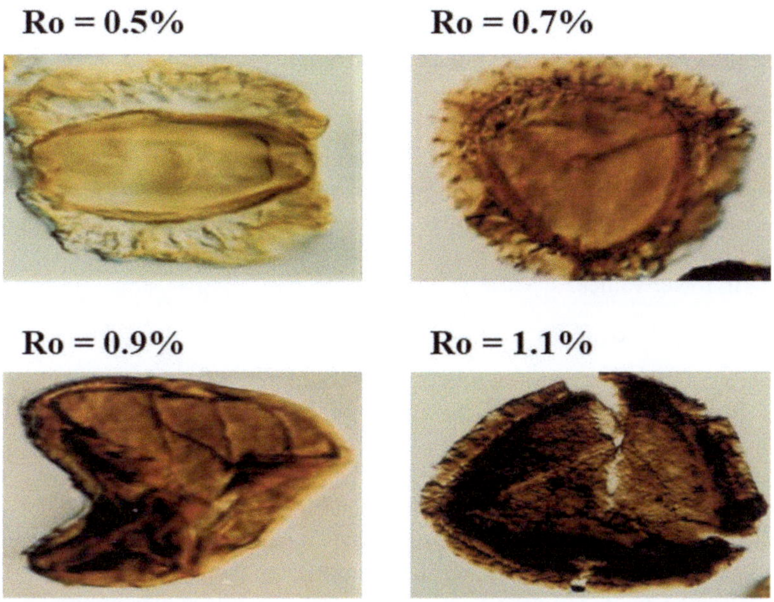

Fig. 2.22 Reflectance of vitrinites of different maturities

Vitrinite reflectance is not only controlled by the changes in paleotemperature but is also affected by time. To reach the same reflectance, the higher the temperature, the shorter the time required, and vice versa, the longer the time needed. Therefore, vitrinite reflectance is an excellent indicator for studying the time–temperature history experienced by source rocks, measuring the maturity of organic matter, and dividing the stages of oil and gas formation. Different types of kerogen have different chemical structures, and the geothermal conditions required to reach various stages

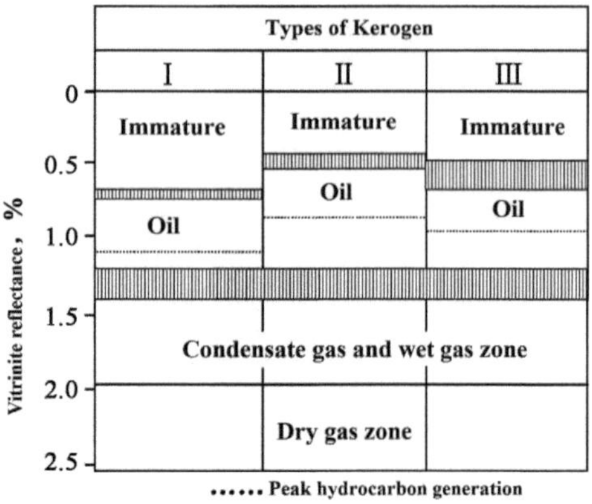

Fig. 2.23 Approximate boundaries of oil and gas zones determined by vitrinite reflectance (according to Tissot & Welte, 1984)

of evolution differ. Thus, when using vitrinite reflectance to judge the maturity of organic matter, distinctions should be made for different types of kerogen (Fig. 2.23).

The boundaries may vary slightly depending on the time–temperature relationship and the mixing conditions of organic matter from different sources. The application of vitrinite reflectance to study maturity has certain limitations. Firstly, the judgment of vitrinite color is subjective. Secondly, it is difficult to find vitrinites in sapropelic samples or those deposited before the Devonian Period (the era of gymnosperms). Thirdly, in samples at the stage of high over-maturity, vitrinite reflectance has reached its maximum value, making it difficult to further distinguish maturity. Finally, vitrinite reflectance is suppressed in samples rich in hydrogen and high organic carbon content, resulting in lower values.

(2) Rock Pyrolysis T_{max}

The higher the thermal evolution degree of a source rock, the less degradable carbon remains in the residual kerogen, the smaller the rock pyrolysis S_2 peak, and the corresponding pyrolysis temperature (T_{max}) of the highest point of the S2 peak is also greater.

However, the value of T_{max} can be influenced by various factors that are not related to maturity. For instance, samples that are highly mature, hydrogen-rich, sulfur-rich, uranium-rich, or contain heavy oils typically exhibit an unusually low T_{max}, whereas samples affected by oxidation and degradation, or those rich in clay minerals, usually show an abnormally high T_{max} (Yang & Horsfield, 2020). Additionally, different types of organic matter correspond to different T_{max} values during various stages of hydrocarbon generation. Consequently, scholars have proposed various schemes for dividing the stages of hydrocarbon generation using T_{max} (Table 2.13), among which the method by Peters and Cassa (1994) is the most widely applied.

(3) Other Maturity Assessment Indicators

In addition to the two most used maturity indicators mentioned above, there are several other techniques such as the Thermal Alteration Index (TAI), kerogen

Table 2.13 Hydrocarbon generation stages indicated by T_{max} values (in degrees Celsius)

Literature source	Immature	Early mature	Oil generation peak	Late mature	Overmature
Espitalié (1986)	<430	430–460			>460
Peters and Cassa (1994)	<435	435–445	445–450	450–470	>470
Baskin (1997)	<430	430–435	445–450	450–460	>460
Dellisanti et al. (2010)	<434	434–465			>465
Killops and Killops (2013)	<430	430–445	445–465		465>

elemental composition, the odd–even predominance ratio of n-alkanes, biomarker compounds, and laser Raman spectroscopy, which can be used to assess the thermal maturity of organic matter.

a. Thermal Alteration Index (TAI): The principle of the thermal alteration index for spores or algae reflects maturity in a similar way to vitrinite reflectance, both utilizing the rule that the color of macerals in hydrocarbon source rocks becomes darker with increasing maturity. However, the change in projection rate of spores and algae under heat is less stable than that of vitrinite, and its correspondence to the hydrocarbon generation stage is poor, thus the application of the thermal alteration index is very limited.
b. Atomic Ratios: The hydrogen to carbon (H/C) and oxygen to carbon (O/C) atomic ratios of kerogen both decrease with increasing maturity. However, other factors such as the type of kerogen and oxidation effects can also influence these two parameters, thus the reliability of inferring maturity through H/C and O/C atomic ratios is relatively low.
c. Odd–Even Carbon Preference (OEC) Ratio: As diagenetic evolution deepens, the n-alkanes in the extractable organic matter of hydrocarbon source rocks gradually lose their odd–even carbon number preference. However, this characteristic is only clearly reflected in the immature or early mature stages and cannot be used for maturity evaluation during the main hydrocarbon generation period.
d. Biomarker Compounds: Biomarker compounds with different configurations have different generation rates or interconversion relationships during the hydrocarbon generation evolution process. Thus, the ratio of related biomarker compounds can be used to indicate the thermal evolution maturity of organic matter. However, biomarker maturity indicators are influenced by multiple factors such as the type of organic matter, sedimentary environment, mineral content, and degradation, lack a unified standard corresponding to the hydrocarbon generation stage, and most biomarkers are difficult to apply to samples with high thermal evolution.
e. Laser Raman Spectroscopy: The condensation degree and structural order of sedimentary organic matter in hydrocarbon source rocks increase with the degree of thermal evolution. The first-order vibration peak parameters of laser Raman

spectroscopy can reflect the information of the crystal structure and chemical bond synthesis and cleavage of carbon materials, and have a good correlation with the thermal evolution degree of sedimentary organic matter in hydrocarbon source rocks, which can indicate the maturity of sedimentary organic matter in hydrocarbon source rocks. However, this method is costly for analysis, the data results are complex to analyze, and it is mainly used for fine characterization of over-mature samples

f. TTI: The Time–Temperature Index (TTI) method was proposed by Soviet scholar Lopatin and further developed by American scholar Waples (1980). This method is based on the theoretical assumption that "the increase in organic matter maturity is linearly related to time and exponentially related to temperature," and it proposes a formula for calculating maturity. However, the temperature experienced by organic matter during the geological burial process is constantly changing, and the TTI method for evaluating maturity is usually applied in basin modeling studies.

2.6 Oil-Source Correlation

Oil-source correlation is the process of determining the genetic connection between hydrocarbons and source rocks based on geological and geochemical evidence. This includes both the comparison between different source rocks and the comparison between hydrocarbons from various reservoirs. Through comparative studies, it is possible to determine the direction and distance of hydrocarbon migration, understand the characteristics of secondary changes in hydrocarbons, and thereby further delineate reliable hydrocarbon source areas, identify exploration targets, and effectively guide the exploration and development of oil and gas.

The petroleum and natural gas formed by the evolution of kerogen in source rocks are partially migrated to reservoir rocks, while the rest remain within the source rock. Therefore, the kerogen and asphalt in the source rock have a genetic relationship with the hydrocarbons from the same rock series. Hydrocarbons from the same source rock have chemical composition similarities, whereas those from different sources exhibit significant differences. This phenomenon exists from macroscopic characteristics to individual compound levels, and this similarity is the basis for oil-source correlation. Appropriate parameters can be selected to identify the "fingerprint" patterns of soluble extract composition in source rocks that are like, identical with, or different from hydrocarbons, and the degree of similarity can be used to prove whether there is a genetic relationship between hydrocarbons and source rocks.

When conducting oil-source correlation, the selection of comparison indicators is the most critical. Typically, the content ratios of characteristic compounds that are common to both hydrocarbons and source rocks and are not affected by migration and thermal metamorphism are used as oil-source correlation indicators. The methods currently used mainly include analysis of the carbon number distribution characteristics of normal alkanes, analysis of biomarker composition characteristics, and analysis of stable carbon isotope composition, among others.

2.6.1 *Application of Normal Alkane Distribution Characteristics for Oil-Source Correlation*

Normal alkanes are the main hydrocarbon components of oil and gas, can serve as markers for the maturity of oil and gas and the source of organic matter, and can also act as "fingerprint" compounds for oil and gas comparison. This analytical method has been widely used in oil-oil and oil-source comparisons.

The carbon number distribution, main peak carbon number, and especially the carbon number distribution pattern of normal alkanes are very useful oil-source comparison parameters. The percentage content of individual components can be calculated from the gas chromatogram of normal alkanes, with carbon number as the horizontal axis and percentage content as the vertical axis, to plot the normal alkane carbon number distribution curve. Hydrocarbons with a genetic relationship have similar characteristics in the carbon number distribution curve of normal alkanes. Figure 2.24 is a comparison diagram made by Williams (1974) using normal alkanes for three types of oil and three sets of potential source rock layers in the Williston Basin. The comparison shows that the three types of oil have a genetic relationship with the three sets of source rocks, respectively.

It should be noted that since normal alkanes are most sensitive to bacterial degradation and thermal effects, and are influenced by selective evaporation and migration fractionation, normal alkane indicators generally work well only for crude oils with medium to low maturity, no obvious biodegradation, and no strong evaporation.

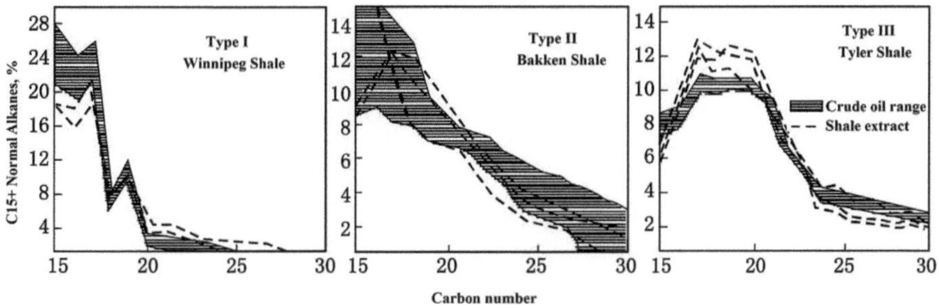

Fig. 2.24 Comparison of C_{15+} normal alkanes from Williston Basin Oil and source rock extracts (Williams, 1974)

2.6.2 Application of Stable Carbon Isotope Composition Characteristics for Oil-Source Correlation

The relationship between the isotopic composition of oil, gas, asphalt, and kerogen is a unique comparison parameter, the importance of which lies in directly linking the oil and gas with the kerogen and asphalt in the potential source rocks. The most used parameter is the stable carbon isotope composition, i.e., $\delta^{13}C$. The isotopic composition of natural compounds is the result of isotopic fractionation, and when the original organic matter and thermal evolution conditions are the same, the carbon isotope composition between the oil and gas and the source rock is comparable.

During the thermal evolution of kerogen, due to thermal decomposition, the carbon isotopes in the products are lighter than those in the residues. Therefore, the carbon isotopes in the asphalt from the same layer are generally lighter than those in the kerogen, but the difference in $\delta^{13}C$ values will not be greater than 2–3‰. The $\delta^{13}C$ values of the oil formed from kerogen are the same as or slightly lighter than those of the asphalt. This difference will also not be greater than 2‰. A large amount of statistical data indicates that $\delta^{13}C$ kerogen > $\delta^{13}C$ asphalt > $\delta^{13}C$ oil, $\delta^{13}C$ kerogen $\geq$ $\delta^{13}C$ asphaltene $\geq$ $\delta^{13}C$ non-hydrocarbons $\geq$ $\delta^{13}C$ aromatics $\geq$ $\delta^{13}C$ saturated hydrocarbons. Since secondary actions such as migration, thermal conversion, and deasphalting affect the isotopic composition of all fractions, various irregular isotopic type curves can often be observed in actual research. Oil samples with the same genesis and secondary alteration effects should still have similar isotopic type curves.

2.6.3 Application of Biomarker Compounds for Oil-Source Correlation

Biomarker compounds have a unique tracing significance due to their characteristics and stable structures. They are widely used to indicate the source input, type of parent rock, and sedimentary environment, and serve as evaluation and research indicators for oil and gas source comparison, oil and gas migration analysis, biodegradation analysis, and description of reservoir fluid heterogeneity. At the same time, certain changes in the geological evolution process, such as configuration isomerization, rearrangement, aromatization, and side chain cleavage, make them sensitive indicators for evaluating maturity, sedimentation rate, and thermal history.

(1) Homoterpene Alkanes

This is a group of isomeric alkane compounds derived from the phytol side chain of chlorophyll or lipid compounds. Structurally, there is a methyl side chain every three methylene groups, resembling a structure formed by the hydrogenation and condensation of several isoprene molecules, hence the name homoterpene alkanes. Since the 1960s, C_9–C_{25} homoterpene alkanes have been successively discovered in crude oil and sediments. Among them, 2,6,10,14-tetramethylpentadecane is called pristane (Pr), and 2,6,10,14-tetramethylhexadecane is called phytane (Ph), both of which are usually the most abundant and stable.

Homoterpene alkanes appear in almost every crude oil and source rock extract. Although they are generally not as abundant as saturated hydrocarbons, their relatively stable structure allows them to better resist microbial degradation compared to normal alkanes, making them an important set of comparison parameters. Generally, the higher the relative content of pristane, the more oxidizing the original depositional environment. An excessively high relative content of phytane indicates a more reducing original depositional environment (Fig. 2.25). Commonly used indicators include Pr/Ph, Pr/n-C_{17}, and Ph/n-C_{18} (where n represents straight-chain normal alkanes). At the same time, biodegradation will cause the content of pristane and phytane to relatively increase compared to normal alkanes, whereas thermal evolution of hydrocarbons will produce more normal alkanes, thereby reducing the relative content of homoterpene alkanes (Fig. 2.25).

(2) Sterane and Triterpane Compounds

Sterane and triterpenoid compounds possess unique structures and stable properties, as well as a strong resistance to microbial degradation, which play a significant role in oil-source correlation. Source rocks and crude oils that are genetically related should exhibit similar relative contents and combination characteristics of steranes and triterpanes. Therefore, the distribution patterns of sterane and triterpane series

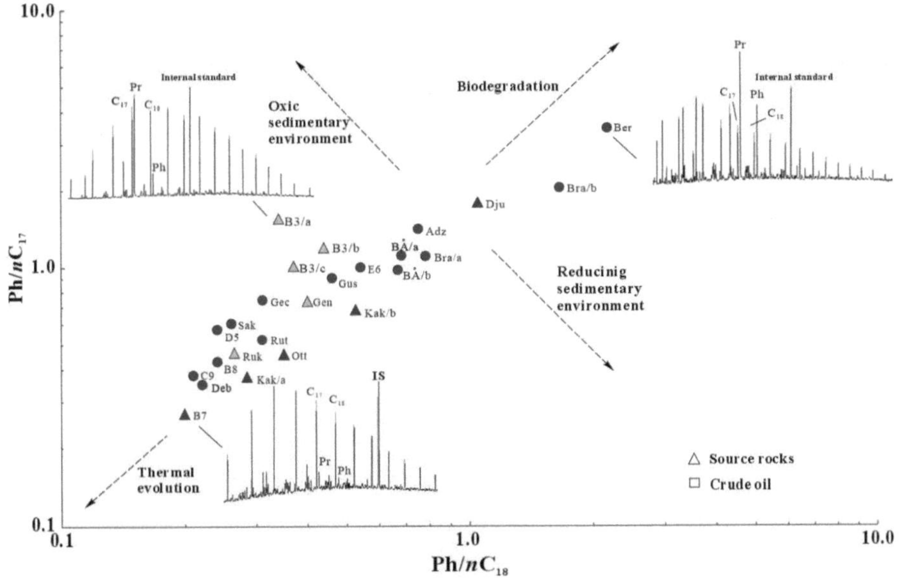

Fig. 2.25 Distribution characteristics of homoterpene alkanes in Paleozoic source rocks and oils of the Baltic Basin in Northern Europe (Yang et al., 2017)

compounds can be utilized for oil-source comparison. Almost all sterane and triterpane parameters used to indicate the sedimentary environment, source input, and maturity of organic matter are also excellent parameters for oil-source correlation. Some characteristic compounds and their distribution curves (such as the C_{27} sterane distribution curve), homologue composition ternary diagrams, and correlation plots are very effective tools for conducting oil-source comparisons.

Utilizing biomarker compositional parameters for oil-source correlation is the most employed method. Given the complexity of factors that influence biomarker composition, relying on any single indicator has its limitations. If there is a genetic relationship between the source rock and the crude oil, both should exhibit similarities in their source nature, sedimentary environment, and maturity. Therefore, when selecting parameters, it is essential to consider the three factors simultaneously. The most frequently used are the mass chromatograms of m/z 217 (steranes) and m/z 191 (triterpenoids). Parameters that reflect the source include the relative content of C_{27}–C_{30} steranes (Fig. 2.26), the ratio of (hopane + moretane) C_{29}/C_{30}, and the oleanane index (OI), among others. Parameters that reflect the sedimentary environment include gammacerane/C_{30}(moretane + hopane) and C_{35} hopane/C_{34} hopane, etc. Parameters that reflect maturity include Ts/Tm, C_{31} hopane S/R, $\alpha\alpha\alpha C_{29}$ sterane S/(S + R), C_{29} sterane $\beta\beta/(\alpha\alpha + \beta\beta)$, etc. These are all parameters that can be used for comparison; one can use a few of them or use them all simultaneously.

The protracted nature of oil and gas formation and their inherent mobility led to a series of transformations during migration, accumulation, and even within the reservoir. Such alterations can obscure or even completely mask the similarities between oil and its source, thereby greatly increasing the ambiguity and complexity of correlation efforts. Therefore, when conducting oil-source correlation, it is essential to thoroughly consider the applicability of the comparison parameters in different regions or strata, to master flexibly, experiment repeatedly, identify the most effective

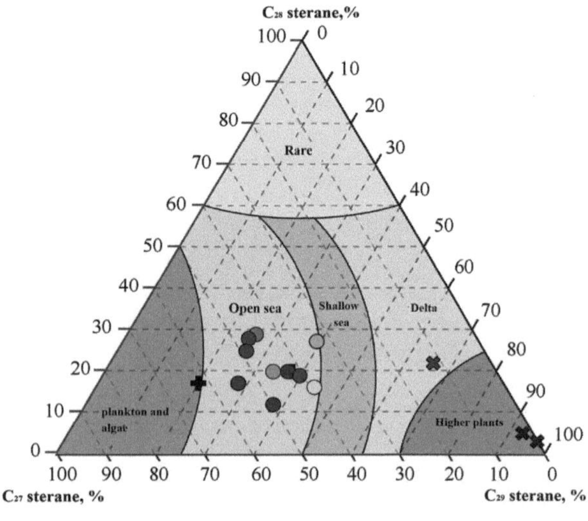

Fig. 2.26 The sedimentary environment and material source indicated by the relative content of steranes (Yang et al., 2022)

methods of comparison, and apply a comprehensive set of indicators. In comparative studies, the more parameters utilized, the more reliable the results tend to be. At the same time, it is imperative to consider the entire causal system of oil and gas formation. Only by fully considering the influences of paleoenvironment, maturity, migration effects, and even biodegradation can we gain a dialectical understanding of the causal connections between oil and its source.

Questions for Discussion

1. What are the core disagreements between the two main schools of thought on the origin of oil and gas? What are the main viewpoints of the inorganic origin theory of oil and gas? What are the main viewpoints of the organic origin theory of oil and gas? What is the primary evidence supporting each?
2. What are the main types of biological organic matter? How is their potential for hydrocarbon formation?
3. What is meant by sedimentary organic matter? What factors determine the abundance of sedimentary organic matter in sediments (rocks)?
4. What is kerogen? How is kerogen classified by type? What are the characteristics of kerogen evolution?
5. What are the main factors influencing the generation of oil and gas? How do they impact the formation of oil and gas?
6. What are the stages of the transformation of organic matter into oil and gas, and what are the characteristics of each stage?
7. What is the threshold of oil generation and the oil generation window?
8. What are biogenic gas, oil-associated gas, and coal-derived gas? How can natural gases of different origins be identified?
9. What are source rocks and source rock systems? What should be the favorable tectonic conditions and lithofacies paleogeographic conditions for the development of source rocks?
10. How to evaluate source rocks based on geological and organic geochemical characteristics?
11. What is oil-source rock correlation? What are its basic principles and purposes? What are the commonly used methods?

References

Baskin, D. K. (1997). Atomic H/C ratio of kerogen as an estimate of thermal maturity and organic matter conversion. *AAPG bulletin, 81*(9), 1437–1450.

Blanc, P., & Connan, J. (1992). Generation and expulsion of hydrocarbons from a Paris Basin Toarcian source rock: An experimental study by confined-system pyrolysis. *Energy & fuels, 6*(5), 666–677.

Botz, R. W., & Von Der Borch, C. C. (1984). Stable isotope study of carbonate sediments from the Coorong Area, South Australia. *Sedimentology, 31*(6), 837–849.

Braun, R. L., & Burnham, A. K. (1990). Mathematical model of oil generation, degradation, and expulsion. *Energy & Fuels, 4*(2), 132–146.

Bray, E. E., & Evans, E. D. (1961). Distribution of n-paraffins as a clue to recognition of source beds. *Geochimica Et Cosmochimica Acta, 22*(1):2–15. https://doi.org/10.1016/0016-7037(61)90069-2.

Connan, J. (1974). Time-temperature relation in oil genesis: geologic notes. *AAPG Bulletin, 58*(12), 2516–2521.

Dai, J., Qi, H., & Hao, S. (1989). Introduction to natural gas geology.

Dai, J. X., Pei, X., & Qi, H. (1992). China natural gas geology.

Dai, J., Wang, T., Song, Y., Zhang, H., Xu, Y., & Zhang, Q. (1997). Formation and distribution of medium-large-sized gas fields in China.

Dai, J. (1992). Identification of various alkane gases. *Science in China (Series B: Chemistry, Life Sciences, and Earth Sciences),* 185–193.

Dellisanti, F., Pini, G. A., & Baudin, F. (2010). Use of Tmax as a thermal maturity indicator in orogenic successions and comparison with clay mineral evolution. *Clay minerals, 45*(1), 115–130.

Difan, H. (1991). Theory foundation of terrestrial facies petroleum generation.

Durand, B., & Espitalié, J. (1976). Geochemical studies on the organic matter from the Douala Basin (Cameroon)—II. *Evolution of kerogen, eochimica et Cosmochimica Acta. 40*,7:801–808.

Espitalié, J., Madec, M., Tissot, B., Mennig, J. J., & Leplat, P. (1977). Source rock characterization method for petroleum exploration. In: Offshore technology conference (pp. OTC-2935). OTC.

Espitalié, J., Deroo, G., & Marquis, F. (1986). La pyrolyse Rock-Eval et ses applications. Troisième partie. *Revue de l'Institut français du Pétrole, 41*(1), 73–89.

Gutjahr, C. C. M. (1966). Carbonization measurements of pollen-grains and spores and their application. *Leidse Geologische Mededelingen, 38*(1), 1–29.

Hao, F., Dong, W. L., Zou, H. Y., et al. (2003). Overpressure fluid flow and rapid accumulation of natural gas in Yinggehai Basin. *Acta Petrol Sinica, 24*(6), 7. https://doi.org/10.1007/BF02974893

Hao, F., Zou, H. Y., Fang, Y., et al. (2004). Overpressure-fault controlled fluid flow and episodic hydrocarbon accumulation. *Acta Petrolei Sinica.* https://doi.org/10.7623/syxb200406008

Hao, F., Jiang, J., Zou, H., Fang, Y., & Zeng, Z. (2004). Differential retardation of organic matter maturation by overpressure. science in china series D: *Earth sciences, 47*(9), 783–793.

Héroux, Y., Chagnon, A., & Bertrand., R. (1979). Compilation and correlation of major thermal maturation indicators. *AAPG Bulletin, 63*(12), 2128–2144.

Horsfield, B., Zou, C., Li, J., Yang, S., Mahlstedt, N., Misch, D., Gross, D., Wei, M., Wang, Y., & Tan, J. (2021). Prediction of the gas-generating characteristics of the Qiongzhusi and Longmaxi Formations, Yangtze Platform, southern China, using analogues. *AAPG Bulletin, 105*, 945–985.

Houfu, Z., & Wanxuan, Z. (1987). *Petroleum Geology* (2nd).

Houfu, Z., Chaoliang, F., Xianzhi, G., Zhihuan, Z., & Youlu, J. (1999). *Petroleum geology*. Beijing: Publishing House of Oil Industry

Huaixian, X., Lihua, C., & Yujin, W. (2001). Petroleum geological test technology and application.

Hunt, J. M. (1979). Petroleum geochemistry and geology. San Francisco: W.H. Freeman and Company.

Jiamo, F., & Jiyang, S. (1975). Theory and practice of petroleum evolution (1)—evolution mechanism and evolution stage of petroleum. *Geochemistry, 2*, 87–107.

Jianzhong, Q., Zhihao, Q., Yin, C., & Meizhen, Z. (2005). New techniques in oil and gas geochemical exploration. *Petroleum Geology & Experiment, 27*(5), 519–528.

Jiawei, Z., Houfa, Q., & Yongsheng, L. (1983). Discovery of Wenliu gas pool generating from coal measures and its significance of prospecting gas in north China basin. *Petroleum Exploration and Development, 1*, 4–12.

Jinxing, D. A. I., Shengfei, Q. I. N., Guoyi, H. U., Yunyan, N. I., Lideng, G. A. N., & Feng, H. O. N. G. (2019). Major progress in the natural gas exploration and development in the past seven decades in China. *Petroleum Exploration and Development, 46*(6), 1100–1110.

Katz, B. J. (2005). Controlling factors on source rock development—a review of productivity, preservation, and sedimentation rate. In N. B. Harris (Ed.), *The deposition of organic carbon rich sediments: Models, mechanisms, and consequences: Society of sedimentary geology* (pp. 7–16).

Killops, S. D., & Killops, V. J. (2013). *Introduction to organic geochemistry*. John Wiley & Sons.

Leventhal, J. S., Daws, T. A., & Frye, J. S. (1986). Organic geochemical analysis of sedimentary organic matter associated with uranium. *Applied geochemistry, 1*(2), 241–247.

Mango, F. D. (1992). Transition metal catalysis in the generation of petroleum and natural gas. *Geochimica et Cosmochimica Acta, 56*(1), 553–555.

Mingyang, Q., Jianhua, G., & Yanran, H. (2019). Sweet spots zone optimization of marine shale gas in complex geolo-gical conditions area out of Sichuan Basin: a case of Paleozoic in northwestern Hunan. *China, 50*(3), 596–606.

National Energy Administration. (2019). Geochemical evaluation methods of source rocks: SY/T 5735—2019. Beijing: Petroleum Industry Press.

Nie, H., Liu, Q., Dang, W., Li, P., Su, H., Bao, H., Xiong, L., Liu, Z., Sun, C., & Zhang, P. (2023). Enrichment mechanisms and resource potential of shale-type helium gas: A case study of the Wufeng Formation–Longmaxi Formation in the Sichuan Basin. *Science China: Earth Sciences, 53*, 1285–1294.

Peters, K. E., Ishiwatari, R., & Kaplan, I. R. (1977). Color of kerogen as index of organic maturity. *AAPG Bulletin, 61*(4), 504–510.

Peters, K. E., & Cassa, M. R. (1994). Applied source rock geochemistry

Qijun, W., & Jianyu, C. (1987). Petroleum Geochemistry. Wuhan: China University of Geosciences Press (in Chinese).

Rice, D. D., & Claypool, G. E. (1981). Generation, accumulation, and resource potential of biogenic gas. *AAPG bulletin, 65*(1), 5–25.

Rigby, D., & Smith, J. W. (1981). An isotopic study of gases and hydrocarbons in the Cooper Basin. *The APPEA Journal, 21*(1), 222–229.

Rongshu, C. (1989). Natural gas geology. China University of Geosciences Press, Wuhan (in Chinese).

Schoell, M. (1983). Genetic characterization of natural gases. *AAPG bulletin, 67*(12), 2225–2238.

Shaochang, W. (1983). A preliminary disscusion on coal-formed gas of Upper Paleozoic in eastern Ordos Basin. *Natural Gas Industry, 8*(2), 1–5.

Stahl, W. J. (1977). Carbon and nitrogen isotopes in hydrocarbon research and exploration. Chem. Geol (Netherlands).

Tissot, B. P., & Welte, D. H. (1978). Petroleum formation and occurrence: A New Approach to Oil and Gas Exploration. Springer-Verlag.

Tissot, B. P., & Welte, D. H. (1984). Petroleum formation and occurrence. 2nd ed. Berlin: Springer-Verlag.

Wanxian, Z., & Houfu, Z. (1981). Petroleum Geology.

Waples, D. W. (1980). Time and temperature in petroleum formation: Application of Lopatin's method to petroleum exploration. *AAPG Bulletin, 64*, 916–926.

Williams, J. A., (1974). Characterization of oil types in Williston Basin. *AAPG Bulletin, 58*(7), 1243–1252.

Xunzhi, S., Jinxing, D., & Zemin, W. (1985). A study of methane carbon isotope of coal-formed gas in FRG and its inspiration to us. *Natural Gas Industry, 5*(2), 1–9.

Yang, S., & Horsfield, B. (2020). Critical review of the uncertainty of Tmax in revealing the thermal maturity of organic matter in sedimentary rocks. *International Journal of Coal Geology*, 225.

Yang, S., Schulz, H.-M., Schovsbo, N. H., & Bojesen-Koefoed, J. A. (2017). Oil–source-rock correlation of the Lower Paleozoic petroleum system in the Baltic Basin (northern Europe). *AAPG Bulletin, 101*, 1971–1993.

Yang, S., Schulz, H. M., Horsfield, B., Schovsbo, N. H., Noah, M., Panova, E., ... & Hahne, K. (2018). On the changing petroleum generation properties of Alum Shale over geological time caused by uranium irradiation. *Geochimica et Cosmochimica Acta, 229*, 20–35.

Yang, S., Horsfield, B., Mangelsdorf, K., Rinna, J., Hartwig, A., di Primio, R., ... & Elias, R. (2022). Corroborative insights into genetically related kerogen, asphaltenes and hydrocarbons using free and bound biomarkers. *International Journal of Coal Geology, 261*, 104080.

Yongchang, X. (1994). Natural gas origin theory and application.

Zeikus, J. G. (1977). The biology of methanogenic bacteria. *Bacteriological reviews, 41*(2), 514–541.

Zhang H., Fang C., Gao X., et al. (1999). Petroleum Geology (3rd ed.). Beijing: Petroleum Industry Press.

Zhongxiang, P. (1986). Petroleum geology.

Zobell, C. E. (1947). Microbial transformation of molecular hydrogen in marine sediments, with particular reference to petroleum. *AAPG Bulletin, 31*(10), 1709–1751.

Chapter 3
Reservoirs and Cap Rocks

Abstract Reservoirs and cap rocks are the fundamental components of oil and gas reservoirs. Reservoirs serve as the storage space for oil and gas, and are the core of the reservoir system. Cap rocks are the geological layers or bodies that overlie the reservoirs and seal in the oil and gas, preventing their upward escape. They are also essential for the formation of oil and gas reservoirs. This chapter introduces the basic characteristics, types, origins, and main factors influencing the properties of reservoirs and cap rocks, with a particular focus on the characteristics, origins, and main research areas of different types of reservoirs, such as clastic rocks and carbonate rocks. The structure, characteristics, origin, and distribution of reservoirs and cap rocks are closely related to the formation of oil and gas reservoirs, the reserves of oil and gas, production capacity, and output. They directly affect the deployment and efficiency of oil and gas exploration and development.

3.1 Overview of Reservoir Rocks

3.1.1 Definition and Classification of Reservoir Rocks

3.1.1.1 Definition of Reservoir Rocks

Rocks that possess certain storage spaces capable of storing and filtering fluids are referred as reservoir rocks. The strata composed of reservoir rocks are known as reservoir layers, or simply reservoirs. If a reservoir layer contains industrially valuable oil or gas flow, it is termed as oil layer, gas layer, or oil and gas layer. Theoretically, any rock can serve as oil and gas reservoir, but over 99% of oil and gas reserves are concentrated in sedimentary rocks, with sandstone and carbonate reservoirs being the primary types. Oil and gas layers are the core of oil and gas reservoirs. Their lithology, development characteristics, internal structure, distribution range, and the laws of physical property variation are closely related to oil and gas reserves, production capacity, and output, directly affecting the deployment of oil and gas exploration and development.

© China University of Petroleum Press 2025

H. Liu and Y. Jiang, *Petroleum Geology*,
https://doi.org/10.1007/978-981-96-7086-4_3

The two essential characteristics of reservoir rocks are porosity and permeability. Porosity refers to the storage space formed by various pores, caves, and fractures within the rock. The quality of porosity directly determines the amount of oil and gas that the rock layer can store, commonly represented by porosity. Permeability refers to the ability of fluids to flow through the rock under a certain pressure difference, and the quality of permeability controls the seepage capacity of the fluid through the reservoir, commonly represented by permeability.

3.1.1.2 Types of Reservoir Rocks

Based on research purposes and the needs of oil field production practices, there are various classification schemes for reservoir rocks:

(1) Classified by Rock Type

Reservoirs can be divided into three major categories: clastic reservoirs, carbonate reservoirs, and special rock reservoirs (including igneous rocks, metamorphic rocks, argillaceous rocks, etc.).

(2) Classified by the Type of Storage Space

Reservoirs can be divided into porous reservoirs, fractured reservoirs, and composite types such as pore-fracture reservoirs, vug-pore reservoirs, pore-cave reservoirs, and pore-fracture-cave composite reservoirs.

(3) Classified by Permeability

According to the magnitude of permeability, reservoirs can be divided into high-permeability reservoirs, medium-permeability reservoirs, and low-permeability reservoirs.

(4) Classified by the Characteristics of Oil and Gas Reservoirs

They can be divided into conventional reservoirs and unconventional reservoirs, with the latter including mudstones, tight rocks, and coal rocks.

3.1.2 Porosity of Reservoir Rocks

Porosity reflects the development degree of pores in rocks, which is the ratio of the volume of pores to the volume of reservoir rock. It is a dimensionless value that can be expressed as a decimal or percentage. Here, the term "pore" is used in a broad sense, referring to the spaces in rocks that are not filled with solid material, including pores (in the narrow sense), caves, and fractures.

3.1.2.1 Types of Pores

Based on the genesis of pores, they are divided into two types: primary and secondary pores. Primary pores are the pore spaces preserved in sedimentary rocks after sedimentation and compaction, such as intergranular pores in sandstone. Secondary pores are those formed by tectonic compression or the circulation of formation water, such as dissolution pores, shrinkage pores, and fractures.

According to the size of the pores in the rock and their different effects on fluid action, the pores can be divided into three types:

(1) Ultracapillary Pores

Pore diameter >0.5 mm, or fracture width >0.25 mm. Under natural conditions, fluids can flow freely in these pores, following the general laws of hydrostatics. Some large fractures, caves, and pores in unconsolidated or loosely cemented sandstone layers mostly belong to this type.

(2) Capillary Pores

Pore diameter between 0.5 and 0.0002 mm, or fracture width between 0.25 and 0.0001 mm. Fluids in this type of pore cannot flow freely due to the action of capillary forces, and can only flow when the external force is greater than the capillary resistance. Micro-fractures and pores in general sandstone mostly belong to this type.

(3) Microcapillary Pores

Pore diameter <0.0002 mm, or fracture width <0.0001 mm. This type of pore is mainly composed of micrometer and nanometer-level pores, where micrometer-level pore throats refer to pores with a diameter greater than 1 μm, and nanometer-level pore throats refer to pores with a diameter less than 1 μm. In these pores, due to the huge attraction between the fluid and the surrounding medium molecules, the fluid cannot flow under normal temperature and pressure conditions. Increasing temperature and pressure can only cause the fluid to diffuse in a molecular or molecular cluster state. Pores in shale and tight sandstone mostly belong to this type.

From the above, for conventional reservoirs, the pores in the rock can be divided into two categories according to their effect on fluid seepage, namely effective pores and ineffective pores. Effective pores are interconnected capillary and ultracapillary pores, while ineffective pores are of two types, one being microcapillary pores, and the other being dead pores or isolated pores (Fig. 3.1).

3.1.2.2 Porosity

Based on the size and connectivity of the pores, porosity can be divided into total porosity and effective porosity.

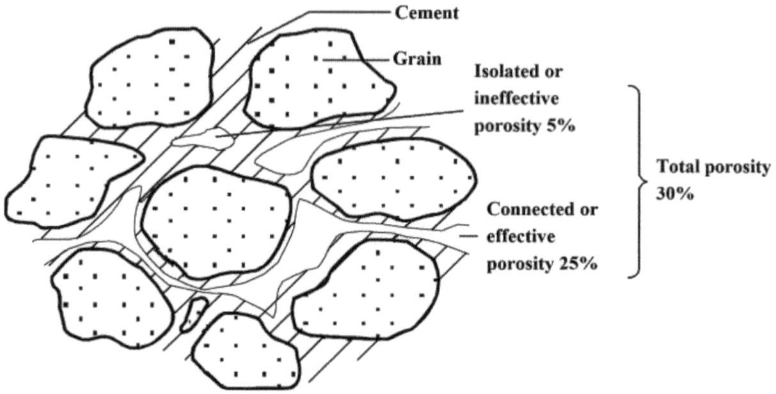

Fig. 3.1 Schematic diagram of connected and isolated pores in clean sandstone

(1) Total Porosity

The ratio of the sum of all pore volumes in a rock sample to the total volume of the rock sample is called total porosity, expressed as:

$$\phi = \frac{\sum V_{\phi}}{V_r} \times 100\% \tag{3.1}$$

where: ϕ—Porosity, %;

$\sum V_{\phi}$—Total volume of all pores in the rock sample, cm^3;

V_r—Total volume of the rock sample, cm^3.

The larger the total porosity of reservoir rock, the greater the space for storing fluids in the rock.

(2) Effective Porosity

Effective porosity refers to the ratio of the volume of interconnected pores (effective pore volume) that can store and filter fluids in a rock sample to the total volume of the rock sample, expressed as:

$$\phi_e = \frac{\sum V_e}{V_r} \times 100\% \tag{3.2}$$

where: ϕ_e—Effective porosity, %;

$\sum V_e$—Total volume of effective pores, cm^3;

In conventional reservoir production practice, only interconnected pores have practical significance. However, in unconventional reservoirs, hydraulic fracturing can enable the flow of oil and gas from microcapillary pores or isolated pores.

Under the same conditions, the total porosity of the same rock sample is greater than the effective porosity, and for sandstones with weak cementation or no cementation, the difference between the two is not significant. In reservoir evaluation and reserve calculation, effective porosity is generally used as the standard, commonly referred to simply as porosity. The porosity of conventional reservoirs is mostly distributed between 20 and 30%, but there are also some, such as fractured limestone reservoirs, with porosity as high as 70%, or as low as less than 5%.

3.1.3 Permeability of Reservoir Rocks

The ability of a rock to allow fluids to pass through under a certain pressure difference is called the rock's permeability. Rocks that allow fluids to pass through their interconnected pores under formation pressure are called permeable rocks. Otherwise, they are non-permeable rocks. The quality of rock permeability is represented by the magnitude of the permeability value. It can be divided into absolute permeability, effective permeability, and relative permeability.

3.1.3.1 Absolute Permeability

It refers to the permeability measured when there is only one type of fluid (single-phase) in the rock pores, the fluid does not have any physical and chemical reactions with the rock, and the fluid flow complies with Darcy's linear seepage law.

When a single-phase fluid flows through the porous medium in a layered manner (Fig. 3.2), the liquid flow rate through the rock cross-section per unit time is directly proportional to the pressure difference and the size of the cross-sectional area, and inversely proportional to the length of the liquid passing through the rock and the viscosity of the liquid:

$$Q = \frac{K(P_1 - P_2)F}{\mu L} \tag{3.3}$$

where: Q—Liquid flow rate through the rock per unit time, cm^3/s;

F—Cross-sectional area of the liquid passing through the rock, cm^2;

μ—Viscosity of the liquid, 10^{-3} Pa s;

L—Length of the rock, cm;

$(P_1 - P_2)$—Pressure difference before and after the liquid passes through the rock, MPa;

K—Permeability of the rock, D (Darcy), $10^{-3}\mu m^2$, 1 mD $= 0.987 \times 10^{-3}\mu m^2$.

For liquids, the calculation formula is:

$$K = \frac{Q\mu L}{(P_1 - P_2)F} \tag{3.4}$$

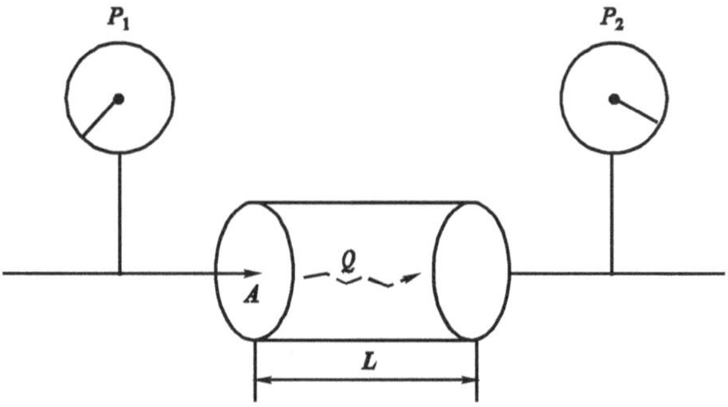

Fig. 3.2 Schematic diagram of the basic setup

According to Darcy's law, it is known that when P_1, P_2, F, L, μ are all constants, the flow rate (Q) is directly proportional to the permeability (K), that is, the amount of fluid passing through depends on the rock's ability to allow fluid passage.

Theoretically, absolute permeability is only related to the properties of the rock and is independent of the properties of the fluid and the measurement conditions. However, in actual measurement, it has been found that the absolute permeability measured for the same rock sample and the same gas is different under different average pressures. The permeability measured with liquid as the medium is always lower than that measured with gas. For gases, due to their different properties from liquids, they are very sensitive to pressure. When the gas flows from high pressure (P_1) to low pressure (P_2), the gas volume expands, and the volume flow rate through each cross-sectional area is a variable. Therefore, the volume flow rate in Darcy's formula should be the average flow rate through the rock. Thus, the formula for gas permeability can be rewritten as:

$$K = \frac{2P_2Q_2\mu_gL}{(P_1^2 - P_2^2)F} \tag{3.5}$$

where μ_g—Viscosity of the gas;

Q_2—Gas volume flow rate after passing through the rock, under the exit pressure P_2.

3.1.3.2 Effective Permeability

In the actual oil layers, the fluid in the pores is often not single-phase but coexists in oil–water two-phase or oil–gas-water three-phase. Under these circumstances, the phases interfere with each other, and the rock's filtration effect on each phase of the fluid is quite different from single-phase flow. To distinguish from the rock's absolute permeability, the concept of effective permeability is proposed.

Effective permeability, also known as phase permeability, refers to the permeability of each phase of the fluid in the rock when multiple phases of fluids exist. The symbols K_o, K_g and K_w are used to represent the effective permeability of oil, gas, and water, respectively. Generally, the effective permeability of the rock to each phase of the fluid is less than the rock's absolute permeability.

3.1.3.3 Relative Permeability

Effective permeability is not only related to the properties of the rock but also to the properties of the fluids and their volume ratios. In practical applications, the ratio of effective permeability to absolute permeability is often used to represent the characteristics of multiphase filtration, called relative permeability:

$$\text{Relative Permeability} = \text{Effective Permeability/Absolute Permeability}$$

The relative permeabilities of oil, gas, and water are K_o/K, K_g/K, K_w/K respectively.

In the pore space of reservoir rock, it is generally occupied by fluids such as water and hydrocarbons. The content of oil, gas, and water in the reservoir pores, accounting for the total pore volume percentage, is called the saturation of oil, gas, and water, commonly represented by S_o, S_w, S_g. If the oil layer contains water, oil, and gas in three phases, then $S_o + S_w + S_g = 1$.

Experiments have shown that effective permeability and relative permeability are not only related to the properties of the rock but also closely related to the properties of the fluids and their saturation. As the saturation of a phase fluid increases, its effective and relative permeability both increase until it is completely saturated with a single-phase fluid, and its effective permeability equals the absolute permeability, while the relative permeability equals 1. On the contrary, as the content of the phase fluid in the rock pores gradually decreases, the effective permeability gradually decreases until a certain limit content is reached, and the phase fluid stops flowing. Therefore, the variation of relative permeability is between 0 and 1.

Relative permeability curves are closely related to the wettability of the rock samples and the heterogeneity of the core. When the wettability of the rock changes from water-wet to oil-wet, the relative permeability of oil tends to decrease, while that of water tends to increase (Fig. 3.3). The heterogeneity of the core affects the direction of permeability, which is very important for guiding oil and gas development.

3.1.4 Pore Structure of Reservoir Rocks

3.1.4.1 Definition of Pore Structure

The space in rocks not filled by particles, cements, or matrix can be divided into pores and throats. Generally, the larger spaces surrounded by rock particles are called pores, while the narrow parts that connect only two particles are called throats. Pore

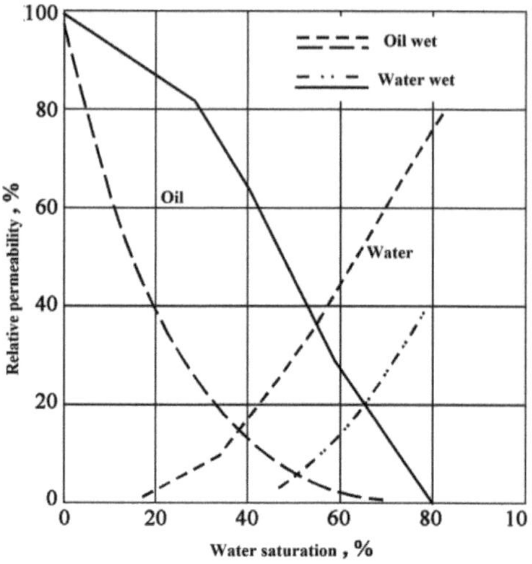

Fig. 3.3 Relationship curves between oil–water saturation and relative permeability in typical water-wet and oil-wet reservoirs (Luca, 2001)

structure refers to the geometric shape, size, distribution, and interconnection of pores and throats. When fluids flow along the complex pore system, they go through a series of alternating pores and throats, which are controlled by the smallest cross-section in the fluid channel (i.e., the throat diameter). Therefore, the shape and size of the throats control the permeability of the pores.

Different configurations of throats and pores can give reservoir rocks different properties. For example, reservoir rocks characterized by thicker throats and larger pore diameters generally exhibit high porosity and high permeability. Those with thicker throats and smaller pores exhibit low to medium porosity and low to medium permeability, those with thinner throats and larger pores exhibit medium porosity and low permeability, and those with both small throats and small pores exhibit low porosity and permeability.

3.1.4.2 Research Methods for Pore Structure

For conventional reservoirs, research methods for pore structure mainly include capillary pressure curve method, cast thin section analysis, scanning electron microscopy, and nuclear magnetic resonance technology. Cast thin sections and scanning electron microscopy observe two-dimensional cross-sections, the capillary pressure curve method is time-consuming, and nuclear magnetic resonance technology can perform three-dimensional imaging of rocks, but its resolution is relatively coarse and is an indirect measurement method. It requires combining nuclear magnetic resonance spectra with capillary pressure curves, studying the relationship and mutual transformation between the two, and then accurately evaluating the size of pores and throats (Fig. 3.4).

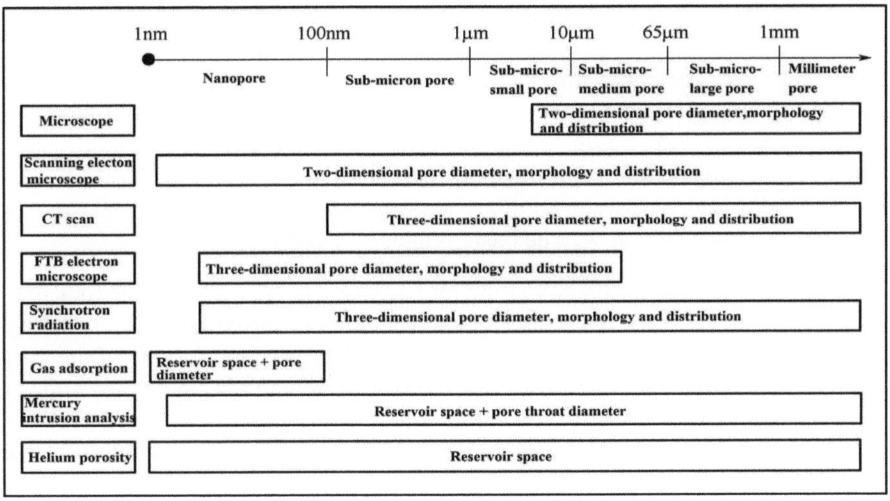

Fig. 3.4 Characterization and testing techniques for reservoir pore-throat structure and their effective range (Zhu et al., 2016)

In terms of pore structure description theory, the fractal geometric characterization of pore structure is currently a hot topic of research. Pore structures of rocks have fractal characteristics, and various reservoirs such as sandstones have good fractal properties within the pore size range. The fractal dimension can characterize the features of the pore structure. The smaller the fractal dimension, the smoother the pore throat surface, the better the homogeneity, and vice versa, indicating a more complex pore structure and more uneven pore size distribution.

Compared to conventional sandstone reservoirs, tight reservoirs are characterized by small pore-throat sizes, strong heterogeneity, and complex pore structures. Therefore, conventional methods for characterizing reservoir pore structures, such as optical microscopy, tungsten-filament scanning electron microscopy, and conventional mercury intrusion techniques, are limited by resolution and cannot fully evaluate the pore structure of tight reservoirs. Consequently, a multitude of technologies with higher resolution and characterization accuracy have been employed to study the pore structure of tight reservoirs. These include field-emission scanning electron microscopy (FE-SEM), focused ion beam field-emission scanning electron microscopy (FIB-FESEM), transmission electron microscopy (TEM), computed tomography (CT) scanning, high-pressure mercury injection, nitrogen adsorption, and small-angle X-ray scattering, among others. These new technologies can systematically characterize the three-dimensional nanoscale pore-throat system.

Digital core characterization and pore network models are two digital rock physics technologies born from the combination of high-precision three-dimensional pore structure imaging and computer technology. They can intuitively reproduce the three-dimensional pore structure of rocks and quantify the statistical parameters of pore structure characteristics.

3.1.4.3 Mercury Intrusion Curve and Pore Structure Parameters

The mercury intrusion curve, also known as the capillary pressure curve, is one of the main methods for quantitatively studying the pore structure of rocks. It is based on the measured mercury intrusion pressure and the corresponding rock sample mercury volume, and by calculating the mercury saturation value and pore throat radius, the relationship curve between capillary pressure, pore throat radius, and mercury saturation is plotted (Fig. 3.5). The basic principle is as follows:

(1) For rocks, mercury is a non-wetting phase. To inject mercury into the rock pore system, the capillary resistance caused by the pore throat must be overcome. Therefore, when the equilibrium capillary pressure and the volume of mercury injected into the rock sample are determined, the relationship between capillary force and rock sample mercury saturation can be obtained.
(2) The capillary pressure is inversely proportional to the pore throat radius R, that is

$$P_c = \frac{2\sigma \cos\theta}{R} \tag{3.6}$$

where: P_c—Capillary pressure, MPa;
σ—Surface tension of mercury, 10^{-3} N/cm;
θ—Contact angle of mercury wetting, degrees;
R—Pore throat radius, cm.

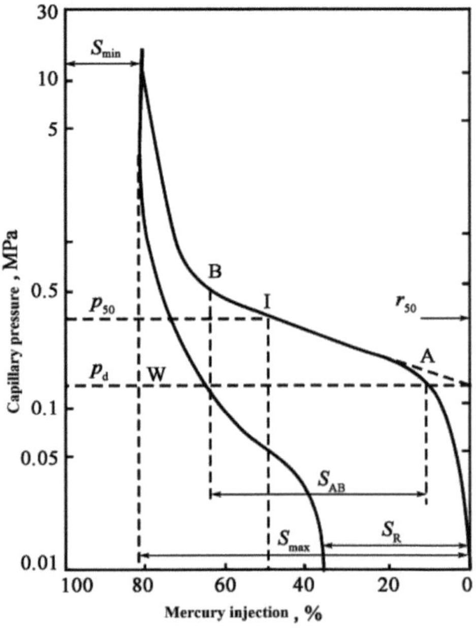

Fig. 3.5 Characteristics of capillary pressure curve. I: Injection curve. W: Withdrawal curve

Therefore, based on the capillary pressure of the injected mercury, the corresponding pore throat radius value can be calculated.

(3) The mercury saturation value can be calculated from the mercury volume:

$$S_{Hg} = \frac{V_{Hg}}{\phi V_f} \tag{3.7}$$

where: S_{Hg}—Mercury saturation, %;
V_{Hg}—Volume of mercury in the pores, cm^3;
V_f—Volume of the rock sample, cm^3;
ϕ—Porosity of the rock sample, %.

The displacement pressure (Pd) refers to the minimum pressure required for the non-wetting phase fluid to displace the wetting phase fluid. The smaller the Pd, the larger the largest interconnected pore throat in the rock sample, the more numerous the large pore throats, and the better the pore structure. The capillary pressure at which the mercury saturation of the non-wetting phase is 50% is called the median saturation pressure (P_{50}), and the corresponding pore throat radius is approximately equal to the median throat radius (r_{50}). When Pc reaches a certain level, the injected mercury volume basically no longer increases. The pore volume not injected by residual water is called the minimum unsaturated pore volume (S_{min}%).

Different capillary curve morphologies reflect different pore sizes and distributions, which can identify the sorting and pore development degree of the rock (Fig. 3.6).

According to the capillary pressure curve, the displacement pressure (Pd), median pore throat radius (r_{50}), median capillary pressure (P_{50}), minimum unsaturated pore volume percentage (S_{min}%), and pore throat radius frequency distribution histogram can be obtained (Fig. 3.7). The more concentrated and larger the r in the pore throat radius histogram, the better the pore structure.

3.1.5 Relationship Between Porosity and Permeability

The relationship between porosity and permeability in rocks is not strictly a functional relationship, but there is a certain correlation. Generally speaking, the absolute permeability increases with the increase of effective porosity, but it is not unlimited, and the relationship between porosity and permeability varies with different reservoir lithologies and types of pore spaces (Fig. 3.8).

(1) Clastic rock reservoirs: There is a good positive correlation between effective porosity and absolute permeability. Permeability increases with the increase of porosity, roughly showing an exponential function relationship (Fig. 3.8).
(2) Carbonate rock reservoirs: The relationship between porosity and permeability varies greatly, with no obvious relationship. For some mudstone, dense limestone reservoirs with developed fractures, the fractures have a much greater

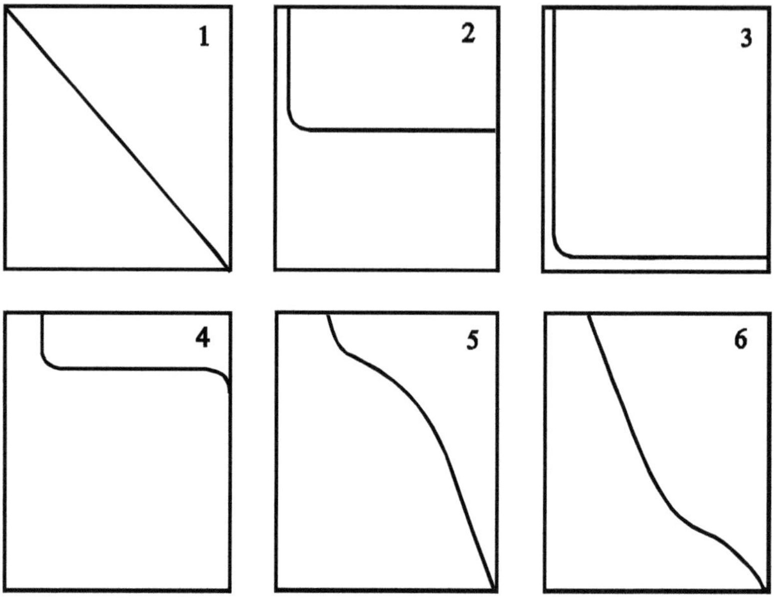

Fig. 3.6 Capillary pressure curves under different sorting and skewness conditions. 1—Unsorted; 2—well sorted; 3—well sorted, coarse skewness; 4—well sorted, fine skewness; 5—poorly sorted, slightly fine skewness; 6—poorly sorted, slightly coarse skewness

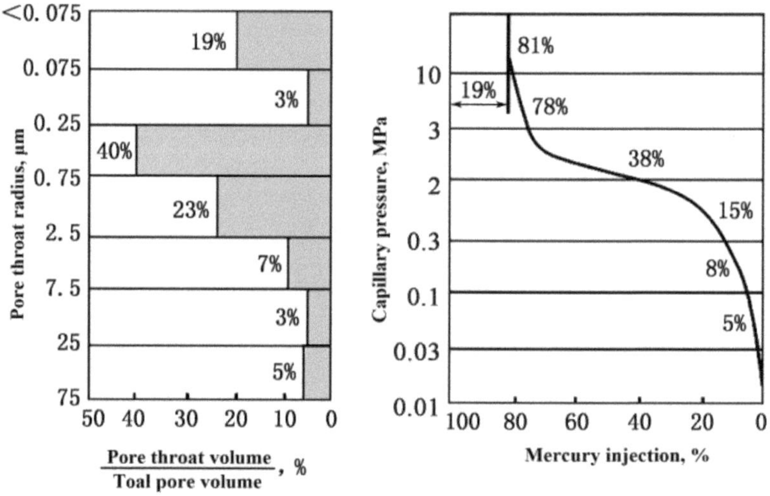

Fig. 3.7 Capillary pressure curve and pore throat distribution histogram

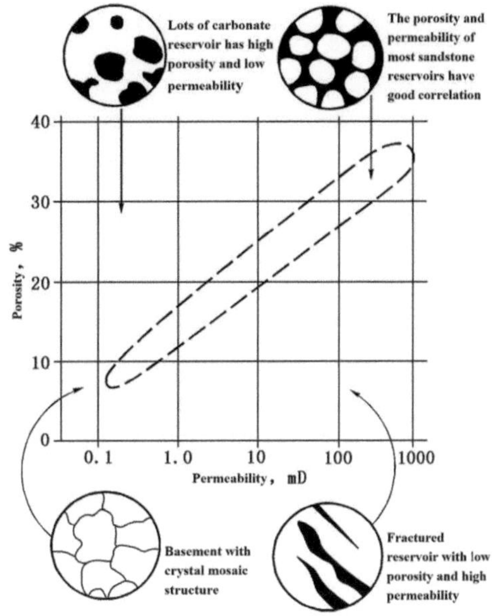

Fig. 3.8 Relationship between porosity and permeability in different reservoirs (Selley, 2015) (Black represents pores)

impact on permeability than pores. Therefore, even though the porosity analyzed in the laboratory is very low, only 5–6%, the permeability is very high. Some biological limestones develop intragranular dissolution pores or body cavity pores, but the pore connectivity is poor or the throats are small, often showing high porosity and low permeability reservoirs (Fig. 3.8). For grainstone limestone, crystalline dolomite with intergranular pores as the main type, caves and fractures are not developed, similar to clastic rocks, permeability increases regularly with the increase of effective porosity (according to Coalson et al., 1990).

(3) Igneous and metamorphic rock reservoirs: The storage space is mainly composed of dissolution pores and fractures, and the impact of fractures is relatively large, and the correlation between the two is poor. If the fractures are well developed, the permeability is very high, if the fractures are not developed, it is an ultra-low permeability reservoir.

Different configurations of pores and throats can also give the reservoir different properties. Under the condition of the same effective porosity, the permeability of rocks with small pore throat diameters is lower than that of rocks with large pore throat diameters, and the permeability of rocks with complex pore throat shapes is lower than that of rocks with simple pore throat shapes.

In general, the larger the effective porosity, the higher the absolute permeability. The reservoir can be classified and evaluated using porosity and permeability (Table 3.1).

Table 3.1 Lithological physical property grading standards (China Petroleum and Natural Gas Industry Standard, SY/T6285-2011)

Parameters	Level					
	Ultra-high	High	Medium	Low	Ultra-low	Extra-low
Permeability ($\times 10^{-3}$ μm^2)	≥2000	500 ≤ ~2000	50 ≤ ~500	10 ≤ ~50	1 ≤ ~10	<1
Porosity (%)	≥30	25 ≤ ~30	15 ≤ ~25	10 ≤ ~15	5 ≤ ~10	<5

3.2 Clastic Reservoir Rocks

Conventional clastic reservoir rocks mainly include various types of sandstone, siltstone, conglomerate, and other clastic sedimentary rocks, and are one of the main types of reservoirs in the world's oil and gas fields. Oil and gas fields such as Kuwait's Burgan Field, Russia's Samotlor Field, and the Prudhoe Bay Field in Alaska, USA, as well as major oil fields in China such as Daqing, Shengli, Dagang, and Karamay, all have their main reservoirs belonging to this category.

3.2.1 Reservoir Space of Clastic Rock Reservoirs

3.2.1.1 Types of Reservoir Space

The reservoir space of clastic rock reservoirs is divided into three categories: pores, fractures, and cavities according to their morphology. According to the genetic mechanism, it is divided into two major categories of primary pores and secondary cavities, with 12 subcategories (Table 3.2). Primary pores mainly include intergranular pores, mineral cleavage fractures, bedding interlayer fractures, etc. (Fig. 3.9). Secondary cavities refer to pores, karst caves, and fractures formed in the diagenesis of rocks after the sedimentation process, including intragranular and intergranular dissolved pores of particles, mold pores, supergranular pores, intercrystalline pores, structural fractures, diagenetic fractures, karst caves, etc. The reservoir space of clastic rocks is mainly intergranular pores (including primary intergranular pores and secondary intergranular pores). Other types of pores are relatively few, but in some reservoir rocks, they can become the main reservoir space.

Affected by the contact type and cementation type of clastic rock particles as well as the shape, size, and roundness of sandstone particles themselves, there are four common types of pore throats in clastic rocks (Fig. 3.10):

(1) The throat is the narrowed part of the pore: It develops in sandstones dominated by intergranular poreslarged intergranular pores, and almost all are effective pores. It is commonly found in types with particle support, contact of floating particles, and without cement.

Table 3.2 Types and characteristics of reservoir space in clastic rock reservoirs

Category	Subcategory			Space size	Characteristics
Primary pores	Pore	Primary intergranular pore		<2 mm	Primary intergranular or residual pore
		Primary intragranular pore			Micropores within lithic fragments, pores within pyroclastic lithic fragments, micropores within matrix, etc.
	Fracture	Mineral cleavage fracture, interlayer fracture			
Secondary cavities	Pore	Inter-granular dissolved pore	Dissolution at particle edges	<2 mm	Partial dissolution at the edges of particles such as feldspar and lithic fragments
			Partial dissolution of cement and within crystals		Local dissolution of cement such as calcite
			Matrix dissolution		Partial dissolution of clay matrix
		Intragranular dissolved pore within components	Intragranular dissolution of particles		Such as feldspar and lithic fragments
			Intragranular dissolved pore within matrix		Partial dissolution of clay matrix
			Intragranular dissolved pore within cement		Partial dissolution within cement such as calcite or its crystal
		Superlarge pore			Caused by the dissolution of cement and particles together

(continued)

Table 3.2 (continued)

Category	Subcategory			Space size	Characteristics
		Mold pore	Grain mold, crystal mold, biological mold		The shape is retained after the dissolution of particles, crystals, or biological debris
		Intercrystalline pore			Pores between crystals such as kaolinite and dolomite formed in the late stage
	Cavity	Karst cave		>2 mm	Mostly related to supergene leaching
	Fracture	Contraction fracture		0.01 to >1 mm	Diagenetic contraction
		Diagenetic fracture and its dissolution			No directionality, thin fracture and small extension range
		Structural fracture and its dissolution			Smooth extension, distinct sets, and mutual cutting

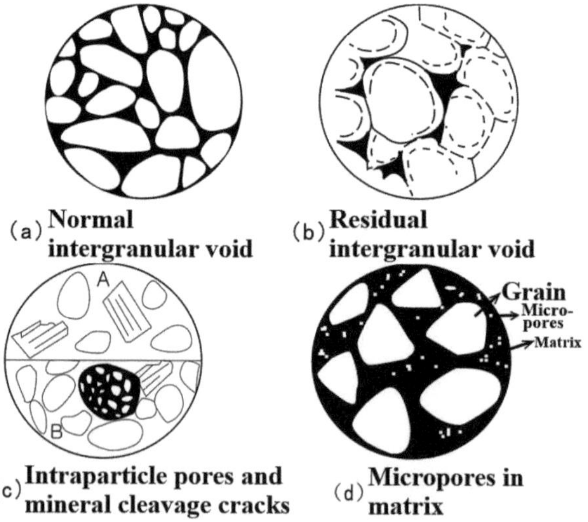

Fig. 3.9 Types of primary pores in clastic rocks

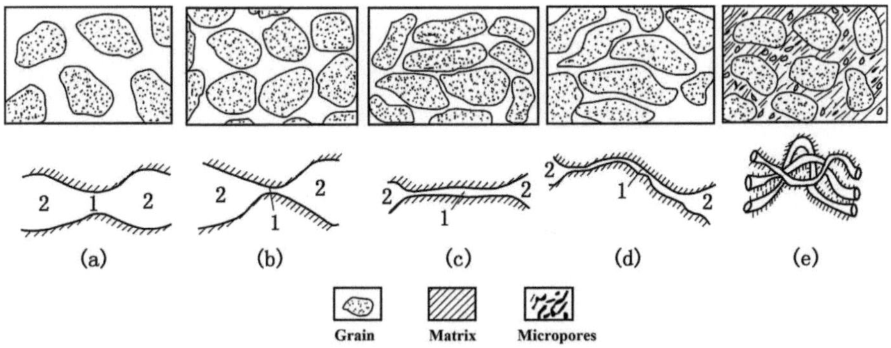

Fig. 3.10 Types of pore throat (Luo, 1986). **a** Throat is the constricted part of the pore, **b** the variable cross-section contraction part is the throat, **c** laminar throat, **d** curved laminar throat, **e** tubular throat, 1—throat, 2—pore

(2) The variable cross-section contraction part is the throat: When the particles are compacted and arranged relatively closely, the throat becomes much narrower, with the characteristics of high porosity and low permeability. It is commonly found in particle-supported, contact, and point-contact types.
(3) Flaky or curved flaky throat: When sandstone is further compacted or when crystals regrow due to pressure solution, the pores surrounded by their regrown edges become smaller, generally less than 1 μm, with the characteristics of small pores and extremely fine throats. It is commonly found in contact, line contact, and concave-convex contact types.
(4) Tubular throat: Many micropores in the matrix and cement are both pores and connecting channels. These micropores are distributed crosswise in the matrix and cement like microcapillaries. Its porosity is low and permeability is extremely low. It is commonly found in matrix-supported, basal, pore, and sutured contact types.

3.2.1.2 Development Characteristics of Secondary Pores

The study of secondary pores in clastic rocks began in the 1930s but has not been taken seriously. Since the 1970s, people have made breakthroughs in the study of sandstone diagenesis, recognized that secondary pores account for a large proportion in sandstone pores, and established identification criteria.

Schmidt and McDonad (1979) believe that the identification of secondary pores should rely on various evidences. Secondary pores are generally larger and more irregular in morphology and distribution than primary pores. Under the microscope, petrological criteria for identifying secondary pores include (1) partial dissolution, (2) mold, (3) heterogeneous filling, (4) superlarge pores, (5) elongated pores, (6) dissolved particle edges, (7) intracomponent pores, and (8) broken particles (Fig. 3.11), etc.

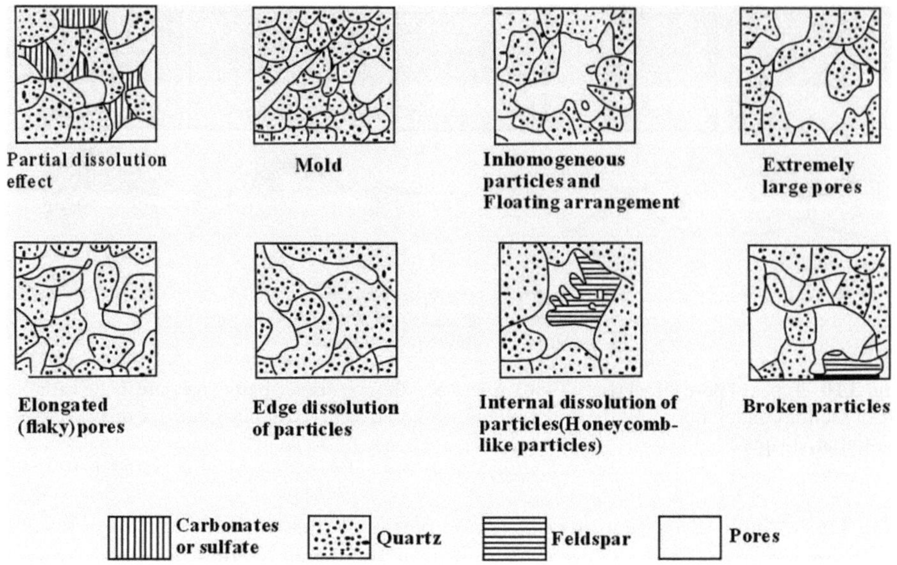

Fig. 3.11 Petrological criteria for identifying secondary pores in sandstone (according to Schmidt and McDonad, 1979)

3.2.1.3 Formation Mechanism of Secondary Pores

Dissolution is the main diagenetic process for the formation of secondary pores. The causes of secondary pores include the dissolution of carbonates or silicates, the rupture and contraction of rock components, etc. (Table 3.3).

Table 3.3 Diagenesis that causes secondary pores in sandstone (Schmidt and McDonad, 1979)

Diagenesis	Amount of secondary pores formed
Rock rupture	Less, and in individual strata there may be more
Particle rupture	Very few
Contraction	Less
Calcite	More
Dolomite	More
Siderite and ankerite	More
Dissolution sulfate	Less
Other evaporites	Less
Other non-silicates	More
	Very few

(1) Dissolution of Rock Components by Organic Acids

Organic acids have two dissolution mechanisms for carbonate minerals. One is that organic acids produce CO_2 through decarboxylation and dissolve in water to form carbonic acid, thereby dissolving carbonates (Fig. 3.12a). The other is that H+ dissociated from organic acids dissolves carbonates (Fig. 3.12b).

(2) Dissolution of Aluminosilicate Minerals by Organic Acids

The complexation of organic acids with Al^{3+} leads to a strong dissolution effect on aluminosilicate minerals. Taking the reaction between potassium feldspar and oxalic acid $H_2C_2O_4$ as an example, the reaction equation is as follows:

$$\underset{\substack{\text{Potassium feldspar} \\ (108.77\ \text{cm}^3)}}{KAlSi_3O_8} + \underset{\text{Oxalic acid}}{H_2C_2O_4} + 2H_2O + 2H+ = \underset{\substack{\text{Quartz} \\ (68.1\ \text{cm}^3)}}{3SiO_2} + \underset{\text{complex}}{[AlC_2O_4 \cdot 4H_2O]} + +K+ \tag{3.8}$$

In reaction Eq. (3.8), each mole of oxalic acid complexes with one mole of aluminum ion, thereby dissolving one mole of feldspar and simultaneously precipitating three moles of quartz. The molar volume decreases by 37% (40.67 cm^3), that is, 1 cubic centimeter of potassium feldspar can be dissolved to form 0.37 cubic centimeters of pore space in the presence of organic acids.

Organic acids can react with aluminosilicates such as feldspar and montmorillonite to cause alteration, reducing volume and increasing pores. Taking potassium feldspar as an example:

$$\underset{(217.45\ \text{cm}^3)}{\text{Potassium feldspar}} + 2H+ + H_2O = \underset{(99.5\ \text{cm}^3)}{\text{Kaolinite}} + \underset{(90.88\ \text{cm}^3)}{\text{Quartz}} + 2K+ \tag{3.9}$$

(3) Formation of Carbonic Acid and Dissolution of Rock Components

The formation of carbonic acid is divided into organic genesis and inorganic genesis. The organic genesis theory believes that organic matter can produce CO_2 through

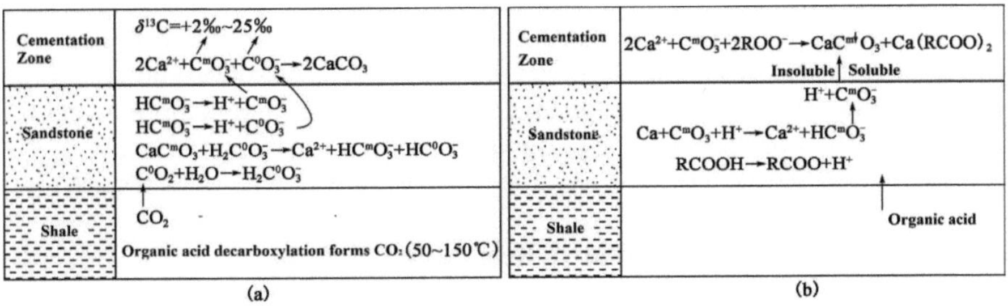

Fig. 3.12 Carbonic acid hypothesis and organic acid hypothesis of the dissolution mechanism of organic matter thermal evolution (MeshrilD, 1986). **a** Carbonic acid hypothesis of organic matter decarboxylation; **b** organic acid hypothesis of organic acid dissociation

oxidation, reduction of sulfate minerals by bacteria, biological fermentation, thermal decarboxylation, mineral redox reactions, and thermal decomposition of carbonate minerals (such as the decomposition of siderite), forming acidic water. Subsequently, acidic water selectively dissolves rock components.

The inorganic genesis theory of CO_2 (Hutcheon, 1980; Maffler & White, 1969) believes that in the deep burial stage of diagenesis, clay minerals react with carbonates to produce CO_2. Reactions such as the conversion of kaolinite to chlorite under high temperature conditions and the alteration of illite to feldspar or chlorite can release some CO_2.

(4) Hypothesis of TSR (Thermochemical Sulfate Reduction)

The main products of the TSR reaction between hydrocarbons and sulfates are organic acids, bicarbonate ions, hydrogen sulfide, and solid bitumen. The total reaction equation can be expressed as:

$$SO_4^{2-} + \text{Hydrocarbons} + H_2O \xrightarrow[R_O > 1.3\%]{>100-135\,^\circ C} H_2S + S + \text{Bitumen} + \text{Organic Acid} + HCO_3^- + \text{Heat} \tag{3.10}$$

Acidic substances such as organic acids and carbon dioxide generated by the TSR reaction have extensive dissolution effects on minerals such as calcite, dolomite, and anhydrite in the layers near where the TSR reaction occurs.

(5) Leaching Effect of Atmospheric Precipitation

The leaching effect of atmospheric water mainly occurs under unconformities formed by tectonic uplift or sea level drop and in fracture zones, causing exposed surface or near-surface rocks to be dissolved and producing secondary pores. Carbonate cements in sandstone undergo dissolution during sandstone uplift and weathering. Surdam's (1997) study of soil confirmed that in addition to the formation of H_2CO_3 from CO_2, the soil also contains oxalic acid. Feldspar in sandstone in the unconformity zone is easily weathered into kaolinite and can also produce a large amount of HCO^{3-}. Atmospheric water has strong dissolution ability, such as the dissolution of carbonates, and feldspar and biotite are leached and altered into kaolinite. The intrusion depth of atmospheric water can reach 2 km. The dissolved substances can be carried away from the dissolution area with large-scale fluid migration, and the generated dissolution space can be preserved.

3.2.1.4 Evolution and Quantitative Restoration of Pores

The pore evolution of clastic rocks is closely related to the diagenetic stage. Taking the Paleogene sandstone in the Jiyang Depression as an example, in the early diagenetic stage, mechanical compaction is the main process, and primary pores decrease sharply. Secondary pores begin to form in the early diagenetic stage B. The middle diagenetic stage A is a secondary pore development zone, which is formed before and during the peak of oil generation and can provide good reservoir space for oil and gas. In the middle diagenetic stage B, due to the thermal cracking of organic acids, a large amount of CO_2 is generated, accompanied by the third transformation of clay minerals, which can form a second secondary pore zone and fractures. In the late diagenetic stage, both primary and secondary pores are rarely developed, and the porosity is very low, but fractures can develop, providing favorable space for the storage of natural gas.

At present, the restoration research of reservoir porosity is mostly semi-quantitative. With the help of rock cast thin sections, we can analyze the diagenetic time sequence of different rock facies. Constrained by the diagenetic evolution sequence, the contents of various authigenic minerals and dissolved pores in different stages can be quantitatively counted. Infer backward in time if very helpful in obtaining the paleophysical properties or paleoporosity of these rock facies in different stages or different oil and gas accumulation periods (Luo et al., 2020). The specific process is: In rock thin sections, we identify the diagenetic products after the last stage of oil and gas charging. Use Eq. (3.1) to estimate the paleoporosity before asphalt charging in this stage:

$$\varphi\text{paleo} = \varphi\text{current} + \Delta\varphi\text{cementation} - \Delta\varphi\text{dissolution} + \Delta\varphi\text{compaction} \quad (3.11)$$

where: φpaleo is the porosity when asphalt is charged in a certain stage; φcurrent is the currently observed porosity; $\Delta\varphi$cementation is the reduction in porosity due to cementation after asphalt charging in the certain stage; $\Delta\varphi$dissolution is the increase in porosity due to dissolution after asphalt charging in the certain stage; $\Delta\varphi$compaction is the reduction in porosity due to compaction after asphalt charging in the certain stage.

Then, let φcurrent = φpaleo and use Eq. (3.1) to estimate the paleoporosity φ'paleo before the second-to-last stage of asphalt charging. In analogy, let φcurrent = φ'paleo again and similarly estimate the porosity φ''paleo before the third-to-last stage of asphalt charging. Repeated to do so until the paleoporosity at the time of asphalt charging in all stages is restored. Use the empirical relationship between porosity and permeability of sandstone in the study area, we can estimate the paleo-permeability in different stages.

3.2.2 *Factors Affecting the Reservoir Physical Properties of Clastic Rock Reservoirs*

The quality of the reservoir properties of clastic rock reservoirs mainly depends on the influences of factors such as sedimentary environment, diagenesis, and diagenetic environment.

3.2.2.1 Sedimentary Environment

Each sedimentary environment forms different sediments and sedimentary rocks. Different rock types have different compositions, structures, and textures. The composition, structure, and texture of rocks affect reservoir physical properties.

(1) Mineral Composition of Clastic Particles

The mineral composition of clastic rocks is mainly quartz, feldspar, and lithic fragments. The influence on the porosity and permeability of reservoir rocks is mainly in two wats: First, the weathering resistance of mineral particles, including the degree of hardness and the degree of dissolution and expansion in water. Second, the wettability of minerals, the adsorption force between mineral particles and fluids, that is, oil repellency and water repellency.

Sandstones composed of clastic particles that are generally hard, insoluble and non-swelling in water, and non-adsorbing to oil usually have good oil storage physical properties, otherwise, they are poor. Quartz and feldspar are the most common particles in clastic rocks and account for more than 95%. Their content has the most significant influence on the reservoir properties of clastic rocks. Under the same diagenetic conditions, quartz sandstone has better oil storage physical properties than feldspar sandstone.

(2) Grain Size and Sorting Degree of Clastic Particles

For unconsolidated wet sand, grain size has little influence on original porosity, while particle sorting has a significant influence on porosity. The relationship is (Beard & Wely, 1973):

$$\text{Original porosity} = 20.91 + 22.9/S_o \tag{3.11}$$

In the formula, S_o is the Trask sorting coefficient, which is obtained by taking the square root of the ratio of the particle size at 25% (R_1) and the particle size at 75% (R_2) on the cumulative curve.

If the particle sizes of the rocks are unequal, particles of different sizes will have complex arrangements. The large pores between large particles will be filled by small particles, making the pores smaller and reducing the porosity and permeability of the rock. Generally, the better the sorting degree of particles, the better the reservoir physical properties, and the greater the porosity and permeability (Fig. 3.13).

When the sorting is constant, the absolute permeability of the reservoir is proportional to the median grain size of clastic particles. The larger the grain size, the greater the porosity and permeability.

(3) Arrangement of Clastic Particles

The assumption is that the clastic particles are equal spheres. In the cubic arrangement stacking way, the particle arrangements is the loosest and the porosity is the largest with the theoretical porosity 47.6%. The pore radius is large, the connectivity is good, and the permeability is also large. In the oblique cube way, the particle arrangement

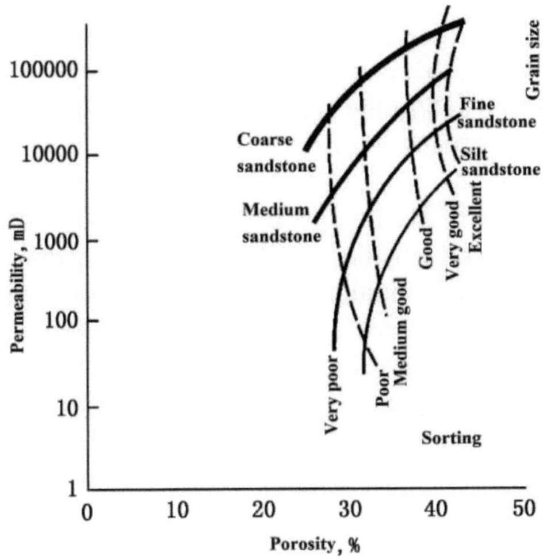

Fig. 3.13 Influences of grain size and sorting on the porosity and permeability (Brayshaw, 1996)

is the tightest and the porosity is the smallest, with the theoretical porosity 25.9% (Fig. 3.14).

Depositional conditions determine the arrangement of clastic particles. If the water during deposition is relatively calm, the particles are mostly arranged in a nearly cubic shape. If the water is more turbulent, such as in rivers, alluvial fans, and shore facies belts, the particles are mostly piled up in an oblique cube shape. In addition, under the pressure of the overlying stratum load, the sediment particles are arranged in a directional manner.

(4) Influence of Matrix Content on the Original Porosity and Permeability of Sand Bodies

The matrix content has a great influence on the original porosity and permeability of sand bodies, and has a greater influence on permeability. Micropores in the matrix are developed, but they contribute little to permeability. Generally, the content of

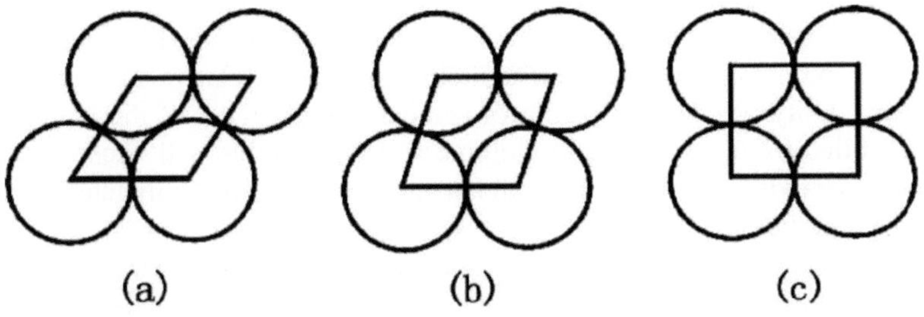

Fig. 3.14 Ideal forms of spherical particle arrangement in rocks. **a** Tightest arrangement form; **b** medium density arrangement form; **c** loosest arrangement form

clay matrix is inversely proportional to porosity and permeability. When the matrix content is less than 5%, the original porosity and permeability are very high. When the matrix content exceeds 15%, the permeability is very low.

(5) Sedimentary Structure

The sedimentary structure of clastic rock reservoirs will also affect physical properties. For fine sandstones and siltstones with horizontal bedding and wavy bedding, they often have a higher shale content, finer particles, poor reservoir properties, and obvious directionality in permeability. The horizontal permeability parallel to the bedding plane is larger, and the vertical permeability perpendicular to the bedding plane is smaller. For sandstones with cross-bedding, the permeability parallel to the cross-bedding plane is the largest, and the permeability in the vertical direction is the smallest. If sandstones contain shale bands, it will also affect reservoir properties, especially vertical permeability.

3.2.2.2 Diagenesis

Diagenesis has a greater influence on the porosity and permeability of sand bodies. The diagenesis that leads to the reduction of porosity and permeability mainly includes compaction and cementation. The diagenesis that improves the reservoir performance of rocks is mainly dissolution.

(1) Mechanical Compaction

As the burial depth increases, the compaction of rocks increases, and the porosity of sandstone significantly decreases. There are many factors affecting compaction, such as the compressive strength of rocks, geothermal temperature, and formation pressure. For rocks with smaller grain sizes and poor sorting, the degree of influence of compaction is extremely great, and it is difficult to become a conventional oil and gas reservoir.

(2) Cementation

Cementation will block pores and reduce porosity. The composition, content, cementation type, and occurrence of cement will all affect reservoir performance.

Generally, sandstones cemented by clay are relatively loose and have better permeability. Calcareous, siliceous, and ferruginous cements are relatively dense and have poor permeability. When the content of cement is high, most of the intergranular pores are filled by them. The pore volume and pore radius will become smaller, and the connectivity between pores will become worse, resulting in deterioration of reservoir properties.

The cementation types of clastic rocks are divided into four types: basal cementation, pore cementation, contact cementation, and chaotic cementation. Generally, contact cementation has high porosity and good reservoir physical properties. Pore cementation has medium physical properties. The relationship between porosity and cementation type of Paleogene clastic rock reservoirs in the Bohai Bay Basin of China can be seen in Table 3.4.

The occurrences of cement are mainly of types such as pore filling, pore lining, pore bridging, and enlarged cementation. In addition to affecting porosity, the occurrence of cement has a greater impact on rock permeability. For example, lining cementation or enlarged cementation, although it does not have a great impact on the total porosity of rocks, their blockage of pore throats will greatly reduce the permeability of rocks (Fig. 3.15). Clay film cementation, especially a complete chlorite clay film, can effectively isolate the surface of quartz particles from pore fluids, prevent the growth of authigenic quartz on the surface of framework particles, and suppress pressure solution. The reservoir physical properties are better. For example, in the deeply buried Middle-Lower Jurassic sandstones on the Norwegian continental shelf, the porosity of sandstones rich in chlorite clay films is 10–15% higher than predicted by the regional porosity-depth trend.

Table 3.4 Relationship between cementation type and porosity of Paleogene sandstones in the Bohai Bay Basin (Zhang et al., 1999)

Cementation type	Contact type	Pore-contact type	Pore type	Pore-cementation type	Basal type
Porosity%	29–34	25–30	24–28	19	<5

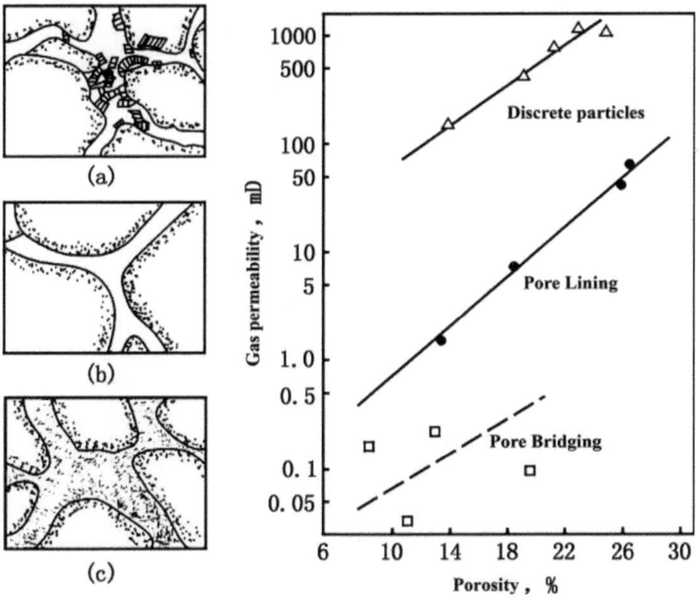

Fig. 3.15 Influence of clay mineral types and occurrences on porosity and permeability (Damsleth, 1992). **a** Dispersed kaolinite, **b** pore-lining chlorite or illite, **c** pore-bridging fibrous illite

(3) Dissolution

In many oilfields in China, clastic rock reservoirs dominated by secondary dissolution pores have been found. The formation of secondary dissolved pores can be manifested as the dissolution of clastic particles, the dissolution of interstitial materials, and the dissolution of authigenic metasomatic minerals (Table 3.5).

Sandstones with secondary pores can have permeability higher or lower than that of sandstones with the same volume of primary pores due to the different properties of secondary pores. When the throats of secondary pores are larger and the shape is more suitable for enhancing pore connectivity, the permeability is higher. On the contrary, if secondary pores are mainly isolated pores such as grain imprints, the permeability is lower.

3.2.2.3 Diagenetic Environment

Diagenesis is affected by media and environments such as lithology, fluid, temperature, and pressure.

(1) Geothermal Gradient

The influence of temperature on diagenesis is mainly in several aspects: ① Geothermal temperature affects the solubility of minerals. The solubility of most minerals increases with increasing temperature. The continuously increasing temperature allows pore water to reach saturation with more ions. ② Affect the transformation of minerals. ③ Affect the reaction between pore fluids and rocks. The rate of many diagenetic reactions doubles every 10 °C and increases by 1000 times every 100 °C (Wilson, 1994). ④ Paleogeothermal temperature controls the diagenetic evolution of organic matter.

Paleogeothermal temperature mainly depends on the geothermal gradient. Figure 3.16 is a graph of the relationship between sandstone porosity and depth of two wells with different geothermal gradients. Wells with a higher geothermal gradient have lower porosity than wells with a lower geothermal gradient. At a depth of 7000 ft, the porosity difference between the two is 10% (Wilson, 1994).

The geothermal gradient is an important factor controlling the compaction of sandstone. The reduction of the geothermal gradient can significantly weaken the compaction of sandstone and is very important for the preservation of pores in deep reservoirs. A lower geothermal gradient corresponds to a smaller amount of sandstone compaction or a larger amount of pore preservation (Fig. 3.17). The porosity reduction rate of sandstone with burial depth or the increase rate of compaction with burial depth in high geothermal gradient areas is higher than that in low geothermal gradient areas. Under the same diagenetic temperature, the porosity of reservoirs is very different. For example, the porosity of Cretaceous (feldspar) lithic sandstones in the Junggar Basin and the Songliao Basin changes greatly under the same diagenetic temperature. Although the burial depth of the Cretaceous reservoirs in the Junggar Basin (about 4,500 m) is much greater than that in the Songliao Basin (about

Table 3.5 Formation environment and basic characteristics of sandstone reservoir bodies

Sedimentary system	Sand body type and characteristics		Oilfield example
Alluvial fan	Planar fan-shaped, cross-sectional wedge-shaped or lenticular; coarse particles, poor sorting; good reservoir properties in the fan center; high permeability of braided channels and central bars		Triassic of Karamay Oilfield, Kongdian Formation of Zaoyuan Oilfield in Dagang
River	Divided into four types: meandering river, braided river, straight river, and anastomosing river. Including sand bodies such as channels, central bars, marginal bars (point bars), and crevasse fans. The cross-section is lenticular. Riverbed sand bodies are long and narrow and irregular, can bifurcate, flat on the top and concave on the bottom in cross-section, and have a large thickness near the river center; there are large changes in structure and grain size, and poor sorting. Serious heterogeneity, large changes in porosity and permeability. The primary pores of channel sandstones are developed and have good porosity and permeability		Permian of Prudhoe Bay Oilfield in Alaska, Jurassic of Changqing Oilfield, Neogene of Jiyang Depression
Aeolian sand	Dune and sand sheet sand bodies are mainly fine-medium grained sandstones with pure sand quality, excellent sorting, and good rounding. Permeability is stable and generally forms high-quality reservoirs and regional conduits. Between dunes are sandstones, siltstones, mudstones, evaporites, limestones, etc. with poor sorting		Chiddian sandstone of Groningen gas field in the North Sea, Jurassic Norphlet sandstone of MaryAnn gas field in Alabama, USA
Lakes	Delta	Distributed in the gentle slope zone of the lake basin, including distributary channel sands, estuary sand bars, frontal sheet sands, etc. Planar bird's foot-shaped or flower-shaped. Lenticular in cross-section, pure sand quality, good sorting, and good physical properties	Cretaceous of Daqing Oilfield, the third member of Shahejie Formation in Dongying Sag, Paleocene of RedWash Oilfield in Utah, USA
	Beach bar	Beach sand bodies are thin and sheet-like, and dam sand bodies are thick, banded or lenticular. Pure sand quality, good sorting, and good physical properties	The fourth member of Shahejie Formation in Chunhua Oilfield of Shengli, the fourth member of Shahejie Formation in Wangjiagang Oilfield of Shengli
	Fan delta	On the steep shore of the lake basin, the subaqueous braided channel sand bodies in the front are developed, and the physical properties are good	The fourth member of Shahejie Formation in Shuguang Oilfield of Liaohe River

(continued)

Table 3.5 (continued)

Sedimentary system	Sand body type and characteristics		Oilfield example
	Braided river delta	Mostly developed in the gentle slope zone or long axis direction of the lake basin, showing forms such as blanket or sheet, flower-shaped (lump-shaped) and branched (bird's foot-shaped), and has good physical properties	Enping Formation in Lufeng Sag of Pearl River Mouth Basin
	Subaqueous fan	Developed in the nearshore steep slope zone, mainly composed of braided channels in the fan center and sheet sands at the fan end	Paleogene of Bonan Oilfield in Shengli, etc.
	Turbidite sand body	Including farshore turbidites, fault trough turbidites, slumped turbidites, etc. The sand body morphology is fan-shaped, banded, lenticular, etc.	The third member of Shahejie Formation in Liangjialou Oilfield of Shengli, the third member of Shahejie Formation in No. 5 Pile Oilfield of Shengli, etc.
Delta	Including channel sands, branch channel sands, estuary sand bars, and frontal sheet sands. Sand bodies in the delta front facies belt are developed. Under different dynamic actions, it can be in the shape of bird's foot, flower, and curved sheet. Pure sand quality, good sorting, and good reservoir physical properties		Cretaceous of aceous of Safaniya Oilfield in Saudi Arabia, Cretaceous of Burgan in Kuwait, Cretaceous of Urengoy gas field in Western Siberia
Coastal	Including onlap and offlap sandstone bodies, coastal sand ridges, and tidal channel sand bodies. High maturity in composition and structure, good sorting and rounding, and good reservoir physical properties. Coastal sand ridges are long and narrow, parallel to the coastline, lenticular in cross-section, flat bottom and convex top, good sorting, and good reservoir physical properties		Paleocene Frio sandstone in East Texas Oilfield, Bisti Oilfield in San Juan Basin, Piper Oilfield in the North Sea
Deepwater submarine fan	Main channel and braided channel sand bodies are developed. Poor maturity in composition and structure, poor sorting. Great changes in reservoir physical properties		Paleocene of Forties Oilfield in the North Sea Basin of the United Kingdom, Miocene of Los Angeles Basin, Oligocene of Marlim Oilfield in Brazil

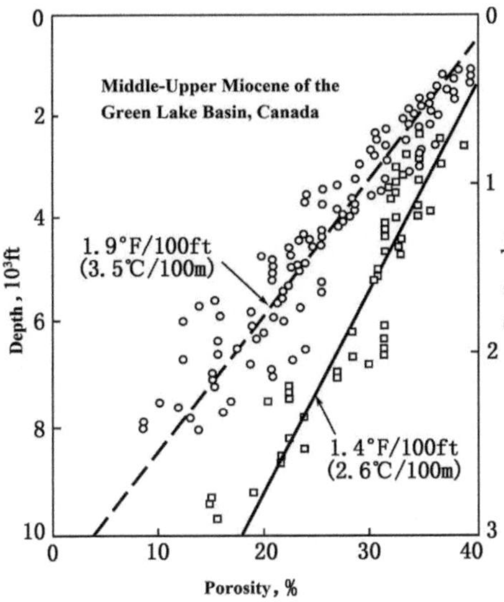

Fig. 3.16 The influences of geothermal gradient on porosity (Wilson, 1994)

2,000 m), under similar reservoir-forming conditions, the porosity of sandstones in the Cretaceous in the Junggar Basin is higher than that in the Songliao Basin.

(2) Abnormal Fluid Pressure

With the increase of overlying sediments, the effective stress on the reservoir increases, the mechanical compaction is strengthened, and the porosity decreases. Overpressure has an important influence on reservoir physical properties, mainly in four ways:

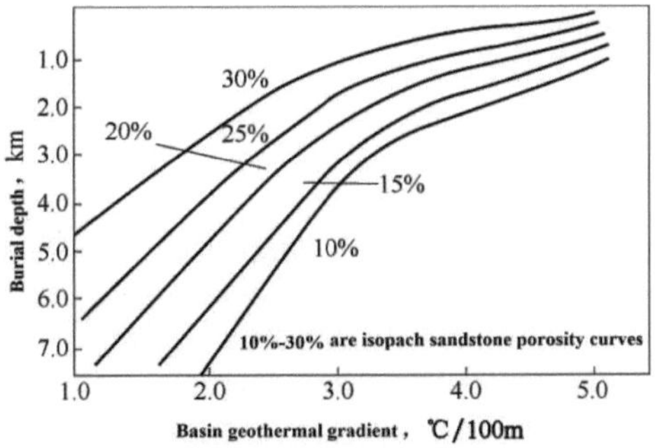

Fig. 3.17 Relationship diagram between basin geothermal gradient and sandstone porosity (Data is taken from western basins, Songliao Basin and South China Sea Basin. The lithology of the reservoir is lithic fine-fine medium sandstone with matrix content less than 2% and cement content less than 5%. It is mainly primary pores.) (Yao, 2019)

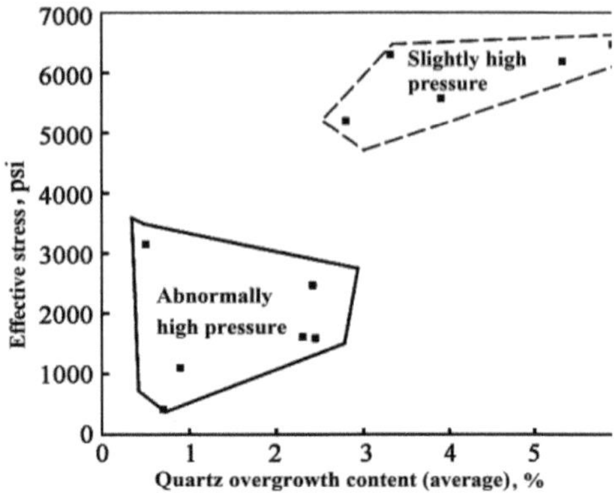

Fig. 3.18 Relationship between fluid pressure and quartz overgrowth (according to Osborne and Swarbrick, 1997)

Overpressure effectively protects the formed pores by slowing down compaction. The reduction in porosity caused by compaction is the result of the effective stress (σ). Effective stress (σ) is the difference between lithostatic pressure and pore fluid pressure. When overpressure exists, the effective stress decreases, and the porosity value at the same depth is larger than that under normal compaction, and the reservoir physical properties are significantly improved. The process of porosity decreasing with burial depth will slow down. Scherer (1987) proposed that about 2% porosity is preserved for every 50 MPa of overpressure.

Overpressure can inhibit or prevent the formation of authigenic minerals. Osborne and Swarbrick (1997) found that the content of enlarged quartz in overpressure zones is significantly lower than that in normal pressure zones (Fig. 3.18).

Overpressure changes the stress conditions when rocks rupture, resulting in overpressure fractures in mudstones and carbonates. At the same time, it can maintain the existing diagenetic contraction fractures and provide good seepage channels and reservoir space for hydrocarbon accumulation.

Overpressure generates secondary pores by enhancing dissolution. Abnormal high pressure inhibits the generation of oil and gas and widens the range of oil generation windows. In the process of thermal evolution of organic matter to generate oil and gas, many organic acids and inorganic acids are produced. Overpressure increases the time and intensity of acid dissolution of carbonate cements and generates many secondary pores (Fig. 3.19).

The influence of pressure on porosity also depends on the stage of compaction that can form overpressure. Although a specific and reliable relationship between overpressure and porosity–permeability has not been established yet, the phenomenon of coexistence of abnormally high porosity and overpressure is very common.

(3) Burial History

Generally, underground temperature increases with depth, and the reaction rate of chemical reactions accelerates with increasing temperature. Therefore, there is a certain corresponding relationship between porosity and thermal maturity. That is, at a certain depth, the porosity in areas with low thermal maturity is much higher than

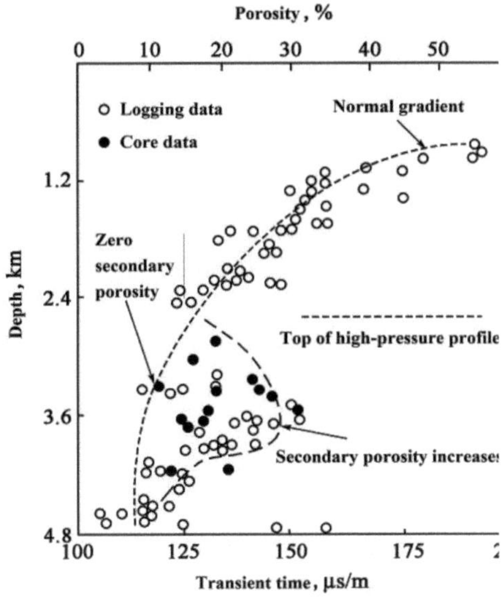

Fig. 3.19 Relationship between porosity and depth of abnormal high-pressure formation in McAllen Ranch oilfield in southern Texas (according to Berg)

that in areas with high thermal maturity. In reservoirs at the same burial depth, even if their current geothermal temperatures are the same, due to differences in burial history, their thermal maturities are very different. This is because thermal maturity is related not only to temperature but also to heating time. Differences in burial history can lead to significant differences in reservoir physical properties (Fig. 3.20). Reservoirs with long-term shallow burial and rapid deep burial in the later stage can preserve many primary pores, which is especially important for deep reservoirs and is one of the important genetic mechanisms for high-quality deep buried reservoirs.

The main oil and gas reservoirs such as the Carboniferous, Triassic, and Jurassic in the Tarim Basin are all buried at a depth of less than 3500 m, and even reach

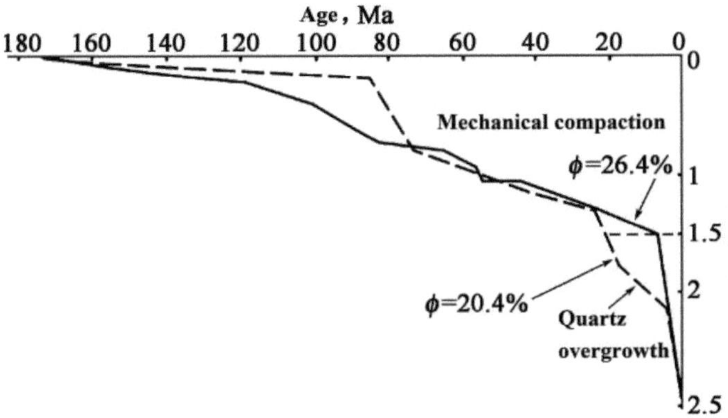

Fig. 3.20 Burial history and pore evolution (Bloch, 1999)

about 6000 m (the Donghe Sandstone of the Carboniferous in the Donghetang area). However, the reservoir quality is very good. The average porosity of the Donghe Sandstone buried nearly 6000 m reaches 15%, and the maximum is 24%. The average permeability reaches $64.04 \times 10^{-3}\ \mu m^2$, and the maximum is $1911 \times 10^{-3}\ \mu m^2$. Having such superior reservoir quality under such deep burial conditions is rare in the history of petroleum geology. The reason is long-term shallow burial, rapid short-term deep burial, and a low geothermal gradient, which are conducive to the preservation of primary pores (Gu, 1998).

(4) Tectonic Factors

Tectonic conditions are very important in reservoir physical properties. In areas with intense tectonic changes, rock formations are prone to fractures, which is conducive to the improvement of reservoir performance. Fractures generated by tectonic movements also have an important influence on improving the reservoir performance of reservoirs.

Sedimentary conditions and diagenesis are the most important geological factors controlling the reservoir performance of clastic rocks. For clastic rocks with different lithologies and burial depths, the relative importance of the two may different. Diagenesis can be the most important controlling factor, and the diagenetic environment also has a greater influence on it. For example, the Tarim Basin with a low geothermal gradient and the Bohai Bay Basin with a high geothermal gradient. At the same depth of 5000 m, the former still develops clastic rock reservoirs with relatively high porosity, while the latter's clastic rock reservoirs have been densified.

Deep reservoirs in basins often undergo multi-stage basin tectonic movement transformations and multi-stage basin fluid activity changes during the burial process, and the physical property evolution process is complex. The development characteristics of deep clastic rock reservoirs are somewhat different from those of middle and shallow layers. They usually show characteristics such as high temperature, high pressure, poor physical properties, complex pore structure and genetic types, strong and diverse diagenesis, and obvious heterogeneity. Generally, middle and shallow reservoirs are mainly dominated by primary and macroscopic secondary pores. In deep layers, the proportion of micropores in the total pores significantly increases. Different scholars have proposed various development mechanisms of deep high-quality reservoirs from different perspectives. Shallow-deep multi-genetic dissolution pore formation controls the development of secondary pores in deep and ultra-deep reservoirs. Tectonic action controls the development of fractures. Early cementation such as chlorite coating, shallow fluid overpressure, early hydrocarbon charging, salt rock development, and low time–temperature index (TTI) type burial history-thermal evolution history control the effective preservation of pores and fractures in reservoirs in deep and ultra-deep layers under different geological backgrounds.

3.2.3 *Types and Characteristics of Clastic Rock Reservoir Bodies*

A sedimentary rock formed in a certain sedimentary environment with certain morphological, lithological, and distribution characteristics and mainly composed of sand is called a sandstone body. It is the main reservoir rock mass of clastic rocks. The differences and heterogeneity of sandstone bodies are largely attributed to the diversity of sedimentary environments. From continental alluvial fans to deep-sea turbidite fans, the main reservoir sandstone bodies formed are alluvial fan glutenite bodies, river sandstone bodies, delta sandstone bodies, shore-shallow lacustrine facies sandstone bodies, coastal sandstone bodies, shallow sea sandstone bodies, deep-water turbidite sandstone bodies, and aeolian sandstone bodies. Due to differences in sedimentary conditions, various types of sandstone bodies have great differences in reservoir physical properties. Table 3.5 summarizes the characteristics of sandstone bodies and representative oil and gas fields in the main formation environments of clastic rocks.

The sandstone reservoirs with the highest oil production in the world are delta distributary channel and estuary bar sandstones, which are often superimposed in multiple stages and have a large longitudinal thickness. Followed by shore-shallow marine facies sand bodies. Submarine fans are a reservoir type with great potential but not fully explored.

The clastic rock reservoirs in major oil and gas basins in China are mostly continental facies, and most of them belong to shallow lacustrine facies, lakeside facies, fluvial facies, delta facies, (fan) braided river delta facies, and semi-deep lacustrine-deep lacustrine facies turbidite fans. Compared with marine basins, continental basins have multiple sources of materials, multiple sedimentary systems, many types of sandstone bodies, and the facies belts are roughly distributed in a ring shape. Vertically, sandstone bodies of different periods and different origins are superimposed on each other. Research on the lithology, lithofacies, thickness, geometry, and paleogeographic restoration of sandstone bodies is particularly important.

3.3 Carbonate Reservoirs

Carbonate reservoir rocks play an important role in the world's oil and gas distribution. The oil and gas reserves in carbonate reservoirs account for about 50% of the total oil and gas reserves in the world, and their oil and gas production accounts for more than 60% of the total world oil and gas production. Oil and gas fields composed of carbonate reservoirs often have large reserves and high single-well production, and are easy to form large oil and gas fields. For example, the Ghawar oil field in the Persian Gulf Basin has recoverable reserves as high as 107×10^8 tons, which is currently the oil field with the largest recoverable reserves in the world. There are a total of nine high-yield wells in the world that once had a daily production of more than 10,000 tons. Eight of them belong to carbonate reservoirs. For example, the reservoir of Well Cerro Azul-4 in the Golden Lane oil area of Mexico is Middle Cretaceous reef limestone. The highest daily production once reached 37,140 tons. In China, carbonate reservoir rocks are also widely distributed, such as the Renqiu Oilfield in North China and the Weiyuan Gas Field in Sichuan.

3.3.1 Reservoir Spaces of Carbonates

Carbonate reservoir rocks usually have three types of reservoir spaces: pores, solution pores (caves) and fractures. Due to the solubility and instability of carbonate rocks, carbonate reservoir rocks have many types of reservoir spaces and great secondary changes. Fractures are often well developed. Compared with sandstone reservoirs, they have greater complexity and diversity (Table 3.6).

There are many types of pore classifications in carbonate rocks. According to the selectivity of fabric, Choquette and Pray (1970) divided them into three major types: fabric-selective, non-fabric-selective and fabric-selective none, which can be distinguished according to pore size, shape, genesis or correlation with other structural components.

Figure 3.21 is a carbonate rock pore genesis classification scheme proposed by Ahr et al. (2008). Carbonate rock pores are produced by three processes, including deposition, diagenesis and mechanical fracture. Each process is independent, but the existence of multi-genetic pore types is due to more than one diagenetic process that can affect the existing pore system in the strata during the diagenetic history. For example, sedimentary pores are changed by diagenesis, but the original sedimentary structure, fabric or bedding can still be recognized. This can be classified as a mixed genetic pore dominated by sedimentary processes. Fractures, especially fracture intensity, are affected by mineral composition, grain size (affected by diagenesis), formation thickness and particle size (sedimentary genesis).

Sedimentary genetic pores transformed by diagenesis can be classified as mixed pore types: sedimentary and diagenetic mixed genetic pores, diagenetic and fracture mixed genetic pores, sedimentary and fracture mixed genetic pores.

Combined with the morphological and genetic classification of pore types commonly used in China, pore types can be divided into two major categories: primary pores and secondary pores (Table 3.7).

3.3.1.1 Primary Pores

The types of primary pores in carbonate rocks include intergranular pores, intragranular pores (body cavity pores), biogenetic framework pores, bird's eye pores and intercrystalline pores. The development of primary pores is controlled by the rock structure and sedimentary structure.

Table 3.6 Comparison of reservoir spaces between sandstone and carbonate rocks (modified according to Choquette & Pray, 1970)

Comparison content	Sandstone	Carbonate rock
Original porosity in sediment	Usually 25–40%	Usually 40–70%
Final porosity in rock	Often half or more of the original porosity, generally 15–30%	Usually not primary porosity. Generally 5–15% in reservoir rocks
Types of reservoir spaces	Intergranular pores are the main ones. There are few fractures and generally no karst caves	There are many types and great changes, and a large number of karst caves and fractures are developed
Size, shape and distribution of reservoir spaces	Evenly distributed. The reservoir space is mainly composed of fabric-selective pores	Greatly variable, ranging from completely depending on the fabric elements of the rock (fabric selectivity) to having no correlation at all
Influencing factors of size and shape of reservoir spaces	Related to grain size and sorting of clastic rocks. The shape of pores depends on the shape of particles	Generally, pore size has nothing to do with grain size and sorting; The shape varies greatly, from depending on the particle shape to completely not depending on the particle shape
Genesis of reservoir spaces	Related to sedimentary environment	Complex, multi-stage and diverse. Strongly affected by diagenesis and later tectonic activities. Fractured reservoir rocks and fracture-cave reservoir rocks have no direct relationship with the environment
Influence of diagenesis	Less influence. Compaction and cementation reduce the primary porosity	Great influence. It can create, eliminate or completely change the porosity. Cementation and dissolution are important
Influence of fracture	Generally not important	Very important in reservoir properties
Relationship between porosity and permeability	Relatively consistent, positive correlation	Greatly variable. Generally, the correlation is poor

Intergranular pore: It is a pore often found in granular limestone such as oolitic limestone, bioclastic limestone and intraclastic limestone. Its characteristics are similar to those of sandstone. The porosity is closely related to particle size, sorting degree, content of argillaceous matrix and content of sparry cement. A considerable

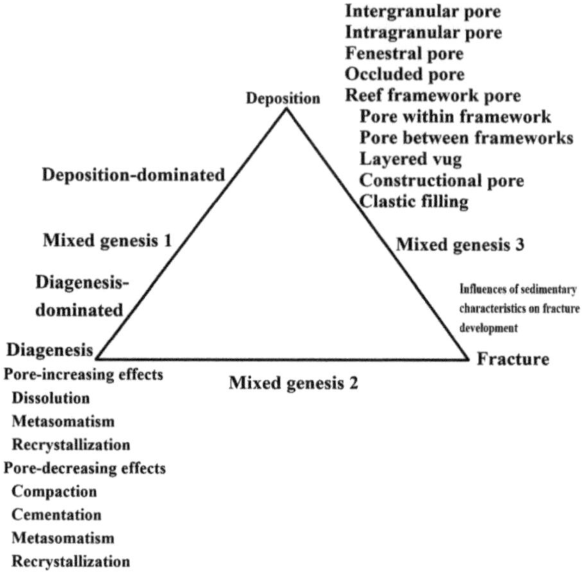

Fig. 3.21 Genetic classification of carbonate pores

Table 3.7 Pore types of carbonate rock reservoirs

Category			Subcategory
Primary pore	Pore	Fabric-controlled	Intergranular pore, intragranular pore, shell masking pore, biogenetic framework pore, intercrystalline pore
		Non-fabric-controlled	Biological drilling pore, intergravel pore, bird's eye pore
	Fissure	Fabric-controlled	Mineral cleavage fissure, shrinkage fissure
Secondary pore	Pore	Non-fabric-controlled	Intergranular or intercrystalline dissolved pore, intragranular dissolved pore, intercrystalline pore
			Mold pores (grain mold, crystal mold, biological mold), fenestral pore, karst breccia pore
	Cave	Non-fabric-controlled	Solution grooves, karst caves, caverns
	Fissure	Non-fabric-controlled	Paleo-weathering fissure, diagenetic shrinkage fissure, pressure-solution fissure, structural fissure, regional fissure

number of carbonate oil and gas reservoirs in the world belong to the intergranular pore type.

Intragranular pore: It refers to the pore inside carbonate particles. Bioclastic limestone often has this kind of pore, so it is also called body cavity pore. For example, the body cavity pore of gastropod shell. Some ooids also have such pores inside.

Biogenetic framework pore: It is the pore left between the firm frameworks formed by reef-building organisms such as colonial corals, stromatoporoids and sponges when they grow in situ. The pore shape varies with the growth mode of organisms. It forms a loose and porous structure between the frameworks. For example, various biohermal limestones often have high porosity and permeability. Second, due to dissolution, the original reservoir space becomes smaller or is completely blocked.

Biological drilling pore: The pore formed by the drilling of certain organisms. It is relatively rare and the pore is often filled.

Bird's eye pore: It is a lenticular or irregular pore that often appears in groups and is distributed parallel to bedding or bedding plane. The pores left by bird's eye structure are often larger in diameter than intergranular pores. They are mostly developed in supratidal or intertidal zones. In the late diagenetic stage, they are formed due to bubbles, shrinkage or dissolution of algal mats. It is a type of reticular or fenestral pore.

Intercrystalline pore refers to the pore between carbonate mineral crystals. For example, sugar-like dolomite has this kind of pore. Intercrystalline pores can be formed during the sedimentary period, but more often they are formed due to recrystallization, dolomitization and other processes in the epigenetic stage.

3.3.1.2 Secondary Pores

The pores formed after carbonate minerals or associated other soluble minerals are dissolved by groundwater or surface water. The types of dissolution pores include intragranular dissolved pores, mold pores, intergranular dissolved pores, solution grooves and karst caves. Karst cave refers to the dissolution that goes beyond the original particle range and is no longer controlled by the original fabric, forming some caves of different sizes and irregular shapes. On the inner wall of dissolved pores or karst caves, crystal clusters of calcite or other mineral crystals are often precipitated, also known as crystal cave pores.

3.3.1.3 Fractures

Fractures are an important type of reservoir space in carbonate rocks. According to the genesis of fractures, they are divided into structural fractures, regional fractures, diagenetic shrinkage fractures, pressure-solution fractures, paleo-weathering fractures and other types (Nelson, 1990).

Structural fracture: A fracture formed after the tectonic stress exceeds the elastic limit of the rock. It is the most important fracture type. Its characteristics are straight edges, long extension and a certain direction.

Diagenetic fracture: A fracture formed due to the pressure of overlying strata and its own water loss, shrinkage, cracking or recrystallization. The distribution of diagenetic fractures is limited by bedding and does not penetrate through layers.

They are mostly parallel to bedding planes and the fracture surfaces are curved and irregular in shape.

Pressure-solution fracture: It is formed by selective dissolution of limestone with uneven composition due to groundwater rich in CO_2 flowing along fractures or bedding. For example, suture line.

Weathering fracture: Also known as dissolution fracture. It refers to the fracture formed or transformed by surface water leaching and groundwater infiltration and dissolution in paleo-weathering crust. The development of such fractures is related to the depth of the phreatic surface. The edges of fractures have obvious oxidation halos. Fractures improve the permeability of strata.

3.3.1.4 Throat Types of Carbonate Rock Reservoirs

The throat is the gap between dolomite or calcite crystals. There are three common types of throats in carbonate rock reservoirs (Fig. 3.22).

Tubular throat: The pores are connected by a thin and long tube. The cross-section is close to circular (Fig. 3.22a). For example, the interconnected channels in the ooid space of oolitic limestone are of this type.

The narrowed part of the pore becomes the throat: There is no obvious boundary between the pore and the throat. The enlarged part is the pore and the narrowed narrow part is the throat (Fig. 3.22b). Due to crystal growth in the pore or various filling materials and other reasons, the pore is narrowed to form a throat. Compared with the pore, the diameter (equivalent) of the throat is not much different.

Sheet-like throat: The gap between dolomite or calcite crystals is generally a sheet-like throat. The sheet-like throat connects polyhedral or tetrahedral pores (Fig. 3.22c). The sheet-like throat is generally very narrow, only a few microns to a dozen microns. This is the most common throat type in carbonate rocks. The genesis of sheet-like

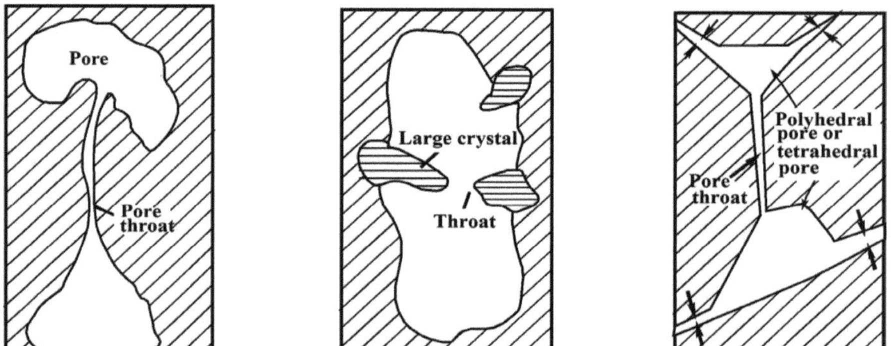

Fig. 3.22 Types of carbonate reservoir throat

throats can be different. For example, the structural fracture type shows that the throat is relatively long and straight, while the sheet-like throat formed by the fractures between dolomite or calcite crystals has the characteristics of narrow, short and flat.

3.3.2 Influencing Factors of Carbonate Reservoir

The formation of carbonate rock reservoir space runs through the entire process of deposition and diagenesis. There are many types of pores and they change quickly. The reservoir space often undergoes the action and transformation of several factors. Primary pores are obviously affected by sedimentation and secondly by diagenesis. The formation of dissolution pores and caves is significantly affected by diagenesis and the formation of fractures is closely related to tectonic activities.

3.3.2.1 Influence of Sedimentary Environment

Each sedimentary environment forms different sediments and sedimentary rocks. Figure 3.23 is the basic facies belt framework of carbonate rocks in a gentle slope continental shelf environment (Sarg, 1988), which is divided into five facies belts, namely the slope and basin facies belts facing the sea direction below the normal wave base surface, the high-energy outer continental shelf facies belt where waves interact with sediments, and the low-energy middle continental shelf and inner continental shelf facies belts facing the land direction. Grainy limestone is mainly developed in the reef and shallow water environments with high energy conditions such as the outer continental shelf. The low-energy basin environment is mainly composed of marlstone. The middle continental shelf and slope belt are mainly developed with grainy marlstone and gravelly limestone.

Different rock types have different compositions, structures and textures, and thus have different reservoir spaces. There is a direct relationship between the development of primary pores and rock types. In carbonate rocks, high-energy sedimentary facies belts are conducive to the development of primary pores. For example, facies belts such as littoral zones, shoals and reefs often develop rock types with relatively coarse structures. Intergranular pores are related to particle size, sorting degree, matrix content. Biogenetic framework is formed in biogenic reef bioclastic rocks and is often related to the size and arrangement of organisms. The size of intercrystalline pores is closely related to grain size and uniformity. Generally, mechanically formed carbonate rocks are characterized by intergranular pores. Biogenically formed rocks are characterized by framework pores and biological drilling pores. Tidal flat deposits are characterized by fenestral pores. The development degree of intergranular pores and biogenetic framework pores determines the quality of primary pores in carbonate rocks.

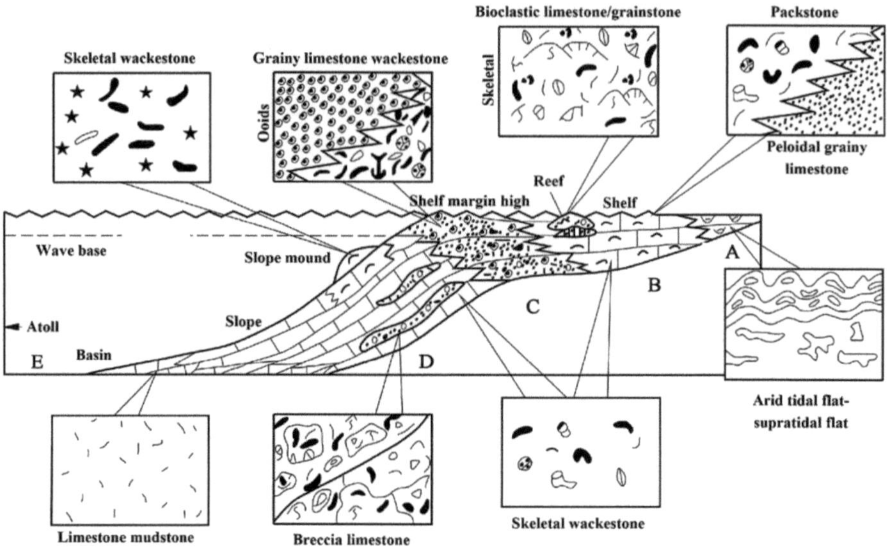

Fig. 3.23 Carbonate Facies Belts and Lithotypes in Shelf Environments A - Inner shelf, B - Middle shelf, C - Outer shelf, D – Slope, E - Basin (Modified from Sarg, 1988)

3.3.2.2 Influence of Diagenesis

Diagenesis has an important influence on the reservoir properties of carbonate rocks. The influence of compaction on the reduction of porosity of carbonate rocks is greater than that of clastic rocks, but there are differences among different lithologies. Generally, compaction has less influence on grainy limestone and dolomite, but has a greater influence on fine-grained rocks such as marlstone (Fig. 3.24). Diagenetic processes conducive to the formation of secondary pores mainly include dissolution and dolomitization. Abnormal high pressure is also conducive to the preservation of primary pores (Fig. 3.25).

Fig. 3.24 Trend of porosity variation on a semi-log plot (according to Brown, 1997)

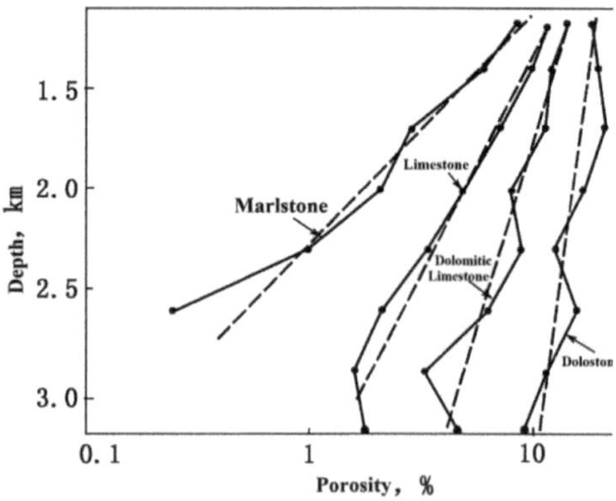

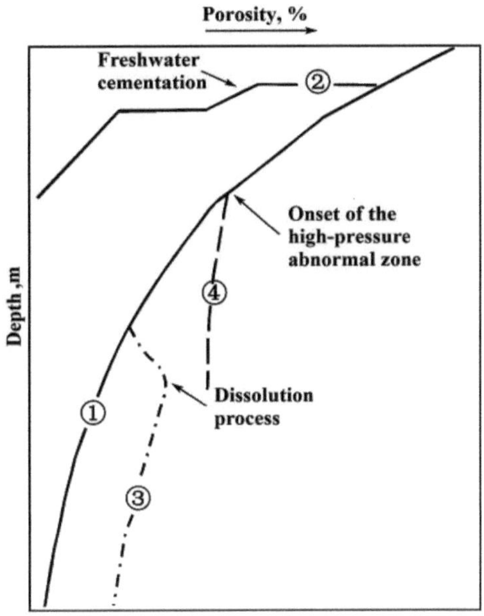

Fig. 3.25 Carbonate diagenesis and porosity evolution (modified from Ma & Tian, 1999). ① "Conventional" compaction curve, ② compaction curve with early freshwater cementation, ③ "ideal" porosity versus burial depth relationship curve formed due to late dissolution, ④ compaction curve under abnormal high pressure (modified from Ma & Tian, 1999)

Curves 1 and 2 represent the normal porosity loss with depth. Curves 3 and 4 represent the changes in porosity under overpressure conditions, when the overburden pressure increases and pore water 3 or hydrocarbons 4 cannot escape, overpressure is formed. Curve 5 represents the porosity under compaction of loose chalk, that is, it is subjected to mechanical compaction before reaching a moderate depth. Curve 6 represents the combined effect of mechanical and chemical compaction. Curve 7 reflects the porosity of trapped pore water under overpressure conditions. Curve 8 reflects the preservation of porosity related to early hydrocarbon injection.

(1) Dissolution of Carbonate Rocks

In a broad sense, dissolution includes dissolution, leaching and karstification. Among them, leaching and karstification mainly occur in supergene environments and are of great significance for the formation of secondary pores.

Dissolution requires unsaturated water (usually water in the vadose zone or mixing zone) to react with carbonate rocks to form a cave system in the vadose zone and phreatic zone.

The carbonate rock reservoirs of the Ordovician large gas field in Ordos, the Sinian-Ordovician paleo-buried hill oil and gas fields in North China, and the Cambrian-Ordovician paleo-buried hill oil and gas fields in the Tarim Basin in China have all been transformed by karstification (Fig. 3.26), and the reservoir properties of the rocks are greatly improved. The development degree of dissolved pores and karst caves is mainly controlled by factors such as the solubility of the rock itself, the dissolution ability of groundwater, and thermodynamic conditions (climate, landform and tectonic changes).

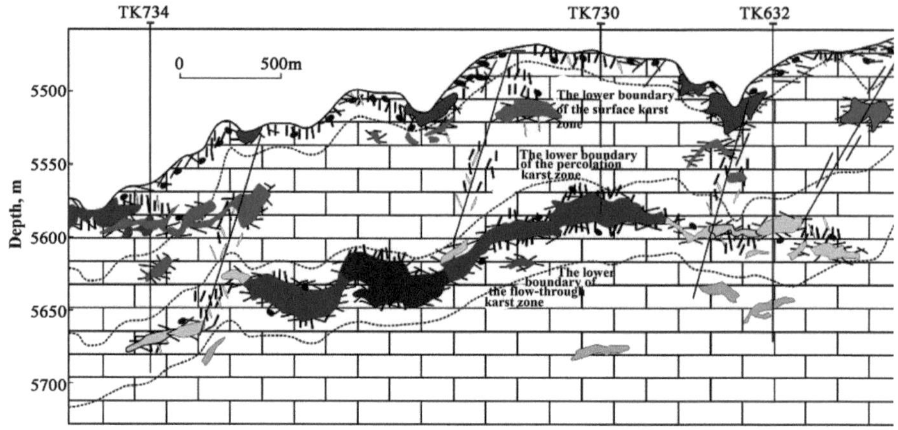

Fig. 3.26 Distribution of Percolation Karst Zone in Block 7 of the Tahe Oilfield (Li et al., 2016)

a. Solubility of Carbonate Rocks

The solubility of carbonate rocks is related to factors such as the Ca/Mg ratio of its composition, the content of clay, particle size, dolomitization, and degree of recrystallization.

Under normal conditions where groundwater is rich in CO_2, the solubility is proportional to the Ca/Mg ratio. That is, limestone is more soluble than dolomite. The content of insoluble residues such as clay in carbonate rocks is inversely proportional to dissolution. That is, the solubility of carbonate rocks decreases with the increase of clay content. Therefore, the solubility of carbonate rocks decreases in the following order:

limestone → dolomitic limestone → calcareous dolomite → dolomite
→ argillaceous limestone → marlstone

The content of calcium sulfate also affects the solubility of dolomite and calcite. Generally, the solubility of dolomite is not significantly related to the increase in calcium sulfate content, while the solubility of calcite decreases noticeably with it.

The structure and texture of rocks also have an impact on the solubility of carbonate rocks. As the particles become smaller, the solubility decreases. In coarse-grained carbonate rocks, there is less clay content and larger intergranular pores or intercrystalline pores. Groundwater can pass through more easily, which is conducive to the formation of dissolution pores. Thick-layered rocks are generally formed in relatively stable environments. They have less clay content, are pure in quality and mostly of medium-coarse grain structure. Therefore, they have high solubility and well-developed pores. Thin-layered rocks are deposited in unstable environments and contain more insoluble residues, which reduces solubility and is not conducive to the development of dissolution pores.

b. Dissolution Ability of Groundwater

When groundwater contains CO_2, the aqueous solution is acidic. As the amount of CO_2 dissolved increases, the acidity of the solution increases, and it has a strong dissolution ability for carbonate rocks. When this acidic water flows in carbonate rock formations, it gradually dissolves the rocks and forms bicarbonate, which is carried away by groundwater. On the contrary, when there is a lack of CO_2 in the water, carbonate precipitation occurs, blocking pores and cementing rocks. The movement status of water also has an impact on the dissolution ability. Good water activity is conducive to the dissolution of rocks to form bicarbonate and be carried away. In addition, the degree of dissolution of rocks is also closely related to the temperature and pressure of groundwater. Generally, it is believed that for every 10 °C increase in ground temperature, the degree of dissolution may double.

c. Influence of Landform, Climate and Structure

The movement of groundwater is an important reason for dissolution. It is related to factors such as landform, climate and structure. In terms of landform, dissolution zones are more developed in areas near river valleys, sea and lake shores. There are often large underground rivers inside carbonate strata in these areas.

In warm and humid areas, dissolution is the most active factor. Karst development is poor under dry and less rainy conditions. In areas with abundant rainfall and warm climate, karstification forms good soil and laterite, a large number of sinkholes and collapse breccia. In warm semi-arid climate conditions, calcrete crusts are often formed.

From a structural perspective, in paleo-weathering crust zones, surface water infiltrates underground along faults and fractures, producing a large number of dissolved pores, karst caves, dissolved fractures and dissolved channels, forming a huge and intricate dissolution space. This is called karst zone. If this area experiences multiple sedimentary interruptions and there are several unconformity surfaces, then several karst development zones can be formed correspondingly.

(2) Dolomitization

Intercrystalline pores related to dolomite can form under many background conditions and can become an important reservoir type. The role of dolomitization in the formation and destruction of pores has long been the focus of debate.

The porosity of dolomite generally changes with the intensity of dolomitization. Powers (1962) pointed out that when the dolomite content exceeds 75%, the porosity increases with the increase of dolomite content. When the dolomite content reaches about 77%, the intercrystalline fractures of dolomite crystals begin to enlarge, and effective intercrystalline pores develop. When the dolomite content reaches 80%, the average porosity can reach 19%. When the dolomite content increases further, the porosity and permeability relatively decay. When the dolomite content reaches more than 95%, the porosity becomes very small.

Dolomitization can strongly change the rock structure. When limestone dominated by mud is transformed into medium or coarse-grained dolomite, the porosity increases. During the dolomitization process of limestone dominated by particles, the porosity changes very little (Fig. 3.27). Generally, the constituent particles of granular limestone are much larger than dolomite crystals. Dolomitization will not have a significant impact on the pore size characteristics of granular limestone. However, when the particles of granular argillaceous limestone are replaced by dolomite crystals larger than 100 μm, its petrophysical properties will be improved.

(3) Other Diagenetic Processes

During the diagenetic and epigenetic stage of carbonate rocks, due to the continuous increase of temperature and pressure, recrystallization will occur. As a result, the crystals become coarser and the pore diameter increases, making the intercrystalline pores larger and increasing the strength of the rock, which is conducive to the formation of dissolution pores.

When groundwater containing calcium sulfate passes through dolomite-developed areas, it will replace dolomite and produce secondary calcite, forming secondary limestone after dedolomitization. Among them, the calcite grains become coarser and the porosity increases, but the distribution is relatively limited. It often appears in dolomite in the form of dendritic or lenticular.

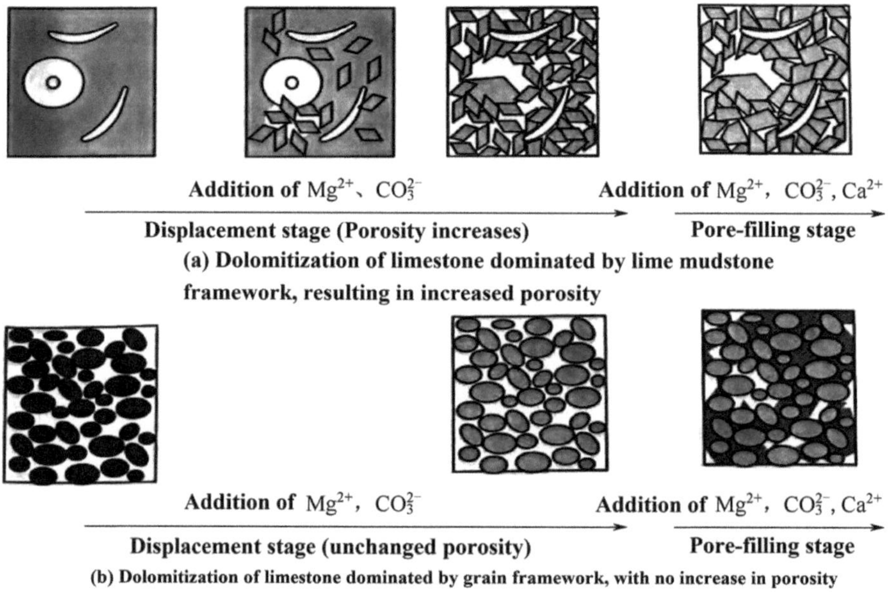

Fig. 3.27 Changes in pore structure of dolomitized limestone dominated by mud (**a**) and limestone dominated by grains (**b**)

3.3.2.3 Influence of Tectonic Activities

The influence of tectonic activities is mainly in the formation of fractures by making rocks rupture. The existence of fractures has a dual role on carbonate rock reservoirs. On the one hand, fractures themselves can serve as reservoir spaces. On the other hand, fractures can serve as seepage channels for diagenetic water, which is conducive to the progress of dissolution, and thus is conducive to the development of dissolution pores and caves. The development degree and distribution pattern of structural fractures are controlled by both lithology and structure.

(1) Controlling Factors of Structural Fractures

When the stratum rocks are under different stresses and cannot withstand a certain stress and rupture, fractures are formed in the entire rock mass. The factors that cause changes in underground stress mainly include folding, faulting, and unconformity erosion (Fig. 3.28). In addition, compaction causes changes in diagenetic minerals and water loss and shrinkage to produce interlayer fractures or shrinkage fractures.

In different structures and different parts of the structure, the fracture development is quite different. The high point, long axis, distortion and fault zone of an anticline are all places where fractures are most developed. On a long-axis anticline with a long and narrow shape, fractures are distributed in zones along the long axis and are most developed at the high point. The fractures are mainly tensile longitudinal fractures. At the high point, there are also tensile transverse fractures and interlayer fractures. If the two wings are asymmetric, the tensile transverse fractures are on the

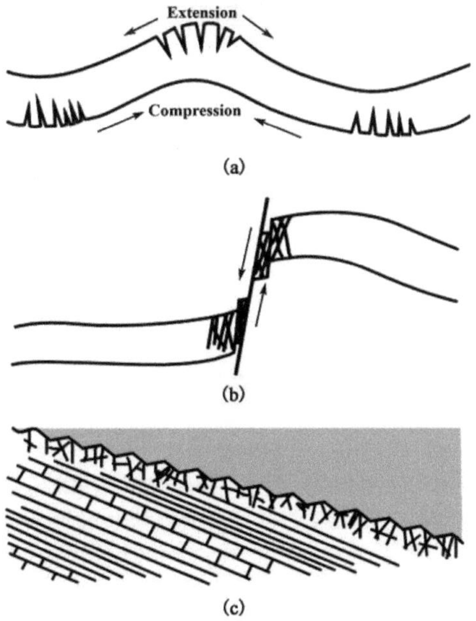

Fig. 3.28 Tectonic setting of fracture development (Selly, 1998)

gentle wing. On a short-axis anticline, fractures are distributed along the axis and are most developed at the high point.

The stress distribution in anticlines and synclines is different, and the types and properties of fractures are also different. In cross-section, tensile-shear fractures are developed in the upper part of the anticline and compressive-shear fractures are developed in the lower part. In synclines, on the contrary, compressive-shear fractures are developed in the upper part and tensile-shear fractures are developed in the lower part. Therefore, fractures are developed in the lower part of the reservoir in synclinal zones.

The strike and characteristics of faults control the development and distribution pattern of fractures. Fractures caused by low-angle faults are more developed than those caused by high-angle faults. Fractures caused by fault groups are more developed than those caused by a single fault. The arch part of fault-induced folds is the most developed in fractures. At the disappearing part of faults, fractures caused by stress release are also well developed.

Understanding the underground structural morphology is the key to improving the drilling success rate. In the practice of searching for high-yield fractured oil and gas fields for a long time, the Sichuan Oilfield has summed up an experience. Drilling on local structures should "occupy high points, follow the long axis, follow distortions, and follow faults."

(2) Internal Causes of Fracture Development

Different lithologies have different brittleness, and the degree of fracture development is also different. Brittle rock strata have developed fractures. Rock brittleness is affected by factors such as rock composition, structure, layer thickness and its combination, and epigenetic changes.

The brittleness of various carbonate rocks and chemical rocks has the following order from large to small: dolomite or argillaceous dolomite → limestone, dolomitic limestone → marlstone → rock salt → gypsum. When the content of argillaceous matter in carbonate rocks increases, it will reduce the brittleness of the rock and weaken the development of fractures. On the contrary, when the content of silica increases, it will increase the brittleness of the rock and be conducive to the development of fractures. Pure and coarse-grained carbonate rocks have high brittleness, are prone to generate fractures, and have more open fractures.

3.3.2.4 Influence of Diagenetic Environment on Pore Transformation

The pore transformation and evolution after sediment deposition occur in three zones, namely the early diagenesis zone, middle diagenesis zone and late diagenesis zone. For the formation and transformation of carbonate rock pores, there are mainly three diagenetic environments: atmospheric fresh water, seawater and burial diagenetic environment, corresponding to the early diagenesis zone, late diagenesis zone and middle diagenesis zone respectively.

(1) Seawater Diagenetic Environment

The seawater diagenetic environment is a potential place to strongly destroy pores through seawater cements. During interglacial periods, seawater is unsaturated with respect to aragonite but supersaturated with respect to calcite. During this period, aragonite particles can be dissolved, leading to the formation of secondary pores and the precipitation of calcite cements. The precipitation of calcite cements is common in atmospheric fresh water environments and deep-sea environments (Palmer et al., 1988).

(2) Atmospheric Fresh Water Diagenetic Environment

This environment is characterized by great variations in the saturation of carbonate mineral phases, ranging from saturated to supersaturated. This is related to the exposure of sediments or strata to the surface environment and the appearance of relatively dilute water. Atmospheric fresh water flowing through the vadose zone is strongly unsaturated with respect to most carbonate mineral phases and will drive the dissolution of aragonite and the precipitation of cements. The flow velocity of relatively fast atmospheric fresh water fluids flowing through the phreatic zone will greatly increase the potential for pore transformation in the atmospheric fresh water diagenetic environment (Back et al., 1979).

(3) Burial Diagenetic Environment

The diagenetic fluid in the burial diagenetic environment is characterized by a mixture of seawater and atmospheric fresh water (Folk, 1974) or basin brine with complex chemical composition formed by long-term rock-water interaction under high temperature and high pressure conditions. Under the high temperature and high pressure state of the burial diagenetic environment, pressure solution is a very important pore destruction process. Since the total amount of fluid in the pores is supersaturated the products of pressure solution can be precipitated in adjacent pores in the form of cements and destroy pores. Finally, locally undersaturated places associated with thermal degradation of hydrocarbons can also form secondary pores through dissolution. Burial diagenesis progresses slowly, but it spans a long geological time and can completely transform rocks. Strata in the burial diagenetic environment after post-orogenic periods can undergo drastic changes due to the influence of atmospheric fresh water recharge.

3.3.3 Types of Carbonate Reservoirs

3.3.3.1 Reservoir Classification Based on Reservoir Space Combination

According to the reservoir space and its combination types, carbonate rock reservoirs can be roughly divided into five basic types:

(1) Pore-Type Reservoir

The reservoir space is mainly composed of various types of pores. Including various intergranular pores, intercrystalline pores, biogenetic framework pores, etc. Such reservoirs are mostly distributed in subtidal zone—open platform, shoal and bioherm. For example, the reservoir of the Ghawar Oilfield is sandy limestone with a porosity of up to 21% and a permeability of 4000 md.

(2) Fracture-Type Reservoir

The reservoir space is mainly fractures, with fewer pores and karst caves. Fractures not only serve as the main oil and gas reservoir space but also as the seepage channel for oil and gas. When fractures form a criss-cross fracture network, it can become a good reservoir.

(3) Fracture-Pore Type Reservoir

The reservoir space is composed of various pores and fractures. The pores of the matrix rock block are the main reservoir space. In addition to providing part of the reservoir space, fractures mainly play the role of connecting the matrix rock blocks and improving the permeability of the reservoir.

(4) Fracture-Karst Cave Type Reservoir

The reservoir space is mainly composed of various karst caves of different sizes. The pores are not developed, but fractures are developed. Karst caves are the main reservoir space, and fractures are the seepage channels. During drilling, there are often phenomena such as emptying and blowout.

(5) Pore-Cave-Fracture Composite Type Reservoir

The reservoir space is composed of pores, dissolution caves and fractures of various origins. Pores and karst caves are the main oil and gas reservoir spaces. Fractures mainly play the role of seepage channels, forming a unified pore-cave-fracture system.

3.3.3.2 Reservoir Classification Based on Genesis and Main Controlling Factors

Carbonate rocks vary greatly and easily. The formation mechanism of reservoir and seepage spaces is diverse, and the main controlling factors for reservoir formation are different. According to reservoir lithofacies, genesis and evolution of reservoir space and its main controlling factors, Ma et al. (2022) divided Chinese carbonate rock reservoirs into five types: grain shoal (reef) type, dolomitization-bioherm type, dissolution pore dolomite (limestone) type, paleo-weathering dissolution type and fracture type.

Combined with previous classifications and China's exploration practice, carbonate rock reservoirs are divided into the following four categories according to reservoir genesis and main controlling factors: (1) reef-shoal limestone reservoir, (2) dolomite reservoir, (3) karst limestone reservoir, (4) fracture-type reservoir (Table 3.8), among which dolomite reservoirs are the most common.

Dolomite reservoirs account for more than half of the world's carbonate oil and gas resources. In North America, it is as high as more than 80%. China's deep carbonate oil and gas are also mainly developed in dolomite reservoirs. The reservoir space develops dolomite intercrystalline pores, intercrystalline dissolved pores, intergranular dissolved pores, karst caves and fractures. The genesis models of dolomite reservoirs include Sabkha dolomite reservoir, reflux infiltration dolomite reservoir, buried dolomite reservoir, hydrothermal dolomite reservoir, microbial dolomite reservoir and structural fracture dolomite reservoir. According to the dissolution mechanism, it is divided into facies-controlled penecontemporaneous dissolution dolomite, fault-controlled dissolution-modified dolomite and epigenic karst dolomite. Favorable facies belts, the stage of dolomitization occurrence, and dissolution modification are the key factors for the development of high-quality dolomite reservoirs.

3.4 Igneous and Metamorphic Rock Reservoirs

Igneous and metamorphic rocks are two types of reservoirs primarily characterized by fractures and secondary dissolution pores as the main storage spaces. Both share similarities in reservoir characteristics and evolutionary processes, yet each has its distinct features. Currently, igneous and metamorphic rock oil and gas reservoirs are important in global oil and gas resource exploration and development.

3.4.1 Igneous Rock Reservoirs

Igneous rocks, also known as magmatic rocks, are formed from the solidification of magma through intrusive or volcanic processes. They include intrusive rocks, volcanic lava (narrow sense volcanic rocks or extrusive rocks), hypabyssal rocks or subvolcanic rocks, and volcaniclastic rocks. Among them, volcanic lava, volcaniclastic rocks, and subvolcanic rocks are collectively referred to as broad-sense volcanic rocks. The discovered igneous rock reservoirs are mainly broad-sense volcanic rocks, with diverse rock types, among which, medium to basic extrusive rocks hold an important position.

3.4.1.1 Distribution of Igneous Rock Reservoirs

Currently, igneous rock oil and gas reservoirs or shows have been discovered in over 300 basins/blocks across more than 50 countries/regions, with industrial oil and gas flows and large-scale reserves obtained in volcanic rocks in 40 basins of

Table 3.8 Main genetic types and characteristics of carbonate rock reservoirs

Characteristic	Reef-shoal limestone reservoir		Dolomite reservoir			Weathered crust karst limestone reservoir	Fracture-type reservoir
	Grainy limestone shoal type	Bioherm type	Facies-controlled penecontemporaneous dissolution dolomite	Fault-controlled dissolution-modified dolomite	Epigenic karst dolomite		
Formation mechanism	High-energy grainy limestone shoal	Bioherm type	Reef-shoal facies dominated, reflux infiltration dolomitization, penecontemporaneous dissolution	Deep hydrothermal dolomitization, dissolution	Karstification; Sabkha dolomitization, reflux infiltration dolomitization	Exposure, epigenic diagenesis (dissolution) effect	Local structural rupture, burial diagenesis; platform facies dominated
Lithology	Oolitic limestone, clastic limestone	Reef limestone	Reef dolomite, granular dolomite	Grainy dolomite, granular dolomite	Grainy dolomite	Various limestones	Various limestones, dolomites

(continued)

Table 3.8 (continued)

Characteristic	Reef-shoal limestone reservoir		Dolomite reservoir			Weathered crust karst limestone reservoir	Fracture-type reservoir
	Grainy limestone shoal type	Bioherm type	Facies-controlled penecontemporaneous dissolution dolomite	Fault-controlled dissolution-modified dolomite	Epigenic karst dolomite		
Reservoir space	Intergranular pore-intergranular dissolved pore, pore-cavity	Biogenetic framework pore, dissolution pore	Intercrystalline pore, intergranular (dissolved) pore	Intercrystalline pore, dissolution pore, fracture, karst cave, etc.	Pore, cave, fracture	Dissolution pore-karst cave-fracture	Micropore, fracture
Example	Puguang Gas Field, Tazhong No.1 belt condensate gas field in Tarim Basin, Paleogene in Bohai Bay, Tertiary in South China Sea, Triassic in Qiangtang Basin	Jiannan Gas Field in Permian, northeastern Sichuan	Weiyuan Dengying Formation in central Sichuan, Jialingjiang Formation of Santengshan in southwestern Sichuan, Longwangmiao Formation in Anyue area of central Sichuan	Permian in Sichuan, Ordovician in Shunnan area of Tarim Basin	Ordovician in Ordos, Cambrian-Ordovician in North China	Sinian -Ordovician in northern Tarim Basin, Sinian-Carboniferous in Sichuan	Feixianguan Formation and Jialingjiang Formation of Triassic in eastern Sichuan

13 countries. Volcanic rock oil and gas reservoirs are mostly concentrated in the Mesozoic-Cenozoic (about 70%), followed by the Paleozoic. Several basins in China have discovered high-yielding oil and gas reservoirs in volcanic rocks (Table 3.9).

Overall, volcanic rock oil reservoirs in China are controlled by deep and large faults, distributed in a string of beads along the fractures, and have a larger thickness at the intersection of faults. There are mainly three sets of volcanic rocks developed: Carboniferous-Permian, Jurassic-Cretaceous, and Paleogene-Neogene. The Mesozoic volcanic rocks in the eastern basins of China are mainly acidic, while the Cenozoic volcanic rocks are mainly medium-basic. The Paleozoic volcanic rocks are developed in the western basins, mainly medium-basic.

3.4.1.2 Volcanic Rock Facies

The combined characteristics of volcanic rock facies can reflect the formation process and genesis of volcanic lava flows. Taking the volcanic rocks in the Songliao Basin as an example, Wang et al., (2007) identified 5 facies and 15 subfacies: I Volcanic Channel Facies, including Volcanic Neck Subfacies (I_1), Subvolcanic Rock Subfacies (I_2), and Cryptobreccia Lava Subfacies (I_3); II Explosive Facies, including Airfall Subfacies (II_1), Thermal Base Surge Subfacies (II_2), and Pyroclastic Flow Subfacies (II_3); III Effusive Facies, including Lower Subfacies (III_1), Middle Subfacies (III_2), and Upper Subfacies (III_3); IV Extrusive Facies, including Inner Zone Subfacies (IV_1), Middle Zone Subfacies (IV_2), and Outer Zone Subfacies (IV_3); V Volcanic Sedimentary Facies, including Volcanic Sedimentary Rock with External Clastics (V_1), Relocated Volcanic Clastic Sedimentary Rock (V_2), and Tuff with Coal Deposition (V_3) (Fig. 3.29).

3.4.1.3 Reservoir Space and Reservoir Types of Igneous Rock Reservoirs

(1) Types of Reservoir Space

The quality of the reservoir physical properties of igneous rocks will determine their oil and gas content. The reservoir space of igneous rocks includes two types: pores and fractures, and is divided into primary pores and secondary pores according to their genesis (Table 3.10). Primary pores include vesicles and residual pores not filled by amygdales, intercrystalline pores, and intergranular pores in volcanic breccia, etc. Secondary pores mainly include devitrification pores in spherulitic rhyolite, dissolution pores in feldspar, volcanic ash, and clay minerals, etc. Fractures mainly include structural fractures and weathering fractures, etc.

The reservoir space of igneous rocks has diverse pores, various geometric shapes, intertwined pores, caves, and fractures. The pores are unevenly distributed with poor connectivity, and fractures play an important role in improving reservoir physical properties. Igneous rock oil reservoirs have the characteristics of high single-well

Table 3.9 Characteristics of igneous rock reservoirs in Chinese oil and gas basins (modified from Zou et al., 2014)

Era	System	Group, formation, member	Basin, depression	Lithology	Porosity/%	Permeability/$\times 10^{-3}$ μm^2
Cenozoic	Neogene	Yancheng Group	Gaoyou Depression	Gray-black, gray-green, gray-purple basalt	20	37
		Bottom of Guantao Formation	Dongying Depression	Olivine basalt	25	80
	Paleogene	Sanduo Formation	Gaoyou Depression	Basalt	22	19
		First Member of Shahejie Formation	Dongying Depression	Basalt, andesite basalt, volcanic breccia	25.5	7.4
		Third Member of Shahejie Formation	Huimin Depression	Olivine basalt	10.1	13.2
		Fourth Member of Shahejie Formation	Zhanhua Depression	Basalt, andesite basalt, volcanic breccia	25.2	18.7
		Xingouzui Formation	Jiangling Depression	Gray-black, gray-green, gray-purple basalt	18–22.6	3.7–8.4
		Kongdian Formation	Weibei Depression	Basalt, tuff	20.8	90
Mesozoic	Cretaceous	Angshenpao Formation of Upper Cretaceous	Lower Liaohe Basin	Trachyte, basalt, andesite basalt	20–25	1–20
		Yingcheng Formation of Upper Cretaceous	Liaohe Basin	Basalt, andesite, dacite, rhyolite, tuff, volcanic breccia	1.9–10.8	0.01–0.87
		Qingshankou Formation	Qijia-Gulong Depression	Medium-acidic volcanic breccia, tuff	22.1	136

(continued)

Table 3.9 (continued)

Era	System	Group, formation, member	Basin, depression	Lithology	Porosity/%	Permeability/× 10^{-3} μm^2
		Suhongtu Formation of Upper Cretaceous	Yingen Basin	Basalt, andesite, volcanic breccia, tuff	17.9	111
	Upper Jurassic	Huoshiling Formation	Erlian Basin	Basalt, andesite	3.5–12.7	1
		Xing'anling Group	Hailar Basin	Basalt, andesite, volcanic clastic rocks	3.6–13	1–214
	Triassic	Budate Group	Hailar Basin	Medium-basic volcanic rocks	Average 5.0	Average 0.03
Paleozoic	Carboniferous-Permian		Junggar Basin	Andesite, basalt, tuff, volcanic breccia	4.2–26.8	0.03–153
	Permian		Tarim Basin	Dacite, basalt, volcanic breccia, tuff	0.8–19.4	0.01–10.5
	Permian		Sichuan Basin	Basalt (Weiyuan, Zhougongshan structure), volcanic breccia lava, tuffaceous breccia lava (Yongtan 1 Well)	6.68–13.22	0.014–0.43

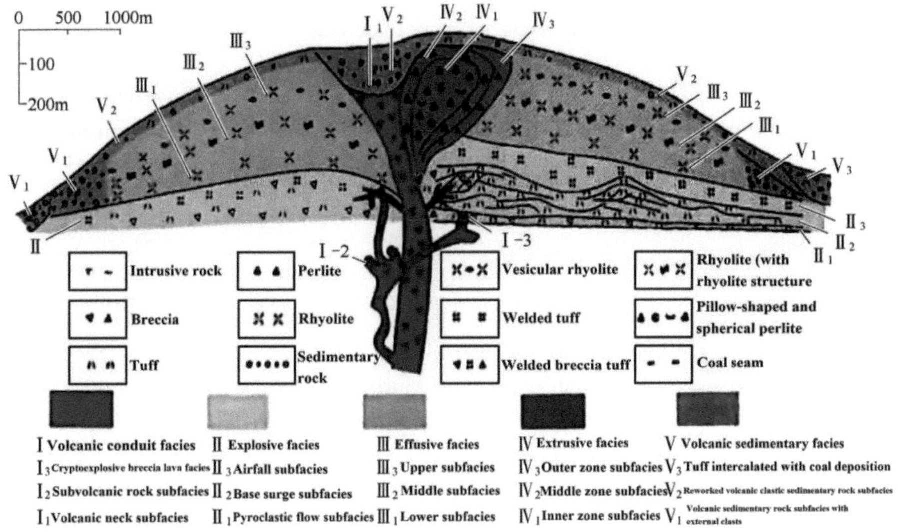

Fig. 3.29 Facies model of acidic volcanic rocks in the Songliao Basin

production in some oil wells, rapid production decline, and large differences in well productivity in the plane, showing the characteristics of fracture seepage.

(2) Genetic Types of Volcanic Rock Reservoirs

The formation of volcanic rock reservoirs is controlled by volcanic, diagenetic, and tectonic processes. According to their genetic characteristics, volcanic rock reservoirs can be divided into three types: primary pore type, secondary pore type, and fracture type. Some scholars also divide igneous rock reservoirs into structural fracture type, weathering and leaching type, primary pore type, and burial dissolution type.

a. Primary Volcanic Rock Reservoirs

The reservoir space is mainly composed of primary pores and fractures, mainly developed in the effusive facies. Generally, the vesicles in volcanic rocks near the volcanic vent are well developed, and the physical properties are good, while the physical properties are poor away from the volcanic vent, such as the volcanic rock reservoirs of the Cretaceous Yingcheng Formation in the Songliao Basin and the rhyolitic welded breccia tuff oil and gas reservoir in Well Xia 72 in the northwestern margin of the Junggar Basin.

b. Weathered Crust Type Volcanic Rock Reservoirs

The reservoir space of volcanic rock is mainly composed of secondary pores formed by weathering, leaching, and structural transformation. The Carboniferous-Permian volcanic rock reservoirs in the Junggar Basin and Santanghu Basin are typical representatives of this type.

Table 3.10 Types of reservoir space in igneous rocks (modified according to Zou et al., 2014)

Type of reservoir space		Corresponding rock type	Characteristics	Oil and gas content
Primary pores	Vesicles	Andesite, basalt, breccia, breccia lava	Mostly distributed at the top and bottom of the lava flow layer, with different sizes and shapes	Better oil and gas content when connected with fractures and caves
	Intergranular pores	Volcanic breccia, agglomerate, volcanic sedimentary rock	Commonly seen in volcanic clastic rocks	Mostly not oil-bearing
	Intercrystalline and intragranular pores	Basalt, andesite, autoclastic breccia lava	Mostly distributed in the middle of the lava flow layer, with small voids	Mostly not oil-bearing
	Condensation shrinkage pores	Basalt	Without a certain direction, and the shape is often irregular	Filled with oil and gas when connected with vesicles
Secondary pores	Devitrification pores	Spherulitic Rhyolite	Microscopic pores, but with good connectivity	Good gas storage space
	Feldspar dissolution pores	Various rocks	Irregular pore shape	One of the main reservoir spaces
	Volcanic ash dissolution pores	Basalt, andesite, breccia lava, breccia	Although the pores are small, due to the large number and good connectivity	Can form a good reservoir
	Carbonate dissolution pores	Various rocks	Larger pores	Good oil and gas content
	Karst caves	Basalt, andesite, breccia lava, breccia	Develop along fractures, self-fragmented clastic rock zones, and structural high positions	Good oil and gas content
Fractures	Explosion fractures	Self-fragmented brecciated lava, subvolcanic rocks	With restorability	Better oil and gas content
	Contraction fissures	Basalt, andesite, self-fragmented breccia lava	Columnar joints, in an open form, planar cracking, but less dislocation	Better oil and gas content

(continued)

Table 3.10 (continued)

Type of reservoir space		Corresponding rock type	Characteristics	Oil and gas content
	Structural fissures	Various rocks	Developed near faults, relatively straight, mostly high-angle fractures	Related to the time of tectonic activity
	Weathering fissures	Various rocks	Connected with dissolution pore-fissure caves and structural fractures	Little significance for reservoir storage

c. Fracture Type Reservoirs

Fracture-type volcanic rock reservoirs are mainly related to tectonic movements, with complex lithology, strong tightness, and brittleness, making the volcanic rocks prone to developing fractures under tectonic stress. The distribution of the reservoirs is controlled by large faults and is distributed in a strip shape. In areas with strong tectonic activities, the fracture density is high, and most of them are fracture—pore composite reservoirs.

3.4.1.4 Influencing Factors of Reservoir Space in Igneous Rocks

The reservoir space of volcanic rock reservoirs is diverse and has strong heterogeneity. Volcanic activity (lithofacies type), tectonic movement, weathering, leaching, and dissolution are the main controlling factors for the development of reservoir space in igneous rock reservoirs.

(1) Lithofacies Type

Different lithofacies of igneous rocks have different rock properties, compositions, rock structures, and abilities to resist weathering and transformation, resulting in different degrees of pore development. Generally, vesicle pores are mostly developed in effusive basalts, and intercrystalline pores are mostly found in lavas of the effusive facies. The facies zone most conducive to pore development is the medium-distance volcanic slope facies (transition facies zone), followed by the near-volcanic vent and far volcanic slope facies. In lava flow, fractures are easily formed in the lower part, and vesicles are developed in the middle and upper parts.

(2) Tectonic Effect

Generally, extrusive rocks have well-developed primary pores, mainly isolated vesicles. Under the influence of later tectonic movements, the originally unconnected vesicles will be connected by crisscrossing and mutually cutting multi-stage fractures, forming a reservoir.

(3) Weathering Effect

Generally, the degree of weathering of igneous rocks is positively correlated with reservoir physical properties. In weathering and erosion zones and structural fracture zones, pores and fractures are well developed. From top to bottom, the weathering intensity weakens, the density of microfissures decreases, and the physical properties decrease accordingly. The thickness of the stratum affected by weathering and leaching is closely related to the ancient landform and the underlying rock fabric. Generally, the influence is deeper in the high positions of the ancient topography and in the areas where the underlying stratum has well-developed primary pores and fractures.

(5) Dissolution Effect

Dissolution, leaching, and hydration can break the rocks and form new minerals. The main forms of dissolution include dissolution, oxidation, hydration, and carbonation. Dissolution can increase the reservoir space and improve the porosity and permeability of the reservoir. The previously formed primary pores, fractures, and structural fractures provide channels for later fluid to act on the rocks, controlling the development of secondary pores.

3.4.2 Metamorphic Rock Reservoirs

Metamorphic rocks are often the basement of sedimentary cover in basins, with dense and hard lithology and no pore space significant for oil and gas storage. If these rocks undergo long-term and intense weathering or structural fracturing, a weathered pore zone appears on the surface, and the porosity and permeability of the rocks greatly increase, making them a good place for oil and gas storage.

3.4.2.1 Rock Types

At present, most of the global metamorphic rock buried hill oil and gas reservoirs are distributed in the Precambrian, and the lithology of the reservoir layers is mainly metamorphic granite, gneiss, and migmatite. Table 3.11 shows the representative metamorphic rock oil and gas reservoirs discovered in China. The rock types are mainly migmatite, followed by slate, phyllite, schist, gneiss, leptynite, and cataclasite.

3.4.2.2 Reservoir Space

According to their genesis and stage, the reservoir space of metamorphic rock reservoirs can be divided into four types: metamorphic crystal, structural, physical weathering, and chemical leaching (Table 3.12), still mainly composed of weathered pores,

Table 3.11 Metamorphic rock oil and gas reservoirs in China

Oilfield	Geological era	Reservoir rock type	Oil reservoir type
Yumen	Paleozoic Silurian	Phyllite Slate	Yaerxia Silurian Buried Hill Oil Reservoir
Liaohe	Proterozoic	Metamorphic Quartz Sandstone	Dujiatai Proterozoic Buried Hill Oil Reservoir
	Archean Anshan Group	Migmatite, Regional Metamorphic Rock	Xinglongtai, Dongshengpu, Jing'anpu, Qijia, Niuxintuo, Ciyutuo, Shuguang, and other Buried Hill Oil Reservoirs
Shengli	Archean Taishan Group	Fragmented Gneiss, Migmatite, Leptynite	Wangzhuang Archean Buried Hill Oil Reservoir, with a single-well daily oil production of over 1,000 tons in Well Zheng 4
			Chengdao Oilfield
Bohai Sea	Proterozoic	Granitic Migmatite	Jinzhou 20-2 Structure Archean Buried Hill Oil and Gas Reservoir
	Archean		Jinzhou 25-1 South, Bozhong 19-6, Bozhong 13-2
Jidong	Archean	Granitic Migmatite	Jidong Archean Metamorphic Rock Oil Reservoir

fractures, and structural fractures. Such reservoirs are mostly developed in the unconformity zone, on the basin edge, slope, and in the paleo-uplifts within the basin, where dissolution pores are developed, and at the same time, fracture-intensive zones with certain directionality and connectivity are formed, providing a good place for oil and gas storage.

The reservoir space of metamorphic rock reservoirs is mainly divided into two types: pores and fractures. Pore-type reservoirs are mainly distributed at the top of the buried hill weathering crust, and fracture-type reservoirs are mainly developed in the interior of the buried hill. Vertically, it is divided into the weathering and leaching zone, the fracture and dissolution zone, and the tight basement zone, and the high-quality reservoirs are mainly distributed in the weathering and leaching zone.

Table 3.12 Common reservoir spaces in metamorphic rock reservoirs

Type	Types of reservoir space
Metamorphic crystal genesis	Intermetamorphic pores, metaremnant intergranular pores, cleavage cracks
Structural genesis	Structural fissures, fractured intergranular pores
Physical weathering genesis	Weathering fissures, weathering fractured intergranular pores
Chemical leaching genesis	Dissolution pores, dissolution cracks

The Archean metamorphic rocks in the Bohai Sea are divided into shallow-buried and deep-buried buried hills according to their burial depth. The shallow-buried buried hills are divided into the clay zone, the sandification zone, the sandified gravel zone, the fracture zone, and the basement zone from top to bottom, with the reservoir space mainly composed of pores and fractures; the deep-buried buried hills are divided into the weathering fracture zone, the interior fracture zone, and the basement zone from top to bottom, with the reservoir space mainly composed of fractures.

3.4.2.3 Influencing Factors

The metamorphic rock buried hill reservoirs are controlled by tectonic movement, weathering and leaching, paleogeomorphology, rock mineral composition, deep thermal fluids, and other factors.

(1) Tectonic Stress Controls Fracture Formation

Tectonic processes play a dual role in the formation of reservoirs in metamorphic buried hills. Tectonic uplift exposes the buried hills to the surface, where they undergo weathering and leaching, leading to the formation of a weathered crust with high-quality reservoir properties at the top of the buried hills. Additionally, multiple episodes and directions of tectonic stress superimpose, significantly altering the metamorphic rocks to produce fault zones and dynamic metamorphic zones, which create many high-quality fracture-type reservoirs. Taking the Bozhong 19-6 oilfield as an example, the Archean metamorphic rocks have been strongly affected by three periods of tectonic movements: the Indosinian, Yanshan, and Himalayan periods. This has led to intense rock fracturing, resulting in a large-scale fracture system with multiple superimposed tectonic fractures in the metamorphic buried hills. The effective reservoir layer of metamorphic rocks with open or partially calcite-filled fractures has developed.

(2) Weathering and Leaching Controls the Formation of Dissolution Pores

Weathering and leaching affect the differential dissolution of minerals and the quality of metamorphic rock reservoirs. Weathering forms many secondary pores and fractures, improving the physical properties of metamorphic rocks. Among them, feldspar dissolution is the main mechanism for the formation of high-quality reservoir layers in felsic metamorphic rocks. For example, in the Archean buried hill reservoir of the Bozhong 19-6 Oilfield, the reservoir space is mainly composed of structural fractures, weathering fractures, followed by dissolution fractures and dissolution pores.

The metamorphic rocks develop feldspar dissolution pores due to weathering and leaching. The clay mineralization of feldspar reduces the mechanical compressive strength of the rocks and also provides a large number of micropore spaces. Weathering also transforms the previously formed fractures. Generally, the rocks are severely weathered and fractured in the fracture sections. In the sections where fractures are not developed, granitic gneiss and other rocks basically maintain the dense appearance of the original rocks.

The metamorphic rock buried hill reservoirs have a vertical zoning. From top to bottom, they are divided into the weathered glutenite zone, the weathered fracture zone, the interior fracture zone, and the bedrock zone (Fig. 3.30). The weathered glutenite zone develops at the top of the metamorphic rock buried hill, with the reservoir space being pores and few fractures, and good physical properties. The weathered fracture zone is controlled by both tectonic stress and weathering, and develops at the bottom of the weathered glutenite zone, with the reservoir space mainly composed of pore-fracture and fracture. The interior fracture zone develops in the interior of the buried hill, distributed near the fault zone, with the lithology mainly composed of fragmented metamorphic rocks and cataclasite, which is usually a high-quality reservoir layer with developed fractures and an interbedded tight reservoir layer. The bedrock zone is basically not affected by weathering and tectonic activities, located at the bottom of the buried hill reservoir, with undeveloped fractures and being a non-reservoir layer.

(3) Paleogeomorphology Controls the Distribution of Different Types of Reservoirs

Paleogeomorphology controls the development degree of the weathering crust. Generally, the thickness of the reservoir is larger in the high positions of the ancient geomorphic structure and decreases in the low positions. In the top of the structural slope and the upper part of the steep slope, the weathered glutenite zone is not easy to be preserved, and the reservoir is mainly composed of the weathered fracture zone. In the wide and gentle platform and the gentle slope, the weathered glutenite zone is easy to be preserved. In the relatively low positions of the structure, the tectonic stress is relatively weak, and the fracture zone is not developed.

(4) Rock Type Controls the Development Degree of Fissures and Pores

Rock type is an important internal factor affecting the development of buried hill reservoirs. According to previous studies, dark minerals such as biotite and hornblende are ductile minerals, which are not easy to form structural fractures, and the higher the content of dark minerals, the more unfavorable it is for the formation of fracture-type reservoirs. Generally, felsic rocks, such as granitic gneiss, metamorphic granite, and migmatite, have strong brittleness and will preferentially generate fractures under the same tectonic stress conditions, and are also more prone to fluid dissolution and transformation in the later stage, forming high-quality reservoirs. Mafic rocks, such as amphibolite, biotite-plagioclase gneiss, biotite-leptynite, etc., have a higher content of dark minerals, a lower content of felsic minerals, and strong plasticity of the rocks, making it difficult to form fractures and undergo dissolution.

(5) Deep Thermal Fluids Have a Dual Control Effect on Reservoirs

The deep thermal fluids in the basin include the thermal fluids associated with geological events such as volcanic activities and organic acids discharged from source rocks,

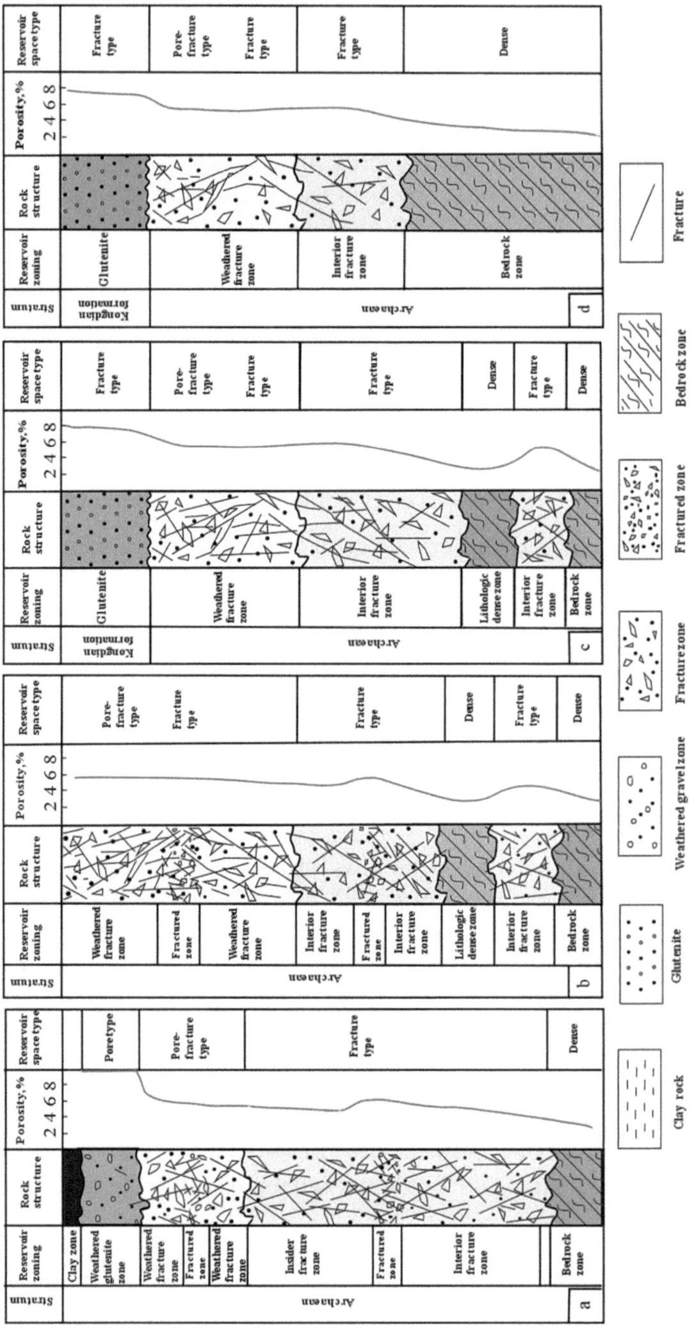

Fig. 3.30 Vertical zonation characteristics of metamorphic buried-hill reservoirs (Changgui et al., 2020)

which have an impact on both the weathering zone and the interior fracture zone of the buried hills. The influence of deep thermal fluids on the reservoir is closely related to the internal fractures of the buried hills, and the deep fluid transformation effect is strong in the fracture-developed areas. On the one hand, acidic thermal fluids containing mantle-derived CO_2 and organic acids discharged from source rocks will dissolve soluble minerals such as feldspar and calcite, forming intra-crystal micropores and dissolution-enlarged pores along the fractures in the reservoir. On the other hand, the deep thermal fluids contain a large amount of silicon-rich, iron-rich, and calcium-rich fluids, which will precipitate in the fractures to form quartz veins, chert veins, siderite, and ferrocalcite, filling the pores and fractures, which has a destructive effect on the reservoir.

3.5 Shale Reservoirs

Shale reservoirs refer to shale formations that develop many micro-nano pore spaces and can serve as effective reservoirs. In the past, such rocks were considered only as "tight" caprocks. However, with the deepening of exploration practices at home and abroad, it has been found that there are also abundant oil and gas resources in shale in sedimentary basins, and shale itself can constitute the reservoir for such oil and gas. In recent years, driven by the demonstration effect of the shale revolution in the United States, there has been a global upsurge in shale oil and gas exploration and development. The total world output of shale gas has exceeded 800 billion cubic meters, and that of shale oil has exceeded 400 million tons. Shale oil and gas are changing the world energy landscape and are currently a new field in oil and gas exploration.

Shale oil and gas resources are worldwidely distributed, such as the vast oil and gas area of the Devonian shale in the Appalachian Basin in the eastern United States, the Paleogene oil and gas area in the western Los Angeles, Ventura, San Joaquin, Santa Maria, and the southern Gulf of Mexico Basin, as well as the Sichuan Basin and adjacent areas in southern China. The micro-nano-scale matrix pores and microfractures within the shale reservoir constitute the reservoir space for shale oil and gas, and the pore-fracture structure controls the occurrence state and exploitation efficiency of shale oil and gas.

3.5.1 *Types of Reservoir Space in Shale*

According to where are the pores developed in shale, the pores can be divided into three basic types: inorganic mineral matrix pores, organic pores, and fracture pores (Fig. 3.31). Inorganic mineral matrix pores can be further subdivided into intergranular pores and intragranular pores. Fracture pores are not classified as matrix pores because they are not controlled by individual mineral particles.

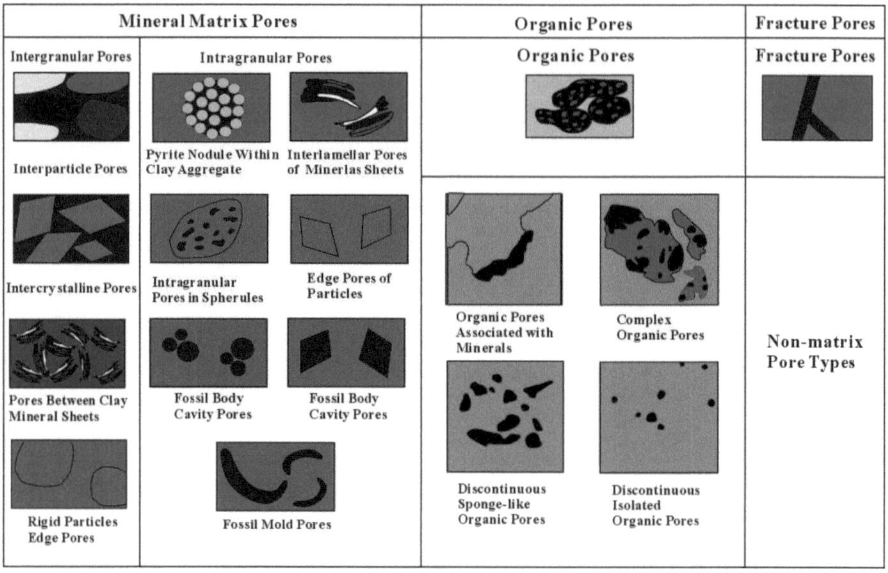

Fig. 3.31 Classification scheme for shale reservoir types (Loucks et al., 2012)

3.5.1.1 Organic Pores

Shale gas exploration practices have confirmed that organic pores are the main pore type in marine shale gas reservoirs in North America and southern China, and are the key factor for the enrichment of shale gas (Gou et al., 2021; Loucks & Reed, 2014). On two-dimensional images, organic pores have various shapes such as irregular, elliptical, and bubble-like (Fig. 3.32). The organic pores within organic-rich shale can be divided into kerogen pores and bitumen pores. Within the oil generation window, kerogen pyrolyzes to form crude oil and is expelled into adjacent intergranular or intragranular pores. During this process, organic pores will form inside the kerogen, but these pores are difficult to detect due to the retention and charging of early crude oil. Within the gas generation window, the bitumen inside the kerogen and the bitumen that has migrated to the outside of the kerogen undergo secondary cracking, transforming into solid bitumen or pyrobitumen rich in nano-pores. The generation and expulsion of hydrocarbons during the thermal evolution of kerogen control the degree of development of kerogen pores. At the same time, some scholars have found through transmission electron microscopy analysis that the morphology and development degree of organic pores within different types of organic matter vary, and solid bitumen pores are the main component of organic pores. There are different views on the origin and source of organic pores, including kerogen hydrocarbon generation—expulsion (i.e., kerogen pores) and retained crude oil cracking (i.e., bitumen pores) (Bernard et al., 2012; Löhr et al., 2015).

Kerogen pores are mostly strongly heterogeneous and unevenly distributed within the organic matter, with irregular angular shapes but relatively uniform sizes

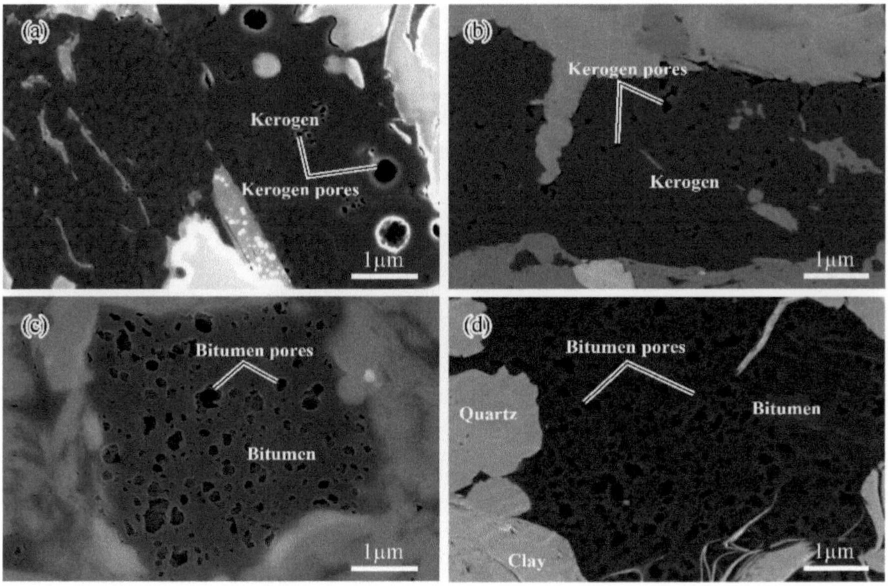

Fig. 3.32 Characteristics of organic pore development in shale

(Fig. 3.33a, b). During the hydrocarbon generation evolution, kerogen is always in a solid state. Due to the differences in structure and composition of different parts, as well as the differences in the intensity of decomposition or condensation reactions, the degree of pore development within the same kerogen particle shows strong heterogeneity. Within the oil generation window, the hydrogen-rich components in kerogen mainly decompose to form liquid hydrocarbons, and aliphatic and heterocyclic functional groups gradually fall off and escape in the form of hydrocarbons and volatiles, generating pores. With the increase in the degree of thermal evolution, the condensation reaction begins to intensify, and the condensation of the aromatic nucleus inside the kerogen can further generate more nano-particles and rearrange to form polygonal organic pores with relatively uniform sizes and mostly angular shapes. The degree of development of kerogen pores depends on the combined effects of the nature of organic matter, hydrocarbon expulsion efficiency, and condensation degree.

Bitumen pores are mostly sponge-like or honeycomb-like, with a composite round or elliptical shape with coexisting sizes, and the pore development characteristics are relatively uniform. The difference in hydrocarbon-generating components is the main reason for the structural difference between kerogen pores and bitumen pores. During the continuous burial and heating of shale, the retained oil begins to crack and convert into gaseous hydrocarbons. This gas generation process occurs simultaneously on the surface and inside the crude oil, and the pore morphology is mainly round or elliptical (Gou et al., 2022; Xu et al., 2020a, 2020b). Since the components and structure of liquid hydrocarbons are relatively uniform, the density and morphology of the generated pores are relatively uniform. In areas where bubbles are densely generated or continuously generated, multiple adjacent small pores often merge to form larger round or elliptical bubble pores, showing the phenomenon of "large pores surrounding small pores, and pores within pores" (Fig. 3.33c, d).

3.5.1.2 Inorganic Pores

(1) Intergranular Pores

Intergranular pores are abundant in shallow-buried sediments and have good connectivity, usually elongated without obvious orientation. These pores continue to evolve with the increase of burial depth, overburden pressure, and diagenesis. During the burial process, plastic particles can deform and squeeze into the pore throats, thereby closing the intergranular pores. Therefore, for deeply buried shale, the intergranular pores are significantly reduced due to compaction and cementation, and many pores are triangular or irregular polygons (Fig. 3.33a), which are considered as the residual pore space between rigid particles after compaction and cementation. Especially when the diagenetic environment (temperature, pressure) changes, clay minerals will undergo corresponding transformations, resulting in the gradual expulsion of interlayer water and the formation of many intergranular pores between clay mineral particles, with most lengths within 1 μm and widths ranging from several tens of nanometers to several hundred nanometers (Fig. 3.33b). In addition, some unstable mineral particles (such as feldspar and calcite) can form ring-shaped intergranular pores at the particle edges due to the influence of dissolution (Fig. 3.33c). Intergranular pores are not only reduced or destroyed by compaction but also controlled by the secondary transformation of mineral particles such as quartz, calcite, feldspar, and clay. Therefore, the development degree and geometric morphology of intergranular pores in shale vary significantly in different regions, strata, and types.

(2) Intragranular Pores

Intragranular pores develop inside mineral particles, and the pore diameter varies widely, ranging from a few nanometers to hundreds of nanometers or several micrometers (Fig. 3.33d–f). Although most of these pores are related to diagenetic transformation, some are primary. According to the location of pore development and the formation mechanism, intragranular pores can mainly be divided into: ① pores formed by the partial or complete dissolution of particles; ② pores preserved inside fossils; ③ pores between crystals inside strawberry-shaped pyrite nodules; ④ cleavage plane (fissure) pores inside clay and mica mineral particles; ⑤ internal pores of particles (such as inside spherules or coprolites).

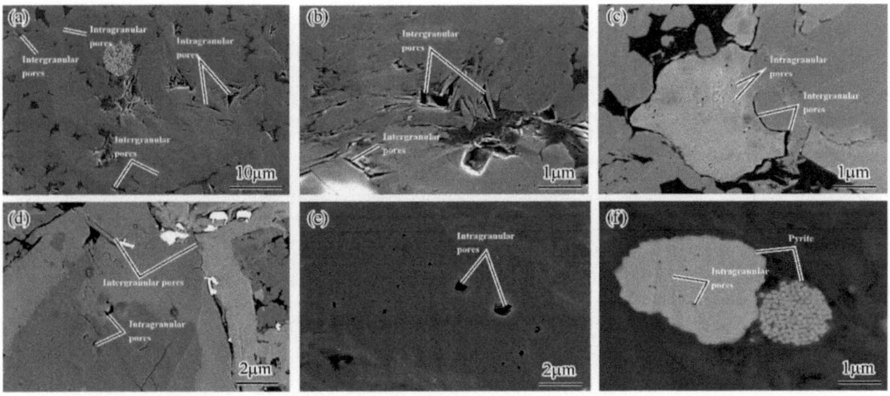

Fig. 3.33 Development characteristics of inorganic pores in shale

3.5.1.3 Micro-fractures

Shale gas mainly exists in two forms in shale reservoirs: free state and adsorbed state. Among them, free gas is mainly stored in larger pores and fracture systems. Therefore, the development characteristics of fracture-related pores in shale reservoirs affect the enrichment and preservation of shale gas to a certain extent. In addition, the pore network related to fractures can serve as a good channel for gas migration, and its development degree is a very important geological indicator for oil and gas exploitation and well pattern deployment. According to whether micro-fractures are affected by tectonic activities, they can be divided into non-structural fractures and structural fractures.

(1) Non-structural Fractures

Non-structural micro-fractures mainly refer to micro-fractures formed by different diagenetic processes during the sedimentation and solidification of shale, including interlayer fractures, shrinkage fractures, and particle edge fractures. Interlayer fractures mainly refer to micro-fractures between bedding layers formed during the deposition of shale (Fig. 3.34a). From the perspective of mechanical properties, interlayer fractures are weak interfaces that are usually easily peeled off. Such micro-fractures have been widely reported in the currently widely exploited laminated shale oil reservoirs. Shrinkage fractures mainly refer to micro-fractures related to tensile stress formed due to the volume changes of organic matter and clay minerals. They usually occur inside organic matter and clay mineral particles, with generally small scales, no obvious directional extension, short extension ranges, and no systematicity, and the fracture widths range from a few nanometers to several hundred nanometers (Fig. 3.34b). Particle edge fractures generally refer to micro-fractures formed between plastic organic matter and detrital particles such as quartz and feldspar. Such micro-fractures often develop around detrital particles, and their morphology is closely related to the size and location of detrital particles (Fig. 3.34c).

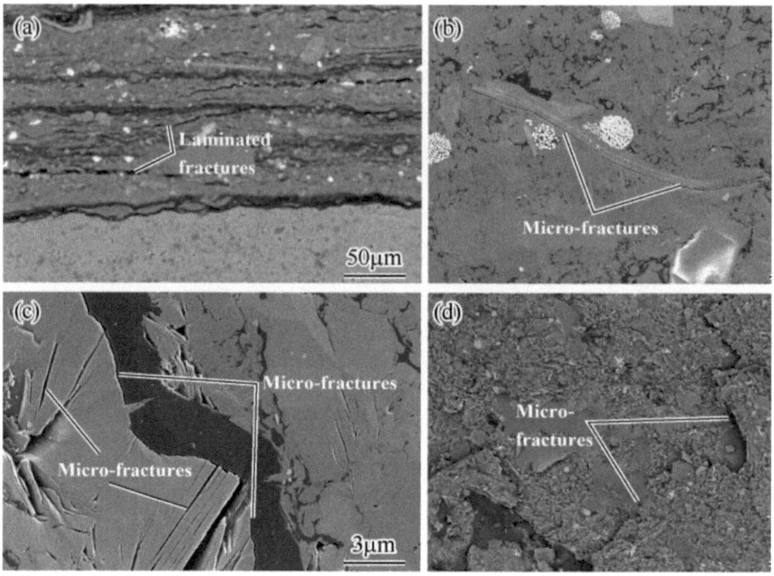

Fig. 3.34 Development characteristics of micro-fractures in shale

(2) Structural Fractures

Structural micro-fractures are usually formed when rocks undergo tectonic activities and the stress in a certain direction exceeds the maximum strength that the rock can withstand. Their direction, distribution, and formation are all related to local tectonic activities. The length of structural micro-fractures is not limited by the size of mineral particles and usually passes through several mineral particles. Structural fractures are easily filled by secondary minerals such as calcite in the later diagenetic process and have no contribution to the storage of shale gas. However, calcite has high brittleness and will have a strong impact on the expansion of induced fractures during production and development, playing a positive role in the exploitation of shale gas (Gou et al., 2019). The intensity of tectonic activity is the key factor for the formation of such fractures. In areas with strong tectonic stress, such as the axis of anticlines, the axis of synclines, and the plunging end of strata, the stratum stress is large and concentrated, and the structural micro-fractures are relatively developed (Xu et al., 2020a, 2020b).

3.5.2 Factors Affecting the Development of Pores in Shale

3.5.2.1 Pore-Forming Ability

(1) Type of Organic Matter

Some scholars used high-resolution scanning electron microscopy to observe the characteristics of typical shale gas reservoirs and found that different organic matter particles undergoing the same degree of thermal evolution and burial transformation history may have completely different pore development situations. In the same field of view, some organic matter develops abundant organic pores, while others do not. In addition, the type of organic matter will affect the production of organic acids, which in turn affects the reaction intensity between acidic fluids and unstable mineral components in the shale reservoir, controlling the quality of the reservoir.

(2) Degree of Thermal Evolution of Organic Matter

Currently, there is still considerable debate about at which stage organic pores first begin to appear. Curtis et al. (2012) observed samples of Woodford shale at different thermal maturation levels using scanning electron microscopy and found that organic matter pores were not developed in shale samples at low thermal maturity stages ($R_o < 0.9\%$). In contrast, at higher maturity stages ($1.0\% < R_o < 1.67\%$), organic matter pores appeared in large quantities, and the porosity of the shale samples reached its maximum at R_o of 3.6%. Subsequently, as the thermal maturity increased, the porosity gradually decreased. Some studies also suggest that the appearance of organic pores corresponds to a R_o of 0.5–0.6%, ~0.7%, or ~0.8%. Additionally, observations based on naturally matured shale samples show that within a maturity range of R_o from 0.65 to 1.15%, the shale pore volume may intermittently decrease and increase with the increase in maturity due to geological processes such as bitumen injection and crude oil cracking. As maturity further increases, the volume of large pores significantly decreases, while the volume of micropores and total pore volume tends to increase (Fig. 3.35). By statistically analyzing the relationship between the thermal maturity of organic matter and porosity in typical shale gas reservoirs both domestically and internationally, it has been found that when the R_o value is greater than 3.5%, residual organic matter generally begins to carbonize, and the internal pores of the organic matter gradually collapse and decrease, with the volume of mesopores and macropores rapidly decreasing. At the same time, interlayer pores of plastic minerals such as clay, and storage spaces such as diagenetic shrinkage fractures, also gradually become compacted or collapse and disappear, leading to a significant decrease in shale porosity (Cheng et al., 2019).

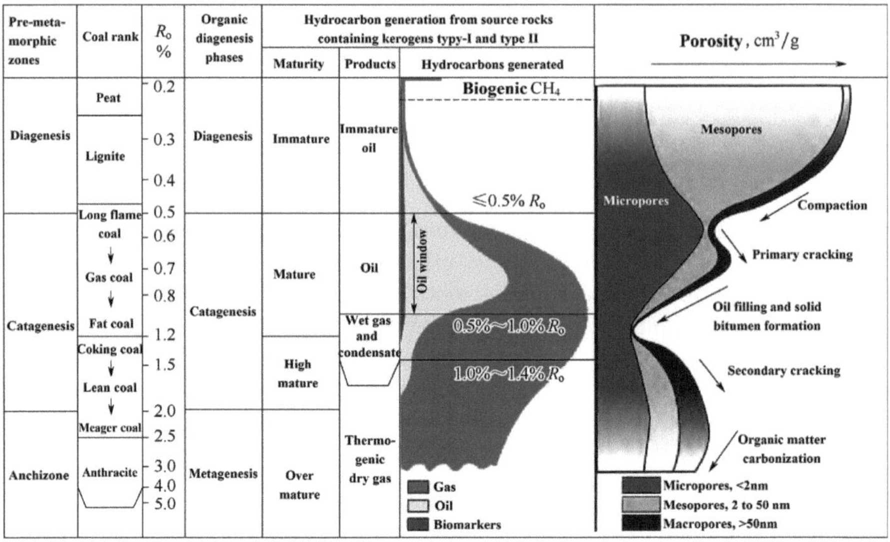

Fig. 3.35 Influences of organic matter maturation on porosity of shale (according to Mastalerz, 2013)

3.5.2.2 Support Capacity

(1) Brittle Mineral Content

Organic-rich shales are primarily composed of minerals such as quartz, feldspar, carbonates, clay, and pyrite. Influenced by mineral composition and structure, these mineral particles exhibit different chemical stability and physical compressive strength. Clay minerals are mostly plastic and chemically unstable. They are easily affected by mechanical compaction and deformation, which in turn influence the quantity and morphology of pores. An increase in the degree of thermal maturation can cause montmorillonite to gradually transform into mixed-layer minerals and illite, thereby releasing silica from clay minerals to form new authigenic minerals. This process subsequently controls the pore structure and rock properties of shales (Sun et al., 2023). Rigid minerals can be mainly divided into two categories. One category includes chemically stable minerals such as quartz and pyrite, which are generally resistant to dissolution and have strong resistance to compaction. Clay minerals and organic matter are ductile particles, they often distribute around the quartz and pyrite. The other category consists of chemically unstable minerals like feldspar and carbonates, which also have strong resistance to compaction but are susceptible to dissolution, often forming dissolution-related pores or fractures around the mineral margins or internally. Shales of different lithologies develop different types and quantities of pore spaces within the reservoir. Quartz, which is widely developed in silica-rich shales, is a chemically stable rigid particle that can resist compaction and effectively protect the pores within the reservoir (Xu et al., 2020a, 2020b).

(2) Support Mechanism of Rigid Particles

Rigid minerals are important supporting particles for protecting the micro-nano pores in organic-rich shale reservoirs. The filling, cementation, and arrangement mechanisms of particles interact to maintain the pore structure and supporting performance of shale reservoirs, which is of great significance for the enrichment of shale oil and gas. When different types of rigid particles are in close and uniform contact, they can form an effective rigid skeleton, forming a support network inside the shale to prevent pore collapse and contribute to the formation of a connected pore network. This connectivity allows hydrocarbon fluids to flow more easily in the shale reservoir and fill these residual pore spaces, thereby improving the enrichment degree of shale oil and gas. On the contrary, if the rigid particles are arranged loosely or irregularly, the incompletely contacted particles may lead to the filling of the pores by ductile minerals such as clay, resulting in the discontinuity of the pore structure, limiting the migration of hydrocarbon fluids in the shale reservoir, and reducing the reservoir performance and oil and gas productivity of shale.

3.5.2.3 Preservation Ability

(1) Protective Effect of Pressure

During the geological history evolution, the formation and evolution of overpressure in a closed fluid system affect the effective stress carried by the reservoir pores and the internal pressure of the pores, thereby controlling the morphology, size, and connectivity of the pores in the shale reservoir. By comparing the porosity characteristics of the high-quality shale sections of the Wufeng Formation-Longmaxi Formation in the southern region of China at different burial depths, it is found that within the same burial depth section, shale reservoirs with high porosity usually have overpressure characteristics, while the porosity of normal-pressure gas-bearing shale reservoirs significantly decrease with the continuous increase of burial depth. Scanning electron microscopy image observation and quantitative statistical results show that under better preservation conditions, there is no large-scale escape of shale gas during the later tectonic uplift process, and the fluid pressure in the pores of the shale gas layer is high, which can effectively prevent or slow down the influence of the later compaction on the plastic pores (mainly referring to organic pores) in the shale, and many pores are preserved. Therefore, the shale layer system with overpressure has a higher organic matter facial porosity, larger pore diameter, and larger pore roundness (Gou et al., 2021). On the contrary, when the preservation conditions of the shale system are poor, the shale gas escapes, resulting in a decrease in the fluid pressure in the shale pores, and the loss of the preservation of pores by overpressure and the strong compaction effect make the size, quantity, and roundness of the organic pores in shale smaller than those in the overpressure area (Fig. 3.36).

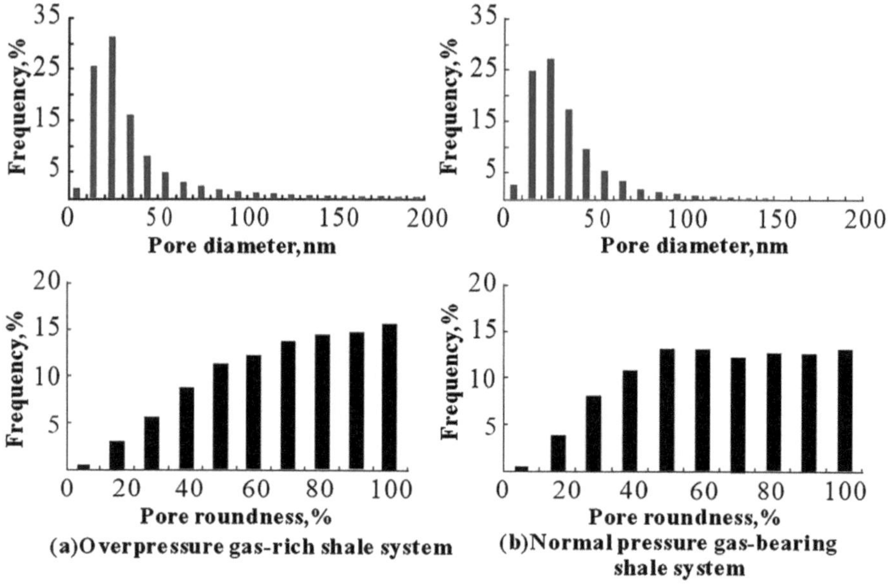

Fig. 3.36 Differences in shale pore structure under different pressure system backgrounds

(2) Destructive Effect of Stress

Shale in the southern China is characterized by old age, deep burial, and has generally experienced multiple stages of tectonic movements after deposition. Under the influence of strong tectonic stress, the organic matter and mineral structure of shale are transformed, which in turn affects the pore and fracture structure characteristics. In recent years, many scholars have conducted many qualitative observations and quantitative descriptions of the deformation characteristics of microscopic pores under different tectonic stress backgrounds. The research results show that the enhancement of effective stress is conducive to the formation of structural fractures mainly dominated by high-angle shear and tensile-shear, leading to the escape of methane gas and the periodic or continuous decrease of pore pressure. The solid bitumen pores where methane gas has been completely or partially leaked will reduce the pore size to increase the pore pressure to prevent being completely flattened and collapsed. Therefore, in areas with relatively strong stress, the organic pores in shale deform and close under the extrusion of stress, and the development degree of organic pores is low. In areas with relatively weak tectonic stress, the development degree of organic pores in shale is high, and the pore morphology is intact (Gou et al., 2022).

3.6 Types and Sealing Mechanisms of Caprocks

The rock layer or geological body that covers the reservoir and can seal the oil and gas to prevent them from escaping upward is called a caprock. Contrary to the reservoir, the caprock has relatively low porosity and permeability, and is a geological body that hinders the escape of oil and gas, affecting the accumulation efficiency and preservation time of oil and gas in the reservoir. Under the condition of sufficient oil and gas sources, the distribution and sealing performance of the caprock control the migration, accumulation, and preservation of oil and gas, and a good caprock can block the seepage and migration of oil and gas, reduce the diffusion and loss of natural gas, and enable it to accumulate and form a reservoir under the caprock, which is a necessary condition for the formation of oil and gas reservoirs.

3.6.1 Types of Caprocks

Caprocks can be classified into different types according to their lithology, distribution range, and positional relationship with the oil and gas reservoir.

3.6.1.1 Classification by Lithology of Caprocks

Caprocks can be divided into argillaceous rock, evaporite, carbonate rock, and other types of caprocks based on their lithology. A statistical analysis of the lithology of caprocks in 34 large and medium-sized gas fields in China shows that about 60% of the large and medium-sized gas fields with mudstone as the caprock, about 18% with gypsum-salt rock, and fewer with shale and limestone, accounting for 11% of the total, respectively (Hu et al., 2009).

(1) Argillaceous Rock Caprocks: Including mudstone, shale, sandy mudstone, calcareous mudstone, etc., with fine grain size, compactness, low permeability, plasticity, adsorption, and expansibility, they are good caprock lithologies. This type of caprock is the most common type in oil and gas fields, with the widest distribution, the largest quantity, and is almost produced in various sedimentary environments. Most of the caprocks in oil and gas fields belong to this type.
(2) Evaporite Caprocks: Mainly including salt rock and gypsum rock, they are the best type of caprocks. Gypsum rock has good sealing ability in a wide depth range, such as the Belle River Mills gas field in the Michigan Basin in the United States (buried depth of 762 m) and the Chatom oil field in Alabama (buried depth of 4900 m), which both have gypsum rock caprocks. The Kuqa foreland basin in China and the Dongpu sag in the Bohai Bay Basin also widely develop gypsum-salt rock caprocks.
(3) Carbonate Caprocks: Mainly including argillaceous limestone, muddy limestone, and dense limestone, such as the Ordos Basin, where carbonate caprocks

are developed. Since carbonate rocks are easily leached by water, structurally damaged, or dissolved to form fractures, which will affect the sealing performance of the caprock, whether carbonate rocks can be used as an effective caprock is closely related to the development degree of their fractures.

(4) Other Types of Caprocks: Rocks such as bauxite, igneous rocks, and coal seams can also serve as caprocks. For example, the Ordos Basin and the Jizhong sag in the Bohai Bay Basin develop bauxite caprocks, the Rongxingtun structure in the eastern sag of the Liaohe Basin develops basalt caprocks, and the Xujiaweizi fault depression in the Songliao Basin develops volcanic rock caprocks. In addition, non-rock substances such as methane hydrate and heavy asphalt formed by the destruction of oil reservoirs can also be caprocks under certain conditions, such as the hydrate gas reservoir in the Blake Ridge in the southeastern waters of the United States (Dillow et al., 1980).

3.6.1.2 Classification by Distribution Range of Caprocks

(1) Regional Caprocks: Refers to caprocks that are widespread in most areas of the petroliferous basin or depression, with large thickness, wide area, and relatively stable distribution. Regional caprocks play an important role in the oil and gas accumulation in the basin or depression, controlling the vertical distribution of oil and gas, the distribution of hydrocarbon facies, and the hydrocarbon abundance (Dai et al., 2012). For example, the natural gas in the Bohai Bay Basin is mainly distributed under the regional caprocks of the First Member of the Shahejie Formation and the Third Member of the Shahejie Formation (Fig. 3.37).

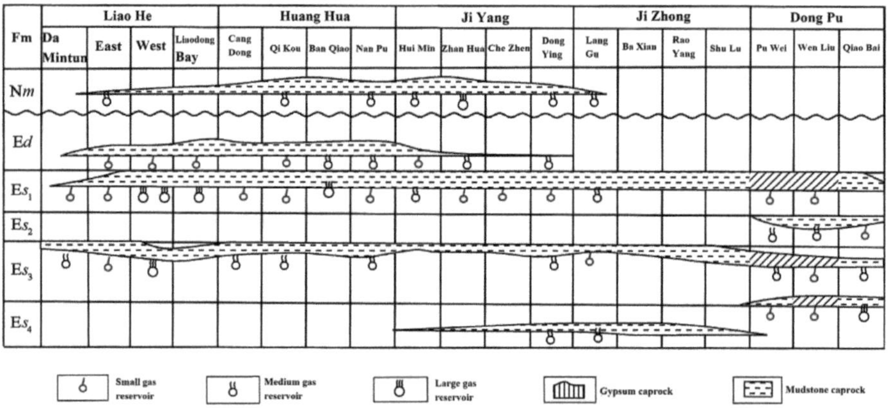

Fig. 3.37 Relationship between natural gas distribution and regional caprocks in the Bohai Bay Basin (according to Wang, 1997)

(2) Local Caprocks: Directly located above the trap reservoir, with less stable lateral distribution and a relatively smaller distribution area than regional caprocks. Local caprocks control the scale and abundance of oil and gas reservoirs.

3.6.1.3 Classification by Positional Relationship with the Oil and Gas Reservoir

Direct Caprocks: The sealed rock layer immediately adjacent to the reservoir. Direct caprocks can be local caprocks or regional caprocks.

Overlying Caprocks: All non-permeable rock layers above the direct caprocks. Generally refers to regional caprocks, which play an important role in the regional accumulation and preservation of oil and gas.

3.6.2 Mechanism of Caprocks in Sealing Oil and Gas

Previously, based on the way caprocks prevent the migration of oil and gas, the sealing mechanism of caprocks was divided into physical sealing, abnormal pressure sealing, and hydrocarbon concentration sealing (Hao et al., 1996). With the in-depth understanding of the sealing mechanism of caprocks, it is believed that physical sealing and overpressure sealing are possible sealing mechanisms, and physical sealing is the most basic and important. Overpressure sealing only plays a role in enhancing the sealing performance when there is overpressure in the sealing layer. For hydrocarbon concentration sealing, it may not exist. This is mainly because diffusion occurs through molecular thermal motion, and the thermal motion of natural gas molecules is constantly occurring. Therefore, it is very difficult to seal the diffusion of natural gas molecules. The so-called hydrocarbon concentration sealed gas reservoirs previously proposed may be that the caprocks of the source rock system or the dense caprocks above them mainly rely on the physical sealing mechanism to seal the underlying gas reservoirs. Therefore, here we mainly introduce the mechanisms of physical sealing and overpressure sealing.

3.6.2.1 Physical Sealing

Physical sealing refers to the sealing of oil and gas by the capillary pressure of the caprock rock, also known as capillary force sealing (Bern, 1975; Howalter, 1979), or membrane sealing (Jon, 2003; Watts, 1988). The caprock rock has a very small pore throat radius (Fig. 3.38), which generates a large capillary pressure directed towards the reservoir. This is the main reason for the capillary pressure sealing.

In fact, since the rocks in the stratum are generally hydrophilic, for the oil and gas covered by the caprock, the forces it receives under static pressure mainly include the buoyant pressure generated by the hydrocarbon column in the upward direction,

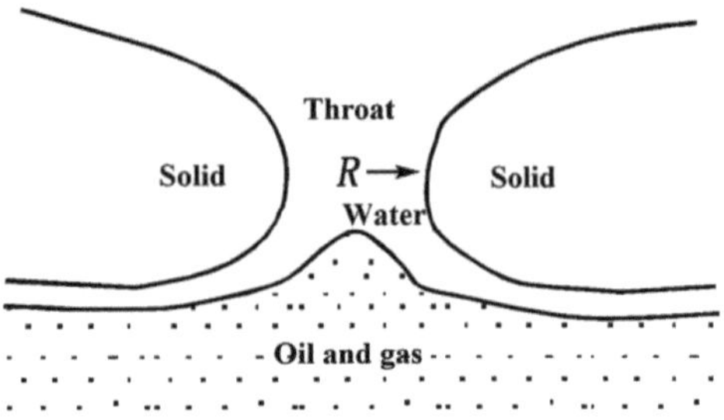

Fig. 3.38 Distribution of fluid in pore throats

the capillary pressure generated by the pore throat of the caprock in the downward direction, and the capillary pressure generated by the pore throat of the reservoir in the upward direction. Whether the caprock can seal the oil and gas in the underlying reservoir by physical sealing is related to the magnitude of the forces in different directions acting on the oil and gas.

(1) Buoyant Pressure

As a rock of the caprock, the pore throat is small, and the rock is like a fine-pore sieve. The surface of the pore is wetted by water (membrane sealing) (Fig. 3.39), and the buoyant force of the underlying hydrocarbon column exerts pressure on the water film, attempting to pass through the rock. When the buoyant pressure cannot overcome the water film, the oil and gas are blocked. The buoyant pressure (Pb) of the hydrocarbon column is expressed as follows:

$$P_b = (\rho_w - \rho_p)gh \tag{3.12}$$

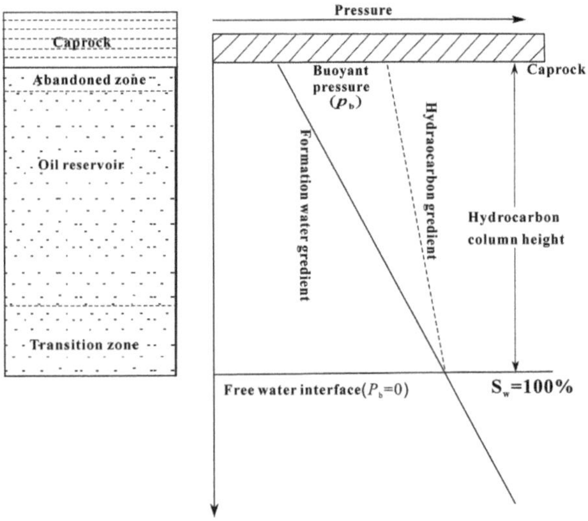

Fig. 3.39 Relationship between Buoyant pressure and depth in pore throats (modified according to Jon, 2003)

where: Pb—Buoyant pressure of the hydrocarbon column (pressure exerted by the hydrocarbon column on the overlying caprock), psi;

ρ_w—Density of water, g/cm^3;

ρ_p—Density of hydrocarbons, g/cm^3;

g—gravitational acceleration, cm/s^2;

h—Height of the hydrocarbon column, m.

The buoyant pressure of the oil (gas) column at any depth point is the difference between the water pressure and the oil (gas) pressure. The magnitude of the buoyant pressure is negatively correlated with the density difference between the hydrocarbon and water, and positively correlated with the height of the hydrocarbon column. Under the same height of the hydrocarbon column, the buoyant pressure generated by natural gas in the reservoir is greater than that generated by oil. For a specific hydrocarbon column, the maximum buoyant pressure is at the top of the oil (gas) column, and the buoyant pressure is 0 at the free water interface (Fig. 3.40).

(2) Capillary Pressure and Membrane Sealing

When the rock is saturated with water, the capillary pressure (P_c) generated is a function of the hydrocarbon-water surface tension (σ), the wetting angle (θ), and the maximum pore throat radius (R):

$$P_c = \frac{2\sigma \cos\theta}{R} \tag{3.13}$$

where: Pc—Capillary pressure, MPa;

R—Pore throat radius of the rock, cm;

θ—Contact angle between the solid and liquid phases, degrees;

σ—Interfacial tension between the two phases, 10^{-3} N/cm.

The capillary pressure (sealing ability) is related to the pore throat radius, wettability, and interfacial tension of the rock. Generally, it increases with the decrease of the pore throat radius of the rock, increases with the enhancement of the water wettability, and increases with the increase of the hydrocarbon-water interfacial tension. It should be noted here that although the wettability of the rock is mostly water-wetting in most cases, under specific rock compositions or the presence of hydrocarbons and other substances, the wettability may change from water-wetting to oil-wetting. The hydrocarbon-water interfacial tension is closely related to the nature of the hydrocarbons and the temperature and pressure conditions of the medium. Under the same temperature and pressure conditions, the gas–water interfacial tension is greater than the oil–water interfacial tension, and both the gas–water interfacial tension and the oil–water interfacial tension show a decreasing trend with the increase of temperature and pressure.

Compared with the reservoir, the pore throat radius of the caprock is usually several orders of magnitude smaller. Therefore, under similar conditions, the downward capillary pressure generated by the caprock is much greater than the upward capillary pressure generated by the reservoir. It is generally believed that the capillary pressure generated by the pore throat of the reservoir can be ignored.

At a specific moment, for oil and gas to migrate through the caprock, they must first displace the water therein and overcome the resistance of the capillary pressure, that is, the buoyant pressure of the oil and gas must reach the minimum pressure (capillary pressure) to enter the capillary. If the buoyant pressure of the oil and gas is less than the resistance of the capillary pressure, the oil and gas will be blocked under the caprock.

In evaluating the sealing ability of the capillary pressure, the concept of displacement pressure is often cited, that is, the capillary pressure of the largest connected pore in the caprock rock. The greater the displacement pressure, the stronger the sealing ability of the caprock. Sediments of different grain sizes often have different displacement pressures, and the larger the grain size, the smaller the displacement pressure. Due to the differences in the physical properties of the reservoir and the sealing fluid, the caprock is considered relative. If there is a physical property difference or displacement pressure difference between the overlying rock layer and the reservoir, a caprock can be formed.

(3) Theoretical Maximum Height of the Hydrocarbon Column

When the buoyant pressure is equal to the displacement pressure (Pb = Pd), the height of the hydrocarbon column accumulated under the caprock reaches the maximum. If the buoyant pressure (Pb) exceeds the displacement pressure (Pd), the hydrocarbons will leak through the caprock. If the buoyant pressure (Pb) is less than the displacement pressure (Pd), as an effective caprock, it may also seal a higher hydrocarbon column. The theoretical height (h) of the maximum hydrocarbon column that the caprock can seal is:

$$\mathrm{h} = \frac{P_d}{P_b} = \frac{\left(\frac{2\sigma\cos\theta}{R}\right)}{(\rho_\mathrm{w} - \rho_\mathrm{p})\mathrm{g}} \tag{3.14}$$

It should be noted that if the hydrocarbons in the reservoir have characteristics such as overpressure, then the overpressure, as one of the driving forces for the migration of oil and gas, can play the same role as the buoyancy. Therefore, in this case, whether the caprock can be sealed depends on the relationship between the displacement pressure of the caprock and the sum of the overpressure and the buoyancy.

According to statistics, the caprocks of large and medium-sized gas fields in China all have relatively high displacement pressures, with a minimum of 8.7 MPa, such as the Wangjiatun, Changde, Shengping, and Xushen 1 Well gas reservoirs in the Songliao Basin. The maximum can reach 28 MPa, such as the Kela 2 Gas Reservoir and Dina Gas Reservoir in the Kuqa Depression of the Tarim Basin. However, in general, with the increase of the displacement pressure of the caprock, the reserve abundance of large and medium-sized gas fields in China gradually increases, indicating that the displacement pressure is an important factor affecting the sealing conditions of large and medium-sized gas reservoirs in China (Lü et al., 2005).

The capillary pressure sealing mechanism is the most common mechanism for caprocks to seal oil and gas. In general, it can only prevent the further migration of free-phase oil and gas, and it is difficult to seal the migration of oil and gas in the water-soluble phase and by diffusion. However, when the permeability is very small and the displacement pressure is so high that the sealing of the caprock can only be broken by structural movements that cause fractures, it can not only prevent the free phase but also the migration of the water-soluble phase of oil and gas.

3.6.2.2 Overpressure Sealing

Abnormal high fluid pressure (overpressure) refers to the pore fluid pressure in the formation being higher than the corresponding hydrostatic pressure. This mechanism of sealing oil and gas by the abnormal high pore pressure of the caprock is called fluid pressure sealing, or simply overpressure sealing. There are many factors that cause overpressure in mudstone, such as undercompaction of mudstone, hydrocarbon generation from organic matter, dehydration of clay minerals, and thermal pressurization of pore fluids, etc.

During the compaction process, at the top and bottom of the mudstone section near the reservoir, the pore water can be fully discharged, forming the upper and lower compaction sections. In the middle, due to the smaller pore throats at the top and bottom, the discharge of pore fluids is hindered, forming an undercompacted zone. The capillary pressure of the upper and lower compaction sections is greater than that of the middle undercompacted section, playing a role in physical sealing. There is an abnormally high pore fluid pressure inside the middle undercompacted section, so that the sum of the capillary pressure and the pore fluid pressure is significantly greater than the capillary pressure of the upper and lower compaction sections, and its sealing ability is stronger (Fig. 3.40) (Hao, 1995). Since the overpressure layer itself is often a dense rock layer and usually has the function of physical sealing, the overpressure layer often has both physical sealing and overpressure sealing functions simultaneously.

The overpressured cap rock is essentially a high potential fluid layer that can prevent the volumetric flow of any fluid, including oil, gas, and water. Therefore, it can not only prevent the movement of free-phase oil and gas but also the flow of water containing dissolved oil and gas. From this perspective, the overpressured cap rock is a relatively effective seal. According to Liu (1992), the height of the gas column sealed by the abnormal pore fluid pressure in undercompacted mudstone with a pressure coefficient of 1.3 is 11 times greater than that sealed by capillary resistance. For example, the mudstone cap rock in the 34/1-35 well in the northern North Sea is overpressured and controlled by the properties and overpressure sealing, and the lower reservoir is rich in oil and gas (Fig. 3.41) (Nordgard Bolas et al., 2005).

It should be noted that overpressure sealing is a dynamic sealing. The sealing ability of the overpressure caprock depends on the sum of the overpressure of the caprock and the capillary pressure. When the capillary pressure is constant, the higher the overpressure the greater the sealing ability. Once the overpressure caprock returns

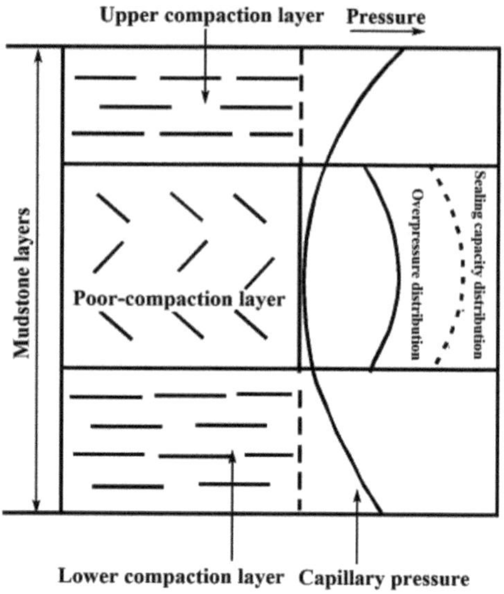

Fig. 3.40 Schematic diagram of sealing capacity of undercompacted mudstone (Hao, 1995)

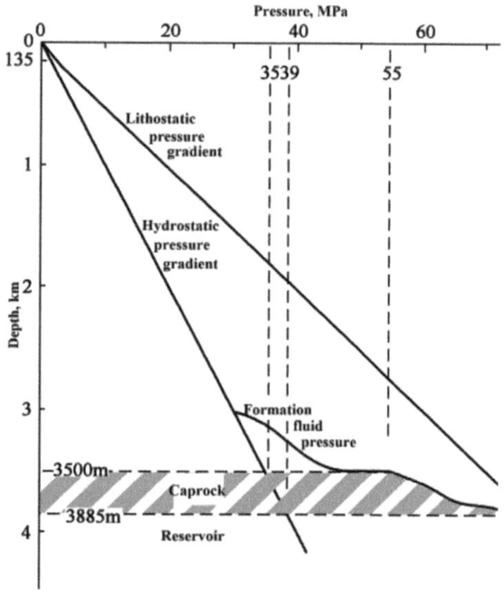

Fig. 3.41 Schematic diagram of overpressure in mudstone caprocks in well 34/1-35 in the Northern North Sea (Nordgard Bolas et al., 2005)

to the normal hydrostatic pressure state for some reason, the sealing of the caprock is entirely due to the capillary pressure sealing.

3.6.3 *Factors Affecting the Effectiveness of Caprocks*

The sealing mechanism of the caprock reflects the basic mechanism of sealing oil and gas, but it cannot fully reflect the characteristics of the caprock as a closed formation. Whether a set of rock layers can serve as an effective caprock is affected by many factors, such as the lithology, toughness, thickness, sedimentary environment and continuity, diagenesis, and tectonic activities of the caprock.

3.6.3.1 Lithology of the Caprock

In theory, any lithology can become a caprock if its displacement pressure is greater than the buoyancy of the hydrocarbon column in the underlying oil and gas reservoir or the upward migration force. In fact, the vast majority of effective caprocks are evaporite, fine-grained clastic rocks, and organic-rich rocks. Shale and mudstone caprocks often coexist with clastic reservoir rocks. Evaporite caprocks are mostly developed in carbonate profiles. In areas with weak tectonic movements and undeveloped fractures, dense marl and limestone can also act as caprocks. This is mainly because different lithologies of the caprock have different sealing abilities. Generally, the order of sealing ability is roughly as follows: salt rock, gypsum rock, various salt-containing mixed rocks, shale, silty shale, calcareous shale, marl, carbonate rock, dense sandstone, etc. Evaporite, fine-grained clastic rocks, and organic-rich rocks have small pore throat radii, large displacement pressures, and strong sealing abilities.

Grunau (1981) statistically analyzed the caprocks of 176 large gas fields in the world, and almost all gas fields rely on their shale or evaporite caprocks. Klemme (1977) statistically analyzed the caprocks of 334 large oil and gas fields in the world, and the large oil and gas fields with shale and mudstone caprocks accounted for 65% of the total, the salt rock and gypsum rock caprocks accounted for 33%, and the dense limestone caprocks accounted for 2%. In the Songliao Basin, Bohai Bay Basin, and other basins in China, clay rocks are the most common caprocks. In the Sichuan Basin, Jianghan Basin, and other basins, evaporite rocks are the most common caprocks.

3.6.3.2 Toughness of the Caprock

Brittle rocks prone to form fractures and ductile rocks usually undergoes plastic deformation. The toughness of the caprock is particularly important for the preservation of oil and gas, especially in environments such as fold belts and thrust fault zones. The toughness of the rock is affected by many factors such as lithology, confining pressure, pore pressure, fluid composition, and temperature (Table 3.13). Different rocks have different toughnesses. Under normal geological conditions, the order of common rock toughness is salt rock > gypsum rock > organic-rich shale > mudstone, limestone.

Mudstone becomes more ductile with increasing depth within a certain depth range (generally around 3000 m). Beyond this depth range, the toughness of mudstone gradually decreases with further increase in depth. The decrease in the toughness of mudstone can easily lead to the formation of microfractures, which will increase the permeability and thus reduce the sealing ability. From this perspective, there should be a favorable sealing depth interval for mudstone caprocks.

The rock mechanical characteristics of gypsum-salt rocks are significantly different from those of clastic rocks, especially salt rocks. The reason is that gypsum-salt rocks themselves have significantly different brittleness and plasticity at different temperatures and confining pressures. Generally, with the increase of burial depth or temperature, the plasticity increases, and it is not easy to break under stress flow, and even the existing faults and fractures can disappear, close, or weld in the salt layer.

The results of triaxial stress–strain physical experiments at different confining pressures and temperatures show that the stress–strain processes of gypsum rock and salt rock can be divided into four stages (Fig. 3.42 left), namely, compaction stage, elastic stage, plastic stage, and failure stage. Due to the small porosity of gypsum rock and salt rock, the compaction stage is very short, that is, the salt rock reaches its compaction strength under a very small stress state. In the elastic deformation stage, the stress and strain have a linear relationship, and the deformation can recover to its original state when the stress is canceled. When the stress exceeds the yield strength, the deformation of gypsum rock and salt rock enters the plastic stage, the rock strain cannot recover when the stress is canceled, and the relationship between stress and strain is nonlinear. When the stress exceeds the compressive strength of the rock,

Table 3.13 Control factors of the toughness of caprocks (Edward, 2001)

Control factor	Control effect on plasticity
Lithology	Brittle caprocks include dolomite, quartzite, anhydrite, and some shales, while ductile caprocks include salt rock, some shales, and some limestones
Composition	Changes in composition such as total organic carbon and clay minerals can change plasticity
Confining pressure	Increasing the confining pressure increases the plasticity
Pore pressure	Increasing the pore pressure reduces the plasticity
Fluid composition	The presence and composition of the fluid affect the plasticity
Temperature	Increasing the temperature increases the plasticity
Strain rate	A high strain rate reduces the plasticity
Time	When the caprock undergoes burial and diagenesis, the plasticity changes over time
Compaction state	As the compaction and diagenesis increase, the plasticity decreases

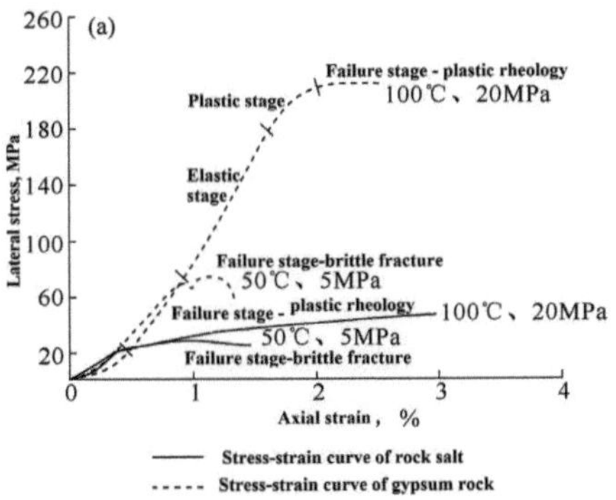

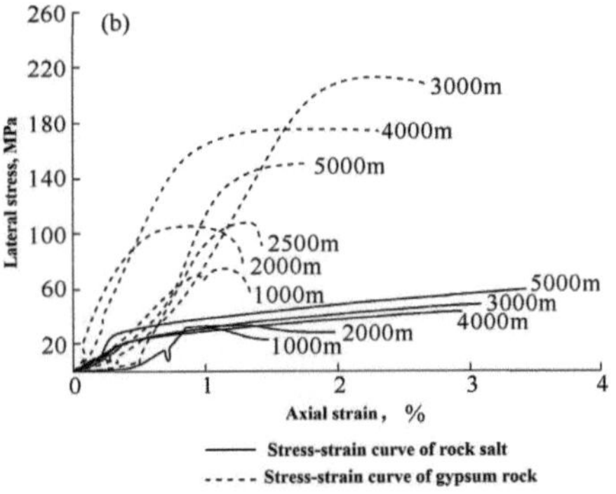

Fig. 3.42 Stress–strain stages of gypsum rock and salt rock under triaxial heating and pressurization (left) and stress–strain curves of gypsum-salt rock in Kuqa Formation of Kuqa foreland thrust belt (right) (according to Zhuo et al., 2014)

gypsum rock and salt rock enter the failure stage, where, when the rock is brittle, the failure stage is manifested as brittle fracture, the axial stress of the rock suddenly drops, and the strain is discontinuous. When the rock is plastic, the failure stage is manifested as plastic rheology, the axial stress of the rock does not decrease, and the strain is continuous. Taking the gypsum-salt rock of the Kuqa Formation in the Kuqa foreland thrust belt as an example, the triaxial stress–strain curves of gypsum rock and salt rock at different burial depths (Fig. 3.42 right) show that both gypsum rock and salt rock have the characteristics of brittle transformation at low temperatures and plastic transformation at high temperatures, but the yield strength and compressive strength of gypsum rock are much greater than those of salt rock, that is, salt rock has strong plasticity, and gypsum rock has high hardness. The stress required for salt rock to undergo plastic deformation is small, and it exhibits plastic flow at a certain temperature.

3.6.3.3 Thickness of the Caprock

There is no simple corresponding relationship between the thickness of the caprock and its sealing ability or hydrocarbon column height. The caprock can be very thin, and a caprock less than 1 m thick can seal a single oil reservoir. In theory, the sealing ability mainly depends on the displacement pressure of the caprock rock. For example, Hubbert's (1953) research suggests that a set of clay rock caprocks several inches thick with a particle size of 10–4 mm has a displacement pressure of approximately 600 psi (1 psi = 6.698 kPa), which is theoretically sufficient to seal an oil column 915 m high. However, from the perspective of sedimentary conditions, when the caprock is thin, the possibility of being continuous, complete, uniform, and free of fractures in the lateral direction is almost zero, and the possibility of forming a large oil and gas field is small. Therefore, from the perspective of preserving oil and gas, there should be an effective lower limit for the caprock that is affected by other geological conditions, and the thicker the better. The reasons are as follows:

Thin caprocks have poor lateral continuity and are prone to rupture. Thick caprocks have good lateral continuity, are less likely to form connected microfractures, thus ensuring the overall continuity throughout the oil-bearing area.

Thick caprocks slow down the filtration and diffusion rate of oil and gas through the caprock, thereby hindering the upward escape of oil and gas.

For fault traps, thick caprocks are less likely to be faulted by small faults. The contact opportunity between mudstones on both sides of the fault is high, and it is easy to form lateral sealing. Moreover, the thickness of the caprock directly affects the height of the trap oil and gas reservoir.

The fluid in the thick caprock is not easily discharged, and it is easy to form an undercompacted layer, thereby forming overpressure in the formation and enhancing the sealing ability of the caprock for oil and gas.

In fact, the thickness of the caprock can generally range from tens of meters to hundreds of meters. The large oil (gas) reservoirs discovered at home and abroad are often associated with relatively thick caprocks, and there is a certain correlation between the thickness of the caprock and the sealing ability (hydrocarbon column height). Many scholars at home and abroad have conducted research and discussions. Ziegler (1992) and others believe that there is no correlation between the thickness of the caprock and the height of the hydrocarbon column, and some scholars believe that there is a good positive linear correlation between the thickness of the caprock and the height of the hydrocarbon column it seals (Lü et al., 1996; Tong et al., 1989). The Soviet scholar Inozemtsev proposed an effective lower limit standard for the thickness of the caprock of 25 m.

Jiang (1998) conducted a study on the relationship between the hydrocarbon column height and the thickness of the overlying direct caprock in 7 pure gas reservoirs in the Neogene gas (oil) fields in the Gudao, Gudong, Yanjia, Feiyantan, etc. in the Jiyang Depression. It is believed that, in general, for an oil (gas) reservoir with a certain hydrocarbon column height, it requires a certain thickness of the overlying caprock to ensure the accumulation and preservation of oil and gas underneath. Due to the great differences in underground geological conditions, the thickness of this caprock can be thick or thin, but it is required to reach a minimum value, which can be called the minimum thickness or critical thickness of the caprock. The height of the hydrocarbon column sealed by the critical thickness of the caprock can be understood as the maximum hydrocarbon column height (critical hydrocarbon column height) that the caprock can seal. There is a positive correlation between the gas column height of the gas reservoir and the minimum thickness (or critical thickness) of the mudstone caprock required (Fig. 3.43). For example, for a gas reservoir with a gas column height of 5 m, the minimum thickness of the caprock is 3 m, and for a gas reservoir with a gas column height of 10 m, the minimum thickness of the caprock is 6 m. With the increase of the gas column height of the gas reservoir, the minimum thickness of the required mudstone caprock also gradually increases. However, the statement that "the thickness of the caprock is directly proportional to the height of the sealed hydrocarbon column" is inappropriate.

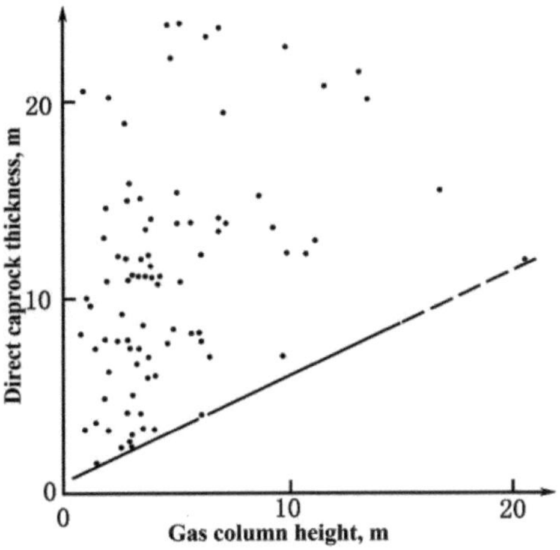

Fig. 3.43 Relationship between the thickness of the caprock and the height of the sealed gas column (taking the shallow gas reservoirs of the Paleogene in the Jiyang Depression as an example) (Jiang, 1998)

3.6.3.4 Sedimentary Environment and Continuity

The influence of the sedimentary environment on the sealing ability of argillaceous rock caprocks is mainly in the fact that the material composition, structure, thickness, lateral distribution continuity, and stability of the caprocks are controlled by the sedimentary environment. Sedimentary environments with deep water bodies, large areas, and relative stability are prone to forming regional argillaceous rock caprocks with pure quality, fine grain size, and good continuity. Argillaceous rocks formed in sedimentary environments with turbulent water bodies have impure quality, high sand content, poor continuity and stability, and can only serve as local caprocks, and the quality of the caprocks is relatively poor.

In the continental sedimentary environment, there is a good correspondence between the sealing ability of argillaceous rocks and the sedimentary environment. Argillaceous rock caprocks in deep lake and semi-deep lake facies have a large distribution area and good sealing performance, and often form regional caprocks. Argillaceous rock caprocks formed in nearshore environments such as shore-shallow lake facies and delta facies have a relatively smaller distribution area and poor sealing performance, and generally form local caprocks. Near-source environments such as fluvial facies and alluvial fan facies generally can only form reservoir layers. In the marine sedimentary environment, argillaceous rock caprocks have a large distribution area and good sealing performance, and often form regional caprocks.

3.6.3.5 Influence of Diagenesis on Caprocks

For common mudstone caprocks, the influence of diagenesis on caprocks is mainly in the compaction of rocks, the transformation of clay minerals, the evolution degree

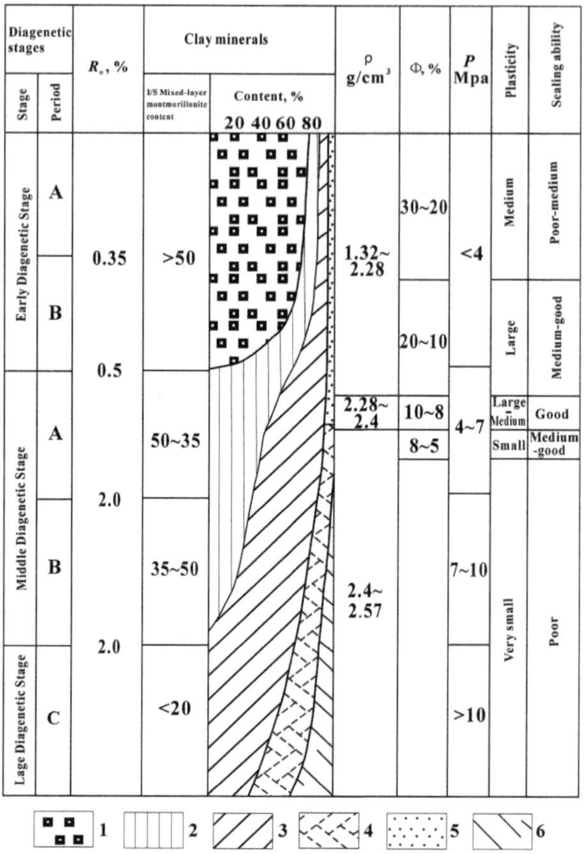

Fig. 3.44 Evolution model of the sealing ability of argillaceous rock caprocks (modified according to Chen, 2003). 1. Montmorillonite; 2. illite/smectite mixed layer; 3. illite; 4. chlorite; 5. kaolinite; 6. chlorite/smectite mixed layer

of organic matter, and the plasticity of rocks. For example, mudstone caprocks at different diagenetic stages have different sealing abilities (Fig. 3.44):

(1) Early Diagenetic Stage A: Buried relatively shallow, the diagenesis of mudstone is poor, and the sealing ability of the caprock is low.
(2) Early Diagenetic Stage B: With the increase of compaction, the particles are arranged more closely, the pore throat radius decreases, the porosity decreases, the displacement pressure increases, and the sealing performance continuously enhances. It has medium to good sealing ability.
(3) Middle Diagenetic Stage A: With the increase of burial depth and geothermal temperature, montmorillonite in the mudstone transforms into illite/smectite mixed layer and illite, and a large amount of interlayer water or crystalline water is precipitated. At the top and bottom of the thick mudstone adjacent to the sand layer, a dense zone is formed, and an overpressure zone is formed inside the thick mudstone. The physical sealing and abnormal pressure sealing work together, and the sealing ability of the argillaceous rock is the strongest.

(4) Middle Diagenetic Stage B—Late Diagenetic Stage: With the further enhancement of compaction, the clay minerals in the argillaceous rock are mainly illite and chlorite, the plasticity of the mudstone decreases, the brittleness increases, and fractures are easily generated under tectonic action, which weakens its sealing performance.

For gypsum-salt rock caprocks, as mentioned earlier, gypsum-salt rock has a dense rock structure and low porosity, and generally, with the increase of burial depth or temperature, the plasticity increases. Taking the Kuqa foreland thrust belt as an example, physical experiments have confirmed that gypsum-salt rocks buried shallower than 3000 m will produce fractures and cracks under strong and rapid compressive stress, or reactivate old fractures, and oil and gas will migrate and dissipate vertically along the salt-penetrating faults. When the burial depth is greater than 3000 m, with the increase of the burial depth of the caprock, the plasticity of the rock increases, and the salt-penetrating faults disappear or weld and close within the caprock section, or the top of the new fractures disappears in the plastic caprock, and oil and gas accumulate under the gypsum-salt rock caprock (Zhuo et al., 2014). Therefore, the sealing ability of gypsum-salt rock caprocks is dynamically evolving.

3.6.3.6 Destructive Effect of Tectonic Activities on the Sealing Ability of Caprocks

The destructive effect of tectonic activities on the sealing ability of caprocks is mainly in:

When the crust is uplifted and the caprock is eroded, with the increase of the degree of erosion, the thickness of the caprock becomes thinner, the sealing ability becomes worse or even disappears.

The faulting has a destructive effect on the sealing ability of the caprock. When the fault completely offsets the caprock, the caprock loses its lateral continuity, and the sealing ability of the caprock is completely lost, which is very unfavorable for the preservation of oil and gas. When the fault does not completely offset the caprock, the caprock does not lose its lateral continuity, but the effective thickness of the caprock for sealing oil and gas is reduced, which can reduce the sealing ability of the caprock. Then the development of the fault itself, including fault rocks, induced fractures, etc., will also affect the sealing ability. When the fault rock is a highly permeable rock or the high-angle fractures in the fault zone are well developed, it will lead to the loss of oil and gas along the fault, thereby indirectly causing the caprock to partially or completely lose its ability to seal oil and gas. For example, in areas with developed faults, the top sealing ability (hydrocarbon column height) determined by the displacement pressure of the caprock is generally greater than the actual height in the oil reservoir. This is because the height of the oil column in the oil reservoir may be controlled by the spill point related to the fault (Allard, 1993).

Magmatic or intrusive activities such as magma or rock bodies have a destructive effect on the sealing performance of the caprock. On the one hand, high temperatures enhance the activity of hydrocarbons and accelerate the diffusion speed, which is conducive to the loss and makes the effectiveness of the caprock worse. On the other hand, the intrusion produces a huge arching effect, which causes the caprock to rupture and reduces the sealing ability or loses the sealing ability.

Faults, magmatic or intrusive activities can cause the caprock to rupture. However, it should be noted that in addition to tectonic activities, if the fluid pressure in the caprock stratum exceeds the rock fracture pressure, the caprock will also generate fractures, which can also make the sealing ability of the caprock worse. This situation is relatively more developed in deep and ultra-deep overpressure formations.

In conclusion, caprocks play an important role in the accumulation and preservation of oil and gas, and the effectiveness of caprock sealing is affected by many factors. The lithology and fabric characteristics of the caprock are the basis for controlling the sealing ability of the caprock, and they are all determined by the sedimentary environment and sedimentary conditions. Diagenesis and tectonic activities are important reasons affecting the effectiveness of caprock sealing.

Questions for Discussion

1. What is a reservoir? What are the basic characteristics of a reservoir?
2. What is total porosity, effective porosity, absolute permeability, effective permeability (or phase permeability), and relative permeability? What is the relationship between porosity and permeability?
3. What is the pore structure of a reservoir? How does it affect the physical properties of the reservoir?
4. What is the principle of studying the pore structure using the capillary pressure curve? What are the commonly used pore structure parameters? What is displacement pressure? What factors does its magnitude depend on?
5. What are the main factors affecting the physical properties of clastic rock reservoirs?

   What are the genetic types of sandstone bodies? What are their main characteristics?
6. What are the main factors affecting the development of reservoir space in carbonate reservoirs?
7. What are the main types of carbonate rock reservoirs?
8. What are the main factors affecting the physical properties of volcanic rock reservoirs?
9. What are the main factors affecting the physical properties of metamorphic rock reservoirs?
10. What are the main factors affecting the physical properties of shale reservoirs?
11. Compare the differences in reservoir space among clastic rocks, carbonate rocks, volcanic rocks, metamorphic rocks, and shale?
12. What is a caprock? What types can it be divided into? What are the mechanisms of caprocks sealing oil and gas?

13. What are the main factors affecting the sealing ability of caprocks? Analyze the relativity of caprocks sealing oil and gas?

References

Ahr, W. M. (2008). Geology of carbonate reservoirs: the identification, description, and characterization of hydrocarbon reservoirs in carbonate rocks. In: Hoboken, N.J., Wiley InterScience, p 277.

Beard, D. C. & Weyl, P. K. (1973). Influence of texture on porosity and permeability of unconsolidated sand. *AAPG Bulletin, 57*, 349–369.

Bloch, J. D., Schroder-Adams, C. J., Craig, J., & McIntyre, D. J. (1999). Sedimentology, micropaleontology, geochemistry, and hydrocarbon potential of shale from the Cretaceous lower Colorado Group in western Canada. *Geological Survey of Canada, Bulletin, 531*, 58.

Brayshaw, A. C., Davies, G. W., & Corbett, P. W. M. (1996). Depositional controls on primary permeability and porosity at the bedform scale in fluvial reservoir sandstones. In: Carling, P. A. & Dawson, M. R. (eds). Advances in Fluvial Dynamics and Stratigraphy. Chichester, England. John Wiley & Sons, 373–394.

Brown, A. (1997). Porosity Variation in Carbonates as a Function of Depth: Mississippian Madison Group, Williston Basin(M). *Aapg Memoir, 69.*

Changgui, X., Xiaofeng, D., Xiaojian, L., Wei, X., & Yiwei, H. (2020). Formation mechanism of high-quality deep buried-hill reservoir of Archaean metamorphic rocks and its significance in petroleum exploration in Bohai Sea area. *Oil & Gas Geology, 41*(2), 235-247.

Chen, Z. (2003). Research on the preservation and destruction of oil and gas reservoirs. Beijing: Petroleum Industry Press.

Choquette & Pray. (1970). Geologic Nomenclature and Classification of Porosity in Sedimentary Carbonates. *The American Association of Petroleum Geologists Bulletin, 54*(2): 207–250.

Curtis, M. E., Cardott, B. J., Sondergeld, C. H., et al. (2012). Development of organic porosity in the Woodford Shale with increasing thermal maturity. *International Journal of Coal Geology, 103*, 26–31.

Dai, J., Ni, Y., & Wu, X. (2012). China's tight sandstone gas and its significant importance in exploration and development. *Petroleum Exploration and Development, 39*(3), 257–264.

Damsleth, E., Tjolsen, C. B., More, H., Haldorsen, H. H. (1992). A Two-Stage Stochastic Model Applied to a North Sea Reservoir. *Journal of Petroleum Technology 44*(4), 402–486.

Edward, A. (2001). Microbiology. Jones and Bartlett Publishers, Massacheustte.

Fu, G., Chen, Z., Lv, Y., et al. (1998). A comprehensive evaluation method for the sealing performance of clayey caprocks. *Petroleum Experimental Geology, 1*, 80–86.

Fu, G., Zhang, B., & Wu, W. (2016). Mechanism and identification method of regional clayey caprocks preventing oil and gas migration along conduits. *Journal of China University of Petroleum (Natural Science Edition), 40*(3), 36–43.

Gou, Q., Xu, S., Hao, F., et al. (2019). Full-scale pores and micro-fractures characterization using FE-SEM, gas adsorption, nano-CT and micro-CT: A case study of the Silurian Longmaxi Formation shale in the Fuling area, Sichuan Basin, China. *Fuel, 253*, 167–179.

Grunau, H. R. (1981). Worldwide review of seals for major accumulations of natural gas. *AAPG Bull, 65*(5), 933.

Hao, S. (1995). The formation and preservation of natural gas reservoirs. Beijing: Petroleum Industry Press.

Hubbert, M. K. (1953). Entrapment of petroleum under hydrodynamic conditions. 37(8), 1954–2026.

Jiang, Y. (1998). Discussion on the relationship between the thickness of reservoir caprocks and the height of the hydrocarbon column they seal. *Natural Gas Industry, 18*(2), 20–23.

Jon, G., & Richard, S. (2003). *Petroleum Geoscience*. Blackwell Pub.
Klemme, H. (1977). World oil and gas reserves from analyses of giant fields and basins (Provinces). Chapter12. In: Meeyer R, (ed). The Future Supply of Nature-made Petroleum and Gas. Oxford: Pergamon Press, 173–256.
Li, Y., Jin, Q., Zhong, J., et al. (2016). Karst zoning and fracture-cavity structure characteristics of the Ordovician in the Tahe Oilfield. *Acta Petrolei Sinica, 32*(6), 928–936.
Loucks, R. G., Reed, R. M., Ruppel, S. C., et al. (2012). Spectrum of pore types and networks in mudrocks and a descriptive classification for matrix-related mudrock pores. *AAPG Bulletin, 96*(6), 1071–1098.
Luca, C. (2001). Integrated Reservoir Studies. Engineering, Environmental Science.
Luo, Z. (1986). Development prospects of the natural gas resources in our country. *Natural Gas Industry, 6*(3), 1–4.
Luo, X., Zhang, L., Zhang, L., et al. (2020). Heterogeneity of clastic rock conduits and oil and gas accumulation. *Acta Petrolei Sinica, 41*(3), 253–272.
Ma, Y., Cai, X., Yun, L., et al. (2022). Exploration and development practice and theoretical and technological progress of the ultra-deep carbonate oil and gas field in the Shunbei area of the Tarim Basin. *Petroleum Exploration and Development, 49*(1), 1–17.
Ma, Y., & Tian, H. (1999). *Carbonate petroleum exploration*. Petroleum University Press.
Mastalerz, M. (2013). Covalent crystal growth. *Nature Chemistry, 5*(10) 810–811.
Meng, Q., Hao, F., & Tian, J. (2021). Origins of non-tectonic fractures in shale. *Earth-Science Reviews, 222*, Article 103825.
Nordgard Bolas, H. M., Hermanrud, C., Teige, G. M. G. (2005). Seal capacity estimation from subsurface pore pressures. *Basin Research, 17*, 583–599.
Osborne, M. J., & Swarbrick, R. E. (1997). Mechanisms for generating overpressure in sedimentary basins: A reevaluation. *AAPG bulletin, 81*(6), 1023-1041.
Powers, R. W. (1962). Arabian upper Jurassic carbonate reservoir rocks. In, W. E. Ham (ed.), Classification of Carbonate Rocks. *American Association of Petroleum Geologists, Memoir 1*, 122–192.
Sales, J. K. (1997). Seal strength VS trap closure: A fundamental control on the distribution of oil and gas. *AAPG Memoir, 67*, 57–83.
Sarg, J. F. (1988). Carbonate sequence stratigraphy. In: C. K. Wilgus, B. S. Hastings, C.G.St. C. Kendall, H. W. Posamentier, C. A. Ross, J. C. Van Wagoner (Eds.), Sea-Level Changes: an Integrated Approach. *Soc. Econ. Paleontol. Mineral. Spec. Publ., 42*, 155–181.
Schmidt, V. & McDonad, D. A. (1979). Texture and recognition of secondary porosity in sandstones. *SEPM Society for Sedimentary Geology*.
Scherer, M. (1987). Parameters influencing porosity in sandstones: A model for sandstone porosity prediction (J). *AAPG Bull., 71*(5): 485–491.
Selley, R. C. (2015). *Elements of petroleum geology*. Academic Press.
Sun, L., Zhao, W., Liu, H., et al. (2023). Discussion on the concept of shale oil "sweet spots" and their applications. *Acta Petrolei Sinica, 44*(1), 1–13.
Surdam, R. C. (1997). A new paradigm for gas exploration in anomalously pressured "tight gas sands" in the Rocky Mountain Laramide basins. Surdam, R. C. Seals, traps, and the petroleum system. *AAPG Memoir, 67*, 283–298.
Tong, X., et al. (1989). The role of regional caprocks in oil and gas accumulation. *Petroleum Exploration and Development, 16*(4), 1–8.
Wang, P. J., Hou, Q. J., Liu, W. Z., Huang, Y. L., Jia, J. T., & Tang, H. F. (2007). Characteristics of volcanic facies and genesis of natural gases in deep Songliao Basin, NE China. *Global Geology, 26*(3), 319–325.
Wilson, I. (1994). Strategic Planning Isn't Dead—It Changed. *Long Range Planning, 27*(4), 12–24.
Xu, C., Du, X., Liu, X., et al. (2020a). Formation mechanism and oil and gas exploration significance of high-quality reservoirs in the Archean buried metamorphic rocks in the Bohai Sea area. *Petroleum and Natural Gas Geology, 41*(2), 235–247, 294.

Xu, S., Gou, Q., Hao, F., et al. (2020b). Shale pore structure characteristics of the high and low productivity wells, Jiaoshiba shale gas field, Sichuan Basin, China: Dominated by lithofacies or preservation condition? *Marine and Petroleum Geology, 114*, Article 104211.

Yao, G. (2019). *Geological theory and exploration practice of deep oil and gas*. Petroleum Industry Press.

Ziegler. (1992). North sea rift system. *Tectonophysics, 208*, 55–75.

Zhai, G., et al. (1989–1996). *Geological atlas of petroleum in China* (Vols. 1, 6, and 10). Petroleum Industry Press.

Zhang, H., Fang, C., Gao, X., et al. (1999). Petroleum Geology. Beijing: Petroleum Industry Press.

Zhu, R., Wu, S., Su, L., et al. (2016). Issues and future development directions in the characterization of pore structures of tight reservoirs in China. *Acta Petrolei Sinica, 37*(11), 1323–1336.

Zhuo, Q., Zhao, M., Li, Y., et al. (2014). Dynamic evolution characteristics of the sealing properties of gypsum and salt caprocks and oil and gas accumulation—A case study of the thrust belt in the Kuqa Foreland Basin. *Acta Petrolei Sinica, 35*(5), 847–856.

Zou, C., Tao, S., Hou, L., et al. (2014). *Unconventional oil and gas geology*. Geological Publishing House.

Chapter 4
Hydrocarbon Migration

Abstract The migration of oil and gas is a crucial link between their generation and accumulation into reservoirs, reflecting the most significant characteristics of oil and gas as fluid mineral resources and is a key focus in petroleum geology. The migration process is complex and can be divided into primary migration and secondary migration based on its characteristics. These two types differ in terms of phase, driving forces, resistance, pathways, timing, and distance of oil and gas migration, which in turn affect the distribution of oil and gas within a basin. Therefore, understanding the content and methods of studying oil and gas migration is essential. This chapter mainly introduces the concepts of oil and gas migration, primary migration, secondary migration, research methods for oil and gas migration, and the relationship between the three fields (structural, fluid pressure, and thermal fields) and oil and gas migration.

4.1 An Overview of Hydrocarbon Migration

4.1.1 The Concept of Hydrocarbon Migration

Oil and natural gas are fluid minerals with mobility. When subjected to driving force, they will flow within the crust. We refer to any movement of hydrocarbons within the crust as hydrocarbon migration. It is known that hydrocarbons are formed in fine-grained sedimentary rocks rich in organic matter and are mostly stored in rocks with good porosity and permeability. The transition of hydrocarbons from a dispersed state in source rocks to an aggregated state in reservoir rocks must involve a migration process. Hydrocarbon migration is an important link connecting hydrocarbon generation and accumulation, and it is one of the essential contents of petroleum geology, which is of significant importance for oil and gas exploration.

Based on migration characteristics, hydrocarbon migration is divided into primary migration and secondary migration. Primary migration refers to the expulsion of hydrocarbons from the source rock, which includes the migration of hydrocarbons from the source rock to the reservoir or conduit layer, as well as the migration within

© China University of Petroleum Press 2025
H. Liu and Y. Jiang, *Petroleum Geology*,
https://doi.org/10.1007/978-981-96-7086-4_4

the source rock itself. For shale oil and gas, some of the hydrocarbons desorbed from the organic matter into the matrix pores and then migrate to the adjacent inorganic shale pores when they reach saturation, which is part of the primary migration. The migration of hydrocarbons gathered in the organic matter pores and inorganic shale pores to the adjacent sandstone layers also falls within the scope of primary migration. Secondary migration refers to all migrations after hydrocarbons enter the reservoir or conduit layer. It includes the migration of hydrocarbons through the conduit layer to the trap and also includes the re-migration of already formed hydrocarbon accumulations due to changes in external geological conditions (Fig. 4.1).

The above definition divides a naturally continuous migration process into two stages in time and space. In fact, it is difficult to determine the order of hydrocarbon migrations which are an almost simultaneous continuous process. For a hydrocarbon molecule, it can be divided into two successive migration processes, but for the overall hydrocarbon migration effect, secondary migration follows immediately after the first hydrocarbon molecule enters the conduit layer.

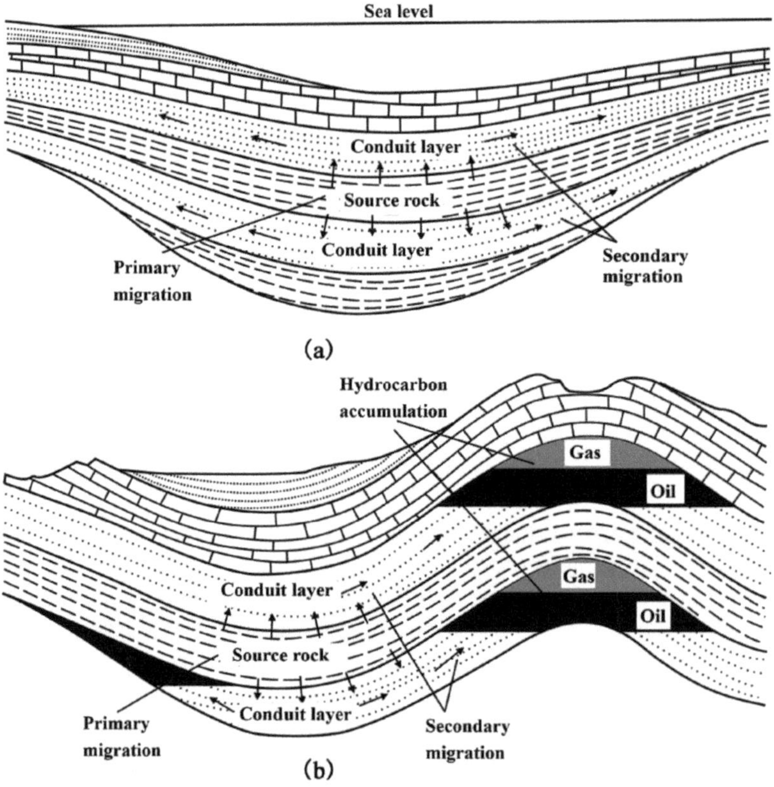

Fig. 4.1 Schematic diagram of hydrocarbon migration. **a** Early stage of hydrocarbon migration; **b** late stage of hydrocarbon migration and the formation of hydrocarbon reservoirs (according to Tissot & Welte, 1984)

Hydrocarbon migration includes two aspects: dynamics and kinematics. Dynamics explores the driving forces of hydrocarbon migration, while kinematics explores the phase state, pathways, direction, main periods, and quantities of hydrocarbon migration. In other words, hydrocarbon migration research is about how hydrocarbons migrate underground, under what driving forces, during what periods, what paths they take, what the main directions are, what geological factors control them, and how much they migrate, etc.

4.1.2 Several Issues Related to Hydrocarbon Migration

4.1.2.1 Pressure

Pressure is a force acting perpendicularly on a unit area of an object, existing within gases, liquids, solids, or at the contact surfaces between fluids and solids, or between solids. Petroleum, natural gas, water, and other subsurface fluids are all in a certain pressure-bearing environment. The international unit of pressure is Pascal (Pa), which can also be represented by atmospheric pressure (atm), where 1 atm = 101.3 kPa.

(1) Lithostatic Pressure and Lithostatic Pressure Gradient

a. Lithostatic Pressure

Lithostatic pressure refers to the pressure caused by the total weight of the overlying rock layers and the fluids in the rock pores, also known as the overburden pressure. It is usually calculated using the following formula:

$$P_r = H\rho_r g \tag{4.1}$$

where P_r—the lithostatic pressure, Pa or MPa; H—the thickness of the overlying sediment or rock layer, m; ρ_r—the density of the overlying sediment or rock layer, t/m^3; g—acceleration of gravity, 9.8 m/s^2.

In fact, the lithostatic pressure is produced by the total weight of the matrix and pore fluids of the overlying sediments, that is:

$$P_r = H\,g\big[(1 - \Phi)\rho_g + \Phi\rho_f\big] \tag{4.2}$$

where Φ—the porosity of the overlying sediment or rock layer, %; ρ_g—the matrix density of the overlying sediment or rock layer, t/m^3; ρ_f—the fluid density in the pores, t/m^3.

In normal compaction, the weight of the overlying sediment matrix and pore fluid that generates the lithostatic pressure acts on the solid matrix and pore fluid of the underlying strata, respectively. The pressure borne by the solid matrix is called the effective stress, i.e., the first term on the right side of Eq. (4.2). The pressure borne by the pore fluid is called the pore fluid pressure, i.e., the second term on the right side of

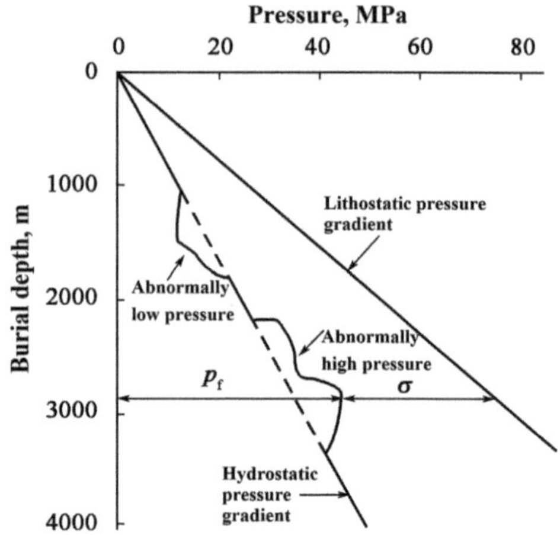

Fig. 4.2 Schematic diagram of hydrostatic pressure gradient, lithostatic pressure gradient, formation pressure, and abnormal pressure

Eq. (4.2). Therefore, the lithostatic pressure can be divided into two parts: effective stress and pore fluid pressure (Fig. 4.2), represented by the following formula:

$$P_r = \sigma + P_f \tag{4.3}$$

where σ—the effective stress, Pa or MPa; Pf—the pore fluid pressure, Pa or MPa.

b. Lithostatic Pressure Gradient

The hydrostatic pressure increases with the increase of the thickness of overlying strata. The pressure increased by each unit thickness increase of overlying strata is called lithostatic pressure gradient. It usually refers to the pressure increased by each 1 m increase in sediment thickness and is expressed in Pa/m. If the average bulk density of overlying sediments is taken as 2.3 t/m^3, then the lithostatic pressure gradient is about 2.3×10^4 Pa/m.

(2) Hydrostatic Pressure and Hydrostatic Pressure Gradient

a. Hydrostatic Pressure

Hydrostatic pressure refers to the pressure caused by the weight of the static water in the pores of the overlying strata, that is, the pressure produced by the connected water column in the strata pores, calculated by the following formula:

$$P_w = H\rho_w g \tag{4.4}$$

where P_w—the hydrostatic pressure, Pa or MPa; H—the height of the overlying water column, m; ρ_w—the density of water, t/m^3; g—the acceleration of gravity, 9.8 m/s^2.

b. Hydrostatic Pressure Gradient

The hydrostatic pressure gradient refers to the increase in pressure when the connected water column in the strata pores increases by a unit height. It is usually represented by the increase in pressure per meter of water column height, expressed in Pa/m. If we take the density of the water as 1 t/m^3, then the hydrostatic pressure gradient is approximately 0.1×10^5 Pa/m.

(3) Formation Pressure and Abnormal Pressure

a. Formation Pressure

Formation pressure refers to the pressure acting on the fluids in the rock pores, also known as pore fluid pressure (P_f). Formation pressure can be composed of various underground pressures such as hydrostatic pressure, dynamic water pressure, buoyancy, capillary pressure, expansion pressure, and permeation pressure. The measured formation pressure is often a comprehensive reflection of the above pressures. If the formation water is in a static state, it generates hydrostatic pressure. When the balance of hydrostatic pressure is disrupted and the formation water flows, it generates dynamic water pressure. In an open formation system, the formation pressure mainly reflects the hydrostatic pressure, which can be represented by the hydrostatic pressure of the formation from the surface to any point underground. In a relatively closed geological environment, the formation pressure will exceed the hydrostatic pressure (abnormally high formation pressure) or be lower than the hydrostatic pressure (abnormally low formation pressure).

b. Abnormal Pressure

Abnormal pressure is the formation fluid pressure in a certain depth range of the strata, which deviates from the normal hydrostatic pressure trend line due to geological factors (Fig. 4.2). When the formation pressure equals the hydrostatic pressure, it is normal pressure. The formation pressure higher than the hydrostatic pressure is called abnormally high formation pressure, and the formation pressure lower than the hydrostatic pressure is called abnormally low formation pressure. The pressure transition zone between normal pressure and abnormal pressure is called the pressure transition zone.

The pressure coefficient is an important parameter for judging abnormal pressure. It refers to the ratio of the measured pressure to the hydrostatic pressure. The pressure coefficient for normal pressure is 0.9–1.1, while the pressure coefficient for abnormally high formation pressure is greater than 1 or 1.1. The pressure coefficient for abnormally low formation pressure is less than 1 (or 0.9).

4.1.2.2 Adsorption and Wettability of Rocks

(1) Adsorption

Adsorption is an interfacial phenomenon between fluids and solids. Adsorption is the phenomenon where fluid molecules adhere to the solid surface under the action of analytical attraction (van der Waals forces) and electrostatic forces. The former is physical adsorption, and the latter is chemical adsorption. During physical adsorption, there is no electron transfer, no generation and destruction of chemical bonds, no rearrangement of atoms. It can form single or multi-molecular layer adsorption, which is unstable and easily desorbed. Chemical adsorption is caused by electrostatic attraction due to chemical bonds, forming only single-molecular layer adsorption, which is relatively stable and not easily desorbed. The two are often transitional. For example, on the surface of clay minerals, multiple layers of water molecules can be adsorbed, with the layer of molecules close to the clay mineral layer being chemically adsorbed, which is difficult to desorb, and it transitions to physical adsorption outwards, with smaller adsorption forces and easy desorption.

The adsorption force of rocks is influenced by rock type, mineral composition, structure, particle size, organic matter content, temperature, formation pressure, water saturation, and hydrocarbon properties and so on. Especially the adsorption capacity of gas, which decreases with increasing temperature, increases with increasing pressure, increases with the increasing specific surface area of mineral particles, and decreases with increasing water saturation. The adsorption of hydrocarbons is also related to the properties and molecular structure of hydrocarbons. The adsorption capacity increases with increasing molecular weight, and the adsorption capacity of normal hydrocarbons is greater than that of isomers. Adsorption is a resistance to hydrocarbon migration, and the adsorption capacity of hydrocarbons is an important aspect of evaluating the resource potential and development engineering of shale oil and gas and coalbed gas.

(2) Wettability

Wettability is an adsorption behavior, referring to the phenomenon where a fluid spreads on a solid surface under the action of surface molecular forces. When two immiscible fluids coexist in the rock pores, due to the different adsorption forces of the rock on different fluids, one type of fluid is preferentially adsorbed and preferentially wets the rock surface, which is called selective wetting. The fluid that easily adheres to the rock is called the wetting fluid (also known as the wetting phase), and the fluid that is not easily adhered to is called the non-wetting phase fluid (also known as the non-wetting phase). For example, when oil and water coexist in the rock pores, if water easily adheres to the rock, water is called the wetting phase, oil is the non-wetting phase, and the rock is hydrophilic (or oleophobic). If oil easily adheres to the rock, oil is called the wetting phase, water is the non-wetting phase, and the rock is oleophilic (or hydrophobic).

The wettability of rocks is related to the mineral composition of the rocks and the properties of the fluids. Inorganic minerals are mostly hydrophilic, while organic matter and metal sulfides are mostly oleophilic. Different components in petroleum have different wettabilities to the same minerals. Surface-active substances can change the wettability of rocks, making them change from hydrophilic to oleophilic, or from oleophilic to hydrophilic. The surface of rock particles is rough, especially at the edges and corners, making wettability more complex. Therefore, the heterogeneity of rocks and fluids inevitably leads to the heterogeneity of wettability.

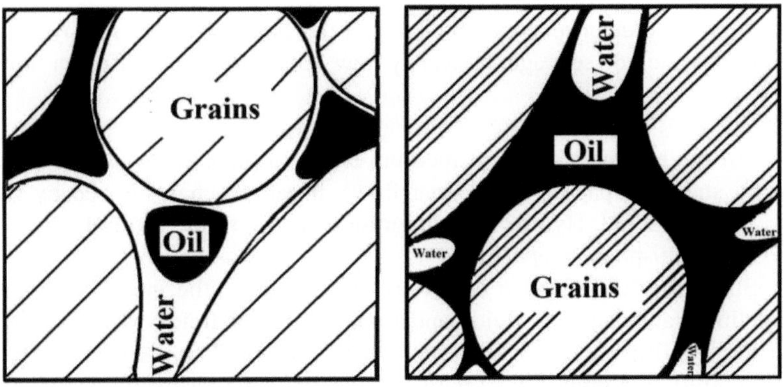

Fig. 4.3 Distribution of oil and water in pore media. **a** Water-wet pore media; **b** oil-wet pore media

Wettability can generally be divided into three types: water-wet, oil-wet, and intermediate wettability. It is generally believed that since most sedimentary rocks are formed in water bodies and water is a polar molecule, rock particles are mostly water-wet and form adsorbed water on the surface of the particles. The original sediments may all be hydrophilic, but later, due to the generation and migration of hydrocarbons, oil can adhere to the surface of the particles, making them oleophilic. Hydrocarbon source rocks themselves have many oleophilic organic particles and can generate hydrocarbons under certain conditions, so they can be considered to have intermediate wettability, which is partly hydrophilic and partly oleophilic.

The wettability of rocks affects the ease of hydrocarbon migration within them. Different wettabilities cause oil and water phases to flow, remain, and be trapped in different ways and quantities in the pores. In hydrophilic media, the pore walls and particle surfaces are wetted by water, and water forms a liquid ring between the particles, while the oil phase cannot remain as a film on the pore walls but is squeezed to the center of the pores. When the oil phase saturation is very low, it forms isolated oil beads (Fig. 4.3). These oil beads can block the pore throats and hinder fluid migration, and a considerable force is required to deform the oil beads to overcome this resistance. In oleophilic media, oil adheres to the pore walls in the form of a film, becoming immobile residual oil. It can be seen that the amount of residual oil in hydrophilic media is less than that in oleophilic media, but the flow of the oil phase in hydrophilic media is more difficult than in oleophilic media.

4.1.2.3 Types of Fluid Flow

Types of fluid flow refer to the dynamic modes of fluid migration in underground rocks under certain external forces. Different types of fluid flow conform to certain fluid dynamics laws and determine the efficiency of fluid flow. The types of hydrocarbon migration flow include seepage, buoyancy flow, surge flow, and diffusion flow.

(1) Seepage

The flow of fluid through a porous medium is called seepage. When the fluid flow is straight, stable seepage (i.e., laminar flow), it can be characterized by Darcy's law and fluid potential. Hydrocarbon migration can be in the form of single-phase seepage and multi-phase seepage.

In multi-phase fluid seepage in porous media, due to the interactions between different fluids and between fluids and rocks, different fluids will have different relative permeabilities and corresponding saturations. Their continuous flow requires a critical migration saturation and relative permeability. Experiments have shown that oil and water in coexistence must reach a certain relative permeability and saturation to occur seepage, that is, they must exceed the residual oil saturation and bound water saturation. We refer to the minimum saturation required for oil (gas) to migrate as a free phase when oil (gas) and water coexist as the critical saturation for oil (gas) migration. It is generally believed that the critical saturation for primary migration of petroleum is 5%, and the critical saturation for secondary migration of oil and water in two phases is 10–20%.

(2) Buoyancy Flow

Buoyancy flow is the upward floating of hydrocarbons in the water medium, which depends only on the density difference between hydrocarbons and water, and is independent of whether the water is flowing or not. It is divided into free upward floating and obstructed upward floating. During the upward floating process, hydrocarbons can be discontinuous, so there is no requirement for critical migration saturation, and Darcy's formula cannot be used to characterize it.

(3) Surge Flow

It refers to the situation where, when the seal of abnormally high-pressure fluid in the formation is broken, or when a closed fault is reopened, oil, gas, and water surge out in a turbulent manner under the action of a strong pressure difference, also known as mixed-phase surge flow, equilibrium flow, jet flow, or rupture flow. It is a special case of multi-phase seepage. After the fluid surges out of the closed system, the pressure is released, and it can be recharged again to cause mixed-phase surge flow again. Therefore, surge flow is intermittent and episodic. Surge flow has no requirements for the saturation of each phase and cannot be calculated using Darcy's formula.

(4) Diffusion Flow

Diffusion flow mainly refers to the diffusion flow of light hydrocarbon molecules under the action of concentration differences. Light hydrocarbon diffusion is the flow of single molecules and can occur in nano-scale micropores, especially in tight sandstone reservoirs, where diffusion flow becomes the main mode of hydrocarbon migration. When hydrocarbons are transported in a water-soluble phase (i.e., hydrocarbons are transported in water as a carrier), under dynamic water conditions, it is mainly seepage, and under static water conditions, it is mainly diffusion flow.

4.2 Primary Migration of Hydrocarbons

4.2.1 Phases of Primary Migration

The phase of primary migration refers to the physical state of oil and gas during their subsurface movement. The main phases of primary migration for oil are aqueous solution, free oil phase, and gas solution, with the free phase being the dominant mode of migration. For natural gas, the primary migration phases include aqueous solution, oil solution, free gas phase, and diffusion phase. All four phases can exist, but their significance varies.

4.2.1.1 Aqueous Phase Migration

Oil migration in an aqueous phase refers to the movement of oil in the form of a true solution or colloidal solution dissolved in water during primary migration. In hydrophilic source rocks, water flowing in a single phase only experiences intermolecular frictional resistance without capillary resistance. When driving water present, the aqueous solution can migrate along fine pore throats. Therefore, from a physical standpoint, the aqueous phase seems to be the most ideal state for migration. However, issues such as the low solubility of oil in water, insufficient water supply during the oil generation period, and the mismatch between the components of oil and their aqueous solubility prevent the aqueous phase from becoming the main phase of oil primary migration.

Unlike oil, natural gas has a higher solubility in water. Under normal temperature and pressure conditions, the solubility of hydrocarbon gases in water is generally 100 times greater than that of oil in water (Bonham, 1978), and even greater under high temperature and high-pressure conditions. Therefore, natural gas can migrate in an aqueous phase. During the immature and early mature stages of source rocks, there is a large amount of interstitial water in the source rocks, and the biogenic gas and low mature gas generated at this time can migrate in an aqueous phase.

4.2.1.2 Free Phase Migration

Free phase migration refers to the migration of oil or natural gas in a free state within the source rock during primary migration. The free phase is the most important phase for oil primary migration. Direct evidence for oil migration in a free phase includes:

(1) Microscopic observations of source rocks, especially studies on shale oil, indicate that most oil exists in pores or fractures in a free state; (2) In thick source rock sections, it has been determined that the content of chloroform extractable substances in the source rock increases with distance from the source rock-reservoir contact interface, with the least content near the contact surface; (3) The free phase can reasonably explain the expulsion of large amounts of oil and gas generated by the source rock, overcoming the various inexplicable phenomena that exist in the aqueous phase migration hypothesis.

Natural gas can also migrate in a free phase, especially when the source rock generates a large amount of gas and there is little formation water, the free phase becomes the main migration phase for natural gas.

4.2.1.3 Gas Solution and Oil Solution Migration

Oil and hydrocarbon gases are mutually soluble. Gas solution migration is the migration of oil dissolved in natural gas in a gaseous state, and oil solution migration is the migration of natural gas dissolved in oil in an oily state.

Generally, the lighter the components of oil, the greater their solubility in natural gas, and this solubility increases with temperature and pressure. Price (1983) conducted a systematic measurement of the solubility of crude oil from the Denver Basin's Spinde oil and gas field in methane and found that under moderate temperatures and higher pressures (100–150 °C, 70–100 MPa), the solubility of crude oil in methane is between 0.5–1.5 g/l. Therefore, the gas solution is also an important mode of primary migration for oil.

The solubility of hydrocarbon gases in oil is much greater than in water, and the solubility of hydrocarbon gases in oil increases with increasing pressure until it reaches the saturation pressure (bubble point pressure). Therefore, under reservoir conditions, the oil solution is also an important mode of primary migration for natural gas.

4.2.1.4 Diffusion Phase Migration

Diffusion is the result of molecular thermal motion, and the diffusion action is effective for gases and light hydrocarbons, but it has little effect on oil, especially for components with more than 10 carbon atoms, which are almost insignificant. The diffusion action of natural gas occurs in the form of single molecules. Since the concentration of natural gas in the source rock during the gas generation peak stage is higher than that outside the source rock, it is an inevitable process for natural gas to migrate in the form of single molecules. It is inferred that in low permeability source rocks, even overpressured source rocks and low mature source rocks, the diffusion phase may be an important mode of hydrocarbon gas migration, and even the only mode.

4.2.1.5 Evolution of Oil and Gas Primary Migration Phases

For sapropelic organic matter source rocks, during the low maturity stage, the burial depth is shallow, the porosity is large, there is a lot of formation water, the amount of hydrocarbons generated is small, and the content of resins and asphaltenes is high. The possibility of oil migration in an aqueous phase is the greatest, the biological methane generated at this stage can migrate in an aqueous phase and a free gas phase. With increasing burial depth, during the mature stage of the source rock, the pore water is not enough to completely dissolve the generated oil and gas, and oil mainly migrates in a free phase, while gas mainly migrates in an oil solution phase. In the high maturity stage of the source rock, liquid hydrocarbons crack into wet gas, and the source rock contains very little water, so a small amount of liquid oil mainly migrates in a gas solution phase, and natural gas mainly migrates in a free phase. In the over-mature stage, the source rock generates a large amount of gas, and at this time, natural gas migrates in a free phase or diffusion phase (Fig. 4.4).

For humic organic matter source rocks, the entire oil and gas generation evolution stage is mainly dominated by gas generation, with limited oil generation potential, and the primary migration of oil and gas shows different phase evolution patterns. The main difference is in the mature stage, because the amount of liquid oil generated by humic source rocks is less than the amount of gas generated. The small amount of liquid oil generated may mainly migrate in a gas solution phase and cannot form a free phase and an oil solution gas phase. However, in the low maturity stage, high

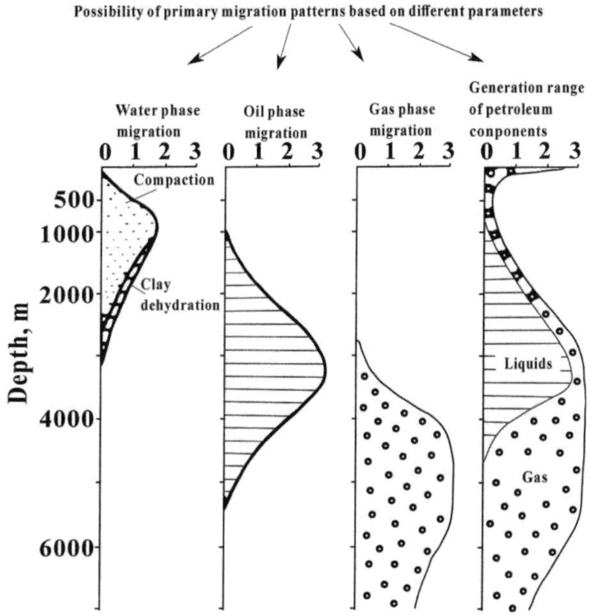

Fig. 4.4 Possible phases during the primary migration process of oil and gas (according to Tissot & Welte, 1978)

maturity stage, and over-mature stage, the primary migration phases are similar to those of sapropelic source rocks.

4.2.2 Primary Migration Dynamics and Direction

Source rocks, due to their fine grain size and large specific surface area, have a strong adsorption capacity for hydrocarbons. The source rocks contain extensive small pore systems, and the migration of oil and gas within the source rocks must overcome the huge capillary resistance between the throats. Therefore, for oil and gas to migrate out of the source rocks, there must be a driving force to overcome adsorption and capillary forces.

It is generally believed that the main driving force for hydrocarbons to be expelled from the source rock layer is the residual pressure generated by various actions within the source rock, which drives the pore fluid to migrate in the direction where the residual pressure decreases. The actions that can generate residual pressure mainly include normal compaction, undercompaction, montmorillonite dehydration, fluid thermal pressurization, hydrocarbon generation, permeation, etc. In addition, the concentration difference between the source rock and the external environment is also a driving force for the primary migration of natural gas.

4.2.2.1 Normal Compaction and Instantaneous Residual Pressure

(1) Mechanism of Instantaneous Residual Pressure Formation

Compaction is one of the most important diagenesis processes of sediments. Compaction leads to a decrease in porosity and the expulsion of pore fluid. If there is oil and gas in the pores at this time, the oil and gas will also be expelled from the pores with the water from the source rock, completing the primary migration of oil and gas. Normal compaction refers to the state where the sediment is in a compaction balance. The skeleton pressure generated by the weight of the rock skeleton of the overlying strata is borne by the rock skeleton, and the pore fluid only bears the pressure of the static water column of the overlying pore water. When new sediments are deposited on the overlying strata, their gravitational load acts on the underlying strata, prompting the particles to re-compact and arrange, and the pore volume decreases. At this moment of changing, the pore fluid bears part of the effective stress that should be borne by the particles, causing the fluid to generate a residual pressure that exceeds the static water pressure. We call this residual fluid pressure. Residual fluid pressure is an abnormal pressure generated during the normal compaction process, and it disappears quickly with the expulsion of the pore fluid, so it is called instantaneous residual fluid pressure.

It is under the action of residual fluid pressure that the pore fluid is expelled. After the expulsion of the fluid, the pore fluid pressure returns to the static water pressure, and the sediment reaches a new compaction balance. Because of this, in a sedimentation and burial process that is continuously sinking, depositing, and compacting, the compaction process of the sediment is a dynamic process from compaction balance—compaction imbalance—compaction balance, and the pore fluid pressure is a change process from static water pressure—instantaneous residual fluid pressure—static water pressure. In this process, the fluid is continuously expelled, and the above process is repeated with the accumulation of new sediments on the overlay.

During the source rocks diagenesis process, the primary migration under compaction mainly occurs in the early stage of rock formation. In a connected pore system, when the hydrocarbon content reaches a critical saturation, hydrocarbons and water will overcome the capillary resistance and be expelled from the source rock in a mixed phase under the action of instantaneous residual pressure. Residual fluid pressure is generated during the normal compaction stage. For most lacustrine fault basins in eastern China, the source rocks of the Paleogene system have not yet or have just entered the oil generation limit, and oil and gas have not been generated in large quantities. At this time, the expelled pore fluid is mainly water and the dissolved hydrocarbons in it. Except for some source rocks with particularly rich organic matter, free oil and gas are difficult to be expelled.

(2) Direction of Compaction Fluid Expulsion

Under the action of residual fluid pressure, the direction of pore fluid expulsion is consistent with the direction of decreasing residual fluid pressure. For new sediments with uniform horizontal thickness, the horizontal residual pressure is equal, all are:

$$E = (\rho_{b0} - \rho_w) g l_0 \tag{4.5}$$

And there is no horizontal residual fluid pressure, only a vertical residual pressure gradient, then:

$$\begin{aligned} dp/dZ &= E_l/l_0 = [(\rho_{b0} - \rho_w) g l_0]/l_0 \\ &= (\rho_{b0} - \rho_w) g \end{aligned} \tag{4.6}$$

Therefore, in a uniform rock layer, the direction of compaction fluid flow is vertically upward.

If the thickness of the new sediment layer changes horizontally, then there will also be horizontal changes in the residual fluid pressure value on the horizontal direction. As shown in Fig. 4.5, when the new sediment is wedge-shaped, since $l_0 > h_0$, then $E_l > E_h$, so the fluid is expelled not only upward but also laterally from l_0 to h_0. If the distance between l_0 and h_0 is X, X, then the horizontal residual fluid pressure gradient caused by the load of the wedge-shaped sediment is:

$$\begin{aligned} dp/dX &= [dP_l - dP_h]/x \\ &= [(\rho b_0 - \rho_w) g l_0 - (\rho b_0 - \rho_w) g h_0]/x \end{aligned}$$

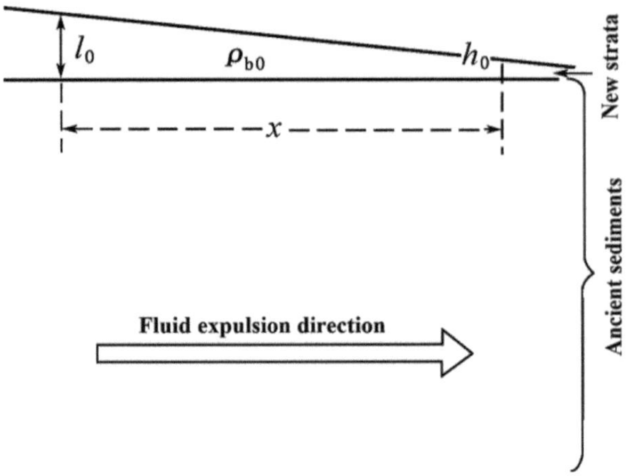

Fig. 4.5 The direction of compaction fluid expulsion under the load of wedge-shaped sediments (according to Magara, 1978)

$$= (\rho b_0 - \rho_w)g(l_0 - h_0)/x \tag{4.7}$$

According to Magara (1978), this value changes between 1/200 and 1/20, that is, the horizontal residual pressure gradient is only 1/200–1/20 of the vertical direction. It can be seen that most of the fluid moves vertically, and only a small part of the fluid moves horizontally. In the source rock, due to the limitations of the migration channel, this lateral migration is also very limited.

In a sandstone-mudstone interbedded formation, due to compaction, the mudstone's porosity decreases more rapidly than that of the sandstone. Under the same load, the instantaneous residual fluid pressure generated in the mudstone is greater than that in the sandstone. Therefore, the direction of fluid migration is from the mudstone to the sandstone. Although the sandstone is also subject to compaction, the instantaneous residual pressure it generates is smaller than that of the overlying and underlying mudstones, preventing the compaction fluid from entering the mudstone and only allowing lateral migration within the sandstone layer (Fig. 4.6). Of course, if there are thickness differences in the mudstone, the compaction fluid may also migrate laterally.

In summary, for a clastic rock sedimentary basin, microscopically, under the action of residual fluid pressure, the compaction fluid always migrates from mudstone to sandstone, and there can also be a certain range of lateral migration within the mudstone. Macroscopically, the compaction fluid always migrates from deeper to shallower areas, from the center of the basin to the basin's edge.

4.2.2.2 Abnormal High Pressure Generated Within the Source Rock

Source rocks in the mature to over-mature stage have very small throat radii, and the source rocks have become closed or semi-closed, providing the prerequisite conditions for the generation of abnormal high pressure. Among them, undercompaction,

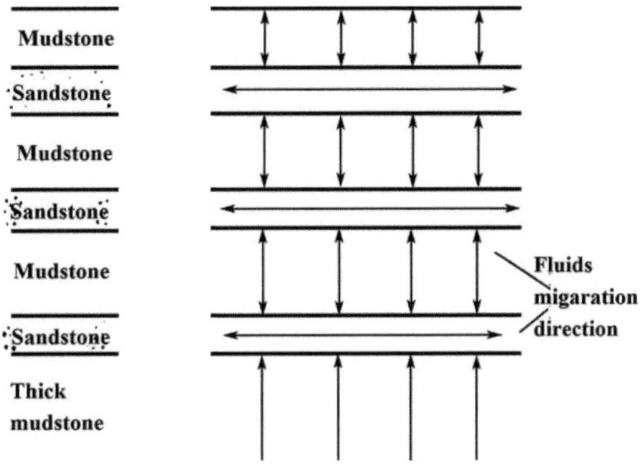

Fig. 4.6 The direction of compaction fluid migration in a sandstone-mudstone interbedded section

clay mineral dehydration, hydrocarbon generation, and thermal pressurization of fluids can all form abnormal high pressure within the source rock, becoming the main driving force for the primary migration of oil and gas.

(1) Undercompaction and Abnormal High Pressure

a. Undercompaction

During the rapid compaction process of argillaceous rocks, many tiny pores are formed, especially the pores at the top and bottom edges that are sealed first, preventing or delaying the expulsion of pore fluid. The pore volume does not decrease with the increase of the overlying load, and the pore fluid bears part of the effective stress from the overlying sediments, causing the pore fluid to have an abnormal high pressure that exceeds its corresponding static water pressure at the same depth. The rock framework, on the other hand, bears a lower effective stress. This phenomenon is called undercompaction.

The load pressure of sediments is borne by both the rock particles and the pore fluid, so the effective stress supported by the particles is inversely proportional to the pore fluid pressure. The changes in pore volume, pore fluid pressure, and particle effective stress in the undercompaction zone can be represented by Fig. 4.7.

Undercompaction is relative to normal compaction, which has not reached the degree of compaction that should be achieved at a certain depth. It is difficult to determine what the normal compaction should be, so undercompaction is not a concept with a strict definition. Undercompaction is manifested as strata having a larger porosity that deviates from the normal compaction trend. Since the compaction curve is influenced by many factors and its normal compaction trend is also difficult to obtain, undercompaction can usually be judged comprehensively based on logging data such as sonic transit time, density, and pore fluid pressure. If the density value is lower, the porosity is larger, or the sonic transit time is higher at a certain depth interval, and the pore fluid pressure is abnormally high, it can be considered that undercompaction has occurred at that depth interval (Fig. 4.8).

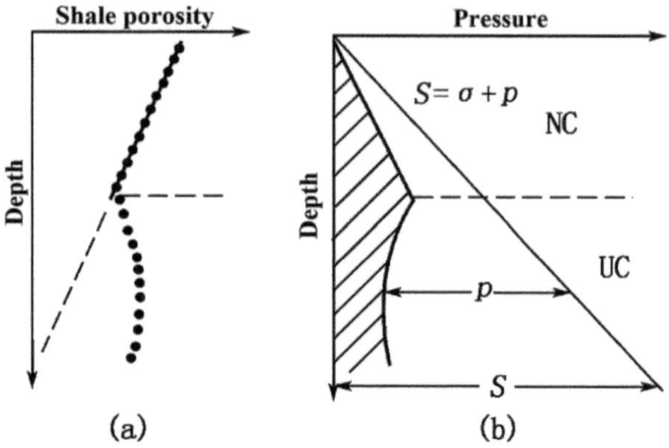

Fig. 4.7 Schematic diagram of the relationship between overlying sediment load pressure (S), fluid pressure (P), and particle-supported effective stress (σ) in normal and abnormal compaction zones (according to Magara, 1978)

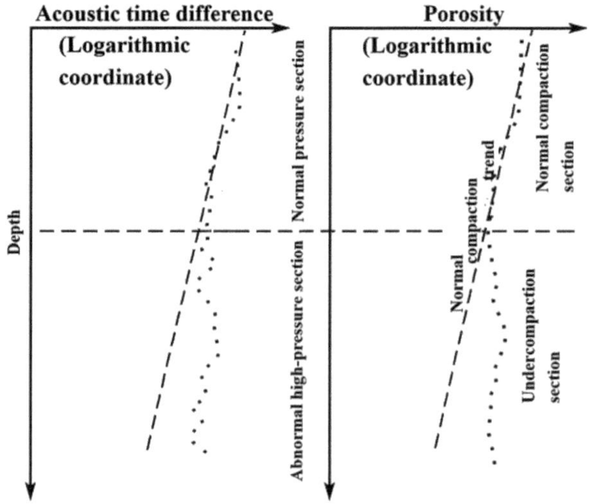

Fig. 4.8 Manifestation of undercompaction on acoustic transit time and porosity curves

b. Mechanism of Hydrocarbon Expulsion by Undercompaction

The abnormal high pressure formed by undercompaction in the source rock can promote the migration of fluids. The principle is that in a closed source rock, when the pore fluid pressure exceeds the rock's rupture strength, micro-fractures are produced, allowing the fluid to be expelled in a surging manner. As the fluid is expelled, the excess pressure is released, the micro-fractures close again, and then the fluid pressure accumulates again, the rock breaks again and expels the fluid, and so on until the undercompaction and abnormal pressure disappear. It can be seen that the formation of abnormal pressure and the release of fluid expulsion have a phased characteristic, but the abnormal pressure produced by undercompaction is much greater in intensity than the instantaneous residual pressure formed by normal compaction.

In a continuously subsiding basin, the depth at which the source rock generates undercompaction is generally below the oil generation limit, at which time hydrocarbons have begun to be generated in large quantities, and the abnormal high pressure is sufficient to overcome various resistances in the source rock to expel oil and gas.

c. Direction of Fluid Expulsion in Undercompaction Zone

The abnormal high pressure in the undercompaction zone drives the expulsion of oil, gas, and water from the center of the undercompaction zone upwards and downwards. Figure 4.9 shows a schematic diagram of the vertical distribution of residual pore fluid pressure and porosity in the undercompaction zone. The △Pa in the figure represents the value of the residual fluid pressure, and △Pmax represents the maximum residual pressure value. The direction of fluid expulsion is from the area of maximum residual pressure upwards and downwards. Since the source rock is more prone to undercompaction in the interior and edges compared to adjacent conductive layers, the direction of fluid migration under the action of undercompaction is from the high residual fluid pressure area to the low residual fluid pressure area, from the interior of the source rock to the edges, and from the source rock to the adjacent reservoir or conductive layers.

(2) Montmorillonite Dehydration and Abnormal High Pressure

Montmorillonite dehydration refers to the role of interlayer water release during the transformation of montmorillonite to illite. Montmorillonite is an expansive clay mineral that contains a large amount of interlayer water, generally with 4 or more layers of water molecules. These interlayer waters can account for 50% of the entire

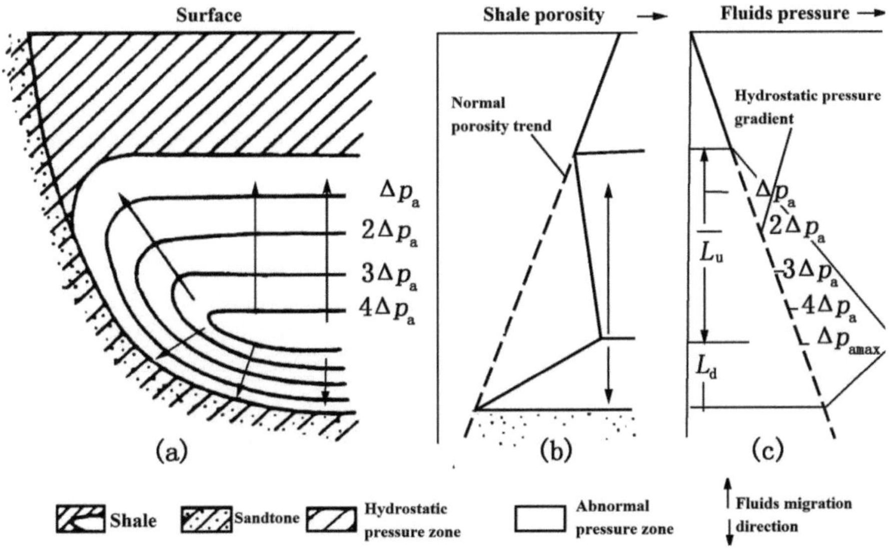

Fig. 4.9 The direction of fluid expulsion in the undercompaction zone (according to Magara, 1968)

mineral by volume and 22% by weight. A portion or all of these interlayer waters can become pore water under compaction and thermal effects. It is generally believed that the temperature range for the transformation of a large amount of montmorillonite to illite is between 99–143 °C. Under the influence of temperature, montmorillonite will lose interlayer water equivalent to 10–50% of its total water content, and such a large amount of water entering the pores will have a significant impact on oil and gas migration.

The role of montmorillonite dehydration in oil and gas migration can be summarized in the following three points:

a. The water released from montmorillonite occupies a certain pore volume in the source rock, generating abnormal high pressure, which in turn produces fractures and provides power and channels for the primary migration of oil and gas.

   The abnormal high pressure is widely present below a depth of 3900 m in the centre of the Junggar Basin. The abnormal pressure has the same starting depth, the main cause is the dehydration effect of montmorillonite (Fig. 4.10). The main clay minerals in this area are mixed layers of illite/montmorillonite—illite—chlorite—kaolinite, and the content of mixed layers of illite/montmorillonite obviously decreases with increasing burial depth, while the content of illite and other minerals obviously increases. For example, in Well Pencan 2, the content of mixed layers of illite/montmorillonite in the high-pressure layer segment decreases to 8%, while the content of illite increases to a maximum of 25%. The corresponding ancient geothermal gradient of this section is 70–110 °C, which is suitable for the dehydration of montmorillonite. This change in clay minerals corresponds well with the abnormal high-pressure segment, and the dehydration of clay minerals not only causes the formation of the abnormal high-pressure zone but also promotes the expulsion of hydrocarbons from the source rock in the abnormal high-pressure zone.
b. When montmorillonite in the source rock dehydrates, due to the phase change and expansion effect of water, hydrocarbons can be detached from the mineral surface to the intergranular pore water and become relatively concentrated. At the same time, because the interlayer water of montmorillonite becomes free water, the volume of mineral particles correspondingly contracts, thereby increasing the effective porosity and permeability. After the dehydration of montmorillonite, its adsorption capacity for hydrocarbons will also be greatly reduced.
c. Montmorillonite dehydration can provide 10–15% of the total volume of mudstone, especially the dehydration section below the oil generation limit depth is more favorable, which can provide a deep water source after the maturity of the source rock.

Since the source rock contains more montmorillonite than the conductive layer, the dehydration effect of montmorillonite will promote the migration of fluids from the source rock to the adjacent conductive layer.

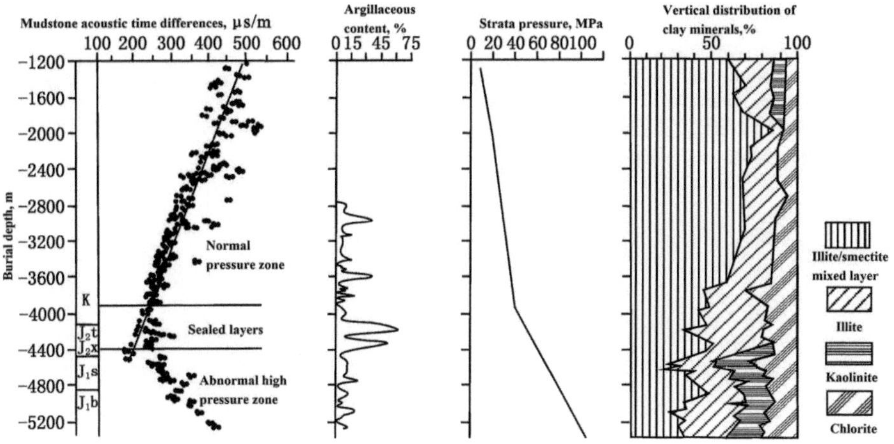

Fig. 4.10 The relationship between clay mineral transformation and abnormal high pressure in Well Pencan 2 of the Junggar Basin

Fig. 4.11 The relationship curve of specific water pressure–temperature–density (according to Barker, 1972)

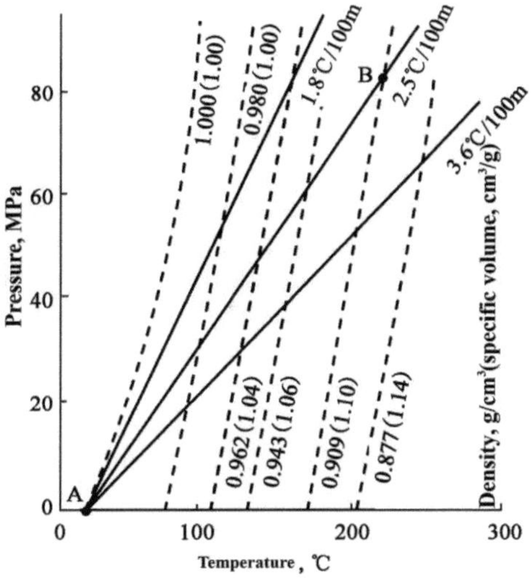

(3) Fluid Thermal Pressurization and Abnormal High Pressure

Fluid thermal pressurization refers to the effect of increasing pore fluid pressure due to the expansion of fluid volume caused by temperature increase in a closed formation pore system. As the temperature increases, the specific volume (i.e., the volume of water per unit mass) of groundwater also increases. As shown in Fig. 4.11, the geothermal gradient line intersects with the water isopycnal line, and the density value at the intersection point decreases with increasing pressure or burial depth (specific volume increases), that is, water expands as the temperature increases.

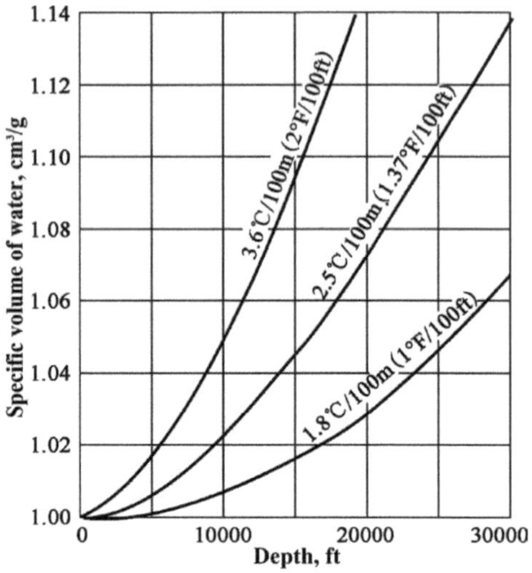

Fig. 4.12 Relationship between volume and depth under there geothermal gradient conditions in the normal pressure zone (according to Magara, 1978)

Different regions have different geothermal gradients, and the expansion of water is also different. Figure 4.12 shows the expansion of water under three geothermal gradients. At a depth of 20,000 ft (6,096 m), when the geothermal gradient is 18 °C/km, the expansion of water is about 3%. When the geothermal gradient is 25 °C/km, the expansion of water is about 7%. When the geothermal gradient is 36 °C/km, the expansion of water is about 15%. This expansion of water will cause the closed-pore system of the source rock to form, and when it accumulates to a certain extent, it will rupture and expel hydrocarbons.

In a sandstone-mudstone section, the fluid in both sandstone and mudstone pores will experience thermal pressurization effects. However, because sandstone has good permeability and is a relatively open system, it often does not lead to abnormally high pressure. The fluid in mudstone is often not well drained, which is prone to abnormal high pressure. Therefore, the fluid in mudstone always migrates towards the adjacent sandstone.

(4) Hydrocarbon Generation by Organic Matter and Abnormal Pressure

The phenomenon of abnormal high pressure caused by the hydrocarbon generation of organic matter is very common and is considered to be the most effective driving force for hydrocarbon expulsion. The essence of this superpressure generation is that the dense solid kerogen is transformed into low-density liquid hydrocarbons or gaseous hydrocarbons, and the volume of oil, gas, and water formed after the maturation of kerogen is much greater than the original volume of kerogen itself, resulting in the expansion of fluid volume in the rock (Osborne & Swarbrick, 1997). Harwood (1977) calculated that for a source rock with an organic carbon content of 1%, the net increase in volume of the generated fluid is about (6–6.8) × 10^{-4} m^3, which is equivalent to 4.5–5% of the total pore volume of shale with a porosity of 10%. These continuously newly generated fluids enter the pores, which will inevitably drive the original fluids in the pores to migrate outward. When the fluids cannot be expelled in time, it will lead to an increase in pore fluid pressure, forming a local range of abnormal high pressure and promoting hydrocarbon expulsion. For carbonate source rocks, pressure generation by hydrocarbon generation is the main driving force for hydrocarbon expulsion. Therefore, the process of hydrocarbon generation in source rocks also breeds the driving force for hydrocarbon expulsion, and the generation and primary migration of hydrocarbons is an inevitable continuous geological process.

The generation of biogenic gas is considered to be one of the main reasons for the development of overpressure in mud shale at low-temperature stages (<50 °C) in the early burial period (Meng et al., 2021). Methanogenic bacteria utilize organic compounds to synthesize and produce biogenic methane, and the continuous generation of gas can lead to oversaturation of methane in the pore water. When additional methane gas enters the fixed pore volume of the rock, a significant increase in fluid pressure may occur. Therefore, the generation of biogenic gas can effectively produce overpressure and is usually manifested as mud volcanoes and shale diapirism (Etiope et al., 2009).

Pressure-volume-temperature simulation calculations show that under a closed system, the volume expansion accompanying the generation of gaseous hydrocarbons will cause a rapid increase in fluid pressure within the source rock, creating extreme fluid overpressure, with pressure scales that can significantly exceed the load of the overlying rocks (Tian et al., 2008), and even reach twice the static rock pressure (Fig. 4.13; Meng et al., 2021).

According to the main controlling factors of overpressure formation, the phased fracturing process of abnormal pressure can be divided into hydrocarbon generation fracturing and hydraulic fracturing: hydrocarbon generation fracturing is mainly

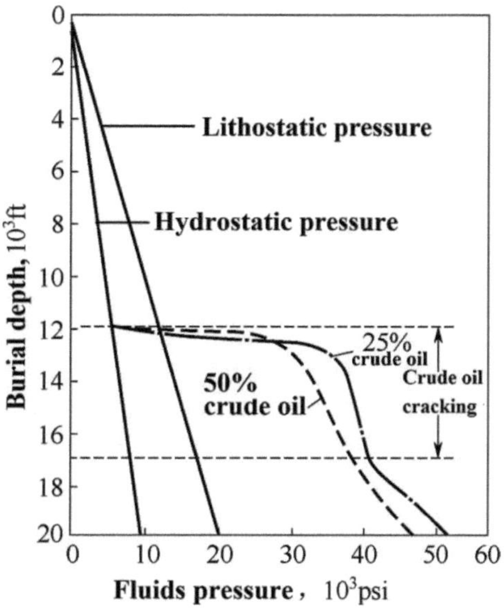

Fig. 4.13 The evolution of pressure during the thermal cracking process of crude oil, with 25% and 50% as the initial oil saturation (according to Meng et al., 2021)

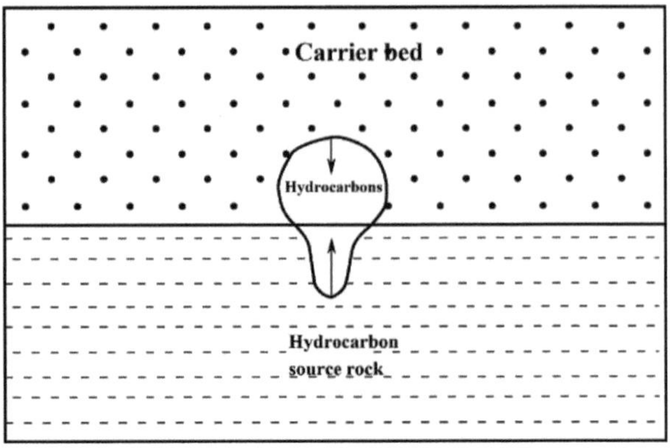

Fig. 4.14 Schematic diagram of capillary pressure difference at the contact surface between hydrocarbon source rock and carrier bed

caused by the generation and thermal cracking of hydrocarbons, leading to overpressure of pore fluid and fluid fracturing. Hydraulic fracturing is mainly caused by the undercompaction of muddy sediments, the dehydration of clay minerals, and the thermal pressurization of water, leading to overpressure of pore fluid and fracturing.

4.2.2.3 Other Driving Forces

(1) Capillary Force Action

At the contact surface between the source rock and the conductive layer, due to the pores of the source rock being much smaller than those of the conductive layer, a capillary pressure difference pointing from the source rock to the conductive layer is formed, driving the hydrocarbons to migrate from the source rock into the conductive layer. For sandstone lenses distributed within the source rock, the action of capillary force is the most direct driving force for hydrocarbon charging.

In hydrophilic source rocks, due to the different curvature radii of capillaries at both ends of the throat, the capillary pressures generated are also different, with the capillary pressure at one end of the throat being greater than that at the pore end. Therefore, under the action of the capillary pressure difference, hydrocarbons are more likely to be expelled from the throat into the larger pores (Fig. 4.14). Hydrocarbons are relatively concentrated in larger pores, which is conducive to the continuous hydrocarbon phase migration.

However, in hydrophilic media, hydrocarbons migrating in a free state are also subject to capillary resistance. If calculated based on the micropore diameter of 10–50 nm in the source rock, the capillary resistance is 1.2–2.4 MPa, and hydrocarbons must overcome such a large resistance to migrate, which seems to make the primary migration of hydrocarbons difficult. In fact, the primary migration of hydrocarbons in the source rock is the final result under the action of various dynamic conditions, such as temperature, pressure, rock fabric, migration channels, and other factors.

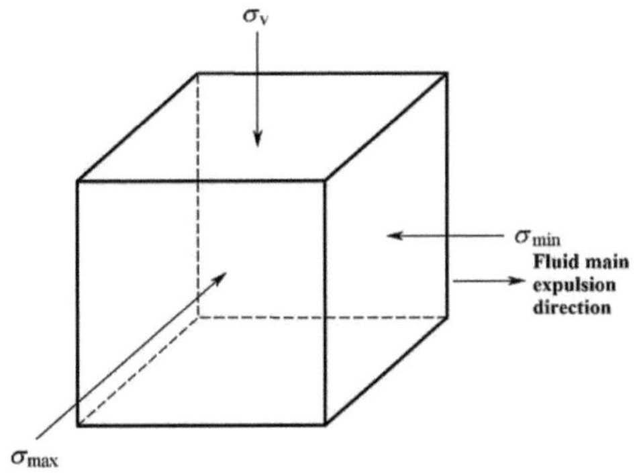

Fig. 4.15 The main direction of hydrocarbon expulsion when the maximum principal stress is a horizontal stress (according to Wang, 1996)

The resistance has limitations and temporariness, while hydrocarbon expulsion is inevitable.

(2) Tectonic Stress

The changes in porosity and fluid pressure of the source rock can not only be produced by the load of the overlying rock layers but also by horizontal tectonic stresses. Field measurements of stress have shown that horizontal tectonic stresses are not only widespread but also very strong. When the horizontal tectonic stress is greater than the vertical load stress, the maximum principal stress is in the horizontal direction, and the fluid will flow along the direction of the minimum principal stress (Fig. 4.15). The action of horizontal tectonic stress on rocks can be understood as a lateral compaction effect, similar to the vertical compaction effect caused by load stress, which can also cause instantaneous residual pressure and rock rupture, only the pressure gradient and rupture direction are different. When the stratum is not significantly inclined, the rupture direction is in the horizontal direction of the source rock. The horizontal permeability of argillaceous source rocks is generally greater than the vertical permeability, and the development of horizontal fractures can greatly improve the efficiency of lateral primary migration of hydrocarbons in the source rock. Tectonic stress can also lead to rock compaction and generate abnormal high pressure. As the stress accumulates and releases periodically, it can also cause phased expulsion of hydrocarbons from the source rock.

(3) Osmotic Action

Osmotic action refers to the movement of water from a side with low salinity through a semipermeable membrane to a side with high salinity, and the osmotic action gradually stops as the salinity difference disappears (Fig. 4.16). Osmotic action is caused by the difference in salinity and is a common phenomenon in nature.

The higher the salinity of the solution, the lower the vapor pressure, and the lower the activity of water; conversely, the higher the activity of water. Therefore, water flows from places with low salinity and high activity to places with high salinity and low activity, forming an osmotic flow. The magnitude of this action is measured

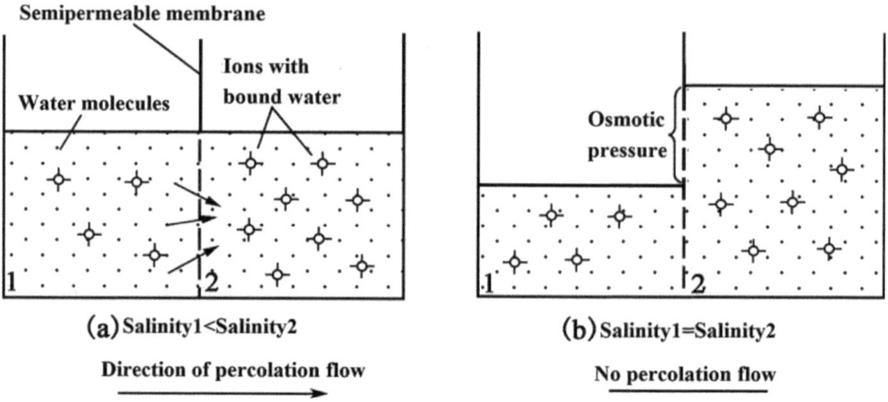

Fig. 4.16 Schematic diagram of osmotic action (according to Magara, 1978)

by the osmotic pressure. Under the influence of the osmotic pressure difference, the fluid flows until the salinity difference disappears.

In a sandstone-mudstone interbedded formation, as the fluid is expelled from the mudstone to the adjacent sandstone under the action of compaction, more salts are left behind in the sandstone, causing the salinity of the pore water in the mudstone to be much lower than that in the adjacent sandstone. Within the mudstone, due to the faster compaction of the edges than the center, pores decrease rapidly, and more salts are filtered down at the edges, causing the salinity at the edges of the mudstone to be greater than in the center. Therefore, under the action of osmosis, the fluid migrates from the mudstone to the adjacent sandstone and from the inside to the edges of the mudstone. In the undercompacted mudstone, which contains more fluids and has lower salinity, the osmotic pressure is higher, ultimately leading to pressure rupture and fluid expulsion (Fig. 4.17).

(4) Diffusion

This refers to the diffusion of hydrocarbon molecules caused by concentration differences. The diffusion action of hydrocarbons in the source rock occurs in the microporous medium, and the concentration of hydrocarbons in the source rock is higher than that in the surrounding rocks, so the direction of hydrocarbon diffusion is from the source rock to the surrounding rocks.

Although the efficiency of hydrocarbon migration by diffusion is relatively low, as long as there is a concentration difference, diffusion action will occur, and even under the conditions of high pressure in the source rock, it can proceed unimpeded. Especially in the early stages of hydrocarbon generation in the kerogen, the hydrocarbons must detach from the organic matter matrix and enter the pores, and when the source rock is deeply buried and becomes extremely dense, and the seepage of fluids is very weak or stops, molecular diffusion is almost the only way for the primary migration of hydrocarbons. The diffusion flow and seepage can convert to each other in the underground pore space, that is, it is a molecular diffusion flow in

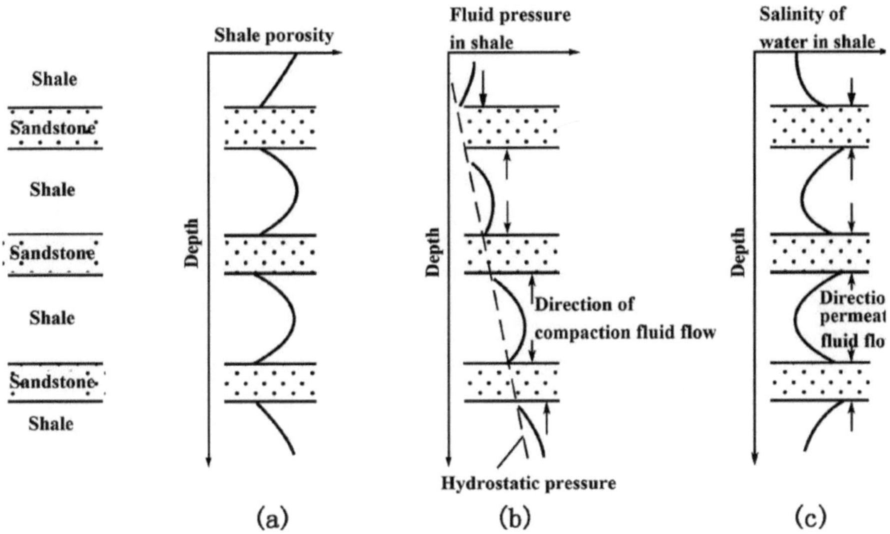

Fig. 4.17 The distribution curves of porosity, fluid pressure, and salinity of pore water in mudstone and the direction of fluid migration in a sandstone-mudstone interbedded formation (according to Magara, 1978)

the tiny pores, and when it reaches the larger micropores, it turns into seepage, and when it encounters tiny pores again, it turns back into molecular diffusion, and so on, continuously converting until it is expelled from the source rock.

(5) Cementation and Recrystallization

During the diagenetic process, cementation and recrystallization can also reduce porosity, block the discharge channels, form diagenetic seals, and the results are similar to compaction actions. As the porosity decreases, the fluids are expelled, and when the fluid discharge is not smooth, it also generates abnormal pressure or intensifies the original abnormal pressure, ultimately leading to rock rupture and hydrocarbon expulsion. Especially for carbonate source rocks that are mainly chemically diagenetic, this action is more important and is considered to be an effective driving force for the primary migration of hydrocarbons in carbonate source rocks.

(6) Buoyancy

In the tiny microcapillary pores of the source rock, the buoyancy force of hydrocarbons relative to the intermolecular forces between hydrocarbons and rocks and capillary resistance is very small. In addition, in the complex pore structure of the source rock, it is difficult for hydrocarbons to form a sufficient length (height) to generate enough buoyancy for primary migration. Therefore, buoyancy is generally not considered in primary migration, but in local larger capillary pores or structural fractures in the source rock, the action of buoyancy still exists and is a driving force for hydrocarbons to be expelled upwards or in the direction of inclination.

4.2.2.4 Evolution of Hydrocarbon Migration Dynamics

The driving forces for hydrocarbon primary migration are diverse, and the main expulsion dynamics vary at different stages of source rock evolution. These differences are mainly influenced by the diagenetic actions and the degree of organic matter evolution of the source rock, as they lead to changes in the conditions of primary migration dynamics, channel conditions, hydrocarbon phases, and quantities. Generally speaking, at medium to shallow depths, compaction is the main driving force, while at medium to deep layers, abnormal pressure is the main driving force. Since the generation of hydrocarbons mainly occurs at medium to deep layers, abnormal pressure is more significant.

Taking argillaceous source rocks as an example: during the diagenetic stage, the porosity is relatively high, there is a large amount of primary formation water, and a small amount of low-mature to immature oil is formed. At this stage, compaction is the main driving force. In the early stage of the maturation process, after a large amount of primary formation water has been expelled and the permeability of the mudstone has become smaller, the fluid seepage is hindered. Then organic matter begins to generate hydrocarbons, and montmorillonite dehydrates in large amounts. At this stage, undercompaction, organic matter hydrocarbon generation, montmorillonite dehydration, and thermal pressurization of fluids all work together to force the expulsion of fluids from the source rock. By the middle stage of the maturation process, organic matter has entered the peak oil generation period, which is also the second stage of montmorillonite dehydration, a large amount of newly generated fluids (oil, gas, water) continuously enter the pores, and the pore pressure continues to increase, eventually leading to the formation of abnormal high pressure, fracture, pressure release, and surging expulsion of hydrocarbons. After entering the overmature stage, organic matter has entered the gas generation peak period, the source rock is relatively compact, and the main driving force for hydrocarbon expulsion mainly relies on the organic matter gas generation and the thermal pressurization of gases (Table 4.1).

Table 4.1 The driving forces for hydrocarbon expulsion at different thermal evolution stages of argillaceous source rocks

Burial depth (m)	Temperature (°C)	Stage of organic matter thermal evolution	Driving forces for hydrocarbon migration
0–1,500	10–50	Immature	Normal compaction, osmosis, diffusion
1,500–4,000	50–150	Mature	Normal to undercompaction, montmorillonite dehydration, organic matter hydrocarbon generation, fluid thermal pressurization, osmosis, diffusion
4,000–7,000	150–250	Highly mature–overmature	Organic matter gas generation, gas thermal pressurization, diffusion

4.2.3 Primary Migration Channels

Migration channels are one of the primary factors affecting the primary migration of hydrocarbons. The type, size, properties, formation period, and distribution characteristics of the channels will influence the phase state, flow type, migration quantity, period, and migration efficiency of hydrocarbon migration, making it an important issue in the study of primary hydrocarbon migration. The main channels for the primary migration of hydrocarbons include larger pores and microlamination surfaces, structural fractures and faults, micro-fractures, and organic matter or kerogen networks.

4.2.3.1 Larger Pores and Microlamination Surfaces

Pores and microlamination surfaces are the main migration pathways during the immature to low-maturity stage of organic matter. Larger pores refer to pores with a diameter greater than 100 nm in the source rock, including large micropores in microcapillaries and a small number of capillary pores (with a diameter less than 2 μm). Although the latter accounts for only a small fraction of the pores in argillaceous source rocks (an average of less than 5%), they not only allow the diffusion flow to pass smoothly but also facilitate Darcy flow, making them the most important expulsion channels.

Understanding the petrophysical characteristics of source rocks is a prerequisite for understanding primary migration channels. Petrophysical research indicates that source rocks possess certain permeability. Argillaceous source rocks are primarily composed of various clay minerals, various clastic and non-clastic minerals, and organic matter, with clay minerals making up the majority and exhibiting strong heterogeneity. Microscopic observation reveals a porous system composed of pores, throats, vugs, and fractures of varying sizes, characterized by a network-like and limited connectivity. Carbonate source rocks such as limestone, argillaceous limestone, and biogenic limestone have a crystalline structure with intercrystalline pores, intracrystalline pores, biological framework pores, and micro-fractures, also exhibiting limited connectivity. Although there is still much debate about the expulsion mechanism of carbonates, the existence of these pores and fractures undoubtedly benefits migration.

Maffhews (1997) compared the pores in argillaceous rocks with the size of hydrocarbon molecules (Fig. 4.18), suggesting that only water and methane molecules can pass through submicropores to ultramicropores (pore diameter of 0.3–1 nm), low molecular weight alkanes, aromatics, and asphaltenes can pass through mesopores to small micropores (pore diameter of 1–10 nm), while large molecular aggregates and colloidal hydrocarbons can only pass through large micropores (pore diameter greater than 50 nm) and open fractures. Microlamination surfaces are the vertical change interfaces of sediments, with good permeability, and are an important lateral migration pathway in argillaceous source rocks. During the mature to over-mature

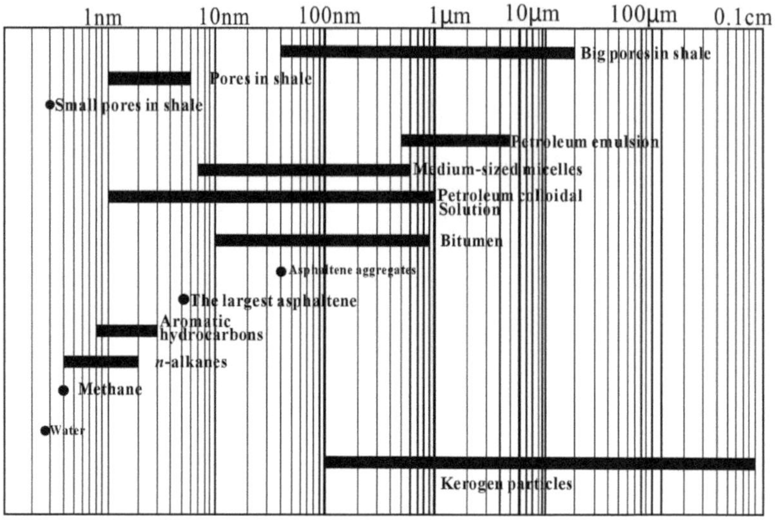

Fig. 4.18 Comparison of pore sizes in shale with hydrocarbon, water, and kerogen molecules (according to Maffhews, 1997)

stage of organic matter, they can form a three-dimensional migration channel system with micro-fractures and kerogen networks.

4.2.3.2 Micro-fractures

Micro-fractures generally refer to fractures with a width less than 100 μm, with most measured widths ranging from 10 to 25 μm, and the smallest widths can be 3–10 nm. They are the main migration pathways during the mature to over-mature stage. At this stage, due to the great burial depth of the source rock, the porosity and permeability are extremely low, and there are essentially no larger pores. The fluid in the pore space can no longer undergo primary migration through the narrow pores, and the internal source rock has become a closed or semi-closed system. At this time, the abnormal high pressure generated within the source rock becomes the main driving force for the primary migration of hydrocarbons, while also promoting the formation of micro-fractures.

The view that abnormal high fluid pressure can lead to the formation of micro-fractures in source rocks has been widely accepted. When the fluid pressure exceeds 1.42–2.4 times the hydrostatic pressure, micro-fractures can be generated in the source rock (Snarsky, 1962). When the pore pressure of the surrounding medium is equal to the pore pressure in the fracture, the fracture can remain open for a long period. When the pore fluid pressure of the surrounding medium is lower than the initial pressure in the fracture, the fracture will close rapidly due to the fluid seepage into the surrounding pores (Rouchet, 1981). After the fracture opens, the previously enclosed fluid is expelled along the fracture, and then the fracture closes under the load of the overlying strata. Later, a new high pressure can be established, leading to the formation of new micro-fractures and expulsion of hydrocarbons (Ungerer et al., 1983, Fig. 4.19).

According to the difference in strike between fractures and the bedding planes of mudstones, fractures can be divided into high-angle fractures, low-angle fractures, and sub-horizontal fractures. These fractures are interconnected in space, forming a three-dimensional expulsion network within the source rock. Due to the fact that the tensile strength of mudstones parallel to the bedding direction is much greater than that perpendicular to the bedding direction, even when the maximum compressive stress is vertical, when the fluid overpressure generated by the hydrocarbon generation first exceeds the sum of the load of the overlying rocks and the vertical tensile strength, the sub-horizontal hydraulic fracture will preferentially form (Wang et al., 2018). Therefore, the formation of sub-layer fractures is more closely related to the thermal evolution and expulsion process of the source rock (Meng et al., 2021).

When the pore fluid pressure is high enough to cause the enclosed source rock to produce micro-fractures, these micro-fractures connect with the pores, forming a micro-fracture-pore system. Under the drive of abnormal high pressure, the fluid flows out of the source rock through the micro-fracture-pore system, and very little flows through the intergranular seepage. When part of the fluid is expelled and the pressure drops, the micro-fractures close, and when the pressure recovers and the micro-fractures reopen, a new surge occurs. Obviously, this process of hydrocarbon expulsion is intermittent and episodic.

Tissot et al. (1971) conducted heating and pressurization simulation experiments on clay rocks with fixed organic matter (Fig. 4.20). The solid line in the figure represents the change in pressure, and the dashed line represents the gas volume

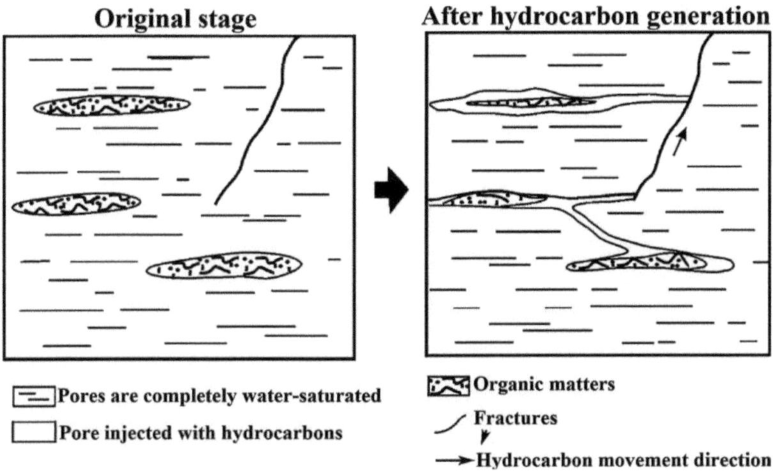

Fig. 4.19 Formation of micro-fractures and injection of hydrocarbons during kerogen generation (according to Ungerer, 1983)

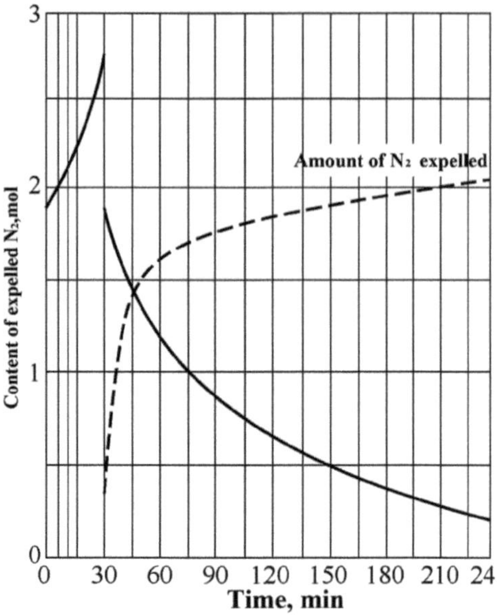

Fig. 4.20 The effect of micro-fractures on hydrocarbon migration in organic clay under pressure (according to Tissot et al., 1971)

expelled. The initial mechanical pressure was 44 MPa, and the amount of N_2 that could be expelled during heating was minimal. It was not until the pressure increased to 54 MPa that the clay rock began to fracture, producing micro-fractures, and the volume of expelled N_2 increased sharply. At the same time, the pressure began to release, and the rate of increase in the volume of expelled N_2 slowed down, indicating that the micro-fractures were gradually closing.

Due to the strong plasticity of argillaceous source rocks, a very high excess pressure is required to cause them to fracture, and when the fluid is expelled and the pressure is released, the micro-fractures close, making them difficult to observe. In addition, it is difficult to distinguish between tectonic fractures caused by stress differences and micro-fractures caused by high pore fluid pressure underground, and there is a phenomenon of mutual superposition. Generally, micro-fractures are larger than micro-pores, and they are more straight and less tortuous, so the capillary resistance to hydrocarbon migration is relatively small, which can greatly improve the expulsion efficiency of the source rock. If hydrocarbon generation is an important reason for the production of abnormal high pressure and micro-fractures, and micro-fractures are important channels for primary migration, it indicates that hydrocarbon generation and expulsion must be a continuous geological process.

4.2.3.3 Organic Matter or Kerogen Networks

The organic matter in source rocks is not dispersed but is mainly distributed along microlamination surfaces. McAuliffe (1979) further confirmed that there is a three-dimensional kerogen network in source rocks. Relatively enriched organic matter can also make microlamination surfaces hydrophilic, which is conducive to hydrocarbon migration. If there is kerogen connection between microlamination surfaces, then during the large-scale oil generation stage, not only can the microlamination surfaces themselves serve as migration channels, but they can also form a three-dimensional interconnected, capillary resistance-free hydrophilic network in space, thus becoming a good channel for primary migration.

4.2.3.4 Structural Fractures and Faults

The structural fractures referred to here are mainly fractures formed in the source rock under the action of tectonic stress differences. However, for a fault basin, it is generally believed that below 2000 m, the vertical stress is dominant, while above 2000 m, the horizontal stress is dominant (Li, 2004), which can produce nearly vertical tension fractures or shear fractures. For a compressional basin dominated by horizontal stress, parallel to the bedding plane tension fractures or shear fractures can be produced. The width of tension fractures is generally greater than 100 μm, which belongs to the capillary diameter, and hydrocarbons can pass through it as long as they overcome the capillary resistance.

Faults that break through the source rock can connect larger pores, microlamination surfaces, and micro-fractures to form a network of primary migration channels. Many studies have shown that hydrocarbon fluorescence is strongest on micro-fractures and microlamination surfaces in the source rock, and it is continuously distributed, with no obvious compositional differentiation in the extension direction, indicating that it is the main channel for primary migration. Faults can also cause the source rock to dock and juxtapose with other strata, allowing hydrocarbon fluids to cross the fault plane into the conductive layer. In addition, the seismic pump effect is beneficial to primary migration. When the fault opens, the volume of the fault zone expands, the fluid pressure decreases, and the fluid in the source rock is sucked in. When the fault closes, the fluid is compressed, and the active fault has a vertical opening, and the fluid will be vertically expelled or enter the adjacent permeable strata.

4.2.3.5 Suture Lines

Suture lines are widely distributed in carbonate rock strata and are the product of pressure dissolution during the post-depositional stage. Suture lines are often distributed along the bedding plane, and there are also many that are oblique or perpendicular to the bedding plane, which are often intertwined with structural fractures in the rock to form the same system. Under microscopic observation, open micro-fractures and micro-pores are also found in the suture lines, and there are oil impregnation and liquid asphalt. Therefore, like various fractures, suture lines are also important channels for primary migration.

In summary, due to the heterogeneity of the source rock itself, it is inevitable to form pores and fracture systems of different sizes and crisscross during the burial process, coupled with secondary pores, micro-fractures, kerogen networks, faults, suture lines, etc., forming a variety of expulsion channels. Among them, micro-pores and micro-fractures are the main ones, and there are also larger pores and fractures, each with different roles. Larger pores and fractures, although fewer, can occur with higher efficiency of volume flow, while in small pores and fractures, diffusion flow can occur, and the two can be converted and supplemented in time and space. Therefore, there are always migration channels in fine-grained source rocks, and as long as there is a driving force, hydrocarbons can always be expelled.

4.2.4 Primary Migration Period and Migration Distance

4.2.4.1 The Main Period of Primary Migration

The period of primary migration refers to the entire period from the beginning of hydrocarbon expulsion from the source rock to the termination of expulsion. As early as the 1980s, Li Mingcheng proposed the period when primary migration began to expel in large quantities, that is, the expulsion threshold. The so-called expulsion threshold refers to the critical point during the thermal evolution process of the source rock, where the generated oil and gas, after satisfying various forms of retention needs such as self-adsorption, pore water dissolution, oil dissolution, and capillary closure, begin to be expelled in a free phase in large quantities. Researchers have tried various methods to determine the expulsion threshold, such as using the depth or period when the source rock reaches the critical migration saturation (5%) as the expulsion threshold (Li, 1989), or using various indicators on the geochemical profile such as hydrocarbon content, chloroform asphalt "A", total hydrocarbon/total organic carbon, source rock average conversion rate ($S_1/(S_1 + S_2)$), and the depth corresponding to the R_o value of different types of kerogen (for example, type I kerogen is set to $R_o = 0.55\%$, type II kerogen $R_o = 0.75\%$).

The main period of primary migration refers to the period when oil or gas is expelled from the source rock in large quantities. For oil, only the period when a large amount of oil is generated, transported in the oil phase, and has the conditions for transportation and channeling is the main migration period. For natural gas, its generation conditions are more extensive than oil, and its migration phase is more diverse than oil. As long as the migration conditions are met, there will be a corresponding main migration period for each gas generation stage. From the perspective of the dynamic conditions provided by the source rock, during the organic matter maturation stage, compaction, undercompaction, clay mineral dehydration, organic matter hydrocarbon generation, and thermal pressure increase have become important driving forces, which are the main periods for episodic fracturing and episodic expulsion. Therefore, the main period of primary migration of oil and gas is the mature stage of organic matter thermal evolution.

Analyzing the main period of primary migration also needs to consider the phase state of oil and gas migration, because the phase state determines the efficiency and quantity of migration, which is an important aspect of migration mechanism research. For oil, since the aqueous solution phase, gas solution phase, and diffusion phase cannot constitute a large-scale migration of oil, only the oil phase migration is the most meaningful. Therefore, the period when oil phase migration conditions can occur should be the main period of oil primary migration. Obviously, oil phase migration mainly occurs during the large-scale oil generation stage, because generally at this time, a higher oil saturation can be achieved to meet the needs for the formation of oil phase migration. For natural gas, it can be transported in various phases, and the issue of the main migration period is not as prominent as oil. From the above analysis, it can be seen that the main migration period of natural gas can occur at any stage after the generation of natural gas.

4.2.4.2 Primary Migration Distance

In fact, it is also a problem of the effective expulsion thickness of the source rock. The distance of primary migration depends on the thickness of the oil-generating layer from a macro perspective. Due to the influence of factors such as permeability, expulsion driving force, homogeneity and thickness of the source rock, the expulsion of hydrocarbons in the source rock is not uniform. Only within a certain thickness range can hydrocarbons be effectively expelled. The expulsion efficiency is high near the contact area between the source rock and the adjacent reservoir (or conductive layer), and the further towards the center of the source rock, the lower the expulsion efficiency, and some may not be expelled at all, becoming dead hydrocarbons. We refer to the thickness within the source rock that can effectively expel hydrocarbons as the effective expulsion thickness.

The effective expulsion thickness of the source rock is related to the combination of the source rock and the adjacent conductive layer, and the permeability of the conductive layer. The larger the contact area between the source rock and the adjacent conductive layer, and the better the permeability of the conductive layer, the greater the effective expulsion thickness. Therefore, the source rock is not necessarily the thicker the better. Only those oil-generating rock systems with small single-layer thickness and frequent interlayering with reservoirs (or conductive layers) have high expulsion efficiency. The overly thick blocky argillaceous source rocks are not the most favorable, and a considerable part of the thickness cannot be effectively expelled.

The Fig. 4.21 shows that the content of hydrocarbons, resins, and asphaltenes in the oil-generating rock above the reservoir in the Algerian region gradually increases with increasing distance from the reservoir, and the closer to the reservoir, the less the content. This indicate that the hydrocarbons in the oil-generating layer within a certain distance from the reservoir have been effectively expelled, and those in the oil-generating layer far from the reservoir cannot be effectively expelled. This distance is the effective thickness of the source rock expulsion, and in this example,

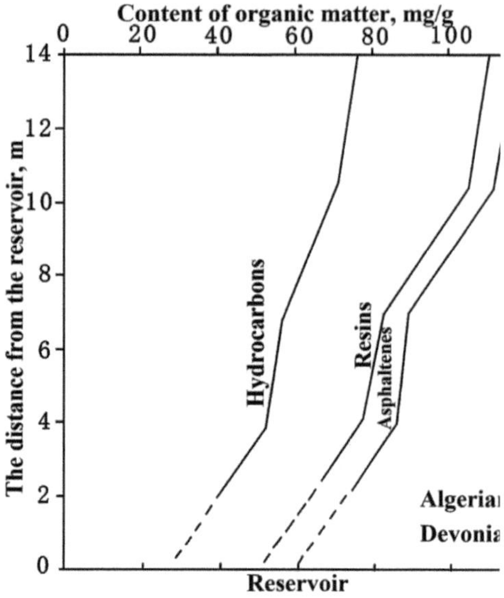

Fig. 4.21 Distribution of hydrocarbons, resins, and asphaltenes in the shale source rock above the reservoir in Algeria (according to Tissot & Pelet, 1971)

the effective expulsion thickness of the oil-generating layer is about 28 m (14 m above and below the reservoir). The curve measured in this figure is a curve where the hydrocarbon content has reached a dynamic equilibrium after a long period of expulsion. This kind of continuous curve with differences in hydrocarbon content also indicates that there is hydrocarbon migration in the middle of the oil-generating layer. Therefore, the view that the oil-generating layer is too thick to be meaningful may not be correct. At present, the micro-fracture expulsion favored by most people does not have the problem of effective expulsion thickness, because micro-fractures are likely to crack open first from the middle of the thick oil-generating layer with the greatest abnormal pressure and expel. It seems that the oil-generating layer is better to be thicker, so that there is a supply and expulsion of hydrocarbons for a long time.

4.2.5 Primary Migration Models

Based on the phase state of oil and gas migration, driving mechanisms, migration channels, stratum temperature, pressure, fluid properties, and other conditions, people have summarized different primary migration models, explaining the migration mechanism from different aspects, and helping us to deeply understand the essence of oil and gas primary migration. Zhang et al. (1999) summarized the primary migration of oil and gas into three models: normal compaction expulsion model, abnormal pressure micro-fracture expulsion model, and light hydrocarbon diffusion auxiliary migration model. The three have differences in migration phase state, driving force, and channels, which can be used to describe the expulsion characteristics of source rocks at different evolutionary stages.

4.2.5.1 Compaction Expulsion Model

This model is used to describe the expulsion action of source rocks in the immature to low-mature stage under normal compaction conditions. At this stage, the burial depth of the source rock layer is not large, the amount of oil and gas generated is small, the source rock has more pore water, the permeability is relatively high, some oil and gas can be dissolved in water in a water-soluble state, and some can be dispersed free oil and gas drops. At this time, the instantaneous residual pressure produced by compaction is the expulsion driving force of the source rock, and it is transported to the conductive layer or reservoir layer through the source rock pores along with the compaction water. This model is based on the impact of compaction on the expulsion of liquid from the source rock.

4.2.5.2 Abnormal High Pressure Expulsion Model

This model is used to describe the expulsion action of source rocks in the mature to over-mature stage under abnormal high pressure conditions. At this stage, the source rock layer has been compacted, the pore water is less, the permeability is lower, the source rock is not easy to expel liquid, a large amount of oil and gas is generated by organic matter, and the pore water is not enough to completely dissolve all oil and gas, and a large amount of oil and gas is in a free state. At the same time, various factors such as undercompaction, montmorillonite dehydration, organic matter hydrocarbon generation, and thermal pressure increase lead to the continuous increase in pore fluid pressure to form abnormal high pressure, becoming the main driving force for expulsion. Micro-fractures formed under abnormal high pressure become the main channels for the primary migration of oil and gas, and the persistence of fracture opening and the connectivity between fractures in various directions control the efficiency of expulsion. When rupture occurs, under the action of abnormal high pressure, the fractures in shale can expand to 1.2–2 times the length of the basic unit, and in mudstone, they can expand to 2–3.5 times the size of the basic unit (Su et al., 2020).

Due to abnormal high pressure, micro-fractures will be produced, so the expulsion process at this stage has periodicity. When the abnormal high pressure in the source rock exceeds the rock's fracture limit, micro-fractures are formed in the source rock, and the high-pressure pore fluid is expelled from the source rock through the micro-fractures. After the fluid is expelled, the internal pressure of the source rock decreases, the micro-fractures close, and the expulsion process is paused. When the internal pressure of the source rock accumulates again and reaches the fracture limit of the source rock, the micro-fractures reopen, and a new expulsion process occurs (Ungerer, 1983). This process can be repeated, and a large amount of hydrocarbons is expelled from the source rock (Fig. 4.22). Therefore, abnormal high pressure expulsion is an episodic expulsion process, and overpressure micro-fracture expulsion is an important way for the primary migration of oil and gas.

When the abnormal high pressure inside the source rock pore is not enough to cause the rock to produce micro-fractures, if the pore throat is not too narrow, or because there is a continuous organic phase and kerogen three-dimensional network, making the capillary force not too large, then oil and gas can slowly be expelled from the source rock without the existence of fractures. In this case, the expulsion of oil and gas driven by compaction should be a continuous process. When the pore fluid pressure is high enough to cause the source rock to produce micro-fractures, these micro-fractures can also connect with the original pores to form a micro-fracture-pore system, and under the drive of abnormal high pressure, oil, gas, and water are expelled from the source rock through the micro-fracture-pore system.

When there is a fault that breaks through to the effective source rock, oil and gas can also undergo primary migration along the fault under the drive of overpressure. At this time, there are two types of expulsion: overpressure-dominated and structure-dominated. The overpressure-dominated type refers to the continuous accumulation of overpressure, and when it reaches the fracture limit of the source rock, shearing rupture occurs, forming ruptured faults and a large number of accompanying fractures, and hydrocarbon fluids are injected into the faults and fractures, and are expelled under the drive of overpressure, forming a repeated episodic expulsion process of closed pressure increase—rupture pressure release—closed pressure increase, and multiple expulsion events along the same fault will lead to an increase in fault displacement (Shao et al., 2019). The structure-dominated type refers to the expulsion mode where the fault itself becomes the pressure relief channel for the overpressure fluid during the large-scale oil and gas generation and expulsion stage, and the multi-stage tectonic activity has played a positive role in the multi-stage expulsion of the source rock.

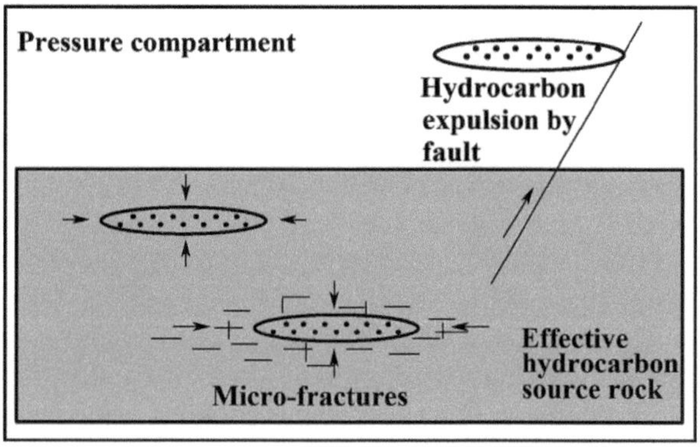

Fig. 4.22 Hydrocarbon expulsion model under abnormal high pressure

4.2.5.3 Light Hydrocarbon Diffusion Expulsion Model

Light hydrocarbons, especially gaseous hydrocarbons, have strong diffusion capabilities, and the diffusion of gases in the source rock is universal. Many scholars believe that the primary migration of gases relying on diffusion only occurs over short distances within the source rock layer (Barker, 1980; Hunt, 1979; Leythaeuser, 1982). After gases enter the nearest conductive layer through short-distance diffusion, they change to other methods for further migration. Diffusion, as a molecular migration process driven by concentration differences, can occur at any stage of the evolution of the source rock. However, during the stages where compaction expulsion and micro-fracture expulsion play the main role, the expulsion efficiency of diffusion is too low to be significant, so the diffusion of light hydrocarbons can be considered as an auxiliary migration model. However, for deep reservoirs that are very tight, or strata in an abnormal high-pressure state, fluid seepage is almost impossible, and the diffusion of natural gas becomes more important, even the only way.

4.3 Secondary Migration

Compared to primary migration, the medium conditions for secondary migration have changed significantly, hence there are many differences between secondary and primary migration.

4.3.1 Phase State, Flow Type, and Critical Saturation of Secondary Migration

4.3.1.1 Phase State of Secondary Migration

Generally speaking, the phase state of oil and gas during the initial stage of secondary migration basically inherits that of the primary migration. Due to the differences in the physical conditions of the dynamic environment between secondary and primary migration, changes and transformations in the migration phase state will occur quickly (Li, 2013). If the research purpose is to gather and form a reservoir, the secondary migration of oil and gas is dominated by the free phase (Mann et al., 1997).

(1) Phase State of Oil Secondary Migration

The free phase is the most effective and important migration phase state for oil secondary migration. There are many favorable factors for oil to migrate in the free phase: First, compared with the source rock, the reservoir has larger pore sizes and pore spaces, allowing hydrocarbon molecules to pass through freely without being restricted by the pore size. Second, the pore diameter in the reservoir is larger, and the capillary resistance that oil experiences during migration in water-wet pores is much smaller than that in the source rock. Third, the aggregation of oil in the trap eventually manifests as the aggregation of the oil phase, without the need for phase transformation. Therefore, the free phase will inevitably become the most important phase state for oil and gas secondary migration. However, the state of free phase oil varies at different times during secondary migration. In the early stage, the oil particles are smaller, with more microscopic and submicroscopic oil particles. As the migration process develops, these dispersed small oil particles gradually connect to eventually form continuous oil beads or oil bodies for migration.

Oil migrating in the form of a water-soluble phase and gas-soluble phase into the conduit layer will, due to the decrease in temperature, pressure, and increase in salinity, reduce the solubility of oil in water and gas, thus becoming a free phase. These exsolved oils slowly aggregate into oil beads of varying sizes dispersed throughout the secondary migration channel. There needs to be an opportunity to connect into larger oil bodies to possibly undergo secondary migration, but overall, the efficiency of oil migration in the form of a water-soluble phase is very low, and it is difficult to gather and form a reservoir in this way.

(2) Phase State of Natural Gas Secondary Migration

Compared to oil, natural gas has two unique physical properties, namely the water solubility and diffusivity of natural gas, leading to a variety of types of natural gas secondary migration phase states. It can migrate in a free phase state, a water-soluble phase state, or a molecular diffusion state.

After water-soluble phase natural gas enters the conduit layer, due to the increase in water quantity, the water solution full of natural gas will first become unsaturated, and natural gas will not immediately exsolve. Due to the influence of changes in temperature, pressure, and salinity, some natural gas will eventually exsolve from the water to become a free phase The free natural gas, due to the increase in burial depth of the strata and the increase in pressure, will also dissolve in water. During the secondary migration process, natural gas can also dissolve in oil and migrate in an oil-soluble phase.

The migration of natural gas in the form of a diffusion phase in the conduit layer is the biggest characteristic different from the migration of liquid oil. As long as there is a concentration difference, the molecular diffusion of natural gas can occur. In both primary and secondary migration, the diffusion phase does not require a phase transformation, especially when fluid seepage is stagnant or in a trap closure state, the diffusion phase migration of natural gas is more important. Similarly, for oil and gas preservation, the diffusion action will lead to the loss of natural gas.

4.3.1.2 Flow Type of Secondary Migration

The flow type of oil and gas secondary migration depends on the properties of the hydrocarbons themselves (the size of oil and gas molecules and the size of the aggregated oil and gas volume) and the migration driving force and channel conditions. Under open secondary migration conduit conditions, oil and gas migrating in the form of a free phase, water-soluble phase, oil-soluble phase, and gas-soluble phase can migrate by seepage, free phase oil and gas, and gas-soluble oil and oil-soluble gas can also migrate by buoyancy flow, and light hydrocarbon molecules migrate by diffusion flow; under closed conduit conditions (such as abnormal pressure sealing boxes, seismic pump effects, etc.), oil and gas mainly migrate by surging flow.

Under normal circumstances, oil, gas, and water coexist and migrate in multiphase in the strata, with common oil–water, gas–water two-phase seepage, or oil–gas–water three-phase seepage. When fluids seep, each phase of the fluid has a certain effective permeability, the size of which depends on its own saturation, and oil and gas must reach a certain critical saturation to migrate. In most oil and gas basins, the migration of oil, gas, and water in three phases is objectively present, but due to the strong mutual solubility between oil and gas, people rarely truly consider it in three phases and often simplify it to oil–water and gas–water two-phase seepage for processing.

4.3.1.3 Critical Saturation of Secondary Migration

Like oil and gas primary migration, oil and gas must reach or exceed their critical saturation to migrate in the form of a free phase in the conduit layer. Because reaching such a saturation can form a connected migration channel, this part of the saturation is mainly lost in the form of residual oil along the migration channel. Regarding the magnitude of the critical saturation of oil and gas secondary migration, the results of different scholars' research are not the same, but the difference is not large. Levorsen (1954) conducted oil–water two-phase drainage experiments on water-wet sandstones, and the results showed that when the oil phase saturation is lower than 10%, the oil phase cannot flow. McAuliffe (1979) believes that the critical saturation for oil and gas to migrate horizontally along the conduit layer in the form of a free phase and to float upwards under the action of buoyancy is both 20–30%. Schowalter (1979) and Sell (1993) and others confirmed through physical simulation experiments that for most conduit layers, the critical saturation for oil phase to undergo secondary migration is suitable for 10–30%.

There is also a similar understanding of the critical saturation for natural gas secondary migration as for oil. Hevorsen (1967) used a two-phase filtration experiment to verify that the gas saturation of the non-wetting phase must reach 5–10% to produce gas phase migration. Schowalter (1979) also believes that the critical saturation for gas migration should reach more than 10%. Hao et al. (1994) believes that taking 10% as the critical saturation for gas is suitable for most reservoir rocks.

4.3.2 Driving Force and Resistance of Secondary Migration

The environment of oil and gas secondary migration is very different from that of primary migration, so the corresponding migration resistance and driving force are also different.

4.3.2.1 Resistance of Secondary Migration

Oil and gas migrating in different phase states will have different resistances. When oil and gas migrate in the form of a free phase, their resistance mainly comes from capillary pressure and adsorption force between rock particles. When migrating in the form of a water-soluble phase, the resistance mainly comes from the friction between water and the pore wall. When migrating in the form of molecular diffusion, the resistance mainly comes from collisions between molecules and between molecules and the pore wall. Under certain stratum conditions, buoyancy can also act as a resistance. Below we mainly analyze the capillary pressure and adsorption force resistance experienced by free phase oil and gas.

(1) Capillary Pressure

The underground rock pore system is mostly water-wet, and the free phase oil and gas migrating in it will inevitably be affected by capillary force. Due to the different radii of the rock pores and throats, the capillary pressure experienced by oil and gas varies, and it will affect secondary migration. As shown in Fig. 4.23, the buoyancy is not enough to deform the surface of the oil bead to enter the throat (a). When the buoyancy or other external forces increase, the oil bead deforms, and its upper end enters the throat (b). Due to the different radii at the upper and lower ends of the oil bead (representing the throat radius and pore radius, respectively), the capillary pressures at both ends are also different. The capillary pressure at the upper end is greater than that at the lower end, and the direction of the capillary pressure difference points to the lower pore. Compared with the rising oil bead, the capillary pressure difference is resistance. The curvature radii at the upper and lower ends of the oil bead are the same, and the capillary pressures at both ends are equal, the capillary pressure difference is zero, and there is no capillary resistance (c). The capillary pressure difference points to the upper pore, and the capillary force is not resistance for the rising oil bead, but an additional driving force for the oil bead to rise (d).

Under stratum conditions, the secondary migration of oil and gas through the rock pore system must go through the above four states, which is a continuous process. Regardless of whether the capillary pressure difference acts as resistance or driving force, the secondary migration of oil and gas is the result of the comparison between capillary force and buoyancy and other migration driving forces. The maximum capillary resistance of oil and gas secondary migration depends on the capillary pressure difference generated by the smallest throat and the largest pore of the rock.

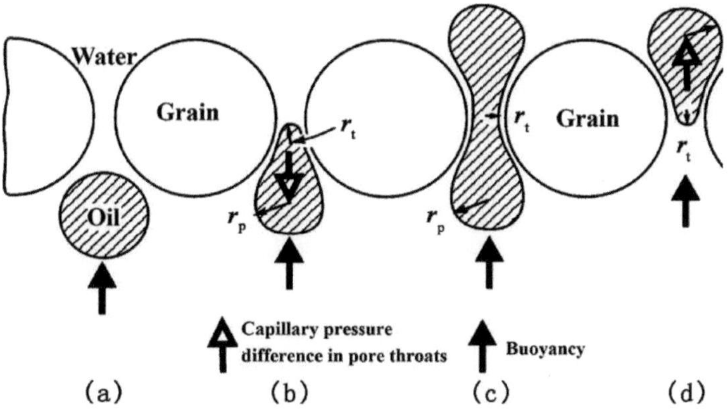

Fig. 4.23 Capillary resistance when oil and gas migrate in reservoirs Rc—throat radius; Rr—pore radius (modified from Berg, 1975)

Oil and gas in rocks will choose the direction of the smallest resistance channel for migration, that is, along the path composed of the largest pores and throats.

At a certain temperature and pressure, the interfacial tension between oil–water and gas–water is different, so the capillary force of oil in rock pores is smaller than that of natural gas (about 1/2 times). From the perspective of capillary force alone, under the same burial conditions, oil is easier to migrate than natural gas.

(2) Adsorption Force

The adsorption of reservoir rocks is generally weaker than that of source rocks. Sandy reservoir rocks are composed of granular quartz, feldspar, rock debris, and cementing materials, etc., and contain little organic matter. They are mainly adsorbed by hydrophilic media on hydrocarbons, and their adsorption force is relatively weak. However, they still have a certain adsorption for oil and gas, especially low permeability or ultra-low permeability reservoirs have stronger adsorption, and there are more residual oil and gas. Therefore, during the secondary migration process, adsorption force is still a migration resistance, but it is much smaller than that of source rocks.

4.3.2.2 Driving Force of Secondary Migration

Under stratum conditions, the driving forces for oil and gas secondary migration include buoyancy, water dynamics, tectonic stress, and molecular diffusion force, etc., which have different driving mechanisms for different fluid types and phase states. When oil and gas migrate in the form of a free phase, the driving force is mainly buoyancy under static water conditions, and in addition to buoyancy, the degree of water dynamics also has an effect depending on its magnitude and direction under dynamic water conditions; when migrating in the form of a water-soluble phase, the driving force is mainly water dynamics. For natural gas, there is also the diffusion of gas molecules. Since secondary migration is most important in the form of a free phase, buoyancy becomes the most important driving force.

(1) Buoyancy

In the stratum water environment, due to the density difference between oil, gas, and water, the free phase of oil and gas will be affected by buoyancy. Taking the buoyancy of oil in water as an example, the magnitude of the buoyancy can be expressed by the formula:

$$F_{w0} = V(\rho_w - \rho_o)g \tag{4.8}$$

where F_{wo} is the buoyancy of oil in water, Pa; V is the volume of the oil phase, cm^3; g is the acceleration of gravity, 980 cm/s^2; ρ_w is the density of the stratum water, g/cm^3; ρ_o is the density of oil, g/cm^3.

Oil and gas must first overcome capillary resistance during migration. As shown in Fig. 4.23b, an oil droplet migrates through the pore throat in the water-wet underground environment. The capillary pressure and buoyancy are opposed to each other until the internal curvature radius of the deformed oil droplet is equal at the upper and lower ends. Only when the buoyancy of oil and gas is greater than the capillary resistance can oil and gas move, which can be expressed by the following formula:

$$V_o(\rho_w - \rho_o)g > 2\sigma(1/r_t - 1/r_p) \tag{4.9}$$

where V_o is the volume of the oil phase, cm^3; σ is the oil–water interfacial tension, 10^{-5} N/cm; r_t is the throat radius, cm; r_p is the pore radius, cm; g is the acceleration of gravity, 980 cm/s^2; ρ_w is the density of water, g/cm^3; ρ_o is the density of oil, g/cm^3.

Regarding this issue, a simple experiment by American scholar K. A. Hill can provide a strong explanation. Figure 4.24 shows a front view of a rectangular box, which is about 1.83 m long, about 10 cm thick, and about 30 cm wide, filled with water-soaked sand, with a transparent glass front for observing the effect of buoyancy. Figure (a), the first stage: three piles of oil are injected into the water-soaked sand, each about 10 cm in size, each occupying a side, not connected, at this time, due to the small volume of the oil pile, the buoyancy is insufficient, and the resistance prevents the oil droplets from floating upwards and they are stagnant. Figure (b), the second stage: more oil is added, causing the three piles of oil to connect and merge, at this time, it can be seen that the upper part of the oil has finger-like oil flows that begin to float upwards, this is because the volume of the oil pile has increased, the buoyancy has increased accordingly, enough to overcome the resistance and float and migrate. Figure (c), the third stage: a few hours later, the entire oil pile has floated and migrated to the top of the box and gathered, with only a few very small oil droplets remaining at the bottom, their diameter is only equivalent to the size of a few pores.

After oil and gas migrate from the source rock to the conduit layer (reservoir layer), they are first stored at the bottom and gradually gather from a dispersed state

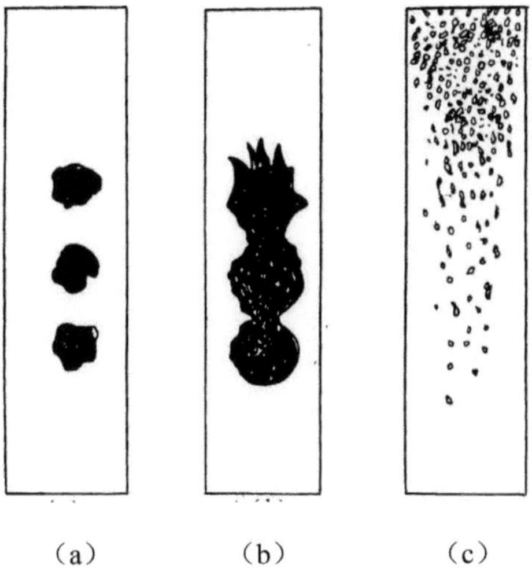

Fig. 4.24 Three consecutive stages of an experiment by Chierman A. Hill, illustrating the relationship between the role of buoyancy and the number of oil droplets (according to Levorsen, 1967)

to a blocky state, becoming an oil and gas body with a certain volume. Under static water conditions, when the gathered volume is large enough, the resulting buoyancy is sufficient to overcome the capillary resistance, and the oil and gas body begins to float vertically upwards and gradually reaches the top of the conduit layer. If the volume of oil V_o is changed to the height of a unit area, the critical height of oil floating (Z_o, unit: cm) can be obtained:

$$Z_o = [2\sigma(1/r_t - 1/r_p)]/[(\rho_w - \rho_o)g] \tag{4.10}$$

Similarly, we can obtain the critical height for gas floating.

It should be noted that the magnitude of the buoyancy experienced by oil and gas is related to their density and temperature and pressure conditions, and the critical height required for oil and gas migration is different. The density difference between oil, gas, and water determines the magnitude of the buoyancy force. Under the same water environment and oil and gas column height, the buoyancy force experienced by gas is greater than that by oil. Therefore, for reservoirs with the same displacement pressure, the height of the gas column required for secondary migration of natural gas is smaller than that of the oil column.

Under static water conditions, if the rock strata are horizontal, after oil and gas reach the top of the conduit layer, they will disperse along the top interface under the seal of the cap rock and will no longer migrate. If the rock strata are inclined, the oil and gas will continue to migrate upwards along the inclined direction under the influence of buoyancy after accumulating to a critical height, until they reach a trap and accumulate. The magnitude of the buoyancy force (F_1) along the inclined direction will be affected by the angle of inclination of the strata (Fig. 4.25). The greater the angle of inclination of the strata, the greater the buoyancy force along the inclined direction, the magnitude of which is represented by Eq. 4.11:

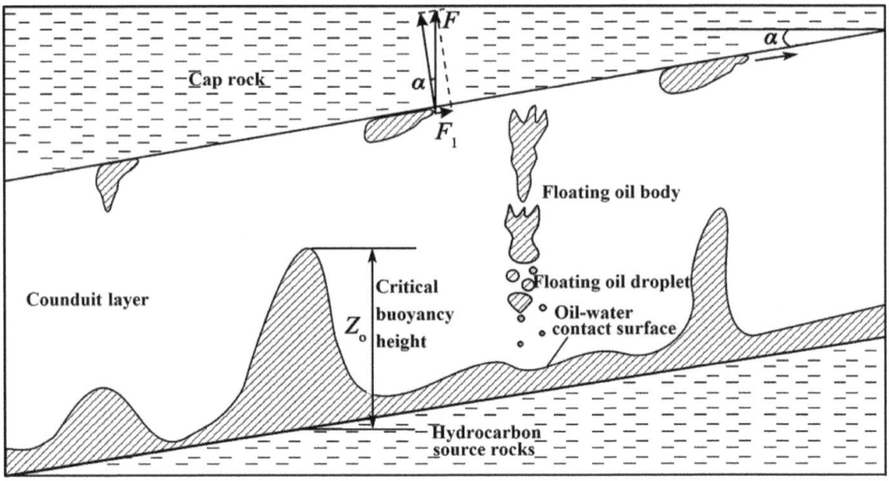

Fig. 4.25 Secondary migration of oil and gas in the conduit layer under hydrostatic conditions

$$F_1 = F\sin\alpha \tag{4.11}$$

The ease of buoyancy as a driving force is related to the reservoir's permeability and the angle of inclination. If the expulsion pressure of the strata is high or the angle of inclination is small, the buoyancy required for the oil and gas column will be greater. Under subsurface conditions, it is very difficult to achieve the height of the oil and gas column needed to drive buoyancy in low-permeability tight reservoirs. Therefore, it is generally said that buoyancy is not the main driving force for the migration of oil and gas in low-permeability tight reservoirs.

(2) Water Dynamics

The reservoir is full of water, and after oil and gas enter the reservoir layer (or conduit layer), they form pore fluids together with water. The flow of water will inevitably affect oil and gas, especially for water-soluble oil and gas, which is an important driving force. The groundwater dynamic conditions of the sedimentary basin have a macro-control effect on the secondary migration of oil and gas. There are mainly two types of groundwater dynamics in sedimentary basins: compaction water dynamics (compaction drive) and gravity water dynamics (gravity drive).

a. Compaction Water Dynamics

Compaction water dynamics mainly come from the compaction drainage of sediments within the basin, which occurs during the early continuous subsidence and differential compaction stage and process of the basin. Usually, during the same period, the strata in the center of the basin are thick and the sediment load is large, while the strata at the edge are thinner and the sediment load is smaller, resulting in differential compaction water flow. In such a basin, the groundwater potentiometric

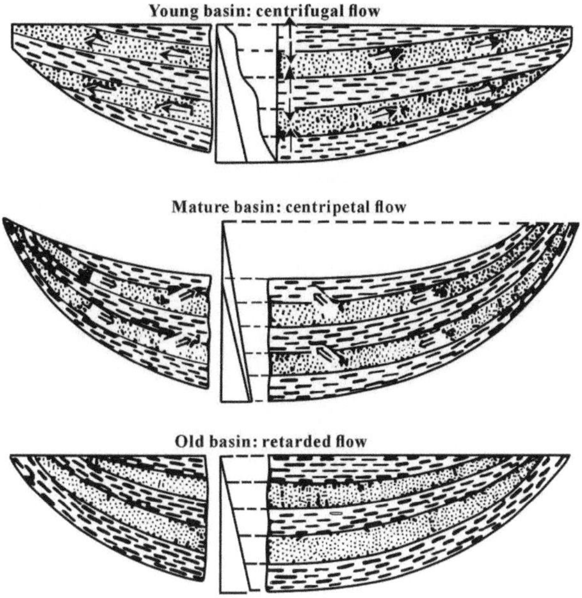

Fig. 4.26 Hydrodynamics during the evolution of the basin (according to Coustau et al., 1975)

surface is highest at the center and deep part of the basin and decreases towards the edge and shallow part, thus forming a regional groundwater dynamic field that points from the depression (depression) area to the edge in a "centrifugal flow" pattern. Basins with this type of groundwater dynamic field are called "compaction flow basins" (Fig. 4.26).

Under the action of compaction water dynamics, the general direction of oil and gas migration is basically consistent with the general direction of oil and gas migration under the action of buoyancy, thus promoting the secondary migration of oil and gas under the action of buoyancy, as well as the original gathering and distribution in the strata. However, in local areas or local structures, the flow of water can move horizontally along horizontal strata or move downward or upward along inclined strata. Therefore, whether water dynamics is a driving force or resistance in the process of oil and gas migration depends on whether the direction of water flow is consistent with the direction of oil and gas buoyancy.

b. Gravity Water Dynamics

Gravity water dynamics mainly occur during the mature stage of basin evolution. With the cessation of basin subsidence and further diagenetic changes, the compaction effect becomes less and less significant. In addition, later crustal movements cause strata to warp and fold, and strata at the edge of the basin are often exposed and connected with atmospheric water, forming gravity flow from the edge of the basin to the center, and regional groundwater shows a "centripetal flow" characteristic, hence it is called a "gravity flow basin" (Fig. 4.26).

Under the drive of gravity water dynamics, the direction of water flow is mainly from the high potential area at the edge of the basin to the low potential area at the center of the basin. The general direction of gravity water flow is exactly opposite to the general direction of oil and gas migration under the action of buoyancy. Although appropriate conditions can form a hydrodynamic trap, excessive gravity water dynamics can also wash out the gathered oil and gas from the trap, causing destruction of oil and gas reservoirs and causing re-migration and re-distribution of oil and gas.

The above two types of water dynamics generally appear in sequence with the evolution of the basin. In the late stage of basin evolution, the groundwater in the basin is basically in a static state, with no fluid energy exchange, and the basin at this time is called a "stagnant flow basin" (Fig. 4.26).

c. Direction of Oil and Gas Migration

In the case of horizontal strata, the direction of water dynamics is perpendicular to the direction of buoyancy, and the oil (gas) body rises to the top interface of the conduit layer and is sealed by the cap rock. At this time, whether oil and gas can migrate laterally depends on the comparison between water dynamics and the capillary force resistance experienced by the oil body.

In the case of inclined strata, such as the transitional zone between the slope zone at the edge of the basin and the deep depression area and the adjacent structural high position, the impact of water dynamics on oil and gas migration depends on the comparison between the magnitude and direction of water dynamics and buoyancy and capillary force (Fig. 4.27).

Under stratum conditions, no matter what the absolute stratum pressure of each point is, the direction of water flow is always from high calculated pressure to low calculated pressure. Figure 4.28 shows that the elevation of the water supply and

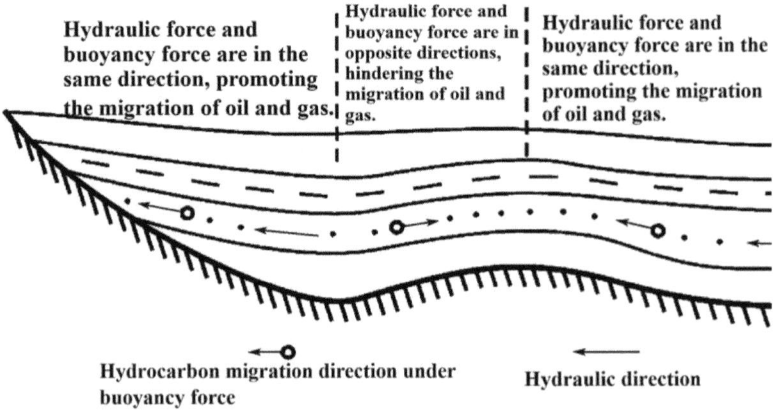

Fig. 4.27 The impact of the combination of water dynamics and buoyancy on the secondary migration of oil and gas

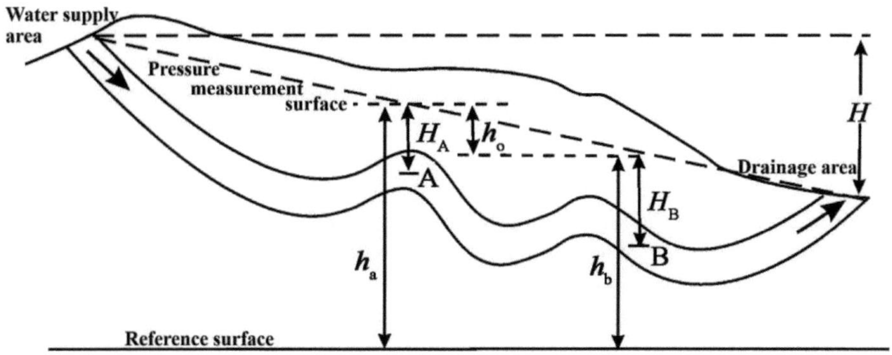

Fig. 4.28 Schematic diagram of calculated pressure and water flow direction (modified from Zhang, 1989)

discharge areas of the conduit layer is different, and the pressure surface is inclined, so the calculated pressure decreases regularly along the inclined direction of the pressure surface, and water flows from the water supply area to the discharge area. The absolute stratum pressure at points A and B in the figure is caused by the static water column at each point. Since the height of the static water column at point B is greater than that at point A ($H_B > H_A$), the absolute stratum pressure at point B is greater than that at point A. However, because the calculated height at point A is greater than that at point B ($H_a - H_b = H_o$), the fluid will flow from point A to point B under the action of the calculated pressure difference.

(3) Tectonic Stress

The stress produced by crustal movement is called tectonic stress. Tectonic stress is the pressure acting on the rock skeleton, while stratum pressure is the fluid pressure in the rock pores. The two interact and transmit to each other, forming a unified pressure system of the rock. As a driving force, the role of tectonic stress in the secondary migration of oil and gas is mainly manifested in the following two aspects:

First, tectonic stress is a direct driving force for secondary migration. Tectonic stress causes strain in rocks, compresses the rock skeleton, and deforms rock particles and pores. This deformation process will inevitably transfer the force to the fluid in the pores, causing the pressure to rise and forming a high potential area, driving oil and gas to migrate to a low potential area. When there is a change in tectonic stress, due to the compression and rebound of the rock skeleton, the fluid pressure will rise or fall, thus producing a stress pump effect, which is also an important mechanism and driving force for the secondary migration of oil and gas.

Second, tectonic stress creates conditions for buoyancy and water dynamics, causing changes in the underground fluid potential. Tectonic stress can form folds and fractures, causing strata to warp, enabling buoyancy to play a role, forming water supply and discharge areas and stratum dynamic water conditions. Changes in the tectonic background and water dynamics conditions will ultimately lead to changes in the underground fluid potential, thus affecting the direction of oil and gas secondary migration. Therefore, although buoyancy and water dynamics are the direct driving forces for oil and gas secondary migration, they are ultimately controlled by the tectonic background, and tectonic stress is the fundamental condition for promoting oil and gas migration.

(4) Abnormal Pressure

Abnormal pressure is the most important driving force for oil and gas primary migration and is an indispensable driving force for secondary migration, especially for secondary migration in low permeability tight reservoirs.

The abnormal pressure in secondary migration mostly comes from the undercompaction and overpressure of adjacent source rocks, and can also come from the strata themselves. Source rocks with abnormal pressure can expel oil and gas into adjacent carrier layers, and the oil and gas that enters the reservoir layer will undergo secondary migration from high-pressure areas to low-pressure areas. For the abnormally high-pressure oil and gas that has gathered in the trap, when its cap rock or fault ruptures or forms a channel under the action of abnormal high pressure, the oil and gas in the trap will enter the overlying or lateral carrier layer with the release of abnormal high pressure, and after the abnormal pressure disappears, the oil and gas will undergo secondary migration mainly by buoyancy. If it is a low permeability tight reservoir, oil and gas will be retained in places where the abnormal pressure is weakened or disappears to form a gathering.

(5) Molecular Diffusion Effect

As long as there is a concentration difference, the molecular diffusion of hydrocarbons can occur, and the molecular diffusion force is an important driving force for both primary and secondary migration. Molecular diffusion is controlled by the concentration gradient, and it always diffuses from the high concentration area to the low concentration area. This kind of diffusion also includes the diffusion of natural gas through the overlying strata after the formation of oil and gas reservoirs, which will lead to the loss of natural gas. Compared with buoyancy, water dynamics, and tectonic stress, molecular diffusion force is only a secondary driving force, and its efficiency is several orders of magnitude lower than that of oil and gas seepage, but it is the main driving force and method for secondary migration in tight strata.

4.3.3 Migration Channels and Migration Modes

Channels are the bridges and links between oil and gas from the "source" to the "reservoir". The oil and gas generated by the source rock can only enter the trap and gather to form a reservoir through effective channels. Their transmission capacity and connectivity characteristics during the migration period directly affect the direction and gathering parts of oil and gas migration. Essentially, oil and gas secondary migration channels are composed of connected pores or fractures, such as the migration of oil and gas along the connected pores and fractures in the fault fracture zone, or along the connected pores and fractures of the upper and lower permeable layers on the unconformity surface, etc.

4.3.3.1 Types of Migration Channels

From a microscopic perspective, oil and gas migrate through the pores, fractures, and cavities in the underground rocks, which are the basic channel spaces for oil and gas secondary migration. From a macro perspective, the geological bodies that can serve as channels for oil and gas secondary migration in the sedimentary basin mainly include permeable strata (conduit layers), faults, and unconformities, which are also the elements that make up the oil and gas conduction system.

(1) Conduit Layers

Conduit layers refer to permeable strata with developed pores, fractures, or cavities and other basic migration spaces. The degree of development and connectivity of the pore space in the conduit layer have an important impact on the difficulty and efficiency of oil and gas migration within it. The common conduit layers in sedimentary basins are mainly permeable reservoir layers, including permeable clastic rock sand bodies, and two types of carbonate rock strata: pore type, fracture type, and dissolution type. From the perspective of conduction channel space, the former is mainly connected pores, sometimes there will be fractures. The latter is mainly fractures, dissolution holes (caves), and connected pores, and its conduction performance is complex and changeable.

As a channel for oil and gas secondary migration, the connectivity of the conduit layer is crucial. From a geological research perspective, the determination of the connectivity of the conduit layer usually includes two levels: one is the geometric connectivity of the conduit layer, which reflects the spatial connection relationship of each part of the conduit layer. The other is the fluid dynamic connectivity, which refers to the connectivity that allows fluid flow within the conduit layer, and the latter truly reflects the characteristics of fluid or oil and gas passing through the conduit layer (Luo et al., 2012).

Under actual geological conditions, the main factors affecting the conduction performance of the conduit layer include lithology and mineral composition, pore structure, cementation and dissolution, heterogeneity, and the degree of fracture development. The association between thickness and conduction performance is uncertain, and the impact of permeability and porosity homogeneity is the most significant. When evaluating the conduction performance of the conduit layer, it is meaningful to consider its permeability performance under the constraint conditions of the fluid connectivity of the conduit layer (Lei et al., 2014). In conduit layers with strong heterogeneity, due to the difference in permeability distribution, it creates a differential migration channel space, and oil and gas always migrate along the channels with good permeability and the smallest capillary resistance. In homogeneous reservoir layers, under the action of buoyancy, oil and gas will concentrate near the top interface of the conduit layer, thus forming a deterministic migration path.

(2) Faults

The practice of oil and gas exploration in China has proved that more than 80% of oil and gas fields are directly or indirectly related to faults (Li, 2013). Faults are an important channel for the secondary migration of oil and gas and are the main way for oil and gas to undergo inter-layer and long-distance vertical migration. The formation of many allochthonous shallow oil and gas reservoirs that are vertically far from the deep source rock layer, and the vertical micro-migration of hydrocarbons caused by fault destruction, all play a role as channels. Faults have an important control effect on oil and gas migration and gathering, showing the dual nature of opening and sealing, and can serve as both channels for oil and gas migration and as sealing layers (Gibson, 1994).

The fault is a zone with a certain width and complex internal structure. Generally, it is composed of a fracture zone and induced fracture zones on both sides of the fracture zone. The fracture zone is often composed of cataclastic rocks, fault breccias and fault gouges (Fig. 4.29). The width of the fault zone is related to factors such as fault scale, fault nature, and lithologic composition. Among them, fault scale and fault nature have a greater impact.

There are two types of conduction for oil and gas along faults, namely lateral migration through the fault and vertical migration along the fault (Fig. 4.30). Lateral migration refers to the situation where oil and gas pass through the fault zone from one side to the other when the permeable strata on both sides of the fault are opposed. The prerequisites are: first, the fault is laterally open (the most common is the opposition of permeable strata on both sides), and second, the displacement pressure of the fault zone and the two permeable strata is comparable, and there is a horizontal fluid potential gradient change. The main factors affecting the lateral conductivity of faults include: the nature of the fault, the opposition relationship between the rock layers on both sides of the fault, the production status of the strata on both sides, shale smearing, fault zone permeability, etc. Extensional faults, permeable strata on

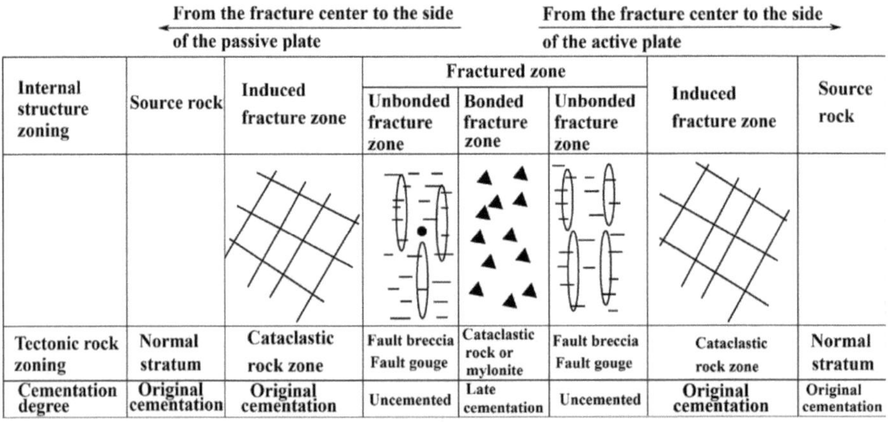

Fig. 4.29 Schematic diagram of fault zone structure (according to Fu et al., 2005)

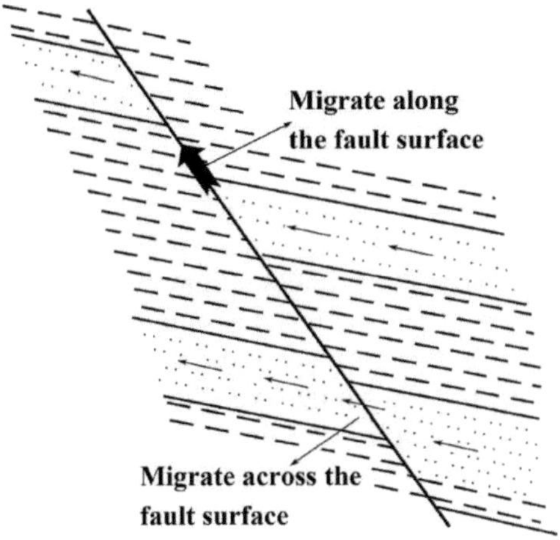

Fig. 4.30 Schematic diagram of oil and gas migration using faults as migration channels (according to Chapman, 1983)

both sides of the fault that are opposed and have similar tendencies, underdeveloped shale smearing, fault zones with certain permeability, and displacement pressures comparable to those of the two permeable strata are all conducive to the lateral conduction of faults.

Vertical migration along faults requires the vertical opening of faults and the existence of vertical fluid potential gradients. The conductivity mainly depends on factors such as fault nature, fault activity, internal structure of the fault zone, normal stress on the fault surface, mud content in the fault zone, development of induced fracture zones, and the degree of late diagenetic filling. Generally, extensional faults have good vertical opening and are conducive to conduction, while compressional and compressional-torsional faults have strong vertical sealing. Active faults are conducive to vertical conduction, and static periods tend to be vertically sealed. Small fault surface inclination, large normal stress on the fault surface, high mud content in the fault zone, underdeveloped induced fractures, or strong late diagenetic filling tend to be vertically sealed, and vice versa is conducive to vertical conduction.

Fault conduction and sealing are relative. It is now a consensus that no matter what the nature of the fault is, the flow of fluids along the fault is periodic. During the activity period, the fault is mostly in an open state with high permeability, serving as a channel for oil and gas migration. During the static period, it often appears to be in a sealed state with reduced permeability, acting as a barrier to oil and gas. Fault activity is often an unbalanced stress and abnormal pressure release process, and the fault itself becomes a channel for fluid migration. After the fault activity stops, because the stress that causes the expansion of the rock disappears, the huge confining stress prompts the fault and the derived fractures to close rapidly, and the corresponding increase in permeability may disappear completely or partially (Luo & Vasseur, 2016). With the continuous activity of the crust and faults, the fault can present a cyclical change pattern of conduction-sealing-conduction-sealing over time, which also indicates that the fault cannot always maintain conductivity or sealing.

(3) Unconformities

The formation of unconformities is usually the result of regional crustal movements, sea (or lake) level changes, or local structural actions, and is of great significance in the study of the burial history and tectonic evolution history of sedimentary basins. Unconformity surfaces in sedimentary basins are generally widely distributed and can connect hydrocarbon source rocks and reservoir layers of different ages, expanding the spatial range and stratigraphic range of oil and gas migration, and playing an important role in oil and gas migration. Many large and super-large oil and gas fields found in the world's oil and gas basins that are related to unconformities are often formed by the migration and gathering of oil and gas through unconformity surfaces.

The conditions for unconformities to serve as channels for oil and gas secondary migration are: the original rock is easily weathered and forms a good weathered leaching zone, has a good unconformity structure type, develops a good oil and gas migration channel space, and has a good cap rock. Due to tectonic uplift causing strata to be exposed at the surface, they are subject to weathering and erosion, surface water leaching, and the degree of weathering varies greatly due to the undulations of the terrain. Moreover, unconformity surfaces often pass through strata of different ages and lithologies, and later diagenetic actions change the permeability of the strata above and below the unconformity surface. Therefore, the pore and permeability characteristics of the strata near the unconformity are complex, and there is still a lot of uncertainty about whether they can form migration channels (Luo et al., 2014; Song et al., 2010).

The difference in the conductive effect of unconformities is closely related to the types of unconformity structures. Generally, a complete unconformity develops an overlying rock layer of the unconformity surface, a weathered clay layer, and a semi-weathered rock layer in three structures. The overlying rock layer of the unconformity surface is usually a basal conglomerate or a water advance sand body, which gradually overlaps in the upward direction along the slope, forming a set of time-stratigraphic deposits, and is a good skeletal sand body conductive layer. The weathered clay layer is the ancient soil layer located at the top of the weathered shell, which is a fine-grained residual material formed by weathering, with a thickness of several meters to more than ten meters, and is locally missing due to later erosion. Because the overlying sediments are compacted and dense, they have good sealing capabilities. Below the weathered clay layer is the semi-weathered rock layer, which can be further divided into a weathered leaching zone and a collapse zone, with developed pore, fracture, or cave systems, with a thickness of several meters to hundreds of meters, and the thickest can reach thousands of meters. The permeability performance of the semi-weathered rock layer depends on the lithology and degree of weathering of the strata being weathered. Generally, the permeability is significantly increased compared to the original rock, and the conductive performance is correspondingly enhanced. It often develops unloading fractures and weathered fractures, which are interwoven and cut, forming a networked pore-cave-fracture system with the connected pores and dissolved holes in the rock, greatly enhancing the conductive performance.

Different depositional, tectonic, and weathering leaching backgrounds form different structural styles of unconformity conductive layers, and their conductive effects on oil and gas are different: those that only develop communication between the upper and lower permeable layers of the unconformity surface become a single-channel type. Those that have conductive capabilities on both the upper and lower sides of the unconformity surface and have developed weathered clay layers are double -channel conductive types. If the weathered clay layer is missing, it can communicate with the upper and lower conductive layers, and can also become a single-channel type. Figure 4.31 shows the structural types of the unconformity surfaces at the bottom of the Neogene and Paleogene in the Jiyang Depression of the Bohai Bay Basin, reflecting the spatial changes in unconformity structures, various configuration relationships between the upper and lower rock layers of the unconformity surface, and the different conductive capabilities of oil and gas for different structural types of unconformities.

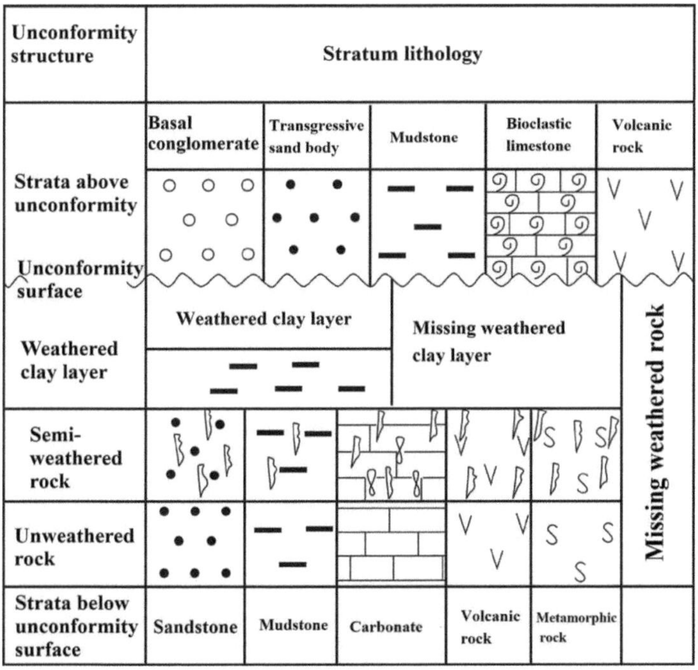

Fig. 4.31 Types of unconformity structures in the Jiyang Depression (according to Sui et al., 2009)

4.3.3.2 Conduction System and Oil and Gas Migration Modes

(1) Conduction System and Oil and Gas Migration Modes

In the secondary migration process of oil and gas from the hydrocarbon source rock to the trap, except for a few that migrate and gather through a single channel type over a short distance, most of them migrate along a system network composed of various channels. We refer to all the paths and their associated surrounding rocks that oil and gas experience from the hydrocarbon source rock to the trap as the conduction system, including connected sand bodies, faults, unconformities, and their combinations (Magoon & Dow, 1994). Conduction layers, faults, unconformities, and other migration channels are the elements that make up the conduction system, called conductive bodies. In the basin, sandstone strata, faults, and unconformities are not isolated conductive bodies. The oil and gas conduction system are often a three-dimensional conduction framework composed of two or several types of conductive bodies combined in space.

When the conduction system is composed of a single conductive body, it is called a simple conduction system, which is further divided into conduction layer conduction systems, fault conduction systems, and unconformity conduction systems according to the type of conductive body, When composed of two or more conductive bodies combined, it is called a composite conduction system, which can be named by combining the types of conductive bodies, such as the fault-sand conduction system composed of faults and skeletal sand body conduction layers.

In continental basins, composite conduction systems are the most common types of conduction systems. Those oil and gas reservoirs that are far from the source and located at the edge of the basin, in the shallow layer, or in non-hydrocarbon-bearing stratigraphic sequences have all undergone long-distance and complex migration paths, and their conduction systems are often composite conduction systems. Li et al. (2004) proposed four types of conduction systems: net-blanket type, T-shaped, stepped, and fractured (Fig. 4.32), based on the spatial combination patterns of various conductive bodies in the Cenozoic of the Jiyang Depression, effectively guiding the exploration of oil and gas in the Neogene, steep slope zone, gentle slope zone, and depression zone.

(2) Modes of Oil and Gas Secondary Migration

The mode of oil and gas secondary migration refers to the pathways and directions of oil and gas migration along the conduction system under certain driving forces, which can generally be divided into three types: lateral migration, vertical migration, and stepped migration. The mode of secondary migration determines the macroscopic migration direction and distance of oil and gas in the basin.

a. Lateral Migration

Lateral migration refers to the migration of oil and gas along the horizontally distributed conduction system, which can transport oil and gas far away from the hydrocarbon generation center to the edge. Lateral migration can be formed by

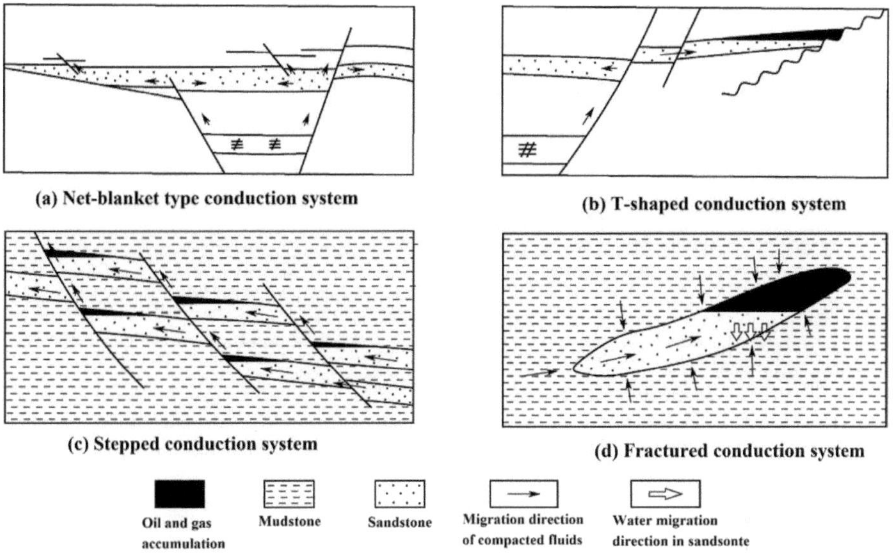

Fig. 4.32 Schematic diagram of the types of conduction systems in the Jiyang Depression (according to Li et al., 2004)

conduction layers or unconformity surfaces alone, or can be completed by conduction layers and unconformity surfaces together. The conduction system that constitutes lateral migration must meet the following conditions: (a) have a good spatiotemporal configuration relationship with the hydrocarbon source rock, (b) have good continuity and stability in planar distribution, (c) have good conductive performance and favorable ancient production status, (d) have well-developed cap rocks and favorable structural shapes. The scale of lateral migration is mainly limited by the continuity of the conduction system itself and is also related to the development of the overlying strata and faults. If the basin is large enough and there is a stable reservoir-cap rock combination, oil and gas can undergo long-distance lateral migration.

Large-scale lateral migration mainly develops in large and craton basins or during the stable development period of the basin, such as the Williston Basin in the United States, where oil and gas migrate laterally upwards along the porous carbonate rocks and unconformity surfaces for more than 100 km (according to Demaison & Huizinga, 1994). The ancient uplift slope on the slope of the depression or the slope of the basin (depression) is also a major area for lateral migration. As shown in Fig. 4.33, fan delta, beach bar and other sand bodies are developed at the southern slope of the Dongying Depression, the upper sub-member of the third member of the Shahejie Formation. The sand bodies are thick, and the front end of the sand bodies are finger-shaped and respectively inserted into the Niuzhuang hydrocarbon depression. They have good connectivity and permeability, forming four advantageous migration channels determined by strong conductive capacity, which respectively form the Bamianhe, Wangjiagang, and Le'an oil fields.

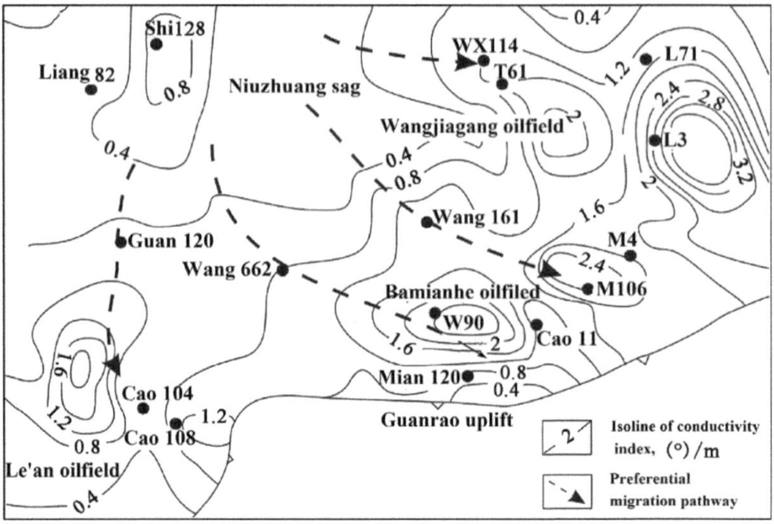

Fig. 4.33 Relationship between contour lines of transport capacity index of framework sand bodies in the upper sub-member of the third member of Shahejie Formation in the eastern section of the southern slope of Dongying Sag and oil and gas show (according to Song et al., 2012)

Lateral migration is the earliest recognized and most easily accepted mode of migration. Due to its wide link range, it is conducive to the formation of large oil and gas reservoirs. However, compared with other migration modes, lateral migration has several shortcomings: First, the long-distance lateral migration causes greater oil and gas loss than other methods. Second, the same amount of oil and gas in low-angle lateral migration produces less buoyancy, so more oil and gas is needed to form sufficient buoyancy for migration. Third, the driving force of lateral migration is small, the migration rate is generally slow, and the efficiency is low.

b. Vertical Migration

Vertical migration mainly refers to the vertical inter-layer migration of oil and gas along faults (or fractures). Under the action of buoyancy, it is generally upward migration, and it can also be upward or downward migration under the action of abnormal high pressure. Vertical migration mainly develops in rift basins and other similar basins in the thrust belts and wrench fault belts. The existence of a large number of shallow oil and gas reservoirs and the display of hydrocarbons near the surface both indicate the existence of vertical migration of oil and gas. The conduction system for vertical migration is mainly composed of oil and gas migration network systems constituted by faults (or fractures), which are mostly composed of one or more main faults and their branch faults, accompanying fractures, and reverse adjustment faults. The conditions for the effectiveness of fault conduction systems are: (a) the fault communicates the hydrocarbon source rock and the trap, and oil and gas migrate directly to the trap through the fault, (b) the fault has good vertical

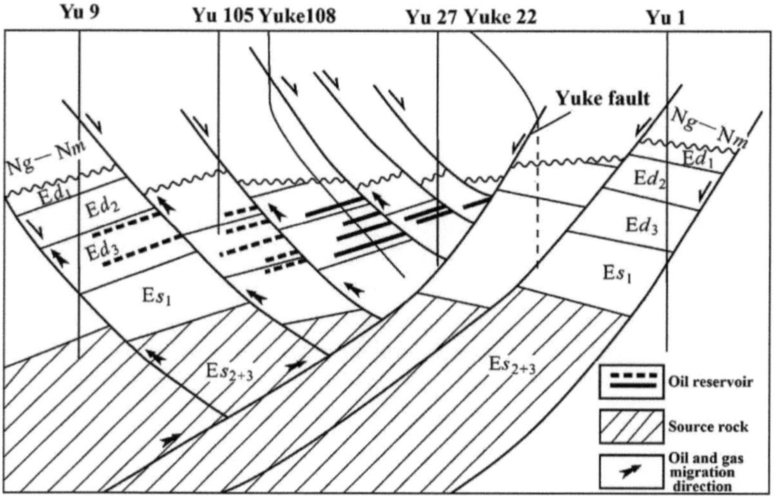

Fig. 4.34 Fault transportation system and hydrocarbon migration of Yuke Oilfield in Shenxian Sag (according to Du et al., 2007)

conductivity, (c) the fault activity period is consistent with the main hydrocarbon generation and expulsion period of the source rock. Continuously active faults and related fractures are the most effective channels for vertical migration of oil and gas.

The eastern continental fault basins in China have developed faults with diverse combinations and significant differences in vertical sealing properties, forming an extremely complex fault conduction system, which has become an important channel for communicating the deep and shallow layers of oil and gas (Fig. 4.34).

In the basin, the vertical migration distance along faults is generally 1–8 km, which is much less than the lateral migration distance. However, compared with other migration modes, vertical migration has the following advantages: First, faults can communicate multiple sets of hydrocarbon source rocks vertically, and the material basis for multi-source oil and gas supply is good. Second, faults are a zone where fractures develop, and their permeability is often better than other conductive bodies, which is conducive to the rapid migration of oil and gas. Third, the driving force of vertical migration is strong, and the conduction efficiency is high. Fourth, fault activity is periodic, and oil and gas can undergo phased migration, reducing the loss of oil and gas during long-term migration. Fifth, faults are conducive to oil and gas crossing different stratigraphic layers vertically, which can form cross-layer migration and is an important migration mode for the formation of shallow allochthonous oil and gas reservoirs. Although the efficiency of vertical migration is high, if the cap rock at the top is destroyed, oil and gas will be lost along the fault in large quantities, forming oil and gas seepages on the surface, so there must be a cap rock seal above the fault conduction system.

c. Stepwise Migration

Stepwise migration refers to the migration of oil and gas along the composite conduction system composed of conduction layers or unconformities and faults in an upward direction in a stepped path. According to the configuration relationship between conductive bodies, the migration modes along the reservoir layer-fault type, unconformity surface-fault type, and reservoir layer-fault-unconformity type composite conduction systems are all stepwise migrations. The stepwise migration mode is most common in basins with developed faults, such as the steep slope zone of continental fault depression basins where fault development is dominant and vertical migration along faults is dominant, with short lateral migration distances along reservoir layers. While in the gentle slope zone, lateral migration along reservoir layers and (or) unconformity conductive layers is longer, and less developed faults play a vertical adjustment role. In the western basins of China, where reverse faults are developed, oil and gas mostly migrate along the composite conduction system composed of skeletal sand bodies and unconformity conductive layers and thrust belts.

The stepwise migration mode can also be explained by the channeling and sealing properties of faults. For example, oil and gas migrating laterally in sandstone layers encounter a fault. If they are blocked by the adjacent rock on the other side, they will migrate upward along the fault. When they enter the overlying sandstone layer, they begin to migrate laterally again. If they encounter a fault and there is no obstruction, they can continue to migrate through the fault surface. If there is an obstruction, they will migrate upward along the fault surface, such as on the Wen'an slope of the Baxian Depression (Fig. 4.35). This forms a stepwise migration mode.

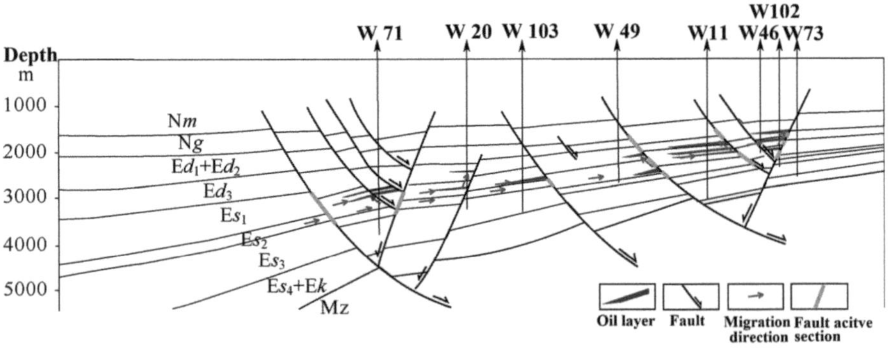

Fig. 4.35 Fault sealing and stepwise migration of oil and gas on the Wen'an slope of the Baxian Depression (according to Liu et al., 2011)

4.3.3.3 Effective Conduction System and Advantageous Migration Channels

Oil and gas secondary migration do not occur throughout the entire conduction system but always takes place along a portion of the path with better conductive properties, stronger migration driving forces, and relatively smaller migration resistance. Li (2004) referred to the channels in the conduction layer that truly underwent the migration effect of oil and gas as the effective migration space of secondary migration, that is, the effective conduction system. Through statistical analysis of oil and gas occurrences in the conduction layers of several basins in China, it was pointed out that the effective migration space accounts for about 5–10% of the entire conduction layer in the main stratigraphic sections where oil and gas migration has occurred.

The paths that oil and gas naturally prefer to flow through during secondary migration are called advantageous conduction systems, that is, advantageous channels, which are part of the effective conduction system of secondary migration and are the main pathways for oil and gas migration (Fig. 4.36). It is emphasized that the advantageous migration channel is the route that oil and gas preferentially choose to migrate, not the trend, nor the potential direction indicated by the fluid potential field. Although the advantageous migration channel only accounts for a very small part of the oil and gas conduction system, it may conduct most of the total oil and gas conducted by the effective conduction system. Therefore, the advantageous migration channel can also be commonly referred to as the "highway" for oil and gas migration. Compared with other migration channels, the advantageous migration channel has more unobstructed conduction conditions, that is, higher permeability, smaller migration resistance, and better connectivity. Once formed, the advantageous migration channel has relative stability, but when geological conditions change significantly, causing the original advantageous migration channel to lose the advantageous conditions for conducting oil and gas, oil and gas will have to change their route and migrate along new advantageous migration channels.

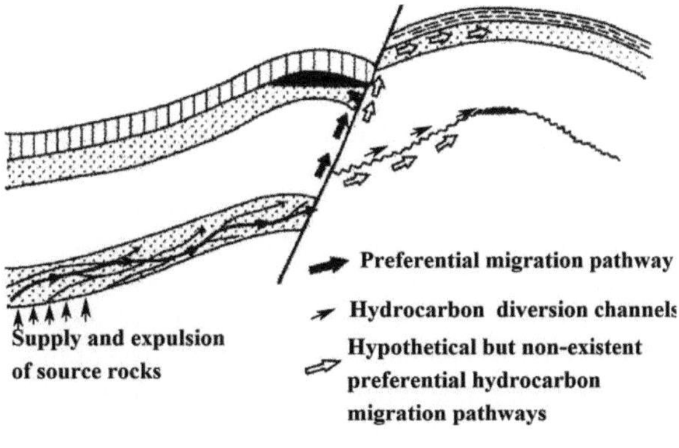

Fig. 4.36 Geological conceptual model of oil and gas migration advantageous channels

4.3.4 Secondary Migration Direction and Influencing Factors

The direction of oil and gas secondary migration is macroscopically controlled by the regional tectonic background and fluid potential field, and microscopically controlled by the migration resistance, conduction system, and the spatial control of migration channel.

4.3.4.1 The Structure and Geometric Shape of the Basin

The geometric shape of the basin is generally described by the paleostructure map of a certain stratum. The trend distribution of structural contour lines in the structural map is almost parallel to the trend distribution of fluid equipotential lines. The direction perpendicular to the equipotential lines is the direction of least effort for fluid migration and is also the advantageous direction for secondary migration of oil and gas. Due to changes in structural shape, oil and gas migration streamlines converge on the concave side and diverge on the convex side. On the side where structural contour lines are dense, the streamlines are also dense, and on the opposite side, the streamlines are sparse. Figure 4.37 shows a schematic diagram of the geometric shape of the basin determining the direction of secondary migration of oil and gas.

The area where streamlines converge is the main direction of secondary migration of oil and gas, and the "partition trough" where the hydrocarbon generation center exists is the "watershed" for the macroscopic direction of secondary migration of oil and gas. Therefore, the tectonic background of the basin determines the macroscopic migration direction of oil and gas, such as the uplift, slope, and inclined stratigraphic layers in the basin, and their upward direction is the favorable direction for secondary migration of oil and gas, while the structural ridges developed on them determine the advantageous migration direction.

4.3.4.2 The Shape and Distribution of the Cap Rock

In the analysis of the combination of secondary migration channels and the direction of advantage, an important factor that should not be overlooked is the cap rock. Oil and gas can only migrate laterally under the action of buoyancy along the top interface of the conduction layer (the bottom interface of the cap rock) when the cap rock is sealed. Once the sealing of the cap rock is lost, lateral migration will terminate and vertical migration will take place. Therefore, the continuity of the cap rock determines the direction and distance of secondary migration.

Hindle (1997) emphasized the importance of the cap rock above the conduction layer, believing that when the cap rock conditions are met, the migration path of oil and gas mainly depends on the structural undulations of the top interface of the conduction layer: the cap rock partition trough determines the macroscopic migration direction of oil and gas, which is controlled by the cap rock, and oil and gas gradually

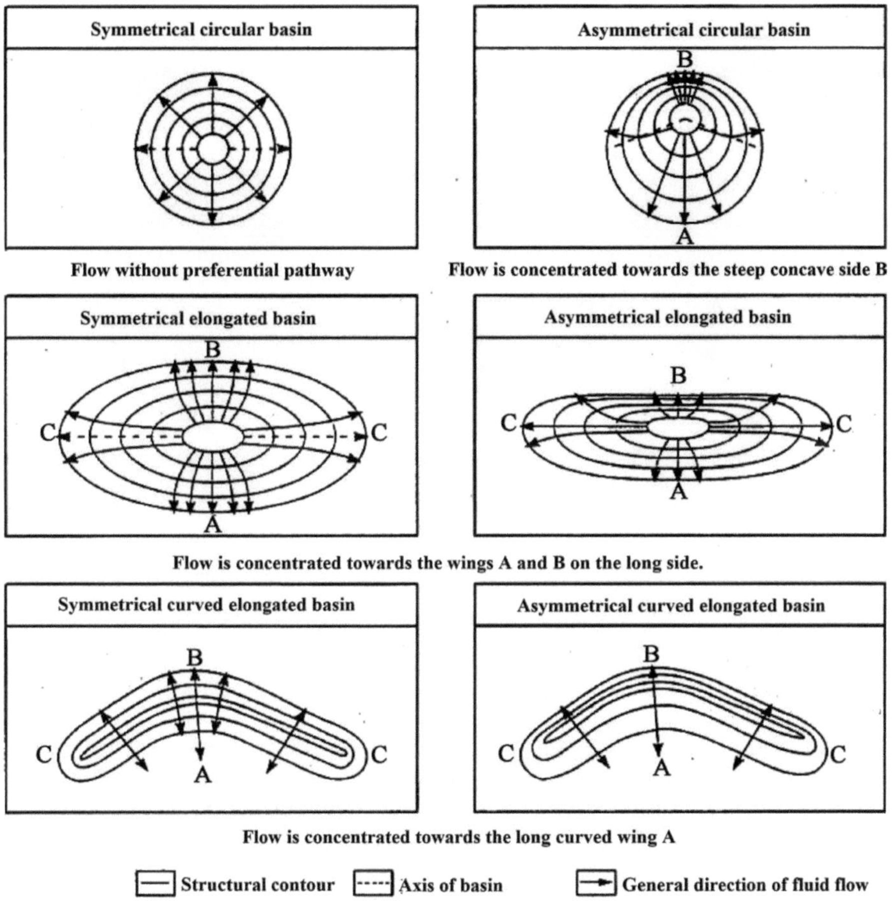

Fig. 4.37 Determining the general direction of oil and gas migration according to the geometric shape of the basin (according to Pratsch, 1982)

converge from being dispersed to form advantageous migration paths. Above the oil and gas generation area, there are many oil and gas migration paths forming a dense network. Far from the oil and gas generation area, oil and gas migration is mainly along a limited number of concentrated paths. It can be seen from Fig. 4.38 that the convex shape of the cap rock is conducive to the formation of advantageous channels, while the concave and flat shapes of the cap rock form dispersed migration paths. For basin-scale oil and gas migration, the undulation shape of the regional cap rock determines the direction of secondary migration of oil and gas, and the distribution of the cap rock determines the range of oil and gas accumulation.

Under normal circumstances, the structure of the top surface of the conduction layer is consistent with the structure of the cap rock. Therefore, in the case of developed cap rocks, the structural ridge line of the top surface of the conduction layer

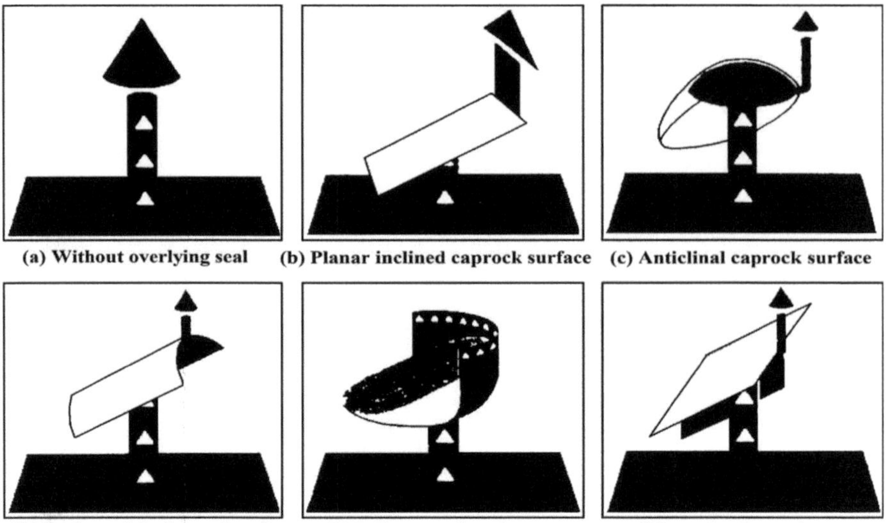

Fig. 4.38 The impact of the shape and occurrence of the sealing layer on the secondary migration path (according to Hindle, 1997)

becomes the direction of the advantageous channel. Jiang et al. (2011) studied the migration path of oil and gas in the Puwei area of the Dongpu Depression by using the coupling relationship between the purpose layer top surface paleostructure ridge and sand body distribution, the relationship between the sand layer top surface paleostructure ridge and oil and gas distribution, and combined with nitrogen-containing compound tracing, and determined that the sand layer top surface paleostructure ridge is the advantageous migration path.

4.3.4.3 Distribution of Sedimentary Facies

Oil and gas are generated, migrated, and accumulated in rocks, and are most closely related to sedimentary rocks. Sedimentary facies are the products of sedimentation in sedimentary environments and are important factors in determining the planar distribution and permeability of conductive layers. For a time-stratigraphic sedimentary facies belt, oil and gas both have the characteristic of migrating from fine-grained sediments to coarse-grained sediments, that is, the macroscopic trend of migrating towards the land source direction. Figure 4.39 shows the delta system developed in the Pliocene of Texas, Gulf Coast, USA. The dark mudstone of the prodelta generates oil and gas, which migrate along the delta front sheet sand towards the land source direction of the estuarine sand bar and the river channel sand body on the delta plain, just like a natural channel for oil and gas migration in the basin. This sourceward direction represents the advantageous direction of secondary migration and also reflects the trend of oil and gas always migrating towards parts with better

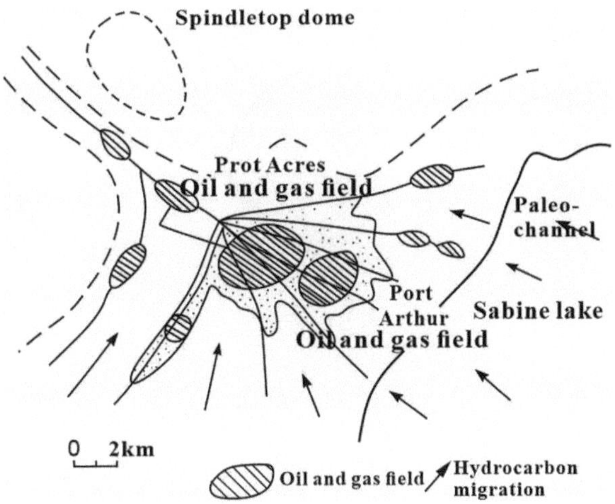

Fig. 4.39 Oil and gas migration direction and oil and gas field distribution of the Pliocene delta in Texas, Gulf Coast, USA (according to Perrodom, 1993)

permeability, which also includes the macroscopic direction of migrating from the thickly deposited sedimentation center to the thinner edge or uplift area.

4.3.4.4 The Shape and Occurrence of Faults

Faults are the most common products of tectonic activity in basins, and the seismic pump effect of the faults themselves and their activity time make faults the main channels for vertical fluid migration. Similarly, the geometric shape and occurrence of faults also have a decisive impact on the migration direction.

The undulation of the fault surface controls the advantageous channel direction for oil and gas migration along the fault. A convex fault surface causes streamlines to converge and form advantageous migration channels, while a concave fault surface causes streamlines to diverge upwards and does not form advantageous channels (Fig. 4.40).

In general, in oil and gas-bearing basins, the secondary migration of oil and gas is along the direction of the best permeability and the smallest resistance, in a certain mode of migration, from high fluid potential areas to low fluid potential areas. From the overall perspective of the basin, the direction of oil and gas secondary migration is always from the center of the basin to the edge of the basin and the ancient uplifts in the basin, and from deep strata to shallow strata. Therefore, the uplifts and slope zones near the hydrocarbon depression often become the main direction of secondary migration of oil and gas, especially the long-term inherited positive structures are the most favorable. The practice of oil and gas field exploration in China has proved that some oil and gas fields with rich oil and gas resources, such as the Daqing Oilfield in the Songliao Basin and the Shengli Oilfield and Dongxin Oilfield in the Dongying

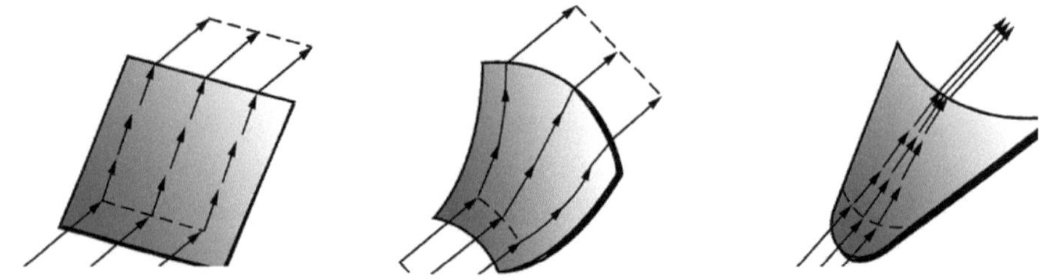

a Planar fault: The aggregation and path remain unchanged from the entry point

b Concave fault: The oil migration path

c Convex fault: The oil migration path

Fig. 4.40 The influence of fault surface morphology and attitude on the distribution of secondary migration pathways (according to Hindle, 1997)

Depression of the Bohai Bay Basin, are all located near the hydrocarbon depression and on the main direction of secondary migration of oil and gas.

4.3.5 Main Period and Distance of Secondary Migration

4.3.5.1 Main Period of Secondary Migration

The secondary migration of oil and gas is a continuation of primary migration, and the two are a continuous process that exists simultaneously. However, the main period of large-scale secondary migration should be after or simultaneously with the first tectonic movement period following the main oil generation period.

Tectonic movement is not only the direct driving force for oil and gas migration but also provides conditions for buoyancy and water dynamics to play a role. More importantly, it can change the attitude of strata, causing them to warp, fold, and produce a large number of fractures, creating conditions for the formation of migration channels. There may be multiple tectonic movements throughout the evolution of the basin. The first large-scale tectonic movement period after the main hydrocarbon generation and expulsion period should be the main period of secondary migration. Because this tectonic movement causes the original strata to incline, and even fold and fracture, destroying the original balance of oil and gas, the oil and gas in the conduction system migrate from high potential areas to low potential areas under the action of buoyancy, water dynamics, and tectonic stress, and eventually gather in the trap. If there are second, third, or even more tectonic movements after the oil and gas have gathered, they will transform, adjust, and even destroy the migration and accumulation of oil and gas.

The main period of secondary migration of oil and gas is also the main period for oil and gas to accumulate and form reservoirs. Therefore, studying the main period of secondary migration of oil and gas is of great exploration significance and is also a focus of petroleum geologists, and many progresses have been made. The content of this part will be introduced in Chap. 5.

4.3.5.2 Distance of Secondary Migration

The distance of secondary migration of oil and gas depends on the migration channel, regional tectonic background, lithology and lithofacies changes, the development of the overlying strata and faults, and the driving force of migration. The vertical distance of oil and gas migration depends on the thickness of the strata in the basin and the vertical extension distance of the fractures, reaching several kilometers at most. The lateral migration distance varies greatly, some are very small or even without lateral migration (such as tight oil and gas reservoirs, shale oil and gas reservoirs, coalbed methane reservoirs, etc.), and some are very large, even reaching hundreds of kilometers. For a basin, the distance of secondary migration of oil and gas is the distance from the hydrocarbon source rock area to the trap, and the maximum distance will not be greater than the distance from the center of the hydrocarbon depression to the edge of the basin in the direction of the long axis.

As long as there is a sufficient oil and gas source, the migration channel is regionally distributed and has good continuity, the conductive cap rock is good, and there are good conditions for the driving force of migration, oil and gas may undergo long-distance migration. Therefore, long-distance secondary migration of oil and gas mainly develops in large cratonic basins or during the stable development period of the basin. The Alberta Basin in Canada is one of the typical long-distance oil and gas migration basins known in the world. The distance of oil and gas migration in the Devonian carbonate rocks and Cretaceous clastic rocks is as long as 500 km and has been confirmed by oil and gas source comparison (Creaney & Allan, 1990; Deroo et al., 1977). The oil and gas migration distance in the Williston Basin in North America may be 160–250 km (Bethke et al., 1991). The oil and gas migration distance in the Persian Gulf Basin may be 100–500 km.

Hu (2005) conducted a statistical analysis of the migration distance of oil and gas in 200 basins or depressions worldwide and found that most areas are characterized by short-distance migration, with only 10–20% of the areas having a migration distance greater than 70 km. Among them, the continental fault depression basins have strong segmentation, large lithology and lithofacies changes, traps are close to oil sources, migration channels are distributed limitedly and change greatly, and oil and gas cannot undergo long-distance migration.

China's oil and gas basins are characterized by multiple tectonic cycles, strong segmentation, mostly continental strata, and large lithology and lithofacies changes.

Table 4.2 Secondary migration distance of oil and gas in some main oil and gas basins in China (modified from Zhang, 1989)

Basins	Distance of migration (km)	
	General	Maximum
Songliao Basin	<40	120
Tarim Basin	<60	130
Ordos Basin	<40	60
Bohai Bay Basin	0–20	30
Jianghan Basin	<10	15
Nanxiang Basin	<10	20
Jiuquan Basin	5–20	30
Zungharian Basin	30–50	80
Zhujaingkou Basin	<20	60

From the current situation of discovered oil and gas fields, they are mostly distributed near the hydrocarbon depression, and the distance of secondary migration of oil and gas is generally not large, usually ranging from a few kilometers to tens of kilometers. Some large basins (such as the Tarim Basin and Songliao Basin) have large areas, and the maximum migration distance is correspondingly larger, exceeding 100 km. For example, in the Hala Hutang Oilfield in the Tarbei area of the Tarim Basin, the oil and gas found in the Cretaceous sandstone reservoir in the southern area comes from the Triassic continental hydrocarbon source rock in the Kuqa Depression. The oil and gas mainly migrate laterally over a long distance through the fault communication with the unconformity surface and the sand body of the Cretaceous Bashi Formation, with a straight-line migration distance of 130 km. Table 4.2 is a statistical data of the secondary migration distance of oil and gas in several main oil and gas basins in China.

4.4 Hydrocarbon Migration Research Methods

The pursuit of the direction and path of hydrocarbon migration is a systematic, multi-dimensional, and complex research task that requires the use of various methods for comprehensive study. Since hydrocarbon migration is controlled by the geological conditions of the basin, geological analysis is the foundation for studying hydrocarbon migration. Through geological analysis, the potential dominant directions and paths of hydrocarbon migration are identified, and then methods such as fluid potential analysis, geochemical methods, laboratory simulation, and numerical simulation are combined for comprehensive judgment.

4.4.1 Fluid Potential Analysis Method

Subsurface fluids, while being subjected to buoyancy, pressure, and capillary forces, also possess their own energy. When fluids flow underground, they follow the principle of energy conservation. As fluids flow and perform work, they inevitably consume their own energy. Therefore, fluids naturally flow from areas of high mechanical energy to areas of low mechanical energy. Hubbert (1940, 1953) first introduced the concept of fluid potential into petroleum geology to depict the energy changes of subsurface fluids and the patterns of fluid migration. Later, Dahlberg (1982) systematically discussed the use of this method to study the direction and accumulation positions of hydrocarbon migration, which has attracted widespread attention from domestic and foreign oil and gas explorers. Fluid potential reflects the combined effects of hydrodynamic, buoyancy, and capillary forces on the motion state of subsurface fluids and is of great significance in the theoretical research of hydrocarbon migration and in solving regional trends and distribution of hydrocarbon migration.

4.4.1.1 Concept of Fluid Potential

Under reservoir conditions, the seepage of subsurface fluids is a mechanical motion process, where oil, gas, and water each have their own potential and migrate under its action. Hubbert (1953) defined the sum of the mechanical energy per unit mass of fluid underground as fluid potential (Φ), represented by the following formula:

$$\Phi = gZ + \int_0^p \frac{dp}{\rho} + \frac{q^2}{2} \tag{4.12}$$

where Z—elevation of the measurement point, m; g—acceleration of gravity, 9.8 m/s; P—pressure at the measurement point, Pa; ρ—fluid density, t/m^3; q—flow velocity, m/s.

The first term on the right side of the equation represents the potential energy caused by gravity, which can be understood as the work done to overcome gravitational changes when moving a unit mass of fluid from a reference surface (elevation = 0) to an elevation Z. The second term represents the fluid's pressure energy (or elastic energy), which can be understood as the work done by a unit mass of fluid due to pressure changes from the reference surface to an elevation Z. The third term represents kinetic energy, which can be understood as the work done to accelerate a unit mass of fluid from a stationary state to a flow velocity q.

When reflecting the variation characteristics of fluid potential on a profile, the concept of a potentiometric surface is often used, similar to a pressure surface. A potentiometric surface refers to an imaginary surface that reflects the variation of potential at different parts of a reservoir, constructed by connecting the fluid potentials at various points within the same reservoir.

4.4.1.2 Potential Gradient and Fluid Migration Direction

Hubbert defined the force experienced by a unit mass of fluid as the intensity of the force field, represented by E:

$$E = -\mathrm{grad}\Phi \tag{4.13}$$

grad Φ refers to the gradient of Φ, and the force field intensity is a vector.

From the above formula, the force field intensity for water, oil, and gas at the same point can be obtained respectively:

$$E_{w,o,g} = g - grad\frac{P}{\rho_{w,o,g}} \tag{4.14}$$

The first term on the right side of the above equation is the gravity force on a unit mass of fluid, which is numerically equal to the gravitational acceleration g. The second term represents the pressure force per unit volume of fluid. The force field intensity is the vector sum of the two. It can be seen that due to the different densities of oil, gas, and water, the force field intensities of the three are different under the same pressure environment.

In a static water environment, the force field intensity for water is 0, while those for oil and gas are not 0. Both force field intensity directions are upward, but since the density of gas is less than that of oil, the force field intensity of gas is greater than that of oil.

In a dynamic water environment, the force acting on a unit mass of oil and gas, compared to a static water environment, is not only subject to the downward gravity g and upward buoyancy $-grad\frac{P}{\rho}$, but also has an additional hydrodynamic force F_w that reflects the flow conditions. Therefore, under the action of hydrodynamic force, due to the different densities of water, oil, and gas, their force field intensities vary in magnitude and direction, and the three flow in their respective directions. The direction of flow varies with the magnitude of the hydrodynamic force. Figure 4.41 shows the force composition diagrams of water, oil, and gas under two conditions of different hydrodynamic forces.

Fluid potential analysis is a common method for studying the direction of secondary hydrocarbon migration and determining hydrocarbon accumulation areas. The prerequisite for applying fluid potential analysis methods is that the conduit layer must be connected. According to the calculation formula of fluid potential, the fluid potential values of a certain conduit layer at different locations can be calculated, and fluid potential contour maps can be made. For different fluids, water potential, oil potential, and gas potential contour maps can be made respectively. As shown in Fig. 4.42, it represents a homogeneous monocline conduit layer, where under the

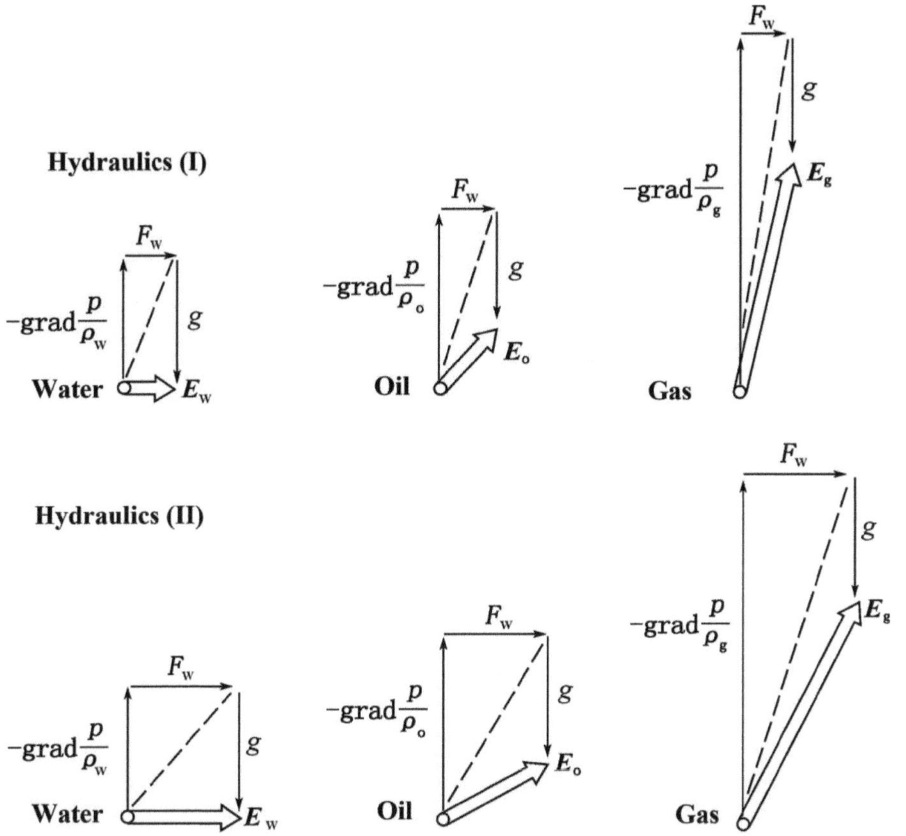

Fig. 4.41 Vector distribution and force field direction of various forces acting on unit mass of water, oil, and gas under different hydrodynamic conditions (according to Dahlberg, 1982)

action of hydrodynamic force, oil and gas migrate along their respective potential directions. In the figure, under the combined action of hydrodynamic force and buoyancy flowing in the downslope direction of the monocline, oil and gas are separated, that is, gas migrates in the upslope direction of the monocline, and oil migrates in the downslope direction of the monocline. If the hydrodynamic force is weak, oil and gas may both migrate in the upslope direction of the monocline. If the hydrodynamic force is strong, oil and gas may both migrate in the downslope direction of the monocline, but due to the differences in their respective force field intensities, the migration directions of oil and gas are different.

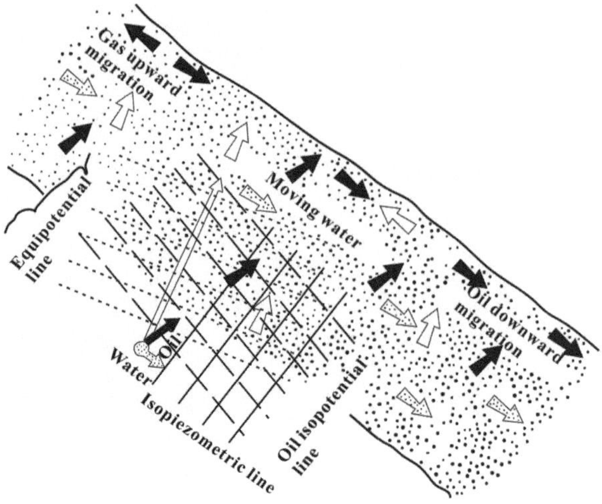

Fig. 4.42 Migration direction of oil and gas in a monocline conduit layer under downslope water flow conditions (according to Hubbert, 1953)

4.4.1.3 Relative Fluid Potential and Hydrocarbon Migration

Dahlberg (1982, 1995), on the basis of the concept of fluid potential, proposed the concept of relative fluid potential and used it to analyze the direction and location of hydrocarbon migration and accumulation, known as the UVZ method, also known as the relative fluid potential method.

Since the migration path of hydrocarbons is determined by the characteristics of the conduit layer, fluid potential cannot solve the specific migration path of hydrocarbons within the conduit layer. Therefore, fluid potential can only analyze the basic pattern and potential migration trend of hydrocarbon migration and accumulation. In addition, hydrocarbon migration occurs in the geological history, and it cannot be fully studied and evaluated using only the current geological conditions and parameters. It is necessary to restore the geological history conditions and paleodynamic characteristics of hydrocarbon migration and trace the migration process of hydrocarbons in each geological historical period. Hydrocarbon migration research must comprehensively consider the impact of various geological backgrounds, such as structural characteristics and tectonic evolution history, sedimentary characteristics and burial history, geothermal distribution and evolution history, and regional petroleum geological characteristics (including the development and history of hydrocarbon source rocks, possible source-reservoir-cap combinations), which control the period and basic migration framework of hydrocarbon migration.

4.4.2 Organic Geochemical Methods

During the migration process, petroleum will inevitably cause its own differentiation and changes in composition and properties with the variation of physical and chemical conditions. We use the regular changes in geochemical indicators and physical properties of petroleum to trace the migration route of oil and gas.

4.4.2.1 Organic Geochemical Methods for Studying Primary Migration

Organic geochemical methods for studying the primary migration of oil and gas mainly include two aspects: using geochemical indicators to determine the depth of expulsion, expulsion time, expulsion efficiency, and effective expulsion thickness. The depth and time of expulsion of hydrocarbon source rocks are mainly determined by studying the abrupt changes in geochemical indicators on the vertical profile. Li Mingcheng, through the analysis of the research by Albrecht and Ourisson (1969), pointed out that due to primary migration, the content of alkanes below the expulsion depth will suddenly decrease, and non-hydrocarbons, asphaltenes, and alkanes are transported together. On the other hand, Leythaeuser (1984, 1986), through the study of hydrocarbon source rocks in the Lower Cretaceous and Paleocene of Svalbard, Norway, pointed out that n-alkanes have a significant differentiation effect during the expulsion process, with low carbon number alkanes being expelled first. According to the decrease of hydrocarbon content in hydrocarbon source rocks from the center to the top and bottom sides adjacent to sandstone, the expulsion rate can be calculated and the effective expulsion thickness of hydrocarbon source rocks can be determined (Fig. 4.21).

4.4.2.2 Organic Geochemical Methods for Studying Secondary Migration

(1) Application of Hydrocarbon Compounds and Crude Oil Properties

During the migration process of oil and gas, chromatographic effects or oxidation effects will occur. Chromatographic effects, also known as geological color layer effects, refer to the phenomenon where physical properties and chemical components of oil and natural gas show regular differentiation and changes along the migration direction due to the selective adsorption effect of minerals. Since the molecules of non-hydrocarbon compounds in oil are large and polar, they are most easily adsorbed on the surface of minerals or dissolved in water. Aromatic hydrocarbons have stronger polarity than n-alkanes and cycloalkanes and are more soluble in water. Therefore, along the migration direction, the relative density, viscosity, wax content, and pour point of crude oil gradually decrease, and the content of high molecular weight hydrocarbon compounds, aromatic hydrocarbons, non-hydrocarbon compounds, and heavy metals (V, Ni, Ca) also correspondingly decrease. In addition, it will also cause changes in some biomarker compounds, such as the sterane compounds, where the 5α, 14β, 17β isomers are relatively enriched compared to the 5α, 14α, 17α isomers, and the rearranged steranes 13α, 17β are relatively enriched compared to the regular steranes 15α, 14α, 17α, and the C_{29} cholestane 20S configuration is relatively enriched compared to the 20R configuration.

Due to the color layer effect, the carbon isotope of methane and condensate components in natural gas will cause significant carbon isotope fractionation during the migration process, which can indicate the direction of oil and gas migration. Along the migration direction, the $\delta^{13}C$ values of methane and condensate components in natural gas gradually become lighter. Oil has stronger adsorption for ^{13}C than ^{12}C, and along the migration direction, the $^{13}C/^{12}C$ ratio decreases, and the $\delta^{13}C/\delta^{12}C$ ratio decreases along the migration direction. For example, the $^{13}C/^{12}C$ ratio in aromatic hydrocarbons is higher than that in n-alkanes and cycloalkanes, and as aromatic hydrocarbons decrease along the oil and gas migration direction, it will inevitably lead to a decrease in the $^{13}C/^{12}C$ ratio.

Through the study of the variation rules of triterpenes in reservoir asphalt, the migration characteristics of natural gas in the Binan-Gubei area of the Jiyang Depression were traced. Figure 4.43 shows that from the Guxi fault step zone of the Gu 22 well to the Gubei buried hill zone of the Gubei Gu 2 and Bo 93 wells, the ratio of triterpenes $(C_{21} + C_{22})/(C_{23} + C_{24})$ increases gradually from 0.37 to 0.5, and then to 0.72, indicating that oil and gas have undergone upward migration along the Guxi fault.

It must be pointed out that the above-mentioned changes in the properties of oil and gas can only occur when the geological color layer effect plays a dominant role along the direction of oil and gas migration. If oxidation plays a dominant role during the migration process, not only the above regularity does not exist, but also the opposite change rule will appear. The Wen'an slope of the Baxian Depression has the characteristic of deteriorating crude oil properties with increasing migration distance of oil and gas, leading to the change of the crude oil property diamond from

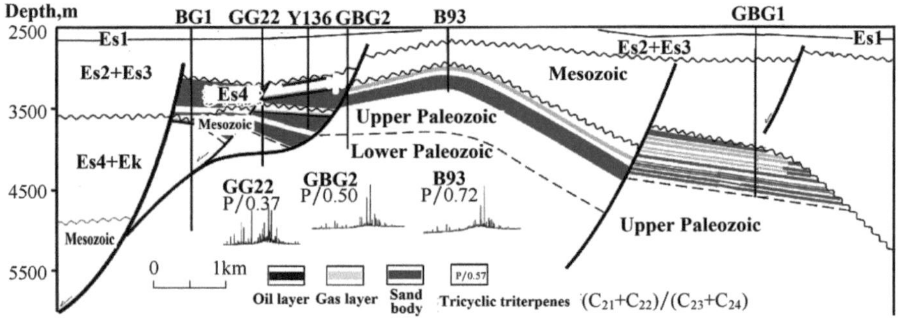

Fig. 4.43 Distribution characteristics of asphaltene triterpanes in the reservoirs of the Binan-Gubei area (according to Jiang & Liu, 2010). *Note* E_{s1}—first section of Shahejie formation, E_{s2}—second section of Shahejie formation, E_{s3}—third section of Shahejie formation, Es4—fourth section of Shahejie formation, Ek—Kongdian formation

a flat type to a square diamond and then to a sharp diamond (Fig. 4.44), and the reason for this change is that oxidation plays a dominant role as the migration distance of oil and gas increases.

In addition to the effects of chromatographic and oxidation effects during the migration process of oil and gas, their chemical composition and physical properties will also be affected by other factors and change, affecting the tracking of the direction and path of oil and gas migration. These factors include thermal alteration effects, deasphalting effects, water washing effects, biodegradation effects, and sulfurization effects. Among them, thermal alteration and deasphalting effects can cause oil to become lighter, while water washing, biodegradation, and sulfurization effects can cause oil to become heavier. Therefore, when analyzing the main direction of oil and gas migration, it is necessary to comprehensively analyze various geological conditions to draw a more correct conclusion.

(2) Application of Nitrogen-Containing Compounds

The application of nitrogen-containing compounds for oil and gas migration tracing has become a more mature technology. The organic nitrogen compounds in crude oil mainly appear in the form of neutral pyrrole-type aromatic compounds. The organic nitrogen compounds applied to oil and gas migration mainly refer to two series: pyrrole-type and pyridine-type. Pyrrole is a five-membered ring compound containing one nitrogen atom. Pyrrole-type compounds refer to condensed pyrroles (or pyrrole benzenoids) and their derivatives, such as carbazole, benzocarbazole, and dibenzocarbazole (Fig. 4.45). According to the substitution of the alkyl group (usually methyl) on the 1–8 positions of the carbazole compounds, the alkyl carbazole is divided into three types: shielded isomers (both C-1 and C-8 are substituted by alkyl groups), semi-shielded isomers (only one of C-1 and C-8 is substituted by

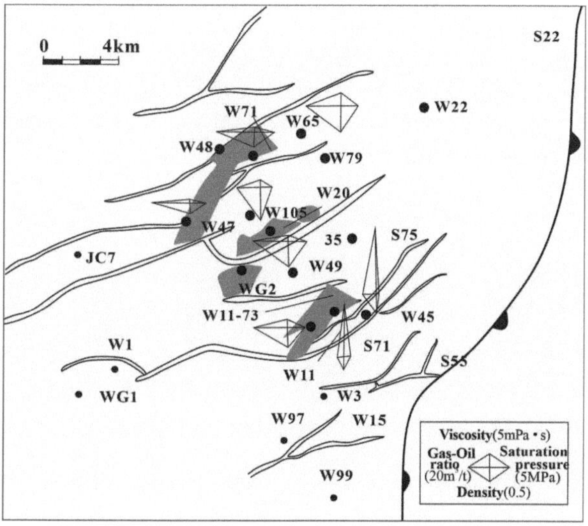

Fig. 4.44 Rhombus diagram of crude oil properties and oil–gas migration direction in the Wen'an Slope of the Ba County Depression (after Liu et al., 2011)

an alkyl group), and exposed isomers (neither C-1 nor C-8 is substituted by alkyl groups).

The basic principle of applying organic nitrogen compounds to evaluate oil and gas migration is that carbazole molecules with nitrogen-containing heterocyclic rings have strong polarity. They can form hydrogen bonds with the negatively charged oxygen atoms on organic matter or clay minerals in the formation through hydrogen atoms bonded to the nitrogen atoms, causing some carbazole molecules to be retained on the migration path during the oil and gas migration process, resulting in the formation of a geological color layer fractionation effect of carbazole. The fractionation effect is mainly manifested in the following points:

With the increase of oil and gas migration distance, the absolute abundance of nitrogen-containing compounds in crude oil decreases.

Nitrogen functional group shielding isomers are relatively enriched compared to semi-shielding isomers or exposed isomers.

Alkyl carbazoles are relatively enriched compared to alkyl benzocarbazoles.

Among benzocarbazole isomers, benzo[a]carbazole is relatively enriched compared to benzo[c]carbazole.

Figure 4.46 is an example of using nitrogen-containing compounds to trace the migration path of oil and gas based on the study of the skeletal sand body conduction system and fault sealing. The Wen'an slope zone of the Baxian Depression is adjacent to the Baxian sag, the center of oil and gas generation. The migration of oil and gas is controlled by the skeletal sand body, fault sealing, and their mutual configuration relationship. Faults play a vertical adjustment role. From the distribution of nitrogen-containing compounds, it can be seen that from the deep sag zone to the upper slope direction, from the Es_3 hydrocarbon source rock series to the overlying Es_1–E_d, the parameters of each nitrogen-containing compound gradually increase, indicating that the migration path of oil and gas is along the skeletal sand body-fault composite conduction system to the edge of the depression in a stepped migration.

The abundance and relative distribution of carbazole-type non-hydrocarbon compounds will also be affected by the original matrix and maturity, which should be distinguished when used, but their impact may be less than that of sterane and triterpane-type biomarker compounds, so they are called the chemical tracers of oil secondary migration abroad.

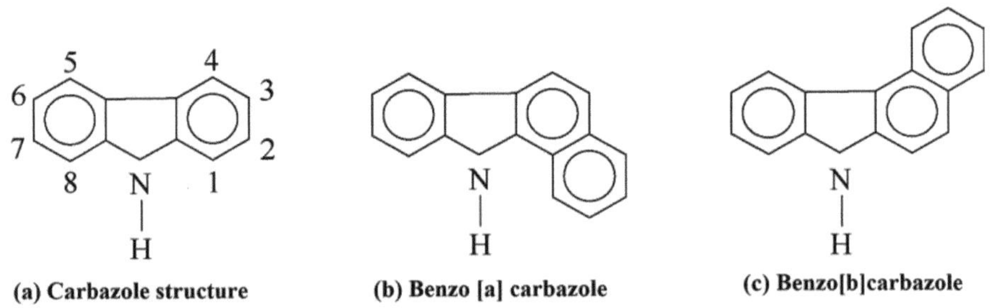

Fig. 4.45 Structural diagram of carbazole and benzocarbazole

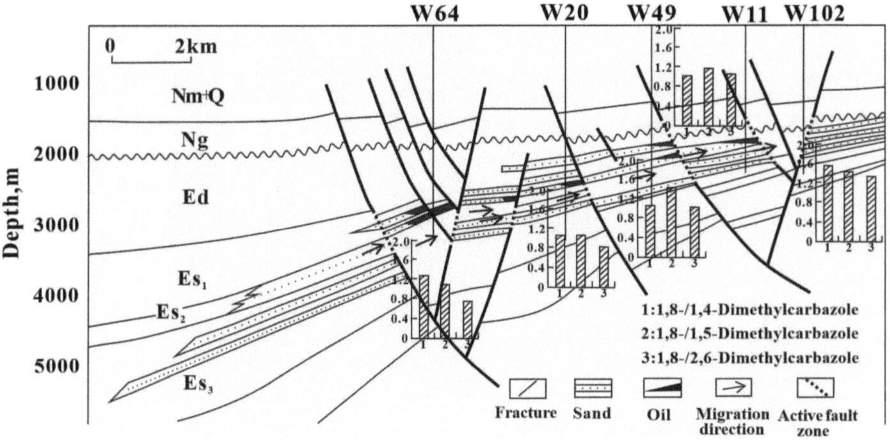

Fig. 4.46 Nitrogen compounds in the Wen'an slope of the Baxian Depression indicate oil and gas migration pathways (according to Liu et al., 2011)

(3) Application of Maturity Parameters

The migration and charging of oil are geological processes that last for a considerable period of time. The maturity of oil injected in the early stage is relatively low, and the maturity of oil injected in the later stage is relatively high (England et al., 1987). During the charging process of the oil reservoir, it is actually the later mature oil with higher maturity that drives the earlier oil with lower maturity, continuously migrating and charging forward in a "wave front" manner until the charging process is completely finished, resulting in a certain maturity difference of oil within the reservoir. Therefore, within an oil reservoir, the migration and charging process of oil can be characterized based on the subtle differences in maturity between early and late injected oils. The gradient change of maturity-sensitive geochemical indicators is often used to judge the direction of oil and gas migration, such as the hopane and sterane indicators of biomarker compounds: C_{29} sterane 20S/(20S + 20R), C_{29} sterane $\beta\beta/(\alpha\alpha + \beta\beta)$, rearranged sterane/regular sterane, C_{31} norhopane 22S/(22S + 22R), Ts/Tm, and C_{27} triaromatic hopane, etc. Non-biomarker parameters such as the methyl phenanthrene index, thiophene, and methyl naphthalene ratio.

Wang et al. (2008) used the ratio of 4-MDBT to 1-MDBT to analyze the direction and path of oil charging in the Ordovician reservoir of the main area of the Tahei Oilfield. The maximum value of 4-MDBT in the Ordovician crude oil of the Tahei Oilfield is in well S_{76}, and the high-value wells are all located on the south side of the Tahei Oilfield, distributed in a linear pattern, showing a trend of gradually decreasing maturity to the north, indicating the migration and charging direction of Ordovician crude oil from south to north. The high-value wells of 4-MDBT can be regarded as charging points. For example, the 4-MDBT isopleths of wells S_{76} and T_{615} delineate two charging points in the western block of the Tahei Oilfield, namely the S–N direction from well S_{76} to T_{607} and the NW direction from well T_{615}. The

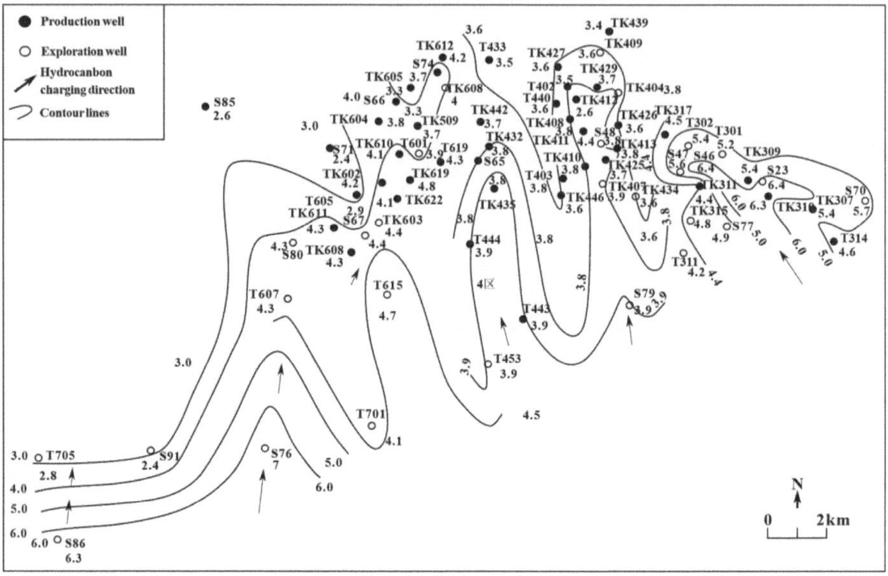

Fig. 4.47 Schematic diagram of the charging direction and path of Ordovician crude oil in the main area of the Tahei Oilfield indicated by the ratio of 4-MDBT to 1-MDBT (Wang et al., 2008)

two merges at well TK_{603} and complete the reservoir charging along the NE direction of wells TK_{610}-TK_{604}-S_{74}-TK_{612} (Fig. 4.47).

4.4.3 Laboratory Simulation Method

The physical simulation method of hydrocarbon migration mainly uses various experimental devices to observe the migration process and state of hydrocarbons in real-time, verify hydrocarbon migration theories, deeply understand the mechanisms of hydrocarbon migration, and provide calculation parameters for quantitative research on hydrocarbon migration and accumulation. It is a direct and effective method for hydrocarbon migration and accumulation research.

As early as 1921, Wang Miers used sandstone samples of various grain sizes to conduct oil migration experiments. Hubbert (1953) carried out corresponding quantitative experiments by taking buoyancy, hydrocarbon dynamics, and capillary force as the controlling factors of migration, and first proposed the concept of fluid potential. In the 1960s, the control effect of micro experimental parameters such as interfacial tension, wettability, and throat radius on the hydrocarbon migration process began to be noticed. In 1989, Lenormand (1988) and others used micro models to study the immiscible displacement process in porous media and summarized the influence of capillary force on hydrocarbon migration with capillary numbers and viscosity ratio coefficients, which are viscous fingering, capillary fingering, and stable displacement.

Chinese petroleum geologists mainly started physical simulation research on hydrocarbon migration in the early 1990s, and have achieved rich results. Zeng and Wang (1999, 2000) made sandstone micro-pore models with core samples and conducted experimental studies on the mechanism of oil displacement of water. The results show that when oil displaces water, phenomena such as breakthrough, jumping, blocking, piston-like displacement, and non-piston-like displacement can be observed, while the gas displacement of water process is obviously controlled by capillary action. Zhang et al. (2004) and others used a tubular glass tube model filled with glass microspheres to simulate the seepage law of oil in a water-saturated skeletal sand body, pointing out that pure buoyancy and relatively small driving force can form an advantageous migration channel, and show strong non-uniformity during the migration process. Once the migration pathway is formed, its form and spatial distribution characteristics are basically consistent until the end of the migration, and the re-injected oil still basically migrates along the original path (Fig. 4.48). Jiang et al. (2005) simulated the influence of sand body homogeneity on migration pathways, pointing out that oil and gas always migrate in the direction where the difference in permeability between the channel medium and the surrounding medium is the greatest (grade difference advantage). Zhao et al. (2009) carried out simulation experiments of gas expulsion and migration of coalbed hydrocarbon source rocks under three-dimensional temperature and pressure conditions, simulating the large-scale charging and migration process of natural gas that may occur under the geological background of sedimentation and uplift.

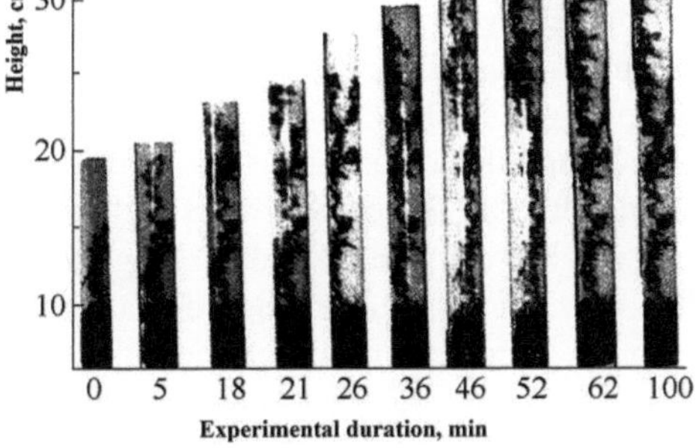

Fig. 4.48 Schematic diagram of the formation process of non-uniform oil migration pathways (according to Luo et al., 2018)

4.4.4 Numerical Simulation Method

Numerical simulation, also known as mathematical simulation, mainly uses computers for operation and is also called computer simulation. Since oil and natural gas are mainly produced in sedimentary basins, numerical simulation in petroleum geology is mainly for basin simulation. Basin simulation is to reproduce the entire process of oil and gas generation, migration, accumulation, and loss. Because simulation technology is the best embodiment of dynamic analysis and comprehensive research, it has become the best method and the only means for quantitative research on hydrocarbon migration.

The main driving force for secondary migration of oil and gas is the buoyancy generated by the density difference between hydrocarbons and water, and the main resistance is the capillary force, which is related to the radius of pore throat, interfacial tension, and rock wettability. Applying basin simulation technology to simulate the secondary migration of oil and gas is to establish a three-dimensional/two-dimensional geological framework, simulate the process of oil and gas being expelled from the hydrocarbon source rock and entering the conductive system, and under the driving force of buoyancy, overcome the surrounding capillary resistance, find the optimal migration path of oil and gas, and accumulate into reservoirs under the sealing of the overlying effective cap rock.

According to the different oil and gas migration simulation algorithms, they can be divided into streamline method, invasion percolation method, and multiphase Darcy flow method. The streamline simulation method is mainly based on the geological elevation characteristics of the two-dimensional structural surface, using the normal method (along the steepest direction), and combining with factors such as sandstone percentage to determine the migration direction of oil and gas. The invasion percolation method is based on dynamic resistance analysis. When the driving force of oil and gas migration is greater than the resistance, oil and gas continue to migrate forward, and when the driving force is less than the resistance, migration stops. It is mainly suitable for simulating oil and gas migration under conditions of complex geological layer geometry and high grid density, with high computational efficiency. The multiphase Darcy flow simulation method is considered the most accurate and complex physical method to describe the flow of fluids in porous media, suitable for simulating oil and gas migration in strata with poor permeability, and can effectively simulate the migration pathways and the amount of hydrocarbon accumulation.

It can be seen that oil and gas migrate from the hydrocarbon source rock stove in the lower left corner of the figure in a divergent trend (Fig. 4.49). As they move away from the hydrocarbon source rock, the migration pathways of oil and gas gradually narrow and converge, forming accumulations in suitable traps.

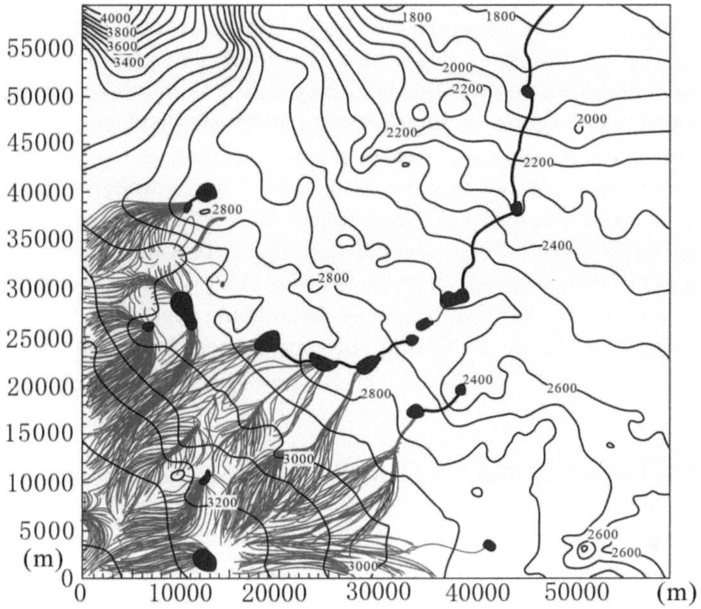

Fig. 4.49 Simulation results of oil and gas migration pathways and accumulation positions based on streamline method (according to Hantschel & Kauerauf, 2009)

4.5 The Role of Geothermal, Geopressure, and Geomechanical Fields in Hydrocarbon Migration

The geothermal field, geopressure field, and geomechanical field (referred to as the "three fields") are manifestations of the Earth's internal energy in different forms within the crust. The geopressure field refers to the distribution of pressure at various points within subsurface fluids. The geothermal field is the manifestation of the Earth's internal thermal energy through rocks with different thermal conductivities within the crust. The geomechanical field refers to the variation of stress states within the crust or the Earth's interior as a function of spatial location. These "three fields" interact and influence each other, controlling the fluid dynamics processes in sedimentary basins, thereby affecting the migration and distribution of hydrocarbons.

4.5.1 The Role of the Geopressure Field in Hydrocarbon Migration

Abnormal high pressure is an important dynamic source that constitutes the regional fluid potential field of a basin and plays a significant role in the primary and secondary migration of hydrocarbons. In compaction flow basins, the center of the basin, depression, or sag is often the main area where abnormal high pressure develops, and on the profile, abnormal high pressure mainly exists within thick layers of shale. During the compaction process, high-pressure shale can conduct pressure to adjacent sandstones and, under the action of pressure differences (or potential), drain water and expel hydrocarbons into the sandstones. This hydrodynamic force is the main factor in forming a regional "centrifugal" fluid potential field. It is under the action of the "centrifugal" fluid potential field that hydrocarbons migrate laterally from the center of oil-generating depressions or sags along pathways such as carrier beds, faults, and unconformities, primarily towards the edge areas for aggregation, forming a "ring-like" distribution pattern of hydrocarbons around the oil-generating depression.

4.5.1.1 The Genesis of Fluid Pressure Seals and Sealing Formations

The formation of abnormal pressure must have a three-dimensional enclosed space, which can be laterally sealed by faults, phase changes, salt masses, etc., and vertically sealed by rock strata. This three-dimensional space, effectively sealed by an impermeable cap rock to prevent pressure from returning to hydrostatic pressure, is called a fluid pressure compartment, or simply a fluid seal or compartment. The compartment is an abnormal pressure unit or pressure seal body divided by sealing formations within a sedimentary basin.

Compartments typically have all the necessary conditions for generation, storage, and sealing, often consisting of a main compartment and secondary compartments. Many sedimentary basins have a normal hydrostatic pressure system in the shallow parts, while deeper parts may contain one or more pressure compartments. The pressure within a compartment can be higher or lower than the normal hydrostatic pressure. Multiple compartments can occur both vertically and laterally. Figure 4.50 illustrates a basin-scale pressure compartment.

Based on the pressure characteristics of the compartments, fluid pressure compartments are divided into two types: one is an overpressured compartment with abnormally high pressure, where the pore fluid supports the cap rock and the overlying rock—fluid weight. The other is an underpressured compartment with abnormally low pressure, where the rock matrix supports the cap rock and the overlying rock—fluid weight. Figure 4.51 expresses the pressure-depth relationship of the two types of compartments. In oil and gas basins, overpressured compartments are commonly found.

Sealing formations are key to forming and separating fluid compartments. Sealing formations are not the same as the cap rock of oil and gas reservoirs. They often cross different stratigraphic and lithologic interfaces and, under certain temperature conditions, form sealing formations with near-zero permeability due to mineralization, filling, and other diagenetic processes. If the sealing formation is carbonate, it is usually caused by silicification. If it is shale (mudstone), it is often related to calcification. When the reflectance of the vitrinite reaches 0.9%, the kerogen has

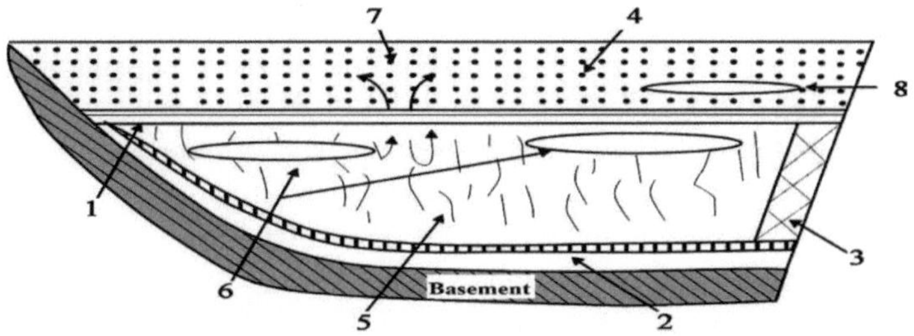

Fig. 4.50 Basin-scale fluid pressure compartment (according to Ortoleva, 1994). 1—The central diachronous formation acts as the top seal across the entire basin, 2—the diachronous formation surrounding the basin base forms the bottom seal, 3—there is a gradual vertical seal change near the fault on the right, 4—above the top seal is the normal pressure compaction zone, 5—a large, fractured compartment is formed within the basin below the top seal; 6—there are also smaller pressure compartments inside, 7—occasionally, some fluids within the basin penetrate the top seal in a pulse-like manner, 8—there are some peripheral pressure compartments within the normal compaction zone

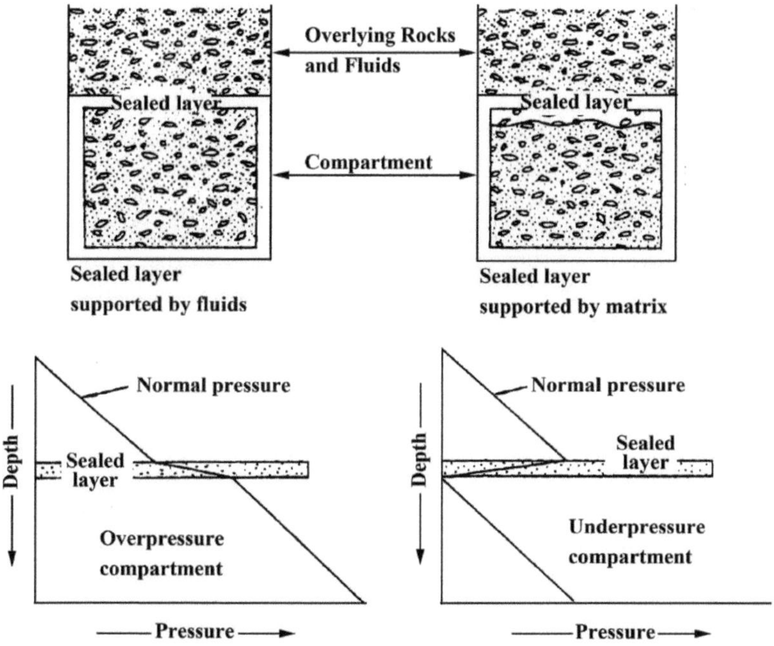

Fig. 4.51 Schematic diagram of overpressured and underpressured compartments (according to Hunt, 1990). **a** Overpressured compartment; **b** underpressured compartment

entered the peak oil generation period, releasing a large amount of carbon dioxide, which helps dissolve a large amount of carbonate to form a secondary pore development zone; when this carbonate solution moves upward to where the reflectance of vitrinite is 0.4–0.5%, the carbonate re-precipitates, forming the top seal, which coincides with the beginning of the oil window. Therefore, oil is often generated within the compartment below the sealing formation.

4.5.1.2 Pressure Compartments and Hydrocarbon Migration

The formation of fluid compartments has a significant impact on hydrocarbon migration. Due to the segmentation of sedimentary basins by compartments, a unified sedimentary basin is divided into many independent hydrodynamic units, which are not interconnected, hindering large-scale hydrocarbon migration within the basin, thus forming many independent hydrocarbon accumulation units. The lateral migration of hydrocarbons can only occur within the compartment, and the principles of fluid potential analysis generally can only be applied within the compartment. In addition, vertical migration is also limited by the compartment, and under certain conditions, hydrocarbons can also migrate outside the compartment through the seal. Therefore, when analyzing the direction and distance of hydrocarbon migration in a basin and studying the pattern of hydrocarbon migration within the basin, the distribution of fluid compartments should be considered.

Due to the episodic opening of fluid compartments, hydrocarbons exhibit a episodic migration characteristic. When the abnormally high pressure in the source rock reaches a pressure that can cause rock fracturing, it can cause the source rock to fracture and produce a large number of micro-fractures, and hydrocarbons and water are expelled along the micro-fractures. After the fluid is expelled, the pressure drops, and the micro-fractures close, and then due to the continuation of the hydrocarbon generation process, the pressure begins to rebuild, and the source rock fractures again. This forms a repeatedly cyclical process composed of "hydrocarbon generation—pressure build-up—fracturing—expulsion—pressure release—closure," until the hydrocarbon generation potential is exhausted and unable to fracture the source rock again.

Based on the relationship between abnormal pressure fluid compartments and hydrocarbon migration accumulation, it can be divided into three hydrocarbon accumulation patterns: extra-compartment accumulation, intra-compartment accumulation, and compartment edge accumulation (Fig. 4.52). The extra-compartment accumulation pattern refers to the abnormal high pressure causing the compartment to fracture, the sealing formation to fracture, or faults cutting the sealing formation to periodically open, allowing hydrocarbons and water to undergo pulse-like mixed-phase flow (episodic migration) outside the compartment, into a lower temperature and pressure stratum, usually the overlying stratum adjacent to the sealing formation, where hydrocarbons and water separate and accumulate in the normal pressure system outside the compartment as temperature and pressure decrease. The intra-compartment accumulation pattern refers to the hydrocarbons generated within the

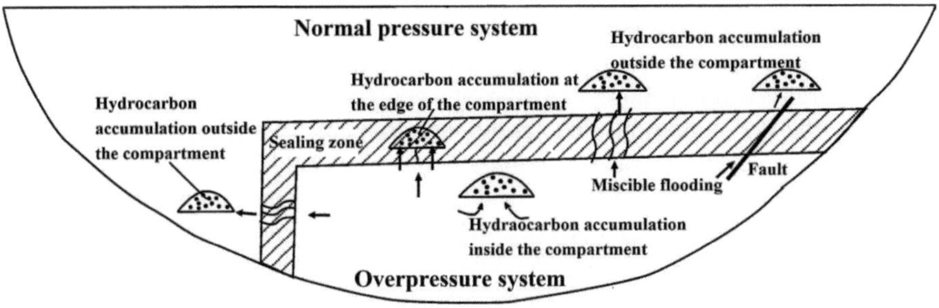

Fig. 4.52 Schematic diagram of compartments and hydrocarbon migration patterns

compartment migrating within the compartment through suitable pathways and accumulating in traps with lower fluid potential, such as sandstone lenses within the source rock formation. If the top sealing formation of the compartment is composed of multiple tight layers interbedded with multiple porous reservoir layers, when the compartment edge fractures, if only the inner sealing layer fractures and the outer sealing layer does not fracture, hydrocarbons can migrate and accumulate at the compartment edge.

4.5.2 The Role of the Geothermal Field in Hydrocarbon Migration

4.5.2.1 Concepts Related to the Geothermal Field

The geothermal field of a sedimentary basin mainly depends on the basement heat flow and its evolution. The intensity of the geothermal field can be measured by the geothermal gradient and the heat flow value.

The geothermal gradient, also known as the geothermal increase rate, refers to the increase in temperature for a certain increase in depth, generally represented by the temperature increase per 100 m or 1 km depth (°C/100 m or °C/1 km). The geothermal gradient reflects the change of ground temperature with depth, and its calculation formula is as follows:

$$G = \frac{T_H - T_0}{H} \times 100 \tag{4.15}$$

where, G is the geothermal gradient, °C/100 m. T_H is the temperature at point H, °C. T_0 is the average surface temperature or the temperature of the isothermal zone (the zone where the influence of the Earth's internal heat and solar radiation heat reaches a balance), °C. H is the depth of the temperature measurement point or the difference in depth between the temperature measurement point and the isothermal zone, m. For a certain area, if the depth and temperature data of the surface isothermal zone have not yet been obtained, the influence of the isothermal zone can also be temporarily ignored, and the geothermal gradient can be calculated directly from the surface, which generally will not cause a significant calculation error.

The geothermal conditions and geothermal gradients of a region are related not only to the characteristics of the regional geothermal field but also to the thermal conductivity of rocks. The thermal conductivity of rocks represents the ability of rocks to conduct heat, and its physical meaning is the amount of heat conducted through a unit thickness of rock per unit time when the temperature difference between the two sides of the rock along the heat conduction direction is 1 °C, with the unit W/(m K). The thermal conductivity of different rocks varies greatly. Generally speaking, the thermal conductivity of crystalline rocks is higher than that of sedimentary rocks. Among sedimentary rocks, coal rocks have the lowest thermal conductivity, followed by shale and mudstone, while rock salt and gypsum have the highest thermal conductivity, and the thermal conductivity of sandstone and conglomerate varies greatly. Pore fluids and pore fillers are both poor conductors of heat, so porosity will greatly reduce the thermal conductivity of rocks. Under the same heat source, regions and strata with low thermal conductivity have higher geothermal gradients.

The heat flow refers to the amount of heat transferred from the Earth's interior to the Earth's surface per unit area over a certain period of time and is a reflection of the deep heat characteristics of the Earth, with the unit mW/m^2. In areas with more radioactive granite bodies or shallow Moho, the heat flow value is generally higher, and the average heat flow value of global sedimentary basins is 60–70 mW/m^2. The heat flow value is not affected by the medium, and it can profoundly reveal the inherent characteristics of the regional geothermal field. Therefore, the heat flow value is the most important geological-geophysical parameter characterizing the characteristics of the regional geothermal field. In numerical value, it is equal to the product of the geothermal gradient and the rock thermal conductivity, that is:

$$Q = K \times G \tag{4.16}$$

where, Q is the heat flow, mW/m^2; K is the rock thermal conductivity, W/m °C; G is the geothermal gradient, °C/100 m.

4.5.2.2 Factors Affecting the Geothermal Field

Actual data show that the geothermal field is heterogeneous, and many factors can directly or indirectly affect the distribution of the geothermal field. Research has shown that the main factors affecting the geothermal field include the nature of the tectonic structure, basement undulations, magmatic activity, lithology, cover folding, faults, groundwater activity, and hydrocarbon accumulation (Yang, 1993).

(1) Tectonic Structure Nature

The nature of the tectonic structure and its position within the structure are the most important controlling factors that determine the basic background of the regional geothermal field. According to plate tectonics, the collision zones between plates and continental rifts form the global seismic belts, volcanic belts, and high-temperature geothermal belts. The thickness of the crust has a significant impact on the geothermal conditions. For example, the crust in the eastern regions of China is generally thinner than in the western regions, hence the geothermal conditions and geothermal gradients in the basins of the east are generally higher than those in the west. The central and western basins, due to their early geological age, higher degree of compaction, and reduced thermal resistance, form a low-heat background with relatively low temperatures and geothermal gradients. The eastern basins are often characterized by mantle uplift, crustal stretching, and rifting, with reduced thickness, and the sedimentary cover is more intensely baked by the mantle, coupled with the upwelling and intrusion of mantle materials, leading to increased geothermal conditions within the basins.

(2) Basement Undulations

In sedimentary basins, the different geothermal conditions in uplifts and depressions are mainly caused by the lateral heterogeneity of the thermal physical properties of rocks. It is essentially the result of redistributing the uniform heat flow from the Earth's interior in the upper crust. Since the thermal conductivity of the basement is often higher than that of the sedimentary cover, the deep heat flow will concentrate towards the uplifted basement, giving it characteristics of high heat flow and high geothermal gradient, while the depressions have lower geothermal characteristics.

For example, in the Dongying Depression of the Bohai Bay Basin, the interior of the depression with thicker cover shows lower geothermal gradient values, generally 2.8–3.7 °C/100 m; the depression margins and buried hill areas with deeper basement and thinner sedimentary cover show higher geothermal gradient values, generally 3.7–4.7 °C/100 m.

(3) Magmatic Activity

Volcanic and magmatic activities have a significant impact on the distribution of geothermal conditions. Many high-temperature geothermal fields in the world are located in modern volcanic areas, where volcanic activity is intense, and magmatic activity in the upper crust is evident, mostly in the edge areas of plate tectonics. The impact of magmatic activity on geothermal conditions is mainly related to the time of magmatic intrusion, the size of the magmatic body, geometric shape, the attitude of the surrounding rocks, and the thermal properties of the surrounding rocks.

(4) Lithology

Lithology has a significant impact on the distribution of the geothermal field. Generally speaking, rocks with high thermal resistance and poor thermal conductivity in the same well have higher geothermal gradients. Rock layers with low thermal resistance and good thermal conductivity have smaller geothermal gradients. According to the concept of heat flow value, in the absence of external heat sources, the product of the geothermal gradient and thermal conductivity for each stratum in the same well is a constant, that is, the heat flow values for each stratum in the well are equal, so there is a compensatory relationship between the geothermal gradients and thermal conductivities of each stratum.

(5) Cover Folding

The folding structure of sedimentary rocks in the cover above the basement has a significant impact on the geothermal field. It has been confirmed by a large amount of temperature measurement data that the geothermal temperature and gradient increase from the wings of an anticline towards its axis or core. The impact of cover folding on the geothermal field is similar to the impact of basement undulations on the geothermal field in the cover.

(6) Faults

Fracture activity can cause the migration of deep high-temperature fluids or shallow low-temperature fluids, leading to local anomalies in the distribution of geothermal conditions along the fractures. When deep high-temperature fluids migrate upward along the fault and intrude into the shallow strata, high geothermal anomalies will appear. If the heat generated by fault friction is not promptly dissipated, the geothermal temperature will rise, and the pore fluid pressure will also increase correspondingly due to the rise in temperature.

(7) Groundwater Activity

Due to the differences in groundwater and hydrogeological conditions, the relationship between the temperature field of groundwater and the surrounding rocks is complex and variable. When lateral groundwater activity is strong and groundwater recharge and flow conditions are good, groundwater activity can cause the temperature of the surrounding rocks to decrease; on the other hand, the rise of deep circulating groundwater can cause local positive geothermal anomalies.

(8) Hydrocarbon Accumulation

A large amount of data indicates that there is often a high geothermal anomaly above hydrocarbon accumulations (oil and gas fields). This phenomenon is due to the fact that hydrocarbon reservoirs themselves provide an additional heat source. The additional heat source of hydrocarbon reservoirs mainly comes from the exothermic reactions of hydrocarbon oxidation under aerobic and anaerobic conditions and the concentration of radioactive elements. In addition, as the fluid seeps upward, it carries the excess heat from the hydrocarbon reservoir to the shallow strata and the surface, increasing a small additional heat flow value above the hydrocarbon reservoir.

4.5.2.3 Geothermal Field and Hydrocarbon Migration

The geothermal field has a multifaceted impact on the migration and accumulation of hydrocarbons. The thermal history of the source rock has a decisive effect on its oil generation status and the process of oil generation and expulsion. The amount of oil generation determines the amount of primary migration and has an important impact on the fluid pressure and fluid potential in the source rock layer. Changes in the geothermal field affect the viscosity of fluids in the strata, the properties and phase state of oil and gas, and the solubility of natural gas in oil and water, thereby affecting the migration capacity of oil and gas. At the same time, the saturation of each phase will also change, and changes in saturation will affect the capillary pressure, buoyancy, and other forces driving the migration and accumulation of oil and gas. Therefore, the spatiotemporal changes of the geothermal field in oil and gas basins are not only important physical conditions in the process of oil and gas migration and accumulation but also indirectly constitute the driving force for oil and gas migration (Song et al., 2002).

In addition, differences in the temperature field can generate convection of formation water. When formation water containing a large amount of natural gas undergoes thermal convection, it can cause the gas to migrate with the water and precipitate out in places with suitable temperature and pressure conditions. This mode is more common in carbonate strata (Fig. 4.53).

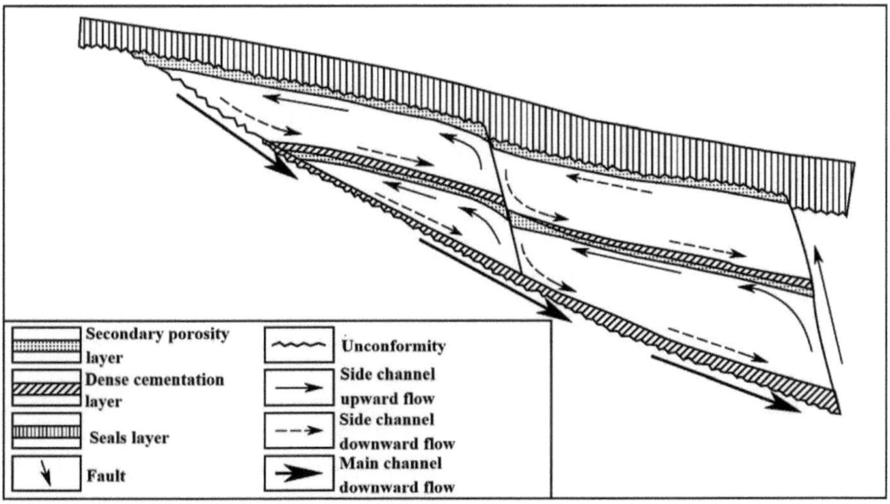

Fig. 4.53 Conceptual geological model of the thermal convection system in carbonate strata

4.5.3 *The Role of the Geomechanical Field in Hydrocarbon Migration*

4.5.3.1 Concept of the Geomechanical Field

Broadly speaking, geological structural phenomena are determined by the total geomechanical stress, which includes both the static geomechanical stress (vertical compressive stress) caused by the weight of the overlying rocks controlled by gravity and the tectonic stress controlled by crustal tectonic movements. The tectonic stress field is variable, while the static geomechanical stress field is relatively constant. It is evident that changes in the total geomechanical stress are mainly caused by changes in the tectonic stress field. Therefore, many scholars are accustomed to using a narrow concept, regarding the static geomechanical stress as the geostatically stress, which belongs to the category of the stress field, and referring to the geomechanical field as the tectonic stress field.

The geomechanical stress field generally changes over time but is relatively stable during certain geological periods. Studying the geomechanical stress field involves investigating the regularity of geomechanical stress distribution, determining the direction, nature, magnitude, and evolutionary characteristics of the stress at a certain point or region on the crust under specific geological eras and conditions.

4.5.3.2 Geomechanical Field and Hydrocarbon Migration

Geomechanical stress is one of the driving forces for hydrocarbon migration and accumulation. Structures such as folds and faults formed by the action of geomechanical stress can serve as pathways and sites for hydrocarbon migration and accumulation. The paleo-stress field influences and controls the direction, pathways, and intensity of primary and secondary hydrocarbon migration during geological history, while the current stress field influences and controls the dynamic changes of oil, gas, and water in oil and gas fields during development.

Earthquakes are a geological phenomenon of geomechanical stress release. It has been observed that earthquake activity can have a certain impact on the oil and gas production of nearby wells. For example, the W-11 well in the Dagang Oilfield, which is 130 km away from the epicenter, has a close relationship with the Tangshan earthquake and strong aftershocks (Fig. 4.54). Sudden increases or decreases in oil and gas well production before and after the occurrence of earthquakes are a manifestation of the effect of stress field changes on hydrocarbon migration.

The relationship between the stress field and hydrocarbon migration is close, mainly manifested in three aspects (Chen et al., 2003): First, tectonic stress can cause deformation and displacement of rock layers, and the deformation of rock layers can lead to changes in the fluid potential field and pressure system, causing hydrocarbon migration. Second, the action of tectonic stress can form pathways for hydrocarbon migration—faults, fractures, and unconformities, while also affecting

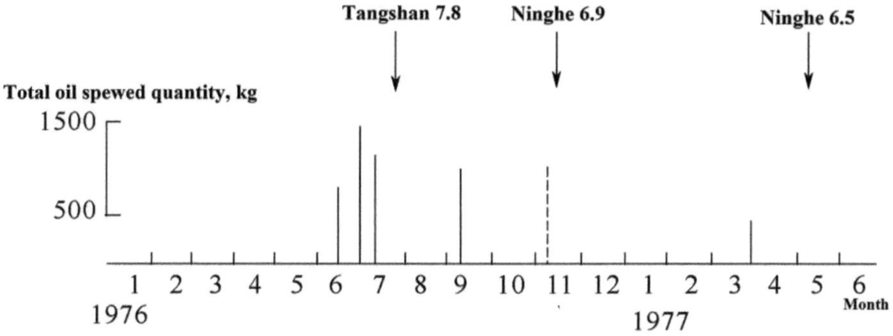

Fig. 4.54 Schematic diagram of the relationship between the W-11 well oil spray and the Tangshan earthquake and strong aftershocks (according to the Science and Technology Office of the National Seismic Bureau, 1983)

the effectiveness of hydrocarbon pathways, which can change the permeability of faults, fractures, and matrix rocks, as well as the sealing properties of faults and fractures. Third, tectonic stress can promote hydrocarbon migration and is a direct driving force for hydrocarbon migration.

The ways in which the stress field affects hydrocarbon migration in the above three aspects are not consistent, and the degree of closeness of the connections is also not consistent. From the degree of their correlation, they can be divided into three levels (Fig. 4.55). The first level of influence is the most direct, the second level is secondary, and the third level of influence is the most indirect. Therefore, when studying the impact of stress on hydrocarbon migration at different levels, different approaches and methods should be adopted.

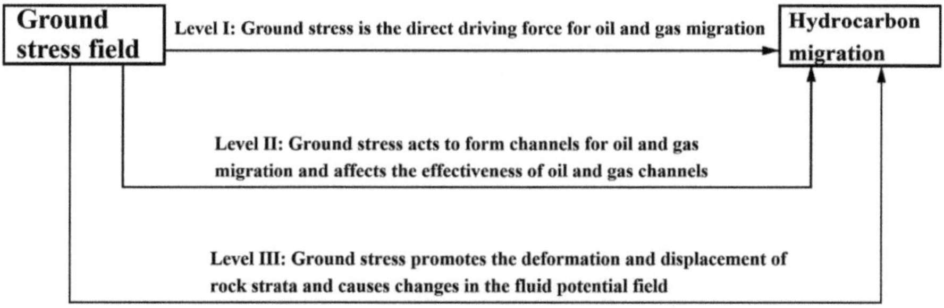

Fig. 4.55 The three levels of stress field influencing hydrocarbon migration (according to Zhai et al., 2003)

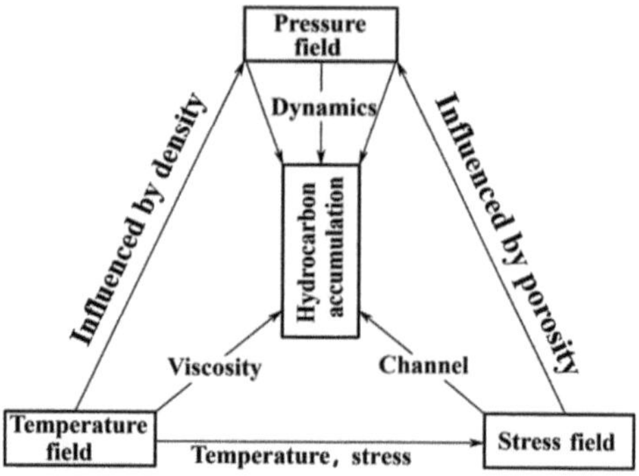

Fig. 4.56 Diagram of the relationship between the coupling of the three fields and hydrocarbon migration

4.5.4 Coupling of the Three Fields and Hydrocarbon Migration

The geothermal field, stress field, and pressure field interact through coupling to form the subsurface fluid potential field, which collectively affects the migration of hydrocarbons (Fig. 4.56). Among them, the pressure field is the main driving force for hydrocarbon migration, the geothermal field can change the viscosity of hydrocarbons, thereby altering the viscous force of hydrocarbon migration, and the stress field can provide or improve the pathways for hydrocarbon migration. The combined action of the three leads to the migration pathways, modes, and directions of subsurface hydrocarbons, thereby completing the migration and accumulation of subsurface hydrocarbons.

Questions for Discussion

1. Compare the differences in geological environment and conditions between primary and secondary migration.
2. Discuss the formation of abnormal pressure in source rocks and its role in primary migration.
3. Compare and discuss the phase state, driving force, pathways, and distances of primary and secondary hydrocarbon migration.
4. How to determine the main periods of primary and secondary migration?
5. What are the types of hydrocarbon secondary migration conduit systems? What factors affect their effectiveness?
6. What is a conduit system? What is an advantageous migration channel? How are the two related?
7. What is the direction of hydrocarbon secondary migration? What factors does it depend on?

8. What are the main research methods for the direction of hydrocarbon secondary migration?
9. What is abnormal pressure? What is a fluid pressure compartment? Analyze the causes of abnormal high pressure in sedimentary basins.
10. What are the three fields? How do they affect the migration of hydrocarbons?

References

Albrecht, P., & Ourisson, G. (1969). Triterpene alcohol isolation from oil shale. *Science, 163*(3872), 1192–1193.

Barker, C. (1972). Aquathermal pressuring—role of temperature in development of abnormal-pressure zones. *AAPG Bulletin, 56*(10), 2068–2071.

Barker, C. (1980). Primary migration: the importance of water-organic-mineral matter interactions in the source rock. *AAPG Studies in Geology, Tulsa, Oklahoma*, 1–13.

Berg, R. R. (1975). Capillary pressures in stratigraphic traps. *AAPG Bulletin, 59*(6), 939–956.

Bethke, C. M., Reed, J. D., Oltz, D. F. (1991). Long-range petroleum migration in the Illinois Basin. *AAPG bulletin, 75*(5), 925–945.

Bonham, L. C. (1978). Solubility of methane in water at elevated temperatures and pressures. *AAPG Bulletin, 62*(12), 2478–2481.

Chapman, R. E. (1983). *Petroleum geology*. Amsterdam: Elsevier Science Publishers B.V.

Coustau, H., Rameau, J. L., Tourisse, C., et al. (1975). Classification hydrodynamique des bassins sedimentaires: Ninth World Petroleum Congress Proc., v. 2. *Geology*, 105–119.

Creaney, S., & Allan, J. (1990). Hydrocarbon generation and migration in the Western Canada Sedimentary Basin. Geological Society, London, *Special Publications, 50*(1), 189–202.

Dahlberg, E. C. (1982). *Applied hydrodynamics in petroleum exploration*. New York: Springer-Verlag.

Dahlberg, E. C. (1995). *Applied hydrodynamics in petroleum exploration* (2nd ed.). New York: Springer-Verlag.

Demaison, G., Huizinga, B. J. (1994). Genetic classification of petroleum systems using three factors: charge, migration, and entrapment. *60*, 73–89.

Deroo, G., Powell, T., Tissot, B. et al. (1977). The origin and migration of petroleum in the Western Canadian sedimentary basin, Alberta, a geochemical and thermal maturation study. *Geological Survey of Canada Bulletin, (262)*, 136.

Du, J., Zhao X., Zhang Y., et al. (2007). Rift Basin Stratigraphic Reservoir in East China. Beijing: Geological Publishing House.

England, W. A., Mackenzie, A. S., Mann, D. M., et al. (1987). The movement and entrapment of petroleum fluids in the subsurface. *Journal of the Geological Society, 144*(2), 327–347.

Etiope, G., Feyzullayev, A., Baciu, C. L. (2009). Terrestrial methane seeps and mud volcanoes: a global perspective of gas origin. *Marine and Petroleum Geology, 26*(3), 333–344.

Fu, X., Fang, D., Lv, Y., et al. (2005). Method for evaluating the vertical sealing of faults based on the internal structure of fault zones. *Earth Science—Journal of China University of Geosciences, 30*(3), 328–336.

Gibson, R. G. (1994). Fault-zone seals in siliciclastic strata of the Columbus Basin, offshore Trinidad. *AAPG bulletin, 78*(9), 1372–1385.

Hantschel, T., & Kauerauf, A. I. (2009). *Fundamentals of basin and petroleum systems modeling* (p. 476). Springer.

Hao, S., Liu, G., Huang, Z., et al. (1994). Simulation model of primary migration of oil and gas. *Acta Petrolei Sinica, 2*, 21–31.

Harwood, J. H. (1977). Non-hydrocarbons of significance in petroleum exploration: Volatile fatty acids and non-hydrocarbon gases. *Mineralogical Magazine, 41*(320), 483–490.
Hindle, A. D. (1997). Petroleum migration pathways and charge concentration: A three-dimensional model. *AAPG Bulletin, 81*(9), 1451–1481.
Hu, C. (2005). Quantitative analysis of the applicability of the "source control theory." *Natural Gas Industry,* (10), 25–27+31+6.
Hubbert, M. K. (1940). The theory of ground-water motion. *The Journal of Geology, 48*(8, Part 1), 785-944.
Hubbert, M. K. (1953). Entrapment of petroleum under hydrodynamic conditions. *AAPG Bulletin, 37*(6), 1954–2026.
Hunt, J. M. (1979). *Petroleum geochemistry and geology.* W. H. Freeman and Company.
Jiang, Y., & Liu, H. (2010). Fault asphalt belts and their geological significance for oil and gas. *Acta Petrolei Sinica, 31*(1), 36–41.
Jiang, Y., Liu, J., Li, X., et al. (2011). Study of oil and gas migration pathways based on structural ridges and geochemical indicators: A case study of the Puwei area in the Dongpu Depression. *Earth Science—Journal of China University of Geosciences, 36*(3), 521–529.
Jiang, Z., Pang, X., Zeng, J., et al. (2005). Types of preferential oil and gas migration pathways and their physical simulation experiments. *Earth Science Frontiers, 12*(4), 507–516.
Lenormand, R., Touboul, E., Zarcone, C. (1988). Numerical models and experiments on immiscible displacements in porous media. *Journal of fluid mechanics, 189*, 165–187.
Lei, Y., Luo, X., Song, G., et al. (2014). Quantitative characterization of connectivity and conductivity of sandstone carriers during secondary petroleum migration, applied to the Third Member of Eocene Shahejie Formation, Dongying Depression, Eastern China. *Marine and Petroleum Geology, 51*, 268–285.
Levorsen, A. I. (1954). *Geology of petroleum.* San Francisco: W. H. Freeman and Company.
Levorsen, A. I. (1967). *Geology of petroleum* (2nd ed.). San Francisco: W. H. Freeman and Company.
Leythaeuser, D., Schaefer, R. G., Yukler, A. (1982). Role of diffusion in primary migration of hydrocarbons. *Aapg Bulletin, 66*(4), 408–429.
Leythaeuser, D., & Schaefer, R. G. (1984). Effects of hydrocarbon expulsion from shale source rocks of high maturity in Upper Carboniferous strata of the Ruhr area, Federal Republic of Germany. *Organic geochemistry, 6*, 671–681.
Leythaeuser, D., & Hu, J. S. (1986). A new method for the identification of hydrocarbon expulsion phenomena and quantitative calculation of hydrocarbon expulsion efficiency. *Foreign Oil & Gas Exploration, 6*, 3–11.
Li M. (2004). *Oil and gas migration* (3rd ed). Beijing: Petroleum Industry Press.
Li, M. (2013). *Migration of oil and gas* (4th ed.). Petroleum Industry Press.
Li, P., et al. (2004). *Formation of concealed oil and gas reservoirs in continental rift basins: A case study of the Jiyang Depression.* Petroleum Industry Press.
Liu, H., Jiang, Y., et al. (2011). Mechanisms and models of oil and gas accumulation in the Neogene of the Jizhong Depression. *Acta Petrolei Sinica, 32*(6), 928–936.
Luo, X., Lei, Y., Zhang, L., et al. (2012). Study of oil and gas migration conduits and quantitative characterization methods. *Acta Petrolei Sinica, 33*(3), 428–436.
Luo X., Zhou L., Shi J., et al. (2014). Dynamical studies on hydrocarbon migration and accumulation in typical superimposed basins in northwestern China. Beijing: Science Press.
Luo X., Vasseur, G. (2016). Overpressure dissipation mechanisms in sedimentary sections consisting of alternating mud-sand layers. *Marine and Petroleum Geology, 78*, 883–894.
Magara, K. (1968). Compaction and migration of fluids in Miocene mudstone, Nagaoka Plain, Japan. *AAPG Bulletin, 52*(12), 2466–2501.
Magara, K. (1978). Geological model predicting optimum sandstone percent for oil accumulation. *Bulletin of Canadian Petroleum Geology, 26*(3).
Magoon, L. B., & Dow, W. G. (1994). The petroleum system: From source to trap. *AAPG Memoir, 60*, 3–24.

McAuliffe, C. D. (1979). Oil and gas migration: Chemical and physical constraints. *AAPG Bulletin, 63*(5), 767–781.

Meng, Q., Hao, F., & Tian, J. (2021). Origins of non-tectonic fractures in shale. *Earth-Science Reviews, 222*, Article 103825.

Office of Academic Research in China Seismological Bureau. (1981). Investigation and study of Tangshan earthquake. Beijing: Seismological Press.

Ortoleva, P. J. (1994). Basin compartments and seals. USA: American Association of Petroleum Geologists.

Osborne, M. J., & Swarbrick, R. E. (1997). Mechanisms for generating overpressure in sedimentary basins: A reevaluation. *AAPG bulletin, 81*(6), 1023–1041.

Pratsch, J. C. (1982). Focused gas migration and concentration of deep-gas accumulation. *Erdoel & Kohle Erdgas Petro-chemie, 35*(1), 59–65.

Price, L. C., Wenger, L. M., Ging, T., et al. (1983). Solubility of crude oil in methane as a function of pressure and temperature. *Organic Geochemistry, 4*(3-4), 201–221.

Rouchet, J. D. (1981). Stress fields, a key to oil migration. *AAPG bulletin, 65*(1), 74–85.

Schowalter, T. T. (1979). Mechanics of secondary hydrocarbon migration and entrapment. *AAPG Bulletin, 63*(5), 723–760.

Sell, O. M. (1993). Experimental Verification of Low-dip, Low-rate, Two-phase (secondary) Migration by Means of R-ray Absorbtion. *Geofluids, 93*, 72–75.

Shao, Z., Wu H., Lang X., et al. (2019). Coupling characteristics between episodic hydrocarbon expulsion and fault throw increase for middle-shallow formations in Songliao Basin. *Earth Science, 44*(2), 551–558.

Snarsky, A. N. (1962). Die primäre Migration des Erdöls. Freiberger Forschungshefte C123.

Song Y., Xia X., Qin S., (2002). Prospect of natural gas exploration in the foreland basins in the middle and western parts of China. *Bulletin of Mineralogy, Petrology and Geochemistry, 21*(1), 26–29

Song, G., Ning, F., Hao, X., et al. (2012). Quantitative evaluation of the conductive capacity of skeletal sand bodies: A case study of the southern slope of the Dongying Depression. *Petroleum Geology and Recovery Efficiency, 19*(1), 4–6.

Song, G., Sui, F., & Zhao, L. (2010). Unconformities in the Jiyang Depression cannot serve as long-distance migration pathways for oil and gas. *Acta Petrolei Sinica, 31*(5), 744–747.

Sui, F., Wang, X., & Zhao, L. (2009). Oil and gas accumulation and exploration in the Jiyang Depression. *Petroleum Geology and Recovery Efficiency, 16*(6), 1–7.

Tian, H., Xiao, X., Wilkins, R. W., et al. (2008). New insights into the volume and pressure changes during the thermal cracking of oil to gas in reservoirs: Implications for the in-situ accumulation of gas cracked from oils. *AAPG Bulletin, 92*, 181–200.

Tissot, B. P., & Welte, D. H. (1978). *Petroleum formation and occurrence*. Springer.

Tissot, B. P., Welte, D. H. (1984). *Petroleum formation and Occurrence* (2nd ed.). New York: springer-Verlag.

Tissot, B. Pelet, R. (1971). Origin and evolution of hydrocarbons in early Toarcian shales, Paris Basin, France. Bull. *AAPG, 55*(12), 2177-2193

Ungerer, P., Behar, E., & Discamps, D. (1983). Tentative calculation of the overall volume expansion of organic matter during hydrocarbon genesis from geochemistry data. Implications for primary migration. *Advances in organic geochemistry, 10*(9), 129–135.

Wang, X., Song, Y., Wang, X. (1996). Physical simulation of petroleum genesis and expulsion-method, mechanism and application. Dongying: Petroleum University Press.

Wang, T., He, F., Li, M., et al. (2005). Methyldibenzothiophene series: Molecular marker to spike the filling pathway of oil. *Chinese Science Bulletin, 50*(2), 176–182.

Yang, X. (1993). *Subsurface temperature and pressure environments in oil and gas regions.* Petroleum University Press.

Zhang, F., Luo, X., Miao, S., et al. (2004). Experimental study and mechanism analysis of the formation process of preferential pathways for secondary oil migration. *Geological Science, 2*, 159–167.

Zhang, H. (1989). *Petroleum geology* (2nd ed.). Beijing: Petroleum Industry Press.
Zhang, H., Fang, C., Gao, X., et al. (1999). *Petroleum geology* (3rd ed.). Petroleum Industry Press.
Zeng, J., & Wang, H. (1999). Simulation experiments on oil migration and accumulation in conduits and lithological traps. *Earth Science, 2*, 85–88.
Zeng, J., & Wang, H. (2000). Simulation experiments on oil migration and accumulation in heterogeneous sand layers. *Journal of China University of Petroleum (Natural Science Edition),* (4), 108–111+132.
Zhai, M., Zhang, Y., Qiu, N., et al. (2003). *Conditions and main controlling factors for oil and gas accumulation.* Petroleum Industry Press.
Zhao, W., Wang, H., & Qian, K. (2009). Development of coal-derived gas theory in China and its role in the development of the natural gas industry. *Petroleum Exploration and Development, 36*(3), 280–289.

Chapter 5
Hydrocarbon Accumulation and the Formation of Hydrocarbon Reservoirs

Abstract Oil and natural gas are initially formed in a dispersed state and must undergo accumulation to form industrial reservoirs that are exploitable. The accumulation of oil and gas, as well as the formation of reservoirs, requires specific geological conditions. Once formed, these reservoirs may still face geological processes such as adjustment and destruction, which influence the final distribution of oil and gas. Therefore, understanding the accumulation of oil and gas and the formation and preservation of reservoirs is essential. This chapter mainly introduces the concepts and measurements of traps and oil and gas reservoirs, the principles of oil and gas accumulation, the conditions for the formation and preservation of reservoirs, the destruction and reformation of reservoirs, the timing of reservoir formation, and the formation mechanisms of condensate gas reservoirs and natural gas hydrates.

5.1 The Concept and Measurement of Traps and Hydrocarbon Reservoirs

5.1.1 The Concept and Measurement of Traps

5.1.1.1 The Concept of Traps

The term "trap" originally means a "snare" or "trap" set by hunters to capture prey. When hydrocarbons migrating through the subsurface conduits encounter a barrier and, under certain conditions, cease to migrate and concentrate, they are said to be trapped, much like game falling into a snare. Thus, a trap can be understood as a site suitable for the accumulation of hydrocarbons and the formation of hydrocarbon reservoirs.

A trap typically consists of three parts: ① Reservoir rock: This is the main part of the trap, providing space for the storage of hydrocarbons. ② Cap rock: Located above the reservoir rock, it prevents the upward migration of hydrocarbons. ③ Barrier: Prevents the lateral migration of hydrocarbons, forming a barrier for their accumulation. Both the cap rock and the barrier are sealing conditions that prevent the loss of

© China University of Petroleum Press 2025
H. Liu and Y. Jiang, *Petroleum Geology*,
https://doi.org/10.1007/978-981-96-7086-4_5

hydrocarbons. The cap rock prevents vertical loss, while the barrier prevents lateral loss. Therefore, a trap is essentially formed by the reservoir rock with pore space and the sealing conditions that can seal the reservoir.

The barrier element of a trap can be the flexure deformation of the cap rock itself, such as an anticline, or it can be constituted by faults, lithological or physical property changes, stratigraphic unconformities, etc. Therefore, the barrier of a trap determines its cause, and based on the sealing conditions of the trap, traps can be classified into structural, stratigraphic, lithological, hydrodynamic, composite, etc., with anticline traps and fault traps being the most important types of structural traps.

A trap is a geological body in the subsurface that can capture dispersed hydrocarbons and form hydrocarbon accumulations, possessing the capacity to store hydrocarbons. It can also be understood as a subsurface container that can store hydrocarbons, but not every trap contains hydrocarbons. Once a sufficient amount of hydrocarbons enter the trap, filling it or occupying a part of it, a hydrocarbon reservoir can be formed. In other words, during the migration process of oil and natural gas, if they encounter barriers that prevent their further migration, they will cease to migrate and accumulate near the barrier to form a hydrocarbon reservoir.

Hydrocarbons, as fluids, inevitably follow the laws of fluid motion during their seepage process. In the process of moving from dispersed to concentrated, hydrocarbons have a tendency to move from high potential areas to low potential areas. When they reach a relatively low potential space within a reservoir that is sealed by an impermeable layer or barrier, the hydrocarbons will accumulate there. Therefore, a trap is actually a low potential area for hydrocarbons in the reservoir, which is jointly sealed by the high potential area of hydrocarbons and the impermeable barrier (He et al., 2010). Especially under dynamic water conditions, this concept of a dynamic trap is more conducive to reflecting the essence of a trap.

5.1.1.2 Measurement of Trap

The size and scale of a trap often determine the size of the hydrocarbon reservoir, and its size is measured by the maximum effective volume of the trap. The maximum effective volume refers to the maximum volume of hydrocarbons that the trap can accommodate and is one of the important parameters for evaluating a trap. In reservoirs with normal hydrostatic pressure conditions, the size of a trap can be described by the following parameters.

(1) Spill Point

The point at which oil and gas begin to overflow from the trap after being filled is called the spill point of the trap (Fig. 5.1). It is the point at which the trap can store the maximum amount of oil and gas, and oil and gas will overflow outward below this point.

Fig. 5.1 Schematic diagram of the main parameters for measuring the volume of an anticline trap

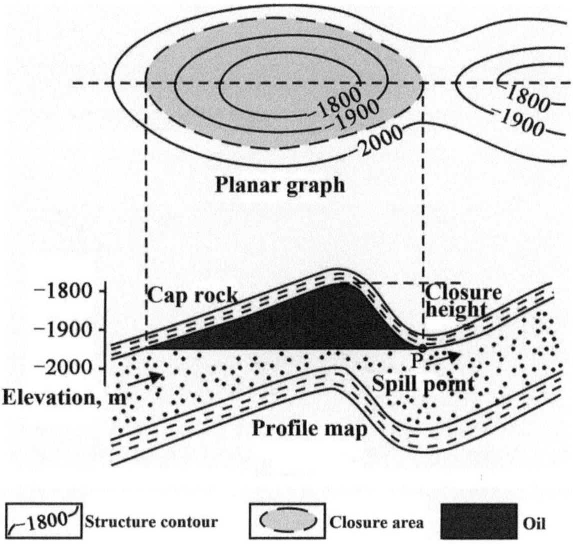

(2) Closure Area

The area enclosed by the structural contour line passing through the spill point is generally referred to as the closure area of the trap. Strictly speaking, the closure area should be the planar projection area of the enclosed space formed by the intersection of the horizontal plane passing through the spill point with the top and bottom surfaces of the reservoir or other sealing faces (such as fault surfaces, stratigraphic unconformity surfaces, lithological sealing surfaces). Obviously, when the thickness of the reservoir is certain, the larger the closure area, the larger the effective volume of the trap.

(3) Closure Height

The elevation difference between the highest point of the reservoir in the trap and the spill point is the closure height of the trap, also known as the closure degree.

As shown in Fig. 5.1, in a typical anticline trap, oil and gas first occupy the highest point of the anticline, then charge downward. When charged to the spill point, that is, the saddle part of the structure, oil and gas begin to overflow and migrate along the updip direction of the reservoir.

It should be noted that the structural closure height and the structural relief amplitude are two completely different concepts. The measurement of closure height is based on the elevation plane of the spill point; while the measurement of structural amplitude is based on the regional dip surface. Anticlinal structures of the same size can have completely different closure heights when the regional dips are different, as shown in Fig. 5.2.

In the fault trap shown in Fig. 5.3, Fault A is sealed, while Fault B is not. The spill point of the trap is point P in the figure. The area enclosed by the top surface

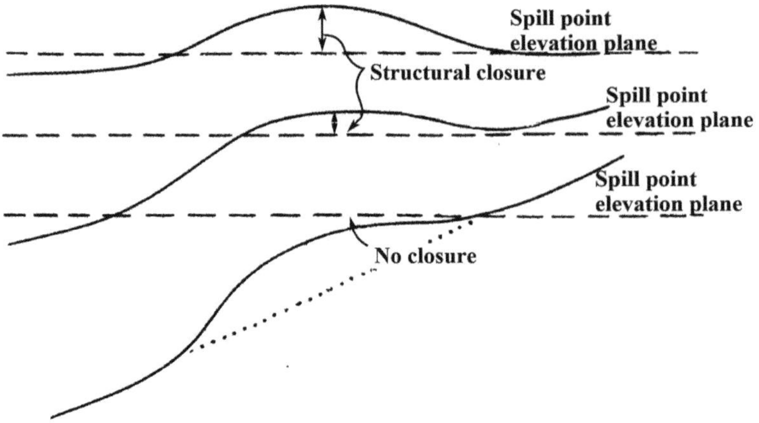

Fig. 5.2 Structural closure height and structural relief amplitude. The same structural relief is represented, but due to different regional dips, the closure height varies

contour line of the reservoir that is tangent to point P and Fault A is the closure area of the trap.

The determination methods for the spill point, closure height, and closure area of lithological and stratigraphic traps are essentially like the above two types. That is, on the plane, the range of the trap is the closed area formed by the top surface contour line of the reservoir and other sealing faces (such as fault surfaces, lithological pinchout lines, stratigraphic unconformity surfaces). The spill point is located on the lowest contour line that can close (Fig. 5.4).

Obviously, the size of a trap (maximum effective volume) is directly proportional to the area, closure height, thickness of the reservoir, and effective porosity. A large-volume trap is the foundation for the formation of a large oil and gas reservoir, usually characterized by a large closure area, thick reservoir, and high effective porosity.

Fig. 5.3 Fault trap spill point, closure height, and closure area schematic

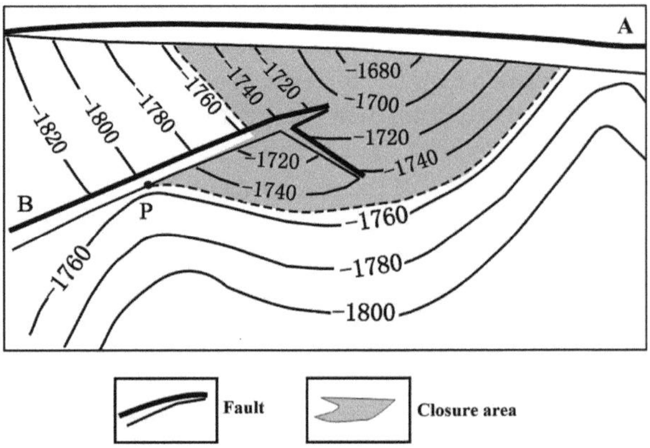

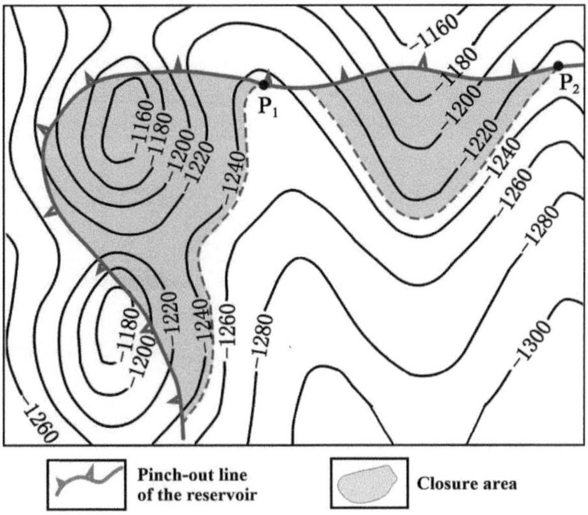

Fig. 5.4 Lithological trap spill point, closure height, and closure area schematic

5.1.2 The Concept and Measurement of Hydrocarbon Reservoirs

5.1.2.1 The Concept of Hydrocarbon Reservoirs

During the subsurface migration process of oil and gas, when they encounter changes in the physical properties and geometric shape of rocks that form a trap, oil and gas enter the trap and accumulate within it, forming a hydrocarbon reservoir. If only oil accumulates in the trap, it is called an oil reservoir. If only natural gas accumulates, it is called a gas reservoir. If both accumulate and form a free gas cap, it is called an oil and gas reservoir.

A hydrocarbon reservoir is the accumulation of oil and gas in a single trap, with a unified pressure system and oil–water contact. It is the basic unit of oil and gas accumulation in the crust and the smallest unit of oil and gas accumulation in a hydrocarbon-bearing basin. In other words, a hydrocarbon reservoir is the smallest oil and gas accumulation unit in the underground rock layers with a unified fluid dynamics system.

In Fig. 5.5, there are three reservoirs in the same anticline, forming three traps with three different pressure systems and three different oil–water contacts, i.e., there are three hydrocarbon reservoirs. In Fig. 5.6, the fault offsets the reservoir, but does not separate it, forming two oil reservoirs blocked by fault mud.

If the amount of oil and gas accumulation in the trap is large enough to have exploitative value, it is called a commercial hydrocarbon reservoir. If the amount of oil and gas accumulation is not large enough to have exploitative value, it is called a non-commercial hydrocarbon reservoir.

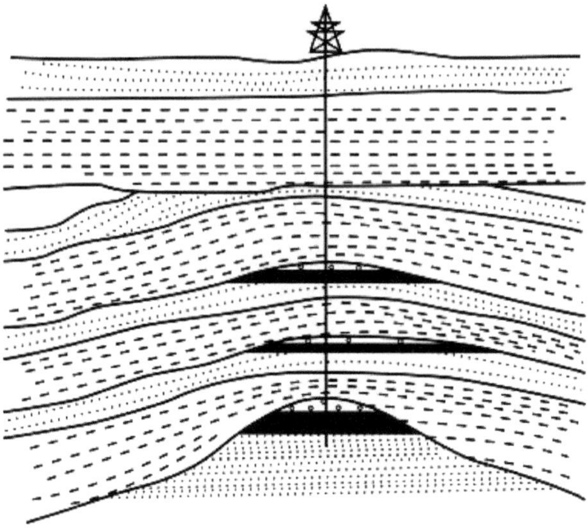

Fig. 5.5 Three hydrocarbon reservoirs composed of three reservoirs

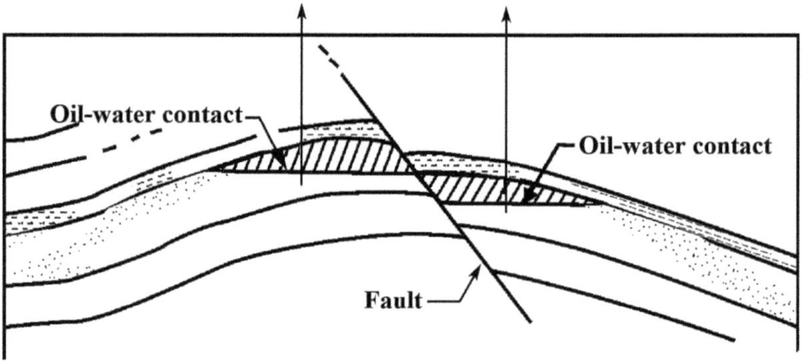

Fig. 5.6 Two oil reservoirs formed by fault barriers

Whether a discovered hydrocarbon reservoir has commercial value mainly depends on factors such as the size of the reservoir's reserves and production, its location and depth, and technical and economic conditions. It is usually measured by the single well daily production of oil and gas. Obviously, the standards for oil and gas layers of different depths are different, and the standards for land and ocean are also different (Table 5.1). A hydrocarbon reservoir with commercial value found in an economically developed area may not have commercial exploitation value in a remote desert area. Hydrocarbon reservoirs that were previously considered to have no exploitative value may become commercially valuable as extraction technology progresses, such as some tight sandstone hydrocarbon reservoirs

Table 5.1 Commercial and non-commercial oil and gas flow standards (according to Zhai et al., 1996)

Category		Oil production (t/d)		Natural gas production ($\times 10^4$ m^3/d)	
Level		Industrial	Non-industrial	Industrial	Non-industrial
Burial depth (m)	<500	0.3–0.5	<0.3	0.05–0.1	<0.05
	500–1,000	0.5–1.0	0.3–0.5	0.1–0.3	0.05–0.1
	1,000–2,000	1.0–3.0	0.5–1.0	0.3–0.5	0.1–0.3
	2,000–3,000	3.0–5.0	1.0–3.0	0.5–1.0	0.3–0.5
	3,000–4,000	5.0–10	3.0–5.0	1.0–2.0	0.5–1.0
	4,000–5,000	>10	5.0–10	>2.0	1.0–2.0

and coalbed methane reservoirs that had no development value under the technical and economic conditions of the past, but have now become important types of commercial hydrocarbon reservoirs. Similarly, a drop in oil prices can turn a profitable commercial hydrocarbon reservoir into a non-commercial one. Therefore, commercial hydrocarbon reservoirs change with political, technical, and economic conditions.

5.1.2.2 Measurement of Hydrocarbon Reservoirs

The size of a hydrocarbon reservoir reflects the amount of oil and gas accumulated in the trap and can be described using parameters such as the area and height of oil and gas content, like trap description parameters. The following uses a typical anticline oil and gas reservoir as an example.

(1) Oil–Water Interface, Oil–Gas Interface, and Pay Zone Height

In a typical oil and gas reservoir, due to gravity segregation, the distribution of gas, oil, and water has a certain pattern: gas occupies the uppermost part, known as the gas cap; oil is in the middle, distributed in a ring shape, known as the oil ring; water is below; thus, forming the oil–gas interface and oil–water interface (Fig. 5.7). Under normal circumstances, these interfaces are nearly horizontal. However, under the influence of water dynamics, the oil–water interface may be inclined. Due to capillary pressure, these interfaces are not abrupt but a transition zone, the thickness of which mainly depends on the properties of the oil and gas-bearing layer and the properties of the oil and gas. Good quality results in a thin transition zone. Under gravity segregation, the oil in the reservoir tends to become heavier from the top of the oil reservoir to the oil–water contact, and due to the oxidation and scouring action of formation water on oil, the oil at the oil–water contact becomes viscous and heavy, even forming a tar mat.

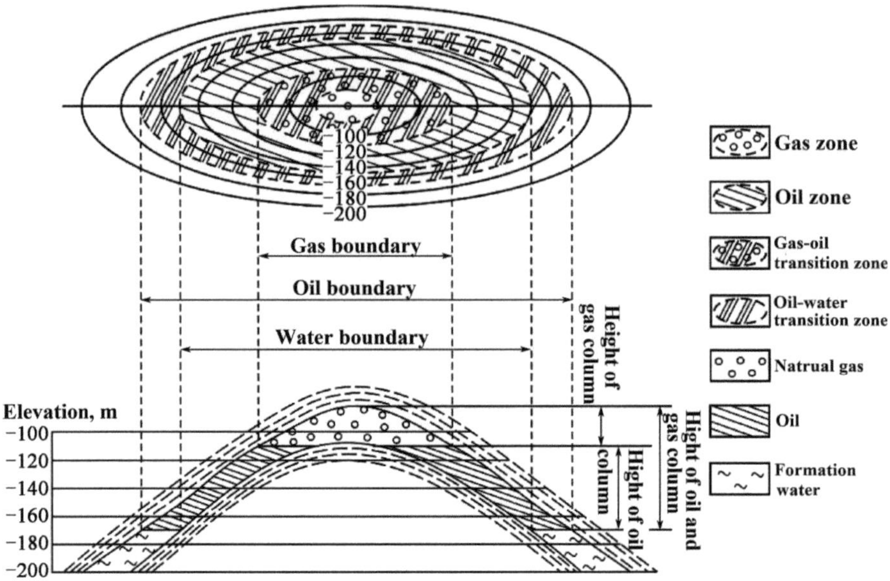

Fig. 5.7 Schematic diagram of the distribution of oil, gas, and water in an anticline hydrocarbon reservoir (contour units in meters)

The vertical distance from the oil–water interface to the highest point of oil and gas content in the reservoir is called the pay zone height, or oil and gas column height, which is an important parameter indicating the size of the oil and gas reservoir. If it is an oil reservoir, it is called the oil column height, and if it is a gas reservoir, it is called the gas cap height, as shown in Fig. 5.7, with the oil column height and gas cap height representing the heights of the oil and gas parts, respectively.

(2) Oil (Gas) Pay Boundary and Oil (Gas) Pay Area

The oil–water interface and oil–gas interface are usually horizontal. The line where the oil–water interface intersects the top surface of the reservoir is called the oil pay boundary, also known as the oil edge, which defines the maximum oil-containing range of the oil reservoir. Outside this boundary, there is no oil; the line where the oil–gas interface intersects the top surface of the reservoir is called the gas pay boundary or gas cap boundary. When the oil height is greater than the thickness of the reservoir, the line where the oil–water interface intersects the bottom surface of the reservoir is called the water boundary, also known as the water edge or inner oil edge, within which there is no free water in the reservoir, and the reservoir is filled with oil. Usually, the oil boundary and water boundary are parallel to the structural contour lines of the top and bottom surfaces of the reservoir. If it is an oil or gas reservoir, the area enclosed by the oil boundary and gas boundary is called the oil area and gas area, respectively. If it is an oil and gas reservoir, as shown in Fig. 5.7, the area enclosed by the oil boundary is called the oil and gas area.

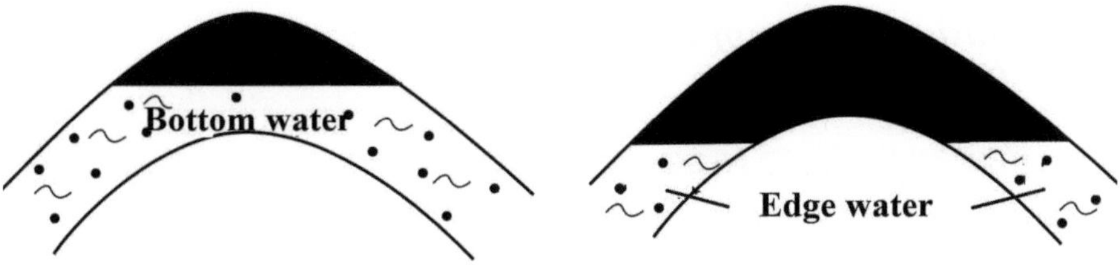

Fig. 5.8 Schematic diagram of the distribution relationship between bottom water, edge water, and oil and gas reservoirs

(3) Bottom Water and Edge Water

If the height of the oil and gas reservoir is less than the thickness of the reservoir, there is no inner oil or gas edge, and the oil and gas accumulate at the top of the trap, with the lower part of the oil and gas reservoir being all water, which is called bottom water. If the thickness of the reservoir is not large, or the structural dip is steep, and the height of the oil and gas reservoir is greater than the thickness of the reservoir, the oil and gas fill the high part of the trap, and the water surrounds the oil and gas reservoir, i.e., outside the inner oil (gas) edge, this water is called edge water. As shown in Fig. 5.8.

In structures with gentle bedding, the oil–water contact surface is relatively wide; in structures with steep dips, the oil–water contact surface is relatively narrow.

(4) Fill Degree

The ratio of the oil (gas) height to the closure height is generally defined as the fill degree or fill coefficient. In addition, some people define the ratio of the oil (gas) area to the closure area as the fill degree; or the ratio of the oil (gas) volume to the closure effective volume as the fill degree. Generally, in oil and gas-rich areas, this coefficient is high; in oil and gas-poor areas, the fill coefficient is low.

(5) Effective Thickness and Oil (Gas) Saturation

The effective thickness is generally the thickness of the part of the reservoir that can produce oil and gas, and its size directly affects the size of the oil (gas) reservoir. Not all of the pores in the oil (gas) layer are occupied by oil (gas), and some formation water is also present. The higher the oil (gas) content, the higher the oil (gas) saturation, and its level is also an important parameter affecting the size of the oil (gas) reservoir.

A typical oil reservoir is roughly divided into four zones from top to bottom in the longitudinal direction, namely the pure oil zone, with an oil saturation of more than 55%, no free water, and pure oil production, the oil–water transition zone, with an oil saturation of 10–55%, producing both oil and water, the residual oil zone, with an oil saturation below 10%, although containing oil, but only producing water, the pure water zone, with only water in the pores, no oil, and pure water production.

5.1.3 Basic Concepts of Unconventional Hydrocarbon Reservoirs

The above concepts of traps and hydrocarbon reservoirs are based on conventional hydrocarbon reservoirs, where the accumulation of oil and gas in traps complies with the principle of gravity segregation, that is, under the action of buoyancy, natural gas and oil accumulate in the upper part of the trap, forming the conventional distribution of upper gas (oil) and lower water. In nature, many oil and gas accumulations do not conform to the principle of gravity segregation, and even do not need the traditional sense of traps. The mode of occurrence and mechanism of oil and gas accumulation in the underground are significantly different from those of conventional hydrocarbon reservoirs, forming unconventional hydrocarbon reservoirs.

The main unconventional oil and gas resources currently studied and developed include tight oil and gas, shale oil and gas, coalbed methane, natural gas hydrates, heavy oil, oil sand, etc. They, together with conventional hydrocarbon reservoirs in oil and gas basins, form a complete series of oil and gas accumulation distribution (Fig. 5.9). Since the formation mechanisms and development methods of natural gas hydrates, heavy oil, and oil sand are very different from the former three, the so-called unconventional oil and gas are usually narrow, mainly including tight sandstone oil and gas, shale oil and gas, and coalbed methane, etc.

Domestic and foreign scholars have given the definition of unconventional oil and gas from geological, development, and economic perspectives. From a geological perspective, based on the geological characteristics of oil and gas reservoirs and the mechanism of oil and gas accumulation, the narrow sense of unconventional oil and gas can be understood as oil and gas accumulations that are widely distributed in low permeability, low porosity reservoir spaces of near-source reservoir bodies or hydrocarbon source rock systems, and are formed under non-buoyancy drive. The distribution is not controlled by conventional traps such as structure, stratigraphy,

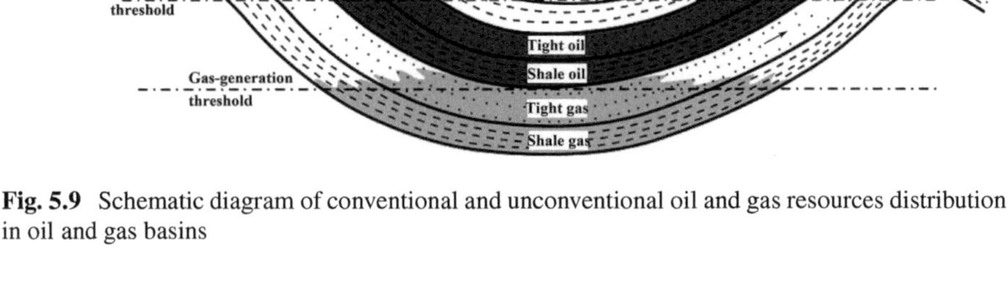

Fig. 5.9 Schematic diagram of conventional and unconventional oil and gas resources distribution in oil and gas basins

or lithology, and the trap boundaries are vague, showing a regional continuous or quasi-continuous distribution.

From an economic perspective, coalbed methane and shale oil and gas with poor economic benefits can be regarded as unconventional oil and gas. From a development perspective, unconventional oil and gas refer to oil and gas resources that cannot be economically developed under existing economic and technical conditions through conventional technology. They are resources that can obtain economic production capacity by changing rock permeability or fluid viscosity through technology, while conventional oil and gas resources do not need to change rock permeability or fluid viscosity to obtain economic production capacity.

Integrating the views of domestic and foreign scholars, unconventional oil and gas reservoirs can be defined as oil and gas reservoirs with characteristics and formation mechanisms different from conventional oil and gas reservoirs, with tight reservoirs, and cannot obtain economic production capacity with traditional technology. They are continuous or quasi-continuous oil and gas accumulations that need to use new technology to improve reservoir permeability or fluid viscosity for economic development.

Compared with conventional oil and gas reservoirs, the two key characteristics of unconventional oil and gas reservoirs are: ① oil and gas are widely distributed in a continuous or quasi-continuous manner, often showing a coexistence of source and reservoir, with unclear trap boundaries and water dynamic effects, ② the reservoir is tight, with no natural industrial stable production, and the Darcy flow characteristics are not obvious. New technology is needed to transform the reservoir or reduce fluid viscosity to obtain economic production capacity.

From a technical and economic perspective, unconventional oil and gas are relative to conventional oil and gas. With technological progress and changes in economic conditions, today's unconventional oil and gas may become conventional oil and gas in the near future. From a geological perspective, there is no distinction between conventional and unconventional oil and gas in terms of their nature. It is only the mode of occurrence of oil and gas in the underground and the geological environment, as well as the mode of accumulation and formation of oil and gas reservoirs, that are significantly different, resulting in differences in their economic and development aspects. Therefore, starting from the concept of a classic trap and oil and gas accumulation, conventional oil and gas reservoirs are the accumulation of oil and gas in classic traps, with oil, gas, and water following the principle of gravity segregation. While unconventional oil and gas reservoirs can be understood as the accumulation of oil and gas in unconventional traps, which is a geological body where oil and gas are accumulated in tight reservoirs, and the oil and gas in it are unconventional oil and gas. Unconventional traps are different from the traditional concept of traps. Due to tight reservoirs, the occurrence of oil and gas in the geological body is different from that in conventional oil and gas reservoirs, and the accumulation of oil and gas does not conform to the buoyancy drive mechanism, showing a continuous or quasi-continuous distribution, and there are no obvious conventional trap boundaries.

5.2 Principles of Hydrocarbon Accumulation

5.2.1 Mechanism of Hydrocarbon Accumulation in Traps

The process of oil and gas displacing pore water in a trap and accumulating to form a hydrocarbon reservoir is called oil and gas accumulation. The combined action of buoyancy and hydraulic dynamics is the main dynamic mechanism for oil and gas accumulation in conventional traps, which mainly includes filtration and displacement. These two actions manifest differently at various stages of oil and gas accumulation.

(1) Filtration Effect

This mechanism (Roberts, 1980) suggests that when hydrocarbon-bearing water or free hydrocarbons enter a trap, the generally hydrophilic and capillary-sealed cap rock does not act as a seal for water, allowing water to continue migrating through the cap rock. Hydrocarbons, however, are filtered down and accumulate in the trap due to the capillary sealing of the cap rock. Under the influence of water dynamics and buoyancy, water and hydrocarbons can be continuously replenished and eventually form a hydrocarbon reservoir in the trap (Fig. 5.10).

This mechanism may only apply to cap rocks with general physical property sealing and occurs in the early stages of hydrocarbon accumulation. When the shale cap rock has abnormally high pressure or salt rock as the cap rock, this mechanism cannot function.

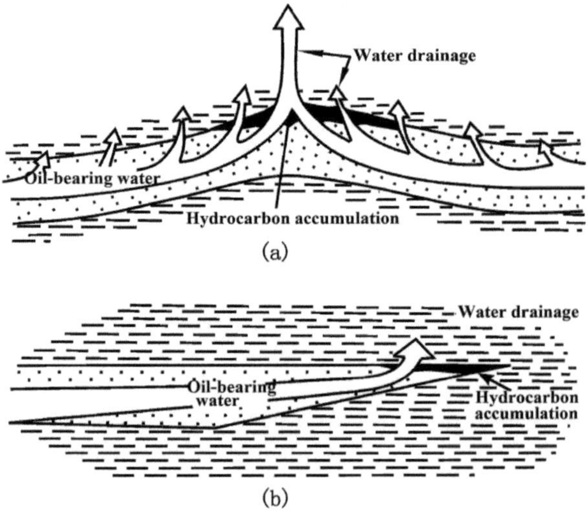

Fig. 5.10 Hydrocarbon accumulation in traps (according to Roberts, 1980). **a** Anticline trap, **b** lithological pinchout trap

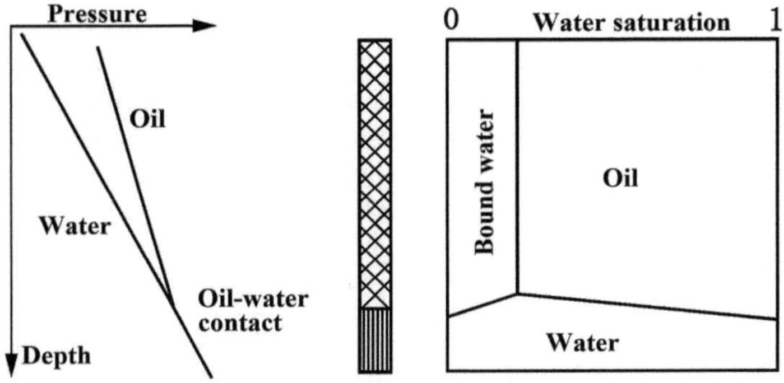

Fig. 5.11 Vertical distribution of oil and water pressure and water saturation in the trap (after Chapman, 1982)

(2) Displacement Effect

Chapman (1982) believed that the fluid pressure in the argillaceous cap rock is generally higher than that in the adjacent sandstone layers, making it difficult for water in the trap to pass through the cap rock. Hydrocarbons entering the trap first accumulate at the bottom, and as the amount of hydrocarbons increases, a continuous hydrocarbon phase of a certain height is gradually formed. At this time, at the oil–water interface, the pressure of oil and water is equal, and at any height above the oil–water interface, due to the density difference, the pressure of oil (or gas) is greater than that of water (Fig. 5.11). Therefore, a downward fluid potential gradient is created, causing hydrocarbons to migrate upwards in the trap while displacing water downwards until the irreducible water saturation is reached. As oil or gas continues to enter the trap, the oil–water or gas–water interface continues to move downward until the hydrocarbons fill the trap, a process mainly dominated by displacement.

In fact, when the overlying rock only has capillary sealing, the accumulation of hydrocarbons in the trap is mostly a mixed mechanism where both effects exist. Since any reservoir is heterogeneous, and the charging of hydrocarbons is gradual, the continuous space first occupied by hydrocarbons may undergo displacement, with water displaced by oil moving downward to the water-bearing part of the reservoir; while the continuous space still occupied by water may undergo filtration, with water being expelled from the overlying rock. According to the principle of two-phase migration, when the oil saturation in the reservoir or at the oil–water interface reaches more than 60%, water can hardly pass through the oil (gas) layer to continue migrating upwards through the overlying rock and can only be expelled downwards (Li, 2013). Therefore, in the early stages of hydrocarbon accumulation, water can seep through the overlying hydrophilic cap rock, with filtration playing a dominant role; when hydrocarbons accumulate to a certain extent, water can hardly pass through the overlying rock, and thus are mainly displaced downwards by hydrocarbons, with displacement playing a dominant role. If the cap rock is an abnormally high-pressure segment or high-quality salt rock and salt rock, the water in the trap cannot seep through the overlying rock and can only be displaced downwards.

In addition to the buoyancy-water dynamics-based hydrocarbon accumulation dynamic mechanisms, abnormal pressure, diffusion, and osmosis, among other actions, may also become important accumulation mechanisms under special circumstances, such as the hydrocarbon charging process caused by fluid curtain migration driven by abnormal pressure.

5.2.2 The Process of Hydrocarbon Accumulation in Traps

In terms of chronological order, the process of hydrocarbon accumulation in traps can be divided into the charging process and the mixing process. These two processes are continuous, sometimes slow, and sometimes rapid.

(1) Charging Process

The modern theory of hydrocarbon migration and accumulation believes that hydrocarbon migration includes both slow, buoyancy-driven gradual migration and rapid, abnormal pressure-driven episodic migration. These two migration modes also manifest as two types of charging in traps: gradual charging and episodic charging.

Taking a typical anticline trap as an example, the highest part of the reservoir in the trap is the low potential area for hydrocarbons. Driven by buoyancy or other forces, hydrocarbons migrate and charge into the trap. Since hydrocarbons migrate and accumulate in porous water-bearing reservoirs, they must initially move upward along the path of least resistance. High permeability and high porosity sections are the areas where hydrocarbons preferentially charge, then continuously expand to relatively lower permeability and lower porosity parts of the reservoir, eventually filling the entire trap.

England & Mackenzie (1989) proposed the process of oil charging in traps from a geochemical perspective (Fig. 5.12). During the process of oil charging and accumulating in traps, oil initially enters the trap through the coarser pores in the reservoir in a dendritic pattern; as the range of migrating oil increases, buoyancy also increases, causing oil to charge into smaller pores and expel residual formation water. If newly generated oil continuously expels from the hydrocarbon source rock and injects from one side of the trap, it advances into the trap like a series of wave fronts, replacing previously generated oil both laterally and vertically, and preventing widespread mixing of the oil column. That is, the charging of an oil reservoir is sequential, with oil first entering the best permeable layers with the lowest pore displacement pressure, and then charging the reservoir in a series of advancing oil wave fronts. The lower maturity oil generated earlier, after entering the trap, is displaced by the higher maturity oil generated later, resulting in heterogeneous maturity and density

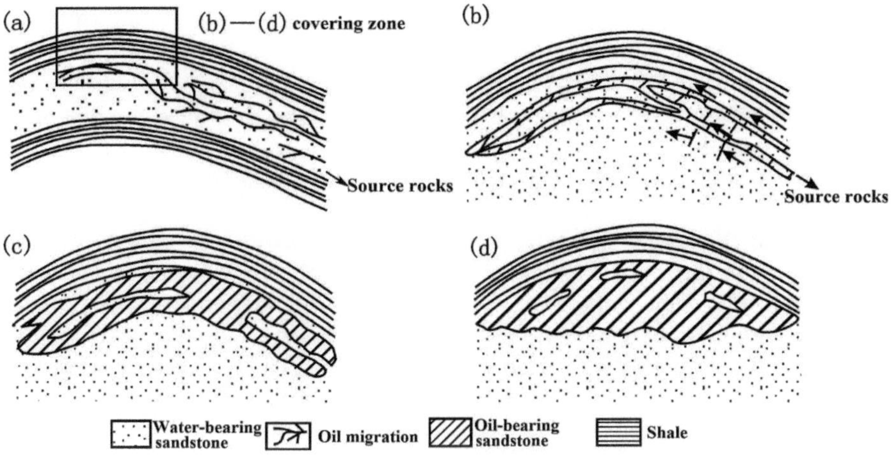

Fig. 5.12 The charging process of a typical oil reservoir (according to England & Mackenzie, 1989)

of oil both laterally and vertically, which is an important reason for the mixing and adjustment of oil in the trap later.

(2) Mixing and Accumulation Process

During the process of hydrocarbon charging in traps, due to the heterogeneity of the reservoir and the differences in the charging process, mainly different charging methods (lateral or vertical) and different charging times and sources of hydrocarbons, there is a strong vertical and planar heterogeneity of hydrocarbons charged in the trap. This fluid distribution heterogeneity is unstable and abnormal, so fluid mixing must occur. The mixing of oil or natural gas in the trap eventually reaches a relatively homogeneous and stable state that conforms to the gravity segregation effect.

England & Mackenzie (1989) believes that there are three main mechanisms for oil mixing in traps: density difference mixing, concentration difference mixing, and thermal convection mixing, with the first two playing an absolute advantage. The reason for density difference mixing is the density difference between the high and low parts of the trap. Oil charged at different times has different maturity and density. The oil with lower maturity and higher density that was charged earlier occupies the top of the trap, while the oil near the wing or bottom of the trap near the charging point has higher maturity and lower density, forming an unstable "heavy on top, light on bottom" density inversion. Therefore, heavier oil sinks, and lighter oil rises. The reason for concentration difference mixing is the concentration difference of oil components in the trap, which is essentially a diffusion mixing effect. The efficiency of diffusion mixing depends on the size of the concentration difference, the size of hydrocarbon molecules, and the physical properties of the reservoir.

The degree of mixing depends on the heterogeneity of the reservoir and the time of mixing. If there are muddy bands or carbonate cementation bands in the reservoir, it will seriously affect the degree of oil mixing. The better the reservoir properties and the lower the degree of heterogeneity, the longer the time after charging, the more fully the density difference mixing effect proceeds, the higher the homogeneity of oil properties, and finally, a concentration gradient formed by gravity segregation is established.

In fact, the charging and mixing processes of oil or natural gas in traps are two processes that proceed almost simultaneously. As hydrocarbons continue to charge into the trap, under the mixing effects of gravity, diffusion, and thermal convection, hydrocarbons keep moving in the trap and also keep accumulating (Li, 2013). Hydrocarbons continuously displace water from the top of the reservoir in the trap downwards, and the oil–water or gas–water interface gradually moves down from the top of the trap, and the pressure in the hydrocarbons gradually increases, until it reaches the spill point of the trap. At the same time, with continuous charging, the height of the hydrocarbon column gradually increases, the charging force gradually increases, and the capillary resistance that can be overcome is enhanced, and hydrocarbons gradually charge from high permeability parts to low permeability parts, increasing oil saturation until it reaches the irreducible water saturation.

5.2.3 Differential Accumulation of Hydrocarbons in Series Traps

In petroliferous basins, when there are multiple hydrodynamically connected traps, and there is an abundant supply of oil and gas from the lower direction, hydrocarbons accumulate in this series of traps, and regular changes in the phase and properties of hydrocarbons occur in each trap along the migration direction. This phenomenon is called differential accumulation of hydrocarbons (Gussow, 1953).

In fact, from the center or lower part of an petroliferous basin or a hydrocarbon-generating depression to a series of traps on the edge of the basin, there are two distinctly different distribution patterns of hydrocarbon phases accumulated in the traps: one is the differential accumulation pattern proposed by Gussow (1953), which can be called the "gas core and oil ring" pattern, that is, natural gas fills the structural traps in the lower center of the basin, while oil fills the structural traps in the higher places. The other pattern is the opposite, which can be called the "oil core and gas ring" pattern, that is, oil occupies the structural traps in the lower center of the basin, and natural gas occupies the structural traps in the higher positions on the edge of the basin. These are two patterns of differential accumulation of hydrocarbons. Both patterns have been found in nature and are the result of differential accumulation of hydrocarbons. The former can be called spill-type differential accumulation, and the latter can be called seepage-type differential accumulation of hydrocarbons. Therefore, differential accumulation of hydrocarbons can be understood as a phenomenon where, when there is an abundant supply of oil and gas from the lower direction, hydrocarbons accumulate in a series of traps, and regular changes in the phase and properties of hydrocarbons occur in each trap along the migration direction.

5.2.3.1 Spill-Type Differential Accumulation of Hydrocarbons ("Gas Core and Oil Ring" Pattern)

It is generally believed that when oil and gas migrate and accumulate from the central deep depression of a basin to the surrounding higher traps along the upper direction, natural gas, being the easiest to flow in the pore medium of rocks, tends to occupy the structural rings on the edge of the basin, while oil occupies the lower structures. People regarded this "oil core and gas ring" distribution pattern as the normal pattern. However, oil and gas exploration practice has found more and more opposite cases, that is, the "oil core and gas ring" distribution pattern appears, prompting people to re-understand this issue. Gussow (1953) systematically studied this phenomenon and proposed the principle of differential accumulation of hydrocarbons. In fact, the emergence of these two patterns is normal, it's just that the geological background and conditions are different.

As shown in Fig. 5.13, under static water conditions, the accumulation of oil and gas in a single trap is divided into three stages. In the early stage, the oil and gas entering the trap, due to gravity segregation, have gas at the top, oil in the middle, and water at the bottom. As the amount of oil and gas continues to increase, the oil–water interface gradually drops to the spill point, and some oil spills from the spill point in the upper direction. Afterwards, oil and gas continue to enter the trap, natural gas gathers at the top of the trap, pushing oil to the spill point, and oil is continuously expelled. When the amount of natural gas is obviously enough to occupy the entire trap, oil can no longer enter the trap but spills along the spill point in the upper direction. In this case, the trap is filled with natural gas.

If there is a series of traps in the same permeable layer in the basin, and the elevation of the spill points increases in turn, and the oil and gas source area come from the lower direction and the amount of oil and gas is relatively sufficient, with long-distance migration conditions of regional inclination, the reservoir is full of water and under static water pressure conditions, oil and gas continue to migrate upwards. The oil and gas differential accumulation situation shown in Fig. 5.14 will be formed: in stage A, oil and gas first charge the trap 1 with the lowest spill point, forming an oil and gas reservoir; in stage B, the oil–water interface of trap 1 reaches the spill point elevation, and natural gas can continue to charge trap 1, but oil spills from trap 1 into trap 2 to form an oil reservoir; in stage C, the gas-oil interface reaches the spill point elevation, trap 1 is completely filled with natural gas, trap 2 is filled with oil and gas to form a small gas cap oil and gas reservoir, and oil spills from trap 2 into trap 3 to form an oil reservoir and natural gas continues to enter trap 2. In stage D, with the continuous charging of oil and gas, trap 1 no longer charges natural gas, trap 2 is filled with oil and gas to form a large gas cap oil and gas reservoir, oil fills trap 3 and spills from the spill point into trap 4 to form an oil reservoir. The result

Fig. 5.13 Accumulation of oil and gas in a single anticlinal trap

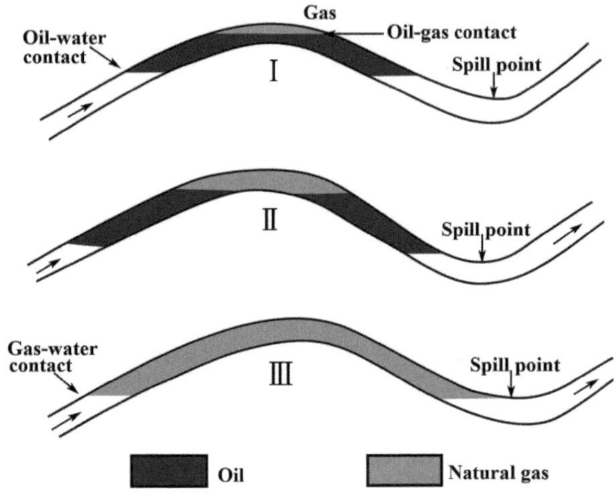

is: trap 1 is a pure gas reservoir, trap 2 is an oil and gas reservoir, traps 3 and 4 are oil reservoirs, among which traps 1, 2, and 3 are all filled, but the properties of the fluids are different, trap 4 is not filled, and trap 5 is a water-containing empty trap. If the oil and gas source is very sufficient and the migration conditions are good, the trap farthest from the oil source can also be filled with oil.

From this, two main conclusions can be drawn: (1) The differential accumulation of oil and gas in a series of traps causes pure gas reservoirs to form in the traps closest to and with the lowest spill points from the oil and gas supply area, while slightly

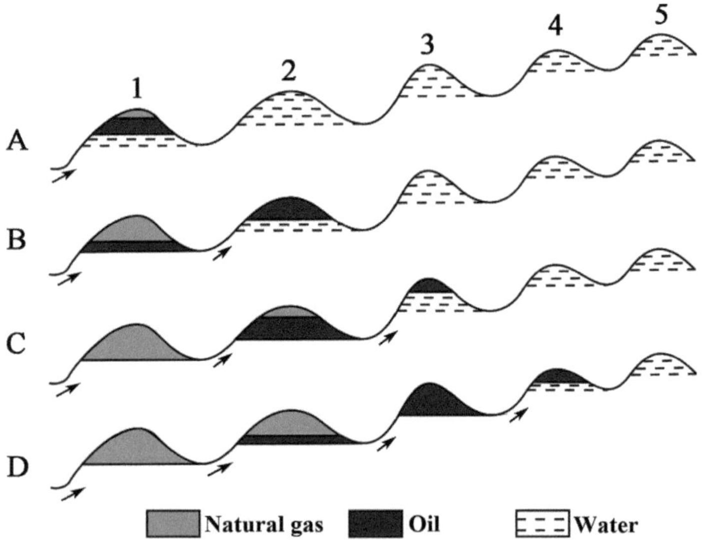

Fig. 5.14 Schematic diagram of differential accumulation

farther and higher spill point traps form oil and gas reservoirs or pure oil reservoirs, and traps farther from the oil source area and with higher spill points only contain water. (2) A trap filled with oil can still accumulate natural gas, but a trap filled with natural gas is ineffective for accumulating oil.

Many oil and gas basins in the world have the characteristics of spill-type differential accumulation of oil and gas. For example, the Appalachian Basin in the eastern United States is a typical example of this pattern. Many gas reservoirs are located around the center of the basin, while the area near the center of the basin is mainly oil reservoirs, as shown in Fig. 5.15.

For example, the differential accumulation of oil and gas in the Silurian reef belt of the Michigan Basin in the United States is shown in Fig. 5.16, and the differential accumulation of oil and gas in the Eromanga Basin in Australia is shown in Fig. 5.17. From the structural low position to the structural high position, gas reservoirs, oil and gas reservoirs, oil reservoirs, and empty traps appear in turn. In many regions such as the Green River Basin in the Rocky Mountain area of the United States, the foredeep of the Zagros Mountains in Iran, and the Alberta Basin in Canada, such examples have been found.

In fact, the differential accumulation of oil and gas is affected by many disturbing factors, mainly the following aspects:

① When there are other branch oil and gas supply sources on the migration path, it will disrupt the original oil and gas distribution pattern.

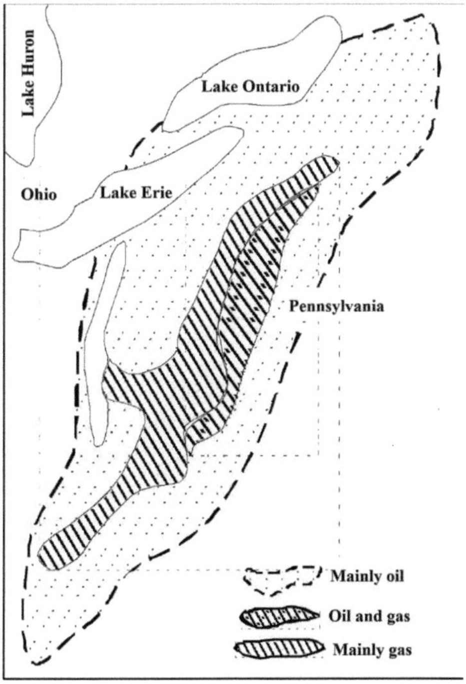

Fig. 5.15 Distribution map of oil and gas reservoirs in the Appalachian Basin, USA (according to A. I. Levorsen, 1967)

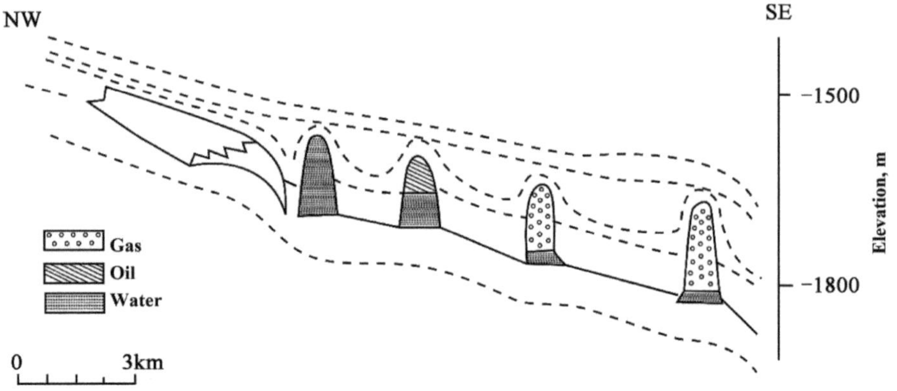

Fig. 5.16 Differential accumulation in the Silurian reef belt of the Michigan Basin, USA (according to Perrodon, 1993)

Fig. 5.17 Differential accumulation diagram in the Eromanga Basin, Australia (according to Bowering, 1982)

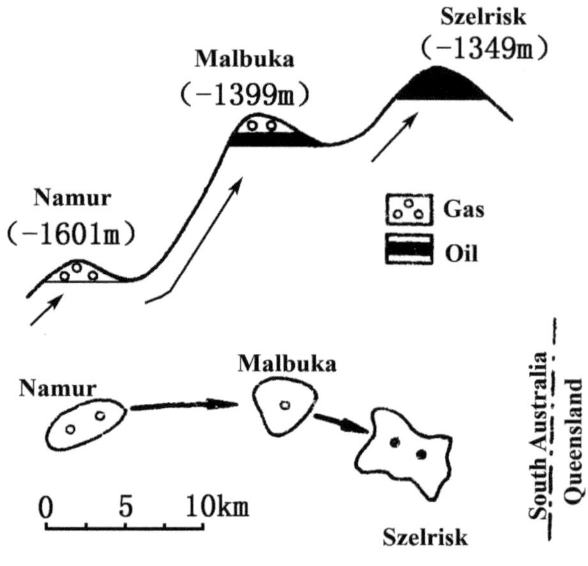

② The dissolution of gas in oil varies with changes in physical conditions (temperature, pressure). It can cause secondary gas caps or lead to the disappearance of the original gas caps, thus affecting the distribution pattern of oil and gas.
③ Later tectonic movements cause changes in trap conditions, which will inevitably lead to the re-distribution of oil and gas.
④ Regional hydrodynamic conditions, the magnitude of the water pressure gradient and the direction of water movement, directly affect the distribution pattern of oil and gas.

5.2.3.2 Seepage-Type Differential Accumulation of Hydrocarbons ("Oil Core and Gas Ring" Pattern)

The principle of differential accumulation of hydrocarbons assumes that the quality of the trap cap is good enough to seal the height of the hydrocarbon column at the spill point, and the hydrocarbons overflow from the spill point after charging the trap, and a series of traps with spill points in turn higher are the same reservoir. In actual geological conditions, only some large oil and gas basins with relatively stable structures may meet these conditions. The more common situation is that hydrocarbons do not necessarily overflow from the lowest part of the trap after filling the trap, but are diverse. For example, due to fault activity, poor cap rock quality, and other factors, hydrocarbons can seep from the top of the trap in the upper direction or the upper trap, or seep before filling to the spill point of the trap, thus forming a distribution of hydrocarbon phases "oil core and gas ring", or a distribution pattern of hydrocarbon phases without obvious zoning. As shown in Fig. 5.18, the fault hydrocarbon reservoirs located in the lower part of the basin (lower step) may lose some oil and gas from the high part of the trap due to fault activity, and multiple activities may cause the trap to only accumulate oil; these lost hydrocarbons continue to migrate to the upper shallow layer or upper direction, forming oil and gas reservoirs in the middle step. The fault hydrocarbon reservoirs in the middle step mainly lose natural gas due to fault activity and form gas reservoirs in the high step. This results in a series of evolution from oil reservoirs, oil and gas reservoirs, gas reservoirs as the burial depth becomes shallower from the center of the basin to the edge of the basin. Many rich oil and gas depressions in the Bohai Bay Basin have an "oil core and gas ring" pattern in the distribution of oil and gas. For example, the distribution of oil and gas phases in the Yong'an Town area in the northeast of the Dongying depression is this pattern, and oil and gas migrate step by step from the low fault block near the oil source to the middle and high fault block area far from the oil source, successively appearing oil reservoirs, oil and gas reservoirs, and pure gas reservoirs (Jiang & Liu, 1997).

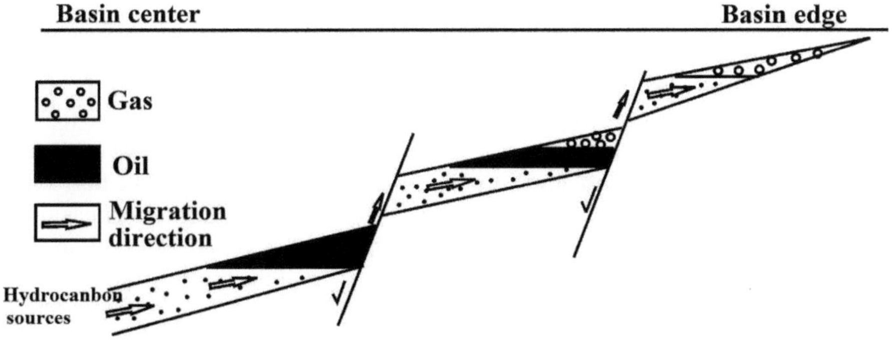

Fig. 5.18 Fault seepage-type hydrocarbon differential accumulation schematic

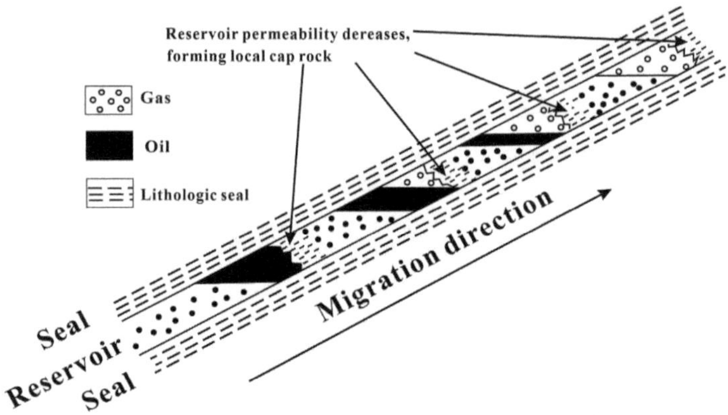

Fig. 5.19 Cap rock seepage-type hydrocarbon differential accumulation schematic

The quality of the cap rock has an important impact on the distribution pattern of oil and gas. If the quality of the cap rock is not high, when the oil and gas accumulated in the trap reach the maximum oil and gas column that the cap rock can seal, some oil and gas can break through the cap rock to seep and migrate upwards, and the accumulation and distribution of oil and gas appear diverse, which can cause gas to occupy the high position, and oil at the bottom, as shown in Fig. 5.19. Therefore, this type of hydrocarbon migration and accumulation pattern can be called seepage-type hydrocarbon differential accumulation (Figs. 5.18 and 5.19).

Sales (1997) believes that the fundamental factor controlling the distribution and differential accumulation of oil and gas underground is the relationship between the sealing strength of the trap and the closure height, which is divided into three situations. The first is that the sealing strength is greater than the closure height, with residual sealing strength, oil and gas can only spill from the bottom of the trap, but will not seep from the top cap rock, and natural gas is preferentially accumulated. The second is with moderate sealing strength, enough to seal the oil column with the same height as the trap closure, but not enough to seal the gas column with the same height, in the dynamic balance process, oil and gas spill from the bottom of the trap and seep from the top cap rock respectively. The third is that the sealing strength is less than the closure height, the sealing strength is not enough to seal the oil column with the same height as the trap closure, and both oil and gas seep from the top cap rock (Fig. 5.20). Due to the differences in lateral spill and vertical seepage of various traps, the differential accumulation and distribution of oil and gas in the horizontal and vertical directions are caused (Fig. 5.21). Therefore, in petroliferous basins, if these three types of cap rocks all play a role, the distribution pattern of oil and gas phases in traps from the center of the basin to the edge of the basin will be diverse.

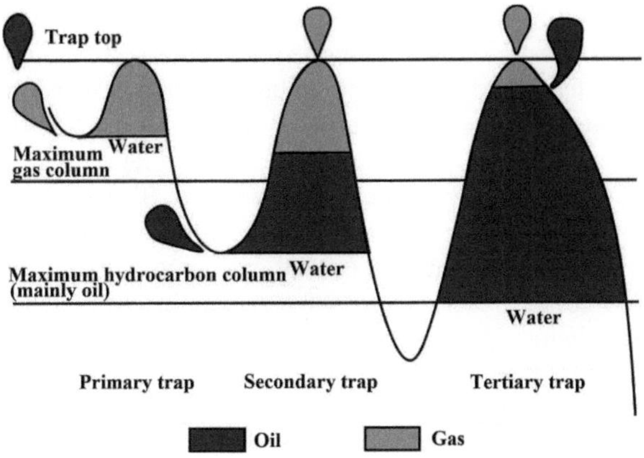

Fig. 5.20 Spill and seepage of oil and gas in three types of traps (according to Sales, 1997)

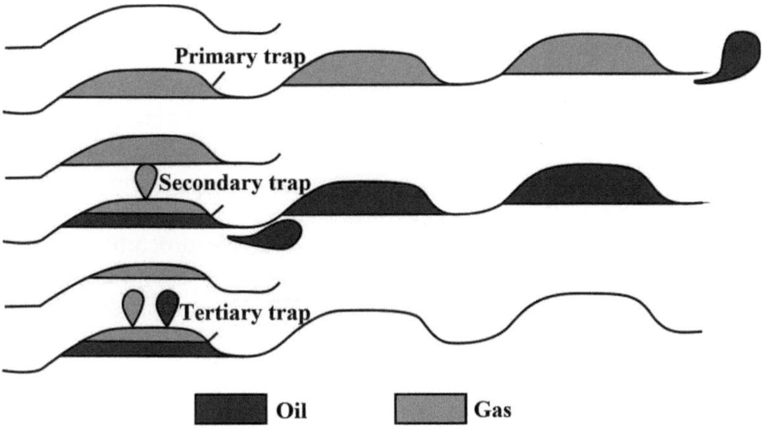

Fig. 5.21 Three types of traps with differences in lateral and vertical oil and gas aggregation (according to Sales, 1997)

5.2.4 Unconventional Hydrocarbon Accumulation Mechanism

The above discussion is based on the hydrocarbon accumulation mechanism of conventional oil and gas reservoirs, that is, under the action of buoyancy, hydrocarbons are charged and accumulated in conventional traps. Different from the formation mechanism of conventional oil and gas reservoirs, the accumulation of unconventional oil and gas represented by tight sandstone oil and gas and shale oil and gas does not conform to the principle of gravity segregation, and the conventional oil and gas accumulation theory cannot explain the accumulation of unconventional oil and gas. Because the unconventional reservoirs are tight and have poor physical properties, the driving force for the transportation and accumulation of oil and gas is mainly the high pressure of hydrocarbon source rocks, mainly controlled by the pressure increase of hydrocarbon generation, undercompaction effect, and structural stress. The nanometer-level pore and throat system limit the role of water column pressure and buoyancy in the transportation and accumulation of oil and gas, so the effect of buoyancy and water dynamics is not obvious. The distance of oil and gas migration is generally short, mainly the first migration or short-distance secondary migration, resulting in unconventional oil and gas being mostly coexistent with the source, or the source and reservoir are adjacent, mainly distributed in the center and slope of the basin near the source, showing a large area of continuous or quasi-continuous accumulation in the source or near the source, and there is no obvious trap. Taking the most common tight sandstone oil and gas and shale gas as examples, their accumulation mechanisms are described below.

5.2.4.1 Tight Oil and Gas Accumulation Mechanism

Tight oil and gas refer to natural gas in tight sandstone reservoirs, and their formation is mainly driven by the original driving force after the generation of oil and gas from hydrocarbon source rocks to tight reservoirs. Tight oil and gas reservoirs are mostly characterized by abnormally high pressure, with formation pressure coefficients generally greater than 1.2, and some tight oil and gas reservoirs are abnormally low pressure, but they still develop abnormally high pressure during the formation period, indicating that the abnormally high pressure of hydrocarbon source rocks during the formation period is the original driving force for the transportation and accumulation of tight oil and gas. It can be divided into two situations: tight oil and tight gas.

(1) Tight Oil Accumulation Mechanism

The formation of tight oil reservoirs is characterized by pressure difference displacement, non-Darcy seepage, and continuous type accumulation. The charging force of tight oil near the source-reservoir interface is mainly driven by the pressure increase of hydrocarbon generation. Oil is expelled from hydrocarbon source rocks into the tight reservoir near the source-reservoir interface driven by the pressure increase of hydrocarbon generation. The greater the pressure increase of hydrocarbon generation, the stronger the ability of oil to enter the small throats in the tight reservoir near the source-reservoir interface, and the smaller the lower limit of pore throat that oil can charge. After oil enters the tight reservoir, it shows non-buoyancy migration and accumulation. In the tight reservoir, due to the small pores and extremely fine throats, the capillary resistance at the throats is large, and the buoyancy of oil beads in the pores of the tight reservoir is much less than the capillary resistance at the throats, making it difficult for oil and water to undergo gravity segregation. Therefore, buoyancy is not the main driving force for the transportation and accumulation of tight sandstone oil, and the driving force for transportation is mainly the overpressure of hydrocarbon source rocks, while the resistance to transportation is the capillary pressure. When the fluid pressure in a tight reservoir is less than the capillary resistance, oil undergoes a retention and aggregation effect, and the coupling relationship between the two controls the oil aggregation process.

(2) Tight Gas Accumulation Mechanism

Under normal circumstances, natural gas in the pores has an upward migration trend. Natural gas in the pores overcomes capillary resistance and migrates upward under the action of buoyancy, while formation water flows in various ways in the opposite direction. This situation only occurs when the formation water at the top and bottom of the natural gas accumulation body is in a free and connected state, that is, it meets the continuity condition of formation water distribution. By the flow of formation water between the top and bottom of the natural gas and the exchange of potential energy, the upward migration of natural gas is achieved until it is blocked by a trap. In this case, the migration of natural gas is achieved through the continuous exchange of relative positions (or potential energy) between formation water and natural gas, showing a "displacement" process, and the result is the accumulation of conventional gas reservoirs at structural high positions. Therefore, the formation of conventional natural gas reservoirs is a buoyancy difference accumulation, and the displacement process of gas and water obeys the displacement principle (Zhang & Wang, 2003).

The accumulation mechanism of tight sandstone gas reservoirs is significantly different from that of conventional gas reservoirs because the reservoirs of tight sandstone gas are tight, and buoyancy hardly plays a role. The displacement process of gas and water is like the piston principle. When the overpressure gradient exceeds the start pressure gradient, natural gas will slowly non-Darcy flow from the lower part of the tight reservoir and upwardly displace the movable pore water. During the migration process, the formation water between the top and bottom boundaries of natural gas cannot achieve potential energy exchange through free flow, showing an overall pushing effect of natural gas on formation water from the bottom, and buoyancy cannot be produced, resulting in an inverted distribution relationship of natural gas below formation water. The capillary pressure between gas and water and the pressure of the formation water column above the natural gas accumulation body constitute the main resistance to the migration of natural gas. Natural gas must overcome the above two kinds of resistance during migration, so the formation process of tight sandstone gas (deep basin gas reservoirs) should show characteristics of abnormal pressure (Zhang & Wang, 2003).

The upward displacement process of natural gas against formation water requires a small driving force for a long period of continuous action. A short period of high-intensity gas injection is easy to cause many micro-fractures, leading to the occurrence of injected gas or gas seepage along the "high-speed channel" during secondary migration. Correspondingly, the expulsion process of natural gas from formation water also requires that the gas–water interface has overall advancement conditions—the relative homogeneity of the tight reservoir in the direction of the main gas migration to form a piston-like displacement effect. When the physical properties of the reservoir gradually improve in the upper direction, the capillary

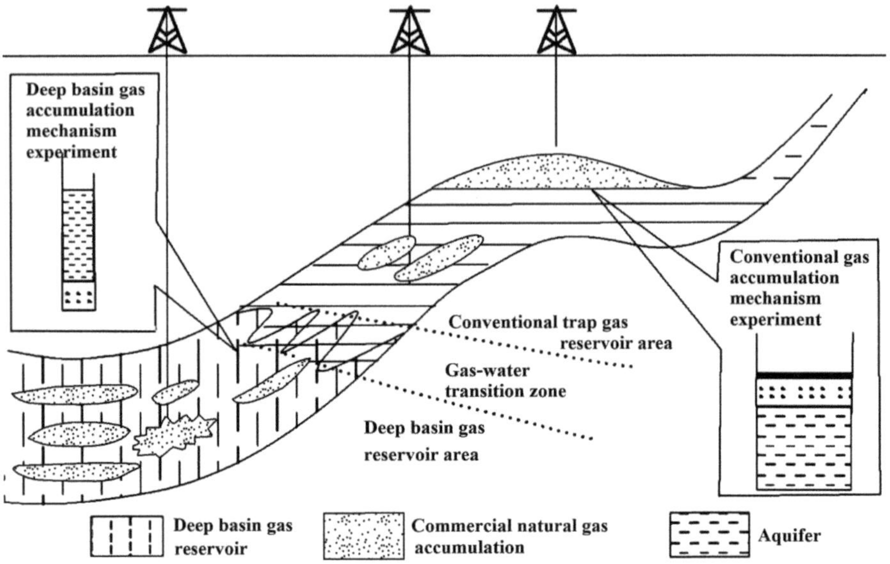

Fig. 5.22 Schematic diagram of the relationship between tight sandstone gas (deep basin gas) reservoirs and conventional gas reservoirs (according to Jin & Zhang, 2003)

also becomes larger and is eventually replaced by large pore throats. When the gas–water interface advances to the critical area where the permeability is good (gas–water transition zone), the relative permeability increases, and the flow resistance of free-phase natural gas becomes smaller and smaller, and can freely exchange mechanical energy with formation water. Gas escapes upward under the action of buoyancy and can accumulate to form conventional gas reservoirs at suitable positions in the upper part (Fig. 5.22).

Li (2013) proposed that the formation of tight sandstone oil and gas reservoirs belongs to the overpressure-retention dynamic mechanism. Under the drive of overpressure, when the start pressure gradient is exceeded, oil and gas will first enter the parts with larger throats in the tight reservoirs in a low-speed non-Darcy flow manner and gradually push into smaller throats, and upwardly displace the pore water. Due to the different flow rates (K/μ) of gas, oil, and water, under the same overpressure gradient, water is the easiest to flow, gas is the next, and oil is the most difficult. Moreover, since buoyancy does not play a role, oil and gas cannot float ahead, so the final distribution state is water on top and oil and gas below. In parts with good permeability where local buoyancy plays a role, oil, gas, and water can show a conventional distribution, but overall, a mixed distribution characteristic appears. Only when the hydrocarbon overpressure is large enough to charge all throats, or when the heterogeneity of the tight reservoir is low, can a more consistent distribution state of water below and oil and gas above be formed. When the overpressure gradient is not enough to overcome the resistance to forward propulsion, this situation may be due to the weakening of hydrocarbon overpressure or the consumption of oil and gas propulsion capabilities, resulting in a decrease in overpressure gradient, and oil and gas stop moving forward and are retained in the tight reservoirs, forming continuous and irregularly edged oil and gas accumulations in three-dimensional space. Among them, at the structural highs and permeability sweet spots, a rich accumulation of oil and gas often occurs.

5.2.4.2 Shale Oil and Gas Accumulation Mechanism

Shale oil and gas are oil and gas contained in organic-rich shale formations, mainly including pure shale type and sandstone interlayer medium type. Since the oil and gas in sandstone interlayers of shale formations are also tight sandstone oil and gas, shale oil is mainly contained in sandstone interlayers of shale formations, so this section mainly discusses the accumulation mechanism of pure shale type natural gas.

Different from the formation mechanism of conventional natural gas reservoirs, shale is both the source rock and the reservoir, so shale gas has no migration or extremely short-distance migration and accumulates in situ, which is a typical "in-situ saturation accumulation". That is, various types of natural gas in the source shale formation are retained and saturated in situ, then migrate upward to the shale formation without hydrocarbon generation capabilities, and after saturation again, they migrate upward to conventional sandstone reservoirs (Fig. 5.23) (Zou et al., 2014).

Shale gas mainly exists in natural fractures and pores in a free state, or on the surfaces of kerogen and shale mineral particles in an adsorbed state, with less shale gas in a dissolved state, that is, the main body of shale gas is free gas and adsorbed gas. Therefore, there are at least two mechanisms for the accumulation of shale gas: one is the accumulation of free-state natural gas, and the distribution of natural

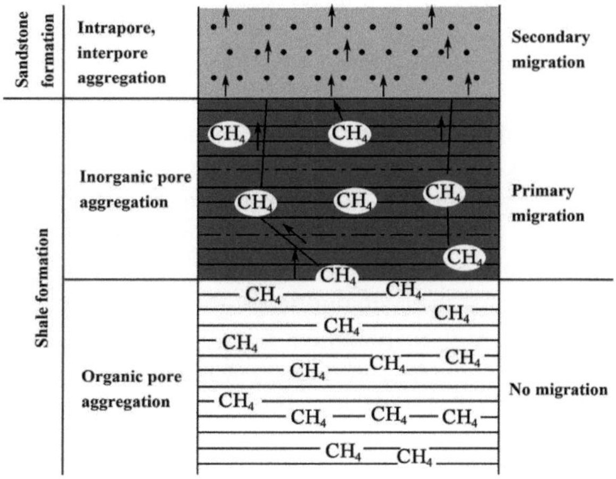

Fig. 5.23 Mechanism of shale gas formation and accumulation patterns (according to Zou et al., 2014)

gas is mainly controlled by the development and distribution of larger pores and fracture spaces in shale. The other is the accumulation of adsorbed-state natural gas. Although the adsorption mechanism of shale gas is like that of coalbed gas, due to the significant differences between shale and coal, the proportion of adsorbed gas in shale and coal is quite different, and its accumulation mechanism is different from that of coalbed gas.

Compared with tight sandstone gas, shale has a more compact porosity and permeability, with almost non-interconnected pores having a smaller radius. Since the generation of hydrocarbons from organic matter has appropriately modified the hydrophilicity of the rock particle surface, the capillary force changes from the resistance to hydrocarbon migration in conventional reservoirs to the driving force for natural gas migration in shale reservoirs, which is different from the typical piston principle in tight sandstone formations. At the same time, diffusion provides another possible pattern for the migration of natural gas in shale, making a significant difference between the accumulation mechanisms of shale gas and tight sandstone gas.

5.3 Formation and Preservation Conditions of Hydrocarbon Reservoirs

The core theory of petroleum geology is the formation and distribution patterns of hydrocarbon reservoirs. Elucidating and mastering the basic principles of hydrocarbon reservoir formation is not only of scientific theoretical significance but also has more important practical significance for the exploration and development of oil and gas resources.

In petroliferous basins, the formation and production state of hydrocarbon reservoirs are controlled by a variety of factors. The formation process of a hydrocarbon reservoir is the transformation process from dispersion to concentration of oil and gas under the effective matching of various reservoir-forming elements. Whether a region can form and preserve rich oil and gas reservoirs depends on whether there are effective static geological elements such as source rocks, reservoirs, seals, and traps, as well as dynamic geological processes such as oil and gas generation, migration, accumulation, and preservation, and their spatiotemporal configuration relationships. Therefore, in petroliferous basins, the formation and preservation of any oil and gas reservoir are the organic cooperation of static elements and dynamic geological processes in oil and gas reservoir formation, and these elements are indispensable. The probability of the occurrence of an oil and gas reservoir can be regarded as the product of the probabilities of these static and dynamic conditions for reservoir formation. Any misjudgment of any condition will lead to the failure of an exploration well, which is the reason why the success rate of wildcat wells is less than 50%.

The formation and preservation of conventional oil and gas reservoirs can be summarized into four basic geological conditions: abundant oil and gas sources, favorable source-reservoir-seal combinations, effective traps, and good preservation. The formation of oil and gas reservoirs is an event that occurs in geological history. Abundant oil and gas sources, good source-reservoir-seal combinations, and effective traps are the basic geological conditions for reservoir formation. The process of oil and gas reservoir formation also requires good preservation conditions. Whether the oil and gas reservoir formed can be preserved to the present, and the continuous preservation conditions are crucial. Therefore, to effectively carry out oil and gas exploration work in a basin or region and improve the success rate of exploration, it is necessary to first study the formation and preservation conditions of its oil and gas reservoirs.

5.3.1 Abundant Oil and Gas Sources

The condition of oil and gas sources is the material basis and prerequisite for the formation of oil and gas reservoirs. Whether the oil and gas sources are abundant is a question that must be clarified first in any exploration new area or new stratigraphic series. If a basin or depression has poor or non-existent conditions for generating oil and gas, no other element matters. The practice of world oil and gas exploration has shown that basins with huge oil and gas reserves first have abundant oil and gas sources, and some basins or depressions that have been explored many times without success are mainly due to poor oil and gas source conditions.

The richness of oil and gas sources mainly depends on the development scale and quality of the source rocks within the basin. The quality includes the abundance, type, and maturity of organic matter, and the scale of the development of source rocks depends on the size of the hydrocarbon depression area and the duration of the depression. In petroliferous basins, the polycyclic nature of crustal movement and the polycyclic nature of sedimentation control the development of source rock series, often forming multiple sets of source rock series. Obviously, a large area and long duration of hydrocarbon depression can form thick polycyclic source rock series and multiple oil generation periods. At the same time, high organic matter abundance and high evolution degree of source rocks are also important conditions for forming abundant oil and gas sources. The type and evolution degree of organic matter determine whether a region is rich in oil or gas, affecting whether a region mainly looks for oil or gas. Source rocks dominated by sapropelic organic matter and basins with moderate evolution degree are mainly rich in oil, while areas rich in humic organic matter and regions with high or low thermal evolution degree of organic matter tend to be rich in gas. For example, the Sichuan Basin is rich in natural gas resources, mainly related to the high thermal evolution degree of its Paleozoic source rocks, while the Bohai Bay Basin is rich in oil and poor in gas, mainly related to the relatively low thermal evolution degree of its Paleogene source rocks.

Large oil and gas fields are distributed in large sedimentary basins. According to statistics, 61 of the world's largest oil and gas fields are distributed in 12 large oil and gas basins, possessing more than half of the world's oil and gas reserves. These basins are all sedimentary basins with inherited stable subsidence, developing large volumes of sedimentary rock systems, with large areas and long durations of hydrocarbon depression, and have abundant oil and gas sources (Table 5.2).

The above statistical data show that oil and gas basins with abundant oil and gas resources all have large areas and large thicknesses of sedimentary rocks, that is, they have a large volume of sedimentary rocks, especially a large volume of source rocks. The distribution area of sedimentary rocks is mostly more than 10×10^4 km^2, and the volume of sedimentary rocks is mostly more than 50×10^4 km^3, with a minimum cumulative thickness of source rock series of 200–300 m, generally more than 500 m, and the thickest can reach more than 1000 m. The latest exploration practice shows that the thickness of effective source rocks may not need to be very large to generate many hydrocarbons. Some basins have effective source rocks with a thickness of only tens of meters, yet they have formed large oil fields, and the quality of source rocks may be more important.

Some basins with small areas but large sedimentary thicknesses, especially those with large thicknesses of source rock series, can also form rich oil and gas accumulations with other favorable conditions. For example, the Los Angeles Basin in the western United States is a small sedimentary basin with an area of only 3,900 km^2. During the late Miocene to the Pleistocene in a short period of time, more than 6000 m of sedimentary rocks were deposited. In the central part of the sedimentary depression, the argillaceous source rock series is 2,000–3,000 m thick, and the oil source is extremely rich. In the oil source area and its vicinity, sandstone reservoirs are well developed, with reservoirs and source rock layers interbedded or finger-like interlaced, and faults are connected. It is very conducive to the migration of oil and gas. There is also a series of anticline structures, with good trap conditions, and the trap area and height are also relatively large. Therefore, many oil and gas fields are formed, and the thickness of oil and gas is particularly large, generally more than 1000 m. In the Long Beach oil field, the thickest oil thickness can reach 1585 m. The amount of oil discovered per unit area in this basin is large, with nearly 20×10^4 m^3 of recoverable oil per square kilometer, leading all oil and gas basins in the world. In addition, the Ploiesti Basin in Romania and the Ventura Basin in California, USA, are all small oil and gas basins with extremely high richness. As shown in Fig. 5.24.

China's eastern Bohai Bay Basin is distributed with dozens of oil and gas-bearing depressions, generally with an area of several thousand square kilometers. For example, the Dongying Depression is 5,700 km^2. They are all relatively independent Paleogene oil and gas-bearing depressions, and each depression's oil and gas generation, migration, and accumulation have relative independence. In these oil-bearing depressions that are not large in area, some have very rich oil and gas resources, forming huge oil and gas fields, such as the Dongying Depression, Bozhong Depression, and the western part of the Liaohe Depression. The oil reserves discovered per unit area in these depressions can be comparable to the world's richest Los Angeles Basin and are known as oil-rich and extremely oil-rich depressions. However, there

Table 5.2 Simplified table of 61 super large oil and gas fields in 12 world oil and gas basins (according to Zhang, 1999)

Basin	Area (× 10^4 km^2)	Development of sedimentary rocks			Development of source rocks		Oil and gas recoverable reserves (t, m^3); number of super large oil and gas fields
		Era	Thickness	Volume (× 10^4 km^3)	Era	Lithology and thickness (m)	
Persian Gulf	240	Pz, Mz, Cz, mainly J, K, E, N	5,000–12,000, average 3,000	704.1, of which J and above is 417	Mainly J_3, K_2, E	Carbonates mainly, up to 4,000, main source rock layer thickness 1,000–1,500	Oil 541 × 10^8; 2^8
Western Siberia	230	Mz, Cz, mainly J, K	Thickest 4,000–8,000, average 2,600	600	J_2~K, mainly J_3, K_1	Mudstone (pro-delta) 500–1,000	Oil 60 × 10^8; 8
U.S. Gulf of Mexico	110	Mz, Cz	Thickest 12,000, average 4,000	545	J_3~N_1, mainly K_3, N_1	Mudstone mainly, some carbonates 1,000–2,000	Oil 53.4 × 10^8; 1
Maracaibo	8.5	Mz, Cz (K~N)	Thickest 10,000, average 4,600	395.7	K~N, mainly E_2	K limestone, mudstone, thickness 150–200, E mudstone, 2,000	Oil 73 × 10^8; 2
Volga-Ural	65	Mainly Pz2	Generally less than 2000, up to 8000 in front of the Ural Mountains, average 3100	218.2	Middle D_2~P_1	Mudstone mainly, total thickness 200–500	Oil 42.7 × 10^8; 2
Libyan Sirte	35	Pz, Mz, Cz, mainly K, E, N	Paleozoic 1500, above K thickest 5000, average 2500	80	K~E, mainly K_2, E	Limestone, dolomite mainly, some mudstone 1000–2000	Oil 40 × 10^8 Gas 7790 × 10^8; 4
Algeria East Gobi	41	Pz, Mz	4000–5000	160	S	Shale 200	Oil 9.9 × 10^8 Gas 29,940 × 10^8; 3

(continued)

Table 5.2 (continued)

Basin	Area ($\times 10^4$ km^2)	Development of sedimentary rocks			Development of source rocks		Oil and gas recoverable reserves (t, m^3); number of super large oil and gas fields
		Era	Thickness	Volume ($\times 10^4$ km^3)	Era	Lithology and thickness (m)	
North Sea	62	P~R	Total thickness 8000 R3000	300	J and R, some C3	Mudstone	Oil 34 $\times 10^8$ Gas 184,080 $\times 10^8$; 4
Niger Delta	6	Cz	Generally 4000–6000 Maximum 12,000	30	N	Mudstone 1000–2000	Oil 27 $\times 10^8$ Gas 11,200 $\times 10^8$; 6 large oil and gas fields
U.S Western Interior	60.2	Pz, Mz	9000	85	∈, C, P	Mainly mudstone, 200–400	1 (gas)
Songliao Basin	22.6	K~N	Thickest 6000 Average 3000	77.5	K	Mudstone 500–1000	1
Bohai Bay	25	Z~Mz, Cz	Cz thickest up to 6000, of which E:4500	125	Mainly E	Mudstone more than 500, thickest 1000–1500	2

are also some depressions that, due to the small volume of source rocks, low organic matter abundance and thermal evolution degree of source rocks, or mainly due to the low thermal evolution degree of organic matter, resulting in insufficient oil and gas generation and relatively poor resource volume. There has been no significant exploration breakthrough with large-scale reserves to date, which can be called poor oil depressions. For example, the Xinxian Depression, Dezhou Depression, and other depressions in the central and western parts of the Bohai Bay Basin have not made significant breakthroughs in oil and gas exploration to date, mainly related to the poor oil and gas source conditions of these depressions, especially related to the low thermal evolution degree of their source rocks. Therefore, the oil and gas content of different depressions in the Bohai Bay Basin is directly related to their oil-generating conditions. Based on this, the depressions can be classified into four levels: extremely oil-rich, oil-rich, oil-bearing, and poor oil (Jiang et al., 2014).

From this, whether a basin is rich in oil and gas mainly depends on the development of the basin's source rock series, among which the volume of the source rock is

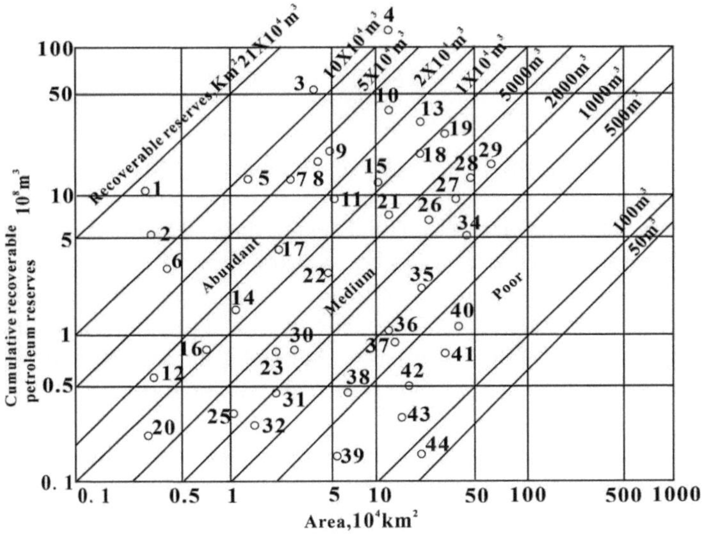

Fig. 5.24 Abundance of some world oil and gas basins (according to Perrodon, 1993). 1—Los Angeles (USA); 2—Ploiesti (Romania); 3—Maracaibo (Venezuela); 4—Persian Gulf (Middle East); 5—Baku (Soviet); 6—Ventura (USA); 7—San Joaquin (USA); 8—Mangystau (Soviet); 9—Sumatra (Indonesia); 10—Louisiana (USA); 11—Tampico Bay (Mexico); 12—Gijón (Spain); 13—Sirt (Libya); 14—Reconquista (Cuba); 15—East Venezuela (Venezuela); 16—Vienna (Austria); 17—Grand Horn (USA); 18—North Slope of Alaska (USA); 19—West Texas (USA); 20—Tirana (Albania); 21—Nigerian Delta; 22—Jambi (Indonesia); 23—Wyoming (USA); 24—Gippsland (Australia); 25—LaHav (Canada); 26—Wyoming (USA); 27—East Algerian Gobi; 28—Texas Coast (USA); 29—Alberta (Canada); 30—Cook Inlet (USA); 31—Paris Basin (France); 32—Sergipe (Brazil); 33—West Gabon; 34—Appalachia (USA); 35—Williston (USA); 36—Palo Duro Basin (USA); 37—Pannonian (Hungary, South); 38—Aquitaine (France); 39—San Juan (USA); 40—Montana (USA); 41—Michigan (USA); 42—Amadeus Basin (Australia); 43—Denver (USA); 44—Radamès (Libya)

the most significant, followed by the organic matter abundance, type, and thermal evolution degree of the source rock. Currently, the hydrocarbon generation intensity is commonly used for evaluation. There are two measurement indicators for the lower limit value of hydrocarbon generation intensity that can form commercial oil and gas flows: one is that the maximum hydrocarbon generation intensity must be greater than 100 × 10^4 t/km^2, and the other is that the average hydrocarbon generation intensity must be greater than 50 × 10^4 t/km^2. Studies have shown that the size of a hydrocarbon-bearing basin's hydrocarbon generation intensity is closely related to the cumulative thickness of its source rock series, the organic matter abundance, type, and maturity of the source rocks.

5.3.2 Favorable Source-Reservoir-Seal Combinations

In stratigraphic sections, the spatial and temporal combination relationship of source rock layers, reservoir layers, and seal layers is called the source-reservoir-seal combination. A favorable source-reservoir-seal combination refers to the ability of abundant oil and gas generated from the source rock layers to be timely transported to good reservoir layers, while the quality and thickness of the seal layers can ensure that the oil and gas transported to the reservoir layers will not escape. The practice of world oil and gas exploration has shown that the formation of large oil and gas fields has a good source-reservoir-seal combination. Therefore, a favorable source-reservoir-seal combination is one of the essential conditions for the formation of rich oil and gas accumulations in oil and gas basins, especially for the formation of large oil and gas fields.

Based on the spatial configuration relationship between source, reservoir, and seal, the source-reservoir-seal combinations can be divided into four types (Zhang, 1999; Fig. 5.25).

(1) Normal source-reservoir-seal combination: refers to the stratigraphic section where the source rock layer is located at the bottom of the combination, the reservoir layer is in the middle, and the seal layer is at the top. This combination type is further divided into continuous and discontinuous types based on temporal continuity or discontinuity. Oil and gas migration from the source rock layer to the reservoir layer is mainly vertical. The normal source-reservoir-seal combination is the most important combination method for many oil fields in China.
(2) Lateral variation source-reservoir-seal combination: This combination is formed by the spatial variation of lithology and sedimentary facies, leading to the lateral

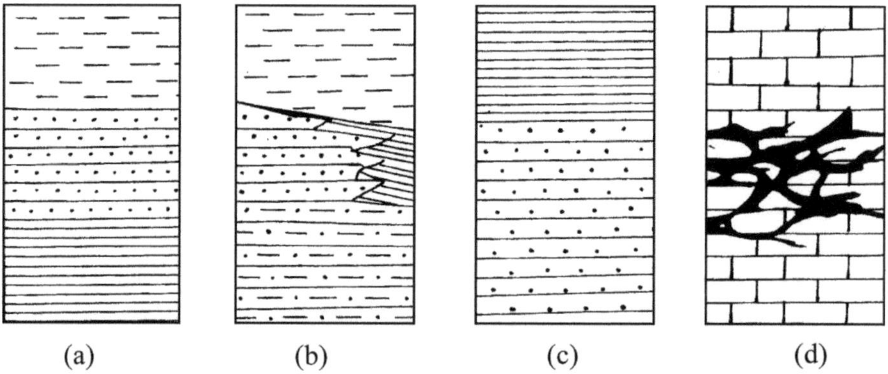

Fig. 5.25 Schematic diagram of source-reservoir-seal combination types. **a** Normal source-reservoir-seal combination, **b** lateral variation source-reservoir-seal combination, **c** top-generated source-reservoir-seal combination, **d** self-generated, self-reservoir, self-seal source-reservoir-seal combination

combination of source, reservoir, and seal layers. This combination is mostly developed on the slope belt of oil-generating depressions or ancient uplift slopes. Due to the horizontal variation of lithology and sedimentary facies, the main characteristic is that the source rock layer and the reservoir layer belong to the same layer, and the two layers contact each other through the horizontal variation of lithology, with oil and gas mainly transported laterally within the same layer. Many oil and gas fields on the northwest edge of the Junggar Basin in Xinjiang, China, belong to this type of combination.

(3) Top-generated source-reservoir-seal combination: The source rock layer and the seal layer belong to the same layer, while the reservoir layer is located below it. For example, in the Renqiu Oilfield in North China, the Shahejie Formation mudstone of the Paleogene is both the source rock layer and the seal layer, directly covering the reservoir layers of the Proterozoic dolomite and the Cambrian and Ordovician carbonate rocks with pores, caves, and fractures.

(4) Self-generated, self-reservoir, self-seal source-reservoir-seal combination: Local oil storage in segments of limestone with developed fractures, sandstone oil storage in thick source rock series, and some oil storage in developed fracture segments of mudstone all belong to this combination type. The biggest characteristic is that the source rock layer, reservoir layer, and seal layer all belong to the same layer. The mudstone oil reservoir in the Zhanhua Depression of the Bohai Bay Basin during the Paleogene is a typical representative of self-generation and self-storage of mudstone. The oil fracture oil reservoir in the Qaidam Basin and other similar combinations.

Based on the temporal relationship between the source rock layer and the reservoir layer, the source-reservoir-seal combinations can be divided into three types: newly generated and ancient storage, ancient generation and newly stored, and self-generation and self-storage. The seal layer of the three types is newer than the reservoir layer.

(1) Newly generated and ancient storage: Oil and gas generated in relatively newer strata are stored in relatively older strata, known as newly generated and ancient storage. For example, in the Renqiu Oilfield in the Jizhou Sag, the oil is generated from the Shahejie Formation of the Paleogene, and the Cambrian, Ordovician, and Silurian carbonates are the reservoir rocks, forming a typical newly generated and ancient storage combination.

(2) Ancient generation and newly stored: Oil and gas generated in older strata migrate to newer strata and accumulate, belonging to ancient generation and newly stored. For example, in the Wenliu Gas Field in the Dongpu Depression of the Bohai Bay Basin, the reservoir layer is the fourth segment of the Shahejie Formation of the Paleogene, and the Carboniferous-Permian coal system is the source rock for gas generation. The coal-formed gas migrates through faults from the depression area to the fourth segment of the Wenliu anticline sandstone reservoir layer and accumulates to form a typical ancient generation and newly stored combination.

(3) Self-generation and self-storage: Refers to the source rock layer and the reservoir layer belonging to the same stratigraphic series, which can be narrow, that is, self-generation and self-storage within the same set of rock layers, and can be broad, that is, the source rock and reservoir rock appear alternately in a continuous sedimentary series, and the oil and gas reservoirs in the source rock series belong to this category, such as the fourth segment above the Shahejie in the Dongying Depression—the third segment below the Shahejie source rock series, where the shale generates oil and gas, and the sandbody in the same series stores oil, and the two belong to the same stratigraphic series deposited continuously.

Based on the sedimentary continuity among the source-reservoir-seal combinations, they can be divided into two major categories: continuous sedimentary source-reservoir-seal combinations and discontinuous source-reservoir-seal combinations separated by faults or unconformities. The former is adjacent in space and continuous in time, and self-generation and self-storage belong to this category; the latter is adjacent or not adjacent in space and discontinuous in time, and newly generated and ancient storage, ancient generation and newly stored all belong to discontinuous storage and cover combinations.

The type of source-reservoir-seal combination affects the expulsion and conduction efficiency of oil and gas. In the source-reservoir-seal combination composed of shale-sandstone, the combination relationship between sandstone reservoir rocks and shale source rock layers is of great significance to the expulsion and accumulation ability of oil and gas. According to Magara's (1978) study, the relationship between the average thickness of sandstone and the total recoverable oil volume in 7241 sandstone oil reservoirs in the United States shows that the contact area between the sandstone body and the surrounding source rock layer is the most important factor controlling oil reserves.

From the statistical results of the thickness ratio of sandstone to shale and the percentage of sandstone thickness in the profile in several oil-producing regions around the world (Table 5.3), the sandstone thickness percentage most favorable for oil accumulation is roughly between 20 and 60%, with a median of 30–40%. Areas with only massive sandstone development or only massive shale development are not favorable for oil accumulation. Only in areas where the sandstone thickness percentage is between 20 and 60%, that is, the single-layer thickness of sandstone reservoirs is about 10–15 m, and the single-layer thickness of shale source rock layers is about 30–40 m, the two are slightly equal in thickness and interbedded, the contact area between sandstone and shale is the largest, which is most conducive to oil accumulation.

Different types of source-reservoir-seal combinations have different channels and conduction capacities for transporting oil and gas, and the conditions for oil and gas enrichment are also different. For example, when the source rock layer and the reservoir layer form a finger-like interlayered combination, due to the large direct contact area between the source rock layer and the reservoir layer, the oil and gas generated in the source rock layer above and below the reservoir layer can be timely

Table 5.3 The optimal sandstone percentage for oil accumulation in some regions (according to Zhang et al., 1999)

Oil-producing region and stratum	Sandstone-mudstone thickness ratio	Sandstone thickness percentage (%)	Researchers
Rocky Mountain Area, Upper Cretaceous, USA	0.25–1	20–50	Krumbein and Nagel (1953)
Peruvian Parinas Sandstone Oil Reservoir	0.60	37	Youngquist (1958)
Salt Creek Area, Cretaceous Ferrant Formation, Wyoming, USA	0.23–0.41	19–29	Dickey and Rohn (1958)
Pennsylvanian System Atoka Formation, Oklahoma, USA	0.50–2.0	33–67	

transported to the reservoir layer, which is most favorable for oil and gas generation and enrichment. When sandstone and mudstone layers exist in an anticline, oil and gas can gather from all around to the anticline, forming a rich oil and gas reservoir. As shown in Fig. 5.27.

When the source rock layer and the reservoir layer form a finger-like interlayered combination, the contact is limited to the finger-like interlayered zone, and the conduction conditions in this zone are like those in the interbedded combination. On the side facing away from the basin and away from the interlayered zone, due to the lack of reservoir layers nearby, the conduction capacity is limited. On the other side, there are only reservoir layers, lacking source rock layers (oil sources), and the source of oil and gas is also limited. Therefore, its conduction conditions and oil and gas enrichment conditions are worse than those in the interbedded combination, as shown in Fig. 5.26.

5.3.3 Effective Traps

Oil and gas exploration practice has shown that not all traps in sedimentary basins accumulate oil and gas. Some traps accumulate oil and gas, some traps only contain water, and some traps with good oil source conditions are often "empty", which raises the issue of trap effectiveness. The effectiveness of a trap can be understood as: under the macro background of having an oil and gas source, the actual ability of a trap to accumulate oil and gas. Under what conditions is a trap effective in accumulating oil and gas? Under what conditions is a trap ineffective?

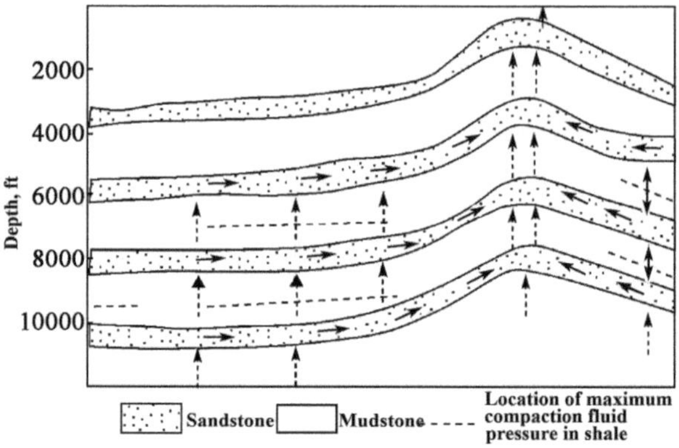

Fig. 5.26 Schematic diagram of the initial migration and accumulation of oil and gas when the source rock layer and the reservoir layer form an interbedded combination (according to R. J. Cordell (1976))

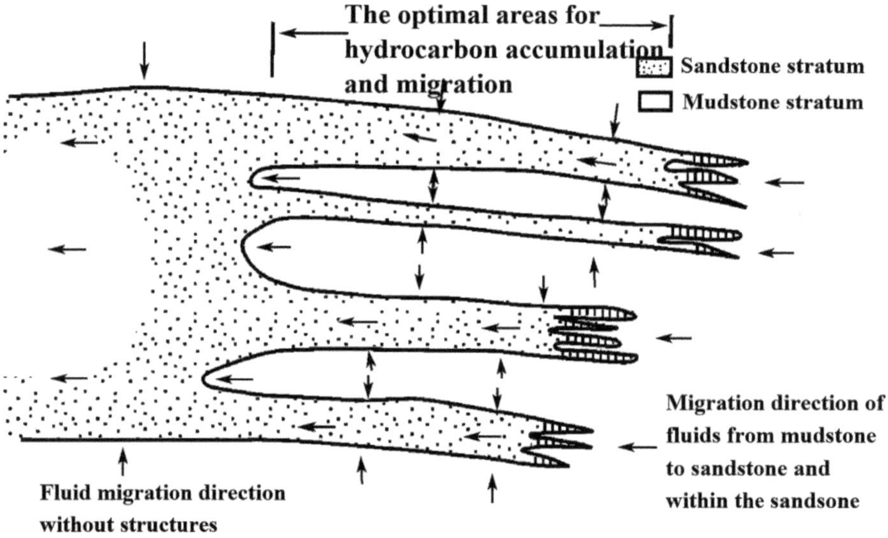

Fig. 5.27 Schematic diagram of the initial migration and accumulation of oil and gas when the source rock layer and the reservoir layer form a finger-like interlayered combination (according to R. J. Cordell (1976))

In general, the main reasons for the occurrence of ineffective traps are the following four aspects: (1) The trap is far from the oil source center and lacks sufficient oil and gas sources. (2) The trap is not on the main migration path of oil and gas, or is shielded by other structures on the migration path. (3) The formation time of the trap is later than the end of oil and gas migration. (4) Due to strong hydrodynamic effects, as well as the quantity and properties of oil and gas, the oil and gas that were originally accumulated in the trap are washed away by water flow. Therefore, the effectiveness of a trap mainly depends on the following conditions: the relationship between the trap's location and the oil source area and the main migration direction, the relationship between the trap's formation time and the main oil and gas migration period, and the impact of hydrodynamic effects and fluid properties on the effectiveness of the trap.

(1) The Relationship Between the Trap's Location and the Proximity to the Oil Source Area

In general, the deep depression areas that develop over a long term in sedimentary basins are the most favorable oil and gas-generating areas within the basin. Many exploration practices have proven that favorable oil and gas-generating areas in sedimentary basins control the distribution of oil and gas. After oil and gas are generated, they first migrate to the traps within and near the oil source area, accumulate to form oil and gas reservoirs, and the excess oil and gas then migrate and accumulate in traps further away in sequence. If the oil source is limited, traps far from the oil source area usually become ineffective traps. Under normal circumstances, the volume of traps that can store oil and gas in a basin is sufficient, and the oil and gas source is always insufficient relative to the capacity of the traps, that is, it is impossible to meet the total effective volume of all traps in the basin. Therefore, under similar conditions, the closer the trap is to the oil source area, the more favorable it is for oil and gas accumulation, and the higher the effectiveness of the trap; the farther away, the worse the effectiveness. Especially in continental sedimentary basins, the lithology and sedimentary facies of the reservoir layers change greatly, and the horizontal migration channels of oil and gas are restricted by many factors. Oil and gas migration distance is short, and traps within and near the oil source area are the most favorable, with high oil and gas enrichment. Traps far from the oil source area are often ineffective. In large marine sedimentary basins where marine sedimentary strata are well developed, the reservoir layers usually have stable lithology and good connectivity, and oil and gas can migrate over long distances. The proximity of the trap's location to the oil source area is not as important as it is in continental sedimentary strata developed sedimentary basins.

The control of oil source centers on the distribution of oil and gas is obvious in many oil source depressions in China's Bohai Bay Basin. Even in oil-rich depressions with good oil-generating conditions, areas with high hydrocarbon generation intensity are more likely to form oil and gas enrichment areas, while areas with relatively lower hydrocarbon generation intensity have lower oil and gas enrichment. For example, the Dongying Depression is the richest oil and gas depression in the Bohai Bay Basin. The richest oil and gas enrichment areas are in the central and northern parts, while the southern areas are relatively poor, mainly due to the distribution of the most favorable oil-generating depressions in the central and northern parts of the depression (Fig. 5.28), which control the distribution of oil and gas.

In continental faulted basins, the oil and gas content of depressions is mainly controlled by the oil-generating conditions of the depressions. The distribution of oil and gas in oil and gas-bearing depressions is obviously controlled by the oil

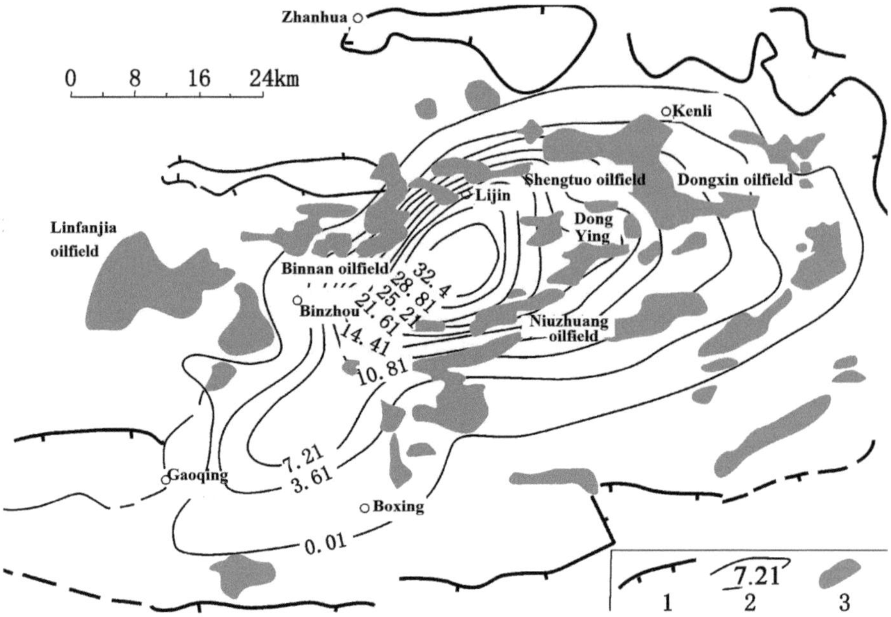

Fig. 5.28 The relationship between the Paleogene oil source center and oil and gas enrichment in the Dongying Depression (according to Shengli Oilfield). 1—Fault, 2—hydrocarbon generation intensity contour lines, $\times 10^6$ t/km^2, 3—oil field

source center (main oil-generating depression), which has become a basic rule for the formation and distribution of oil and gas in continental basins. Therefore, the source control theory has become an important theory guiding the oil and gas exploration in continental basins. For example, in the Songliao Basin, the central deep sag is rich in oil sources, and the Daqing Changyuan is located within the deep sag. After the generation of oil and gas, they migrate and accumulate nearby, forming a super-large oil field, while several structures far from the central sag have obvious poor oil and gas content. In the Bohai Bay Basin, the sedimentary depression acts as the basic hydrocarbon supply unit within the depression. The scale of hydrocarbon source rock development, organic matter abundance, type, and maturity, and other best oil-generating conditions are all distributed in the central area of the depression. However, the sedimentary subsidence characteristics of different depressions vary, leading to significant differences in their oil-generating capabilities. The oil and gas content of the depression and its surroundings mainly depends on the oil-generating conditions of the main sedimentary depression within the depression. Therefore, to a large extent, the depression controls the distribution of oil and gas, which is controlled by the oil-generating depression. The basic hydrocarbon supply unit of the "source control theory" is the oil-generating depression (Jiang et al., 2019).

Oil and gas enrichment requires a certain area to converge oil and gas. The larger the oil and gas field, the larger the area of oil and gas supply required. Therefore, the formation of large oil and gas fields often requires a larger oil supply area, and large-scale inherited positive structures in the oil-generating depression often become oil and gas enrichment areas. This is like the growth and development of trees. A large tree has a wider and deeper root system than a small tree, so-called "big trees have deep roots," which can provide sufficient nutrients needed for the growth and development of trees. Therefore, traps on the boundary of the oil source area have a larger area to converge oil and gas and are often the most favorable traps. It is difficult to form large oil and gas fields solely relying on the hydrocarbon source rocks within the trap range (Fig. 5.29).

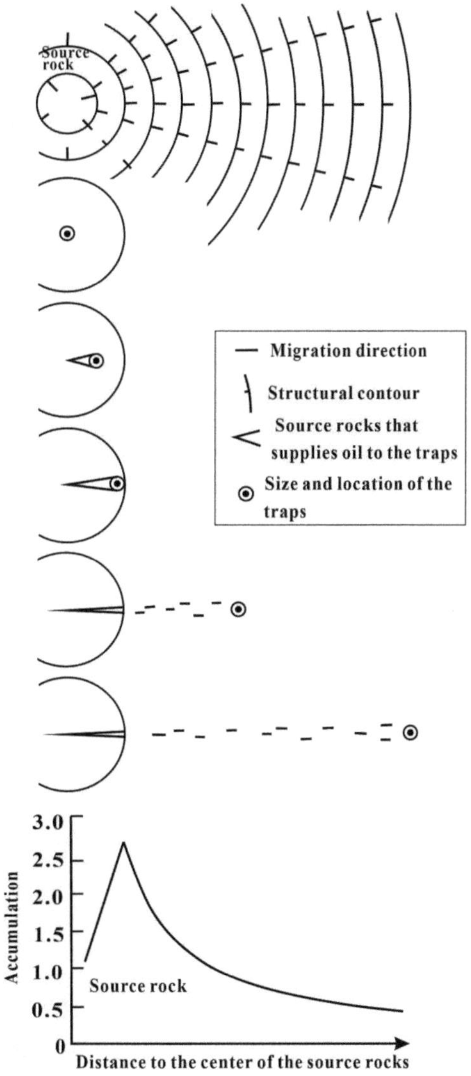

Fig. 5.29 Oil and gas enrichment section of a circular basin, traps on the edge of the source rock are highly enriched in oil and gas (according to Hindle, 1997)

(2) The Relationship Between the Trap's Location and the Main Migration Route of Oil and Gas

The migration of oil and gas from the oil-generating depression is not uniformly divergent, but some directions are relatively concentrated. Oil and gas migrate along the dominant migration pathways, while the amount of oil and gas in other directions is less, or even no oil and gas pass through. According to numerous studies, the space of oil and gas migration is less than 10% of the volume of the reservoir space. Therefore, both horizontally and vertically, the actual migration pathways that oil and gas pass through are limited. This characteristic of oil and gas migration along dominant pathways inevitably leads to some directions being rich in oil and gas, while others are relatively poor. Obviously, the probability of traps located on the main migration pathways accumulating oil and gas is much greater than that of traps on non-main migration pathways, so the former is often effective, and the latter is often ineffective.

As shown in Fig. 5.30 (A), due to the actual oil and gas migration route occurring along the structural ridge, near the hydrocarbon depression, the oil and gas migration pathways form a dense network. Far from the hydrocarbon depression, oil and gas converge towards the structural ridge under the control of the three-dimensional geometric shape of the top surface of the conduit layer. Trap A on the migration pathway and Trap F first accumulate oil and gas. After oil and gas fill Trap A, they migrate towards Trap D and accumulate in Trap D. After oil and gas fill Trap D, they migrate towards Trap E and accumulate in Trap E. Oil and gas do not migrate towards Trap B, causing Trap B to not accumulate oil and gas. Although Trap C is the closest to the oil source area, it becomes an empty trap because it is not on the oil and gas migration route, and traps on the oil and gas migration route become effective traps. If a cross-section from the oil source area of the hydrocarbon depression to the structural high part, S-S, is made, and only looking at the C-D-E structural section, as shown in Fig. 5.30 (B), it is often mistakenly believed that oil and gas migrate along this section, and Trap C is the most favorable, and if oil and gas accumulate in Trap D and Trap E, they also come from Trap C.

According to many scholars' research, most of the oil and gas reservoirs in many foreign basins are concentrated on the main migration pathways. For example, in the Gulf of Mexico Basin, more than 75% of oil and gas are concentrated on the main migration pathways that account for less than 25% of the basin area (Pratch, 1996). In the Paris Basin, more than 81% of oil and gas are concentrated on the main migration pathways that account for 13% of the basin area (Hindle, 1997). Therefore, studying and determining the main oil and gas migration routes is of great significance for evaluating the effectiveness of traps.

(3) The Matching Relationship Between Trap Formation Time and the Main Oil and Gas Migration Period

Traps, as containers for accumulating oil and gas, can only accumulate oil and gas after they are formed. Therefore, the relationship between trap formation time and the time of large-scale oil and gas migration is of great significance for evaluating

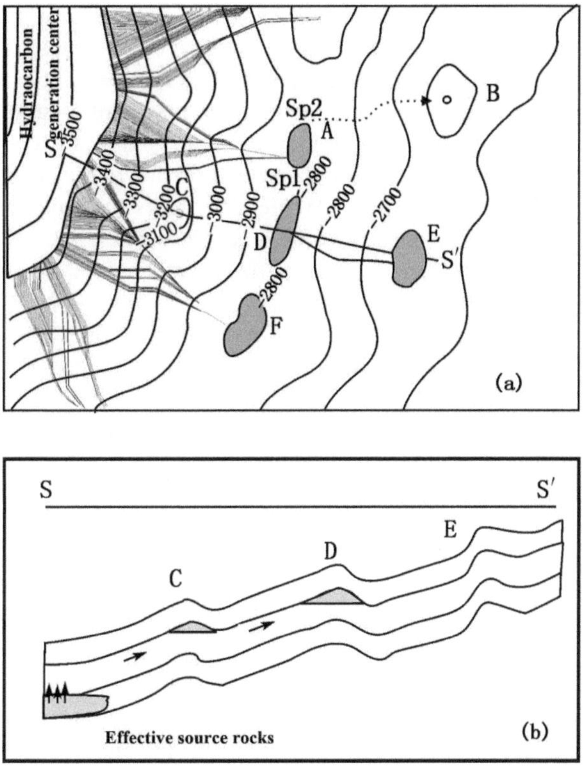

Fig. 5.30 3D ray tracing of oil and gas migration pathways (**a**) and comparison with 2D analysis (**b**) (according to Hao et al., 2005)

the effectiveness of traps. Only those traps that are formed before or at the same time as the large-scale oil and gas migration are effective for oil and gas accumulation. If a trap is formed after the last large-scale oil and gas migration, missing the time to capture oil and gas, such a trap is ineffective for oil and gas accumulation.

In petroliferous basins, the formation time of different types of traps varies greatly. Primary lithological traps developed within the source rock series are formed earlier, and oil and gas accumulate in the traps after short-distance primary and secondary migration, if other conditions are also favorable, these traps often become effective traps. For most traps formed by crustal movement and structural changes, such as anticlines, faults, and stratigraphic unconformities, the relationship between their formation time and the regional oil and gas migration time has a great impact on the effectiveness of the trap. If a basin only experiences one large-scale oil and gas migration, traps formed before this are favorable for oil and gas accumulation, otherwise, they are ineffective.

Many basins often develop multiple sets of effective source rocks and multiple periods of structural movement, accompanied by multiple periods of oil and gas migration and trap formation. In this case, the last tectonic movement that determines the current structural characteristics of the basin controls the last regional oil and gas migration time. Structural traps formed before this and not destroyed, and inherited structural traps, are favorable for oil and gas accumulation. While newly formed traps often become ineffective empty traps. If the crustal movement is very intense, changing the original structural appearance of the basin and the conditions of early traps, breaking the original balance of oil and gas accumulation, oil and gas undergo regional migration and redistribution, then new traps with good oil and gas migration pathways on the up-dip direction of the original oil and gas reservoirs often become effective traps. While early traps that accumulated oil and gas, if the trap conditions are destroyed, oil and gas escape, may become ineffective traps. Therefore, the effectiveness and ineffectiveness of traps cannot be completely determined solely by the time of formation.

The phenomenon of traps becoming ineffective due to the trap formation time being later than the oil and gas migration time can be represented by the Qingcao Bay structure in the Jiuquan Basin. As shown in Fig. 5.31, both the Laojun Temple and Qingcao Bay anticlines in the Jiuquan Basin are in the southern structural zone, and they have similar anticline traps in the Tertiary strata. The drilling results show that the Laojun Temple anticline has rich oil and gas reservoirs, while no oil and gas accumulation was found in the Qingcao Bay anticline. After comparing the geological development history of the two anticline structures, it was found that in addition to being related to lithological changes, the correspondence between the trap formation time and the regional oil and gas migration time is a very important reason. The last regional oil and gas migration time in the Jiuquan Basin was the Pliocene, at which time the Laojun Temple anticline had already formed, and oil and gas accumulated in it, forming rich oil and gas reservoirs. However, the Qingcao Bay anticline trap was formed only at the end of the Pliocene, when the regional oil and gas migration had ended, lacking oil and gas sources, and its elevation is also lower than that of the Laojun Temple anticline, which cannot cause oil and gas to migrate again; therefore, the Qingcao Bay anticline trap is ineffective for oil and gas accumulation and no oil and gas reservoirs are formed (Zhang et al., 1999).

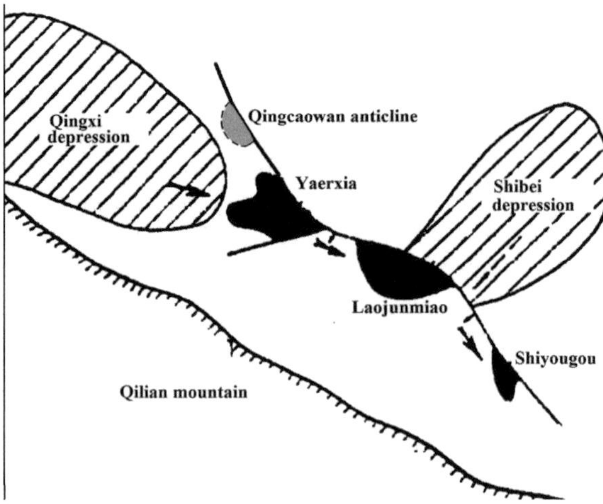

Fig. 5.31 Oil and gas accumulation regional schematic diagram of Qingcao Bay-Laojun Temple in Jiuquan Basin

(4) The Impact of Hydrodynamic Effects and Fluid Properties on Trap Effectiveness

Under normal hydrostatic pressure conditions, the oil–water (or gas–water) interface in the trap is horizontal. Under hydrodynamic conditions, formation water flows along the tilt direction of the potentiometric surface, and the oil–water (or gas–water) interface in the trap also tilts in the direction of water flow. The degree of tilt depends on the intensity of hydrodynamic effects and the density difference of the fluids. As the intensity of hydrodynamic effects increases, the tilt angle of the oil–water (or gas–water) interface gradually increases. When the tilt angle exceeds the tilt angle of the rock layer on the down-dip side in the direction of water flow, the trap that originally accumulated oil or gas becomes ineffective (Fig. 5.32).

In addition to the intensity of hydrodynamic effects, the tilt angle of the oil–water (or gas–water) interface is also related to the properties and density difference of the fluids. As shown in Fig. 5.32, under the action of hydrodynamic effects, the oil–water interface tilts in the direction of water flow.

L is the distance between Well 1 and Well 2. Δh is the pressure surface height difference between Well 1 and Well 2. ΔZ is the height difference of the oil–water interface between Well 1 and Well 2. α is the dip angle of the reservoir layer on the side towards the water flow. β is the dip angle of the potentiometric surface. γ is the dip angle of the oil–water interface.

For oil reservoirs, the dip angle of the oil–water interface can be calculated by the following formula:

$$\tan\gamma_o = \frac{\rho_w}{\rho_w - \rho_o}\cdot\tan\beta = \frac{\rho_w}{\rho_w - \rho_o}\cdot i$$

where ρ_w and ρ_o are the densities of water and oil, respectively, and i is the hydraulic gradient.

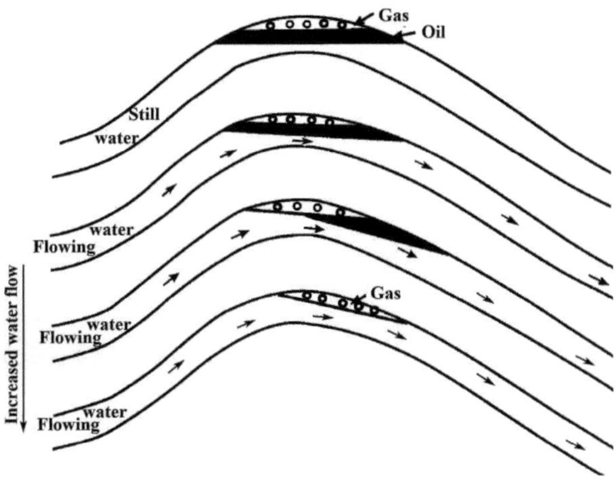

Fig. 5.32 Schematic diagram of the inclined oil–water interface under hydrodynamic conditions

For gas reservoirs, the dip angle of the gas–water interface is calculated by the following formula:

$$\tan \gamma_g = \frac{\rho_w}{\rho_w - \rho_g} \cdot i$$

where ρ_g—is the density of natural gas.

From the above two formulas, under the action of hydraulic gradient and fluid density difference, the effectiveness of a trap for oil accumulation is different from that for gas accumulation, as shown in Figs. 5.32 and 5.33. Assuming the hydraulic gradient remains unchanged, the fluid density directly affects the effectiveness of the trap. Assuming the densities of gas, oil, and water are $\rho_g = 0.001$ g/cm^3, $\rho_o = 0.8$ g/cm^3, $\rho_w = 1$ g/cm^3, then according to the above two formulas, the calculation results are: $\tan\gamma_g = I$, $\tan\gamma_o = 5i$. That is, under the same hydraulic gradient conditions, because the density of natural gas is smaller than that of oil, the dip angle of the oil–water interface is equivalent to 5 times the dip angle of the gas–water interface.

Originally, effective traps that have accumulated oil and gas under hydrostatic conditions, under the action of hydrodynamic effects, whether accumulating oil or gas, can accumulate oil or gas in the trap only if the dip angle of the rock layer on the side towards the water flow is greater than the dip angle of the oil–water (or gas–water) interface. Under the same hydrodynamic conditions, for the same trap, the gas–water interface dip angle may be less than the dip angle of the rock layer on the side towards the water flow ($\gamma_g < \alpha$), and natural gas can accumulate to form a gas reservoir, which means the trap is effective for gas accumulation. However, the oil–water interface dip angle may be equal to or greater than the dip angle of the rock layer on the side towards the water flow ($\gamma_o \geq \alpha$), and oil will be washed away by water, resulting in the trap being filled with water and ineffective for oil accumulation, and the oil reservoir is completely destroyed.

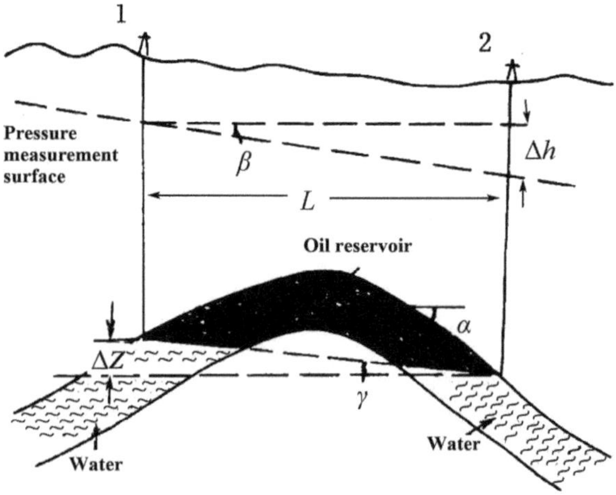

Fig. 5.33 The relationship between hydraulic gradient and trap effectiveness

In addition, in some structural positions that cannot form traps under hydrostatic conditions, such as flexural zones, under the action of strong water dynamics, hydrodynamic traps can be formed.

5.3.4 Good Preservation Conditions

From the perspective of the oil and gas reservoir formation process, in geological history, if a region has sufficient oil and gas sources, favorable source-reservoir-seal combinations, effective traps, and other basic geological conditions, oil and gas reservoirs may be formed. However, as a fluid mineral resource, oil and gas are easily migrated and lost in the crust. The formation of oil and gas reservoirs is a temporary stable state in the geological environment where they are located, and once the environment changes, oil and gas will migrate and be lost or their composition will change. Therefore, whether the formed oil and gas reservoirs can be preserved, especially whether the oil and gas reservoirs formed earlier can be preserved, depends on the changes in the geological environment after the formation of the oil and gas reservoirs, that is, the degree of adjustment and destruction.

Oil and gas reservoirs formed in geological history will inevitably undergo corresponding changes in the hydrocarbons as time and the environment change. Crustal movements causing folding changes, structural uplift, and magmatic activities play a key role in the preservation and destruction of oil and gas reservoirs, and many other factors are derived from the folding and uplift caused by crustal movements.

After the formation of oil and gas reservoirs, folding and uplift caused by crustal movements break the equilibrium state of the oil and gas reservoirs, causing oil and gas to migrate again, resulting in partial or complete loss of oil and gas. The formation and reservoir-forming strata of oil and gas are generally below 250 °C. If they are invaded by high-temperature magma, oil and gas will be burned off, so large-scale magmatic activities are not conducive to the preservation of oil and gas reservoirs. Oil and gas reservoirs are usually in a relatively closed reducing environment, and hydrodynamic conditions are relatively stagnant. Active hydrodynamic environments not only wash oil and gas out of the traps but also carry more oxygen in the flowing formation water, causing oxidation of oil and gas, forming viscous oil and asphalt, thus causing oil and gas reservoirs to lose their commercial value. Therefore, after the formation of oil and gas reservoirs, relatively stable crustal movements, no large-scale uplift, not intense faulting and folding, relatively stable magmatic activities, and relatively closed and stagnant hydrodynamic environments are all important conditions for the preservation of oil and gas reservoirs.

5.3.5 Differences in Formation Conditions Between Gas Reservoirs and Oil Reservoirs and Key Conditions for the Formation of Unconventional Oil and Gas Reservoirs

(1) Differences in Formation Conditions Between Gas Reservoirs and Oil Reservoirs

Petroleum and natural gas are both fluids and are easily migrated. Their generation and migration have the same origin, and the formation and preservation conditions for gas reservoirs and oil reservoirs are similar. Whether it is a gas reservoir (referring to gas reservoirs dominated by hydrocarbon gases) or an oil reservoir, their formation first requires hydrocarbon sources provided by source rocks. Oil and gas migrate in porous and permeable conduit layers, undergo the process of enrichment from dispersion to concentration, form hydrocarbon deposits, and require relatively stable later preservation conditions. Therefore, when studying the formation and preservation conditions of oil and gas reservoirs, oil reservoirs and gas reservoirs are often discussed together. However, there are both similarities and many differences between the mechanisms of petroleum and natural gas reservoir formation, and there are certain differences between traditional oil reservoir conditions and natural gas reservoir conditions (Table 5.4).

Overall, compared with the formation conditions of oil reservoirs, the formation of gas reservoirs has the following main characteristics: (1) multiple and widespread hydrocarbon sources, often multi-source composite reservoirs. (2) Low requirements for reservoirs but high requirements for seals. (3) Easy to migrate and various migration and accumulation modes, in addition to filtration, diffusion migration is also an important mode, and can also form water-soluble gas reservoirs or form water-soluble gas reservoirs through water dissolution of gas. (4) High preservation conditions are required, low accumulation efficiency, and large diffusion loss, the formation of gas reservoirs is always in a dynamic balance of aggregation and dispersion, if the diffusion loss of natural gas in the trap is greater than the injection volume, the gas reservoir will gradually depletion, so late-stage gas reservoir formation is more conducive to the preservation of gas reservoirs (Jiang & Zhang, 2000).

(2) Key Conditions for the Formation of Unconventional Oil and Gas Reservoirs (Tight Oil and Gas, Shale Oil and Gas)

The above discussion is about the basic geological conditions for the formation and preservation of conventional oil and gas reservoirs, among which sufficient oil and gas sources and effective traps are the most important conditions for reservoir formation, and the accumulation of oil and gas in traps complies with the principle of gravity differentiation. The accumulation of oil and gas in unconventional oil and gas reservoirs does not comply with the principle of gravity differentiation, and even does not need traditional traps. The state of oil and gas storage underground and the mechanism of reservoir formation are significantly different from conventional oil and gas reservoirs.

Table 5.4 Differences in formation and preservation conditions between natural gas reservoirs and oil reservoirs

Content	Natural gas reservoirs	Oil reservoirs
Hydrocarbon source	Wide, multi-source, multi-stage. Both organic and inorganic gases. Various organic matters generate natural gas at different stages of evolution. Multi-source natural gas composite reservoirs	From sapropelic and humic-sapropelic organic matters. Mainly generated in the oil window at a certain burial depth
Reservoir and seal conditions	Low requirements for reservoirs, high requirements for seals. Various sealing mechanisms for seals, hydrocarbon concentration sealing can play an important role	High requirements for reservoirs, low requirements for seals. Sealing mechanisms for seals are physical property sealing and abnormal pressure sealing
Migration mode	Easy to migrate and various modes: filtration, pulse-like mixed-phase flow, diffusion, water-soluble convection, among which diffusion and water-soluble convection are important migration mechanisms	Mainly filtration and pulse-like mixed-phase flow
Accumulation mechanism	Diverse: free natural gas directly displaces formation water to form a reservoir, oil-accumulated traps are displaced by natural gas to form a reservoir, water-soluble gas exsolution forms a reservoir, and gas-rich formation water can form a water-soluble gas reservoir	Relatively simple. Free-phase petroleum displaces formation water to accumulate and form a reservoir
Evolution and preservation conditions	Easy to disperse, diffusion loss is important. The formation of gas reservoirs is always in a dynamic balance of aggregation and dispersion, and late-stage gas reservoir formation is conducive to the preservation of gas reservoirs. Low accumulation efficiency	Mainly filtration loss. Diffusion loss is not very important, and the accumulation efficiency is relatively high

For the formation of tight sandstone and shale oil and gas reservoirs, due to the tight reservoirs and poor physical properties, technical transformation is required to have production capacity. The reservoir conditions are low, and the characteristics of short-distance migration and accumulation make source-reservoir coexistence and near-source accumulation favorable conditions for the large-scale accumulation of tight rock oil and gas. Therefore, the most critical conditions for the formation of tight sandstone oil and gas and shale oil and gas are sufficient oil and gas sources and good sealing preservation conditions, while the reservoir rock has good brittleness, which is easy to transform to produce fractures. The spatial combination relationship between source rocks and reservoirs is crucial for the enrichment of unconventional oil and gas. The main controlling factors for oil and gas reservoir formation have evolved from the "source control" of conventional oil and gas to the "advantageous source-reservoir combination and preservation conditions" under source control for unconventional oil and gas, as well as the "dual enrichment" theory of shale gas, that is, the development of high-quality mudstone is the basis for "hydrocarbon control of reservoirs" of shale gas, and good preservation conditions are the key to "reservoir control of production" of shale gas (Guo, 2014).

High-quality source rocks are the material basis for the formation of large-scale unconventional oil and gas reservoirs. High-quality source rocks have high organic matter abundance, good quality, moderately high thermal evolution, and high total hydrocarbon generation. Generally, effective source rocks require TOC > 2%, for oil, the maturity is moderate (R_o is 0.7–1.3%), and for natural gas, usually Ro is greater than 1.5% or 2.0%, with sufficient gas supply and high gas supply rate, so that natural gas can enter the reservoir with poor physical properties on a large scale and displace the pore water that originally occupied the reservoir, forming a large area and low abundance natural gas enrichment area. Of course, the standards for source rocks vary among different basins. As a fluid mineral resource, good sealing preservation conditions are a key condition for the formation of unconventional oil and gas reservoirs. Good top and bottom sealing layers, not intense fractures, can reduce the loss of natural gas, which is conducive to the enrichment and preservation of natural gas. In addition, as a commercial oil and gas accumulation, the development of tight sandstone and shale oil and gas requires fracturing transformation, and the rock properties (such as brittleness) often control the size of oil and gas production capacity, usually requiring the content of brittle minerals in shale (such as quartz, feldspar, etc.) to be more than 40%.

5.4 Destruction and Reformation of Hydrocarbon Reservoirs

In geological history, formed oil and gas reservoirs may change with environmental variations, such as crustal movements, fault activities, and magmatic activities, all of which can cause destruction to the original oil and gas reservoirs. Some oil and gas reservoirs are completely dispersed after being destroyed, while others are partially destroyed and become residual oil and gas reservoirs, with some hydrocarbons re-accumulating in new strata to form so-called secondary oil and gas reservoirs. Therefore, studying the destruction and reformation of oil and gas reservoirs is an essential part of the research on the formation and distribution of oil and gas reservoirs in hydrocarbon-bearing basins and is also an important aspect of oil and gas exploration evaluation in a region.

5.4.1 Destruction of Oil and Gas Reservoirs

The destruction of an oil and gas reservoir refers to the process where the oil and gas within a reservoir, due to changes in the geological environment, partially or entirely dissipate or become viscous oil and asphalt. The main geological factors causing the destruction of oil and gas reservoirs include: the integrity of traps being compromised by crustal movements, faults cutting through oil and gas reservoirs causing upward migration of hydrocarbons, structural uplift leading to the erosion and destruction of the cap rock, shallow burial depth causing oxidation and biodegradation of oil, hydrodynamic erosion and water washing, magmatic activity, and so on.

Macgregor (1996) classified the destruction of oil reservoirs into three major categories: vertical leakage, lateral leakage, and compositional changes, further divided into eight mechanisms: fault leakage, erosion, overpressure leading to ineffective cap rock sealing, trap tilting, hydrodynamic flushing, gas washing, biodegradation, water washing, and cracking (Table 5.5). Different types of destruction occur at different depths; for example, shallow destruction is mainly due to fault leakage, erosion, hydrodynamic flushing, biodegradation, and water washing. The primary destruction mechanisms include fault activity, tectonic changes and erosion, hydrodynamic flushing and water washing, and biodegradation.

(1) Fault Activity

Faults generated by crustal movements are the most significant geological cause of oil and gas reservoir destruction. If an originally formed oil and gas reservoir is cut by an active fault, the sealing conditions are destroyed, the equilibrium conditions of the oil and gas reservoir are disrupted, and hydrocarbons migrate upward along the fault into other reservoirs or escape to the surface, causing partial or complete destruction of the crude oil and gas reservoir. Surface oil and gas seeps indicate the presence of underground oil and gas reservoirs and are a direct sign for early oil exploration.

(2) Tectonic Changes and Erosion

Tectonic changes and other movements caused by crustal movements can alter the trap conditions of originally formed oil and gas reservoirs, such as the integrity being compromised or the spill point of the trap being raised, causing the volume of the trap to decrease and hydrocarbons to laterally overflow from it. These overflowing hydrocarbons may re-accumulate to form oil and gas reservoirs or directly expose on the surface to form oil and gas seeps. Crustal movements change the inclination direction of strata, resulting in the destruction of the integrity of original oil and gas reservoirs and their traps, leading to the re-distribution or re-formation of oil and gas reservoirs.

Intense magmatic activity caused by crustal movements is detrimental to the preservation of oil and gas reservoirs. When high-temperature magma invades oil and gas reservoirs, it can burn off the hydrocarbons. Therefore, large-scale magmatic

Table 5.5 Destruction mechanisms affecting oil reservoir preservation (according to Macgregor, 1996)

Destruction type	Mechanism	Evidence	Depth range	Example
Vertical leakage	Fault leakage: faults destroy the original cap rocks, causing hydrocarbons to migrate and accumulate at higher positions or escape to the surface	Oil seeps related to faults	Shallow to medium	Iran/Iraq Zagros (Beydoun et al., 1992; Dunnington, 1958)
		Hydrocarbon-bearing layers that are trapped and sealed under good sealing conditions		
	Erosion: surface erosion related to evaporation, oxidation, and leakage through surface fault systems	Surface oil seeps, oil stains	Very shallow or surface	Athabasca tar sand exposure zone (Wilson et al., 1973); Pre-Cretaceous of Zagros (Bedoun et al., 1992)
		Outcropping bituminous sand		
		Modified bituminous sediments		
	Overpressure: failure of physical or hydrodynamic sealing	Mud volcanoes and associated oil seeps	General depth	Mud volcanoes, such as S. Caspian; many oil fields in North Sea/ Haltenbanken
		Residual oil and gas shows within the trap under fracturing pressure conditions		
		Oil and gas show within the caprock		
		Young trapped oil reservoirs, and re-trapped hydrocarbon-bearing strata		
Lateral leakage	Trap tilting: original traps tilt due to compression or tilting, causing hydrocarbons to leak at spill points	Evidence of re-migration (hydrocarbons in younger traps)	Shallow or deep	Alaska (Carman & Hardwick, 1983); Papuan thrust fault zone (Earnshaw et al., 1993)
		Exposure of the conduit layer		
		Remaining oil in a compromised trap		

(continued)

Table 5.5 (continued)

Destruction type	Mechanism	Evidence	Depth range	Example
	Hydrodynamic flushing: the action of atmospheric water	Residual oil in reservoirs with freshwater in effective traps	Shallow (requires atmospheric water)	Various examples in North America (Dahlberg, 1982); Illiza, Algeria (1996)
		Tilted oil–water interface		
		Hydrodynamic trap		
	Gas washing: the proliferation or expansion of gas caps	Leaky gas fields with gas cap development	Deep (occasionally shallow)	Alberta (Gussow, 1953); Timan Pechora, Russia (Bogatsky & Pankratov, 1993); Illizi, Algeria (1996)
		An increasing tendency in gas volume within oil fields close to hydrocarbon source rock kitchens		
Compositional changes	Biodegradation, hydrodynamic flushing: bacterial decomposition or dissolution of certain components, separating light fractions	Presence of heavy oil or tar sand	Shallow (requires atmospheric water)	Orinoco tar belt (Demaison, 1977); Alberta tar sand (Creany & Allan, 1990; Masters, 1984); Ugnu, Alaska (Carman & Hardwick, 1983)
		During the oil field charging period, crude oil degradation leads to the widespread presence of heavy oil in the oil field		
		Changes in oil accumulation leading to a decrease in API values towards OWC		
		Geochemical evidence of light hydrocarbon loss		
	Cracking: conversion of crude oil to gas or condensate oil under high-pressure conditions	Gas or condensate gas associated with pitch asphalt	Deep	Elmworth/Rainbow, Alberta (Masters, 1984); Deep Permian/Delaware Basin (Holmquest, 1965)
		The absence of crude oil above a certain temperature		

activity is not conducive to the preservation of oil and gas reservoirs and ultimately leads to their destruction.

Crustal uplift and erosion cause the overlying strata of oil and gas reservoirs to be strongly eroded, and even the cap rock is eroded, exposing the hydrocarbon-bearing strata directly on the surface, leading to the complete destruction of oil and gas reservoirs, with a significant loss of light components and the remaining oil being oxidized and degraded into heavy oil and tar sand. The heavy oil and tar sand of Athabasca are typical examples. Similarly, the oil sand mountains in the Qaidam Basin are due to crustal movements that severely damaged the original oil and gas reservoirs, exposing the Paleogene-Neogene oil-bearing strata on the surface and causing erosion and weathering; the tar sand of the Silurian in the Tarim Basin is also the result of crustal movements damaging ancient oil reservoirs. The petroleum ditch oil field in the western Jiuquan Basin, its Paleogene-Neogene Baiyanghe Formation oil and gas reservoir, was severely damaged by the Himalayan orogeny, causing a massive loss of crude oil to the surface.

(3) Hydrodynamic and Water Washing Effects

Hydrodynamic environments have a significant impact on the preservation of oil and gas reservoirs. Active hydrodynamic environments can wash hydrocarbons out of traps, leading to the destruction of oil and gas reservoirs (Fig. 5.33). At the same time, during the flow of groundwater along the oil–water interface, water washing occurs, where groundwater dissolves some easily soluble components of oil, especially light hydrocarbon components, thereby thickening and weighting the crude oil. The stronger the hydrodynamic force and the shallower the burial, the greater the destruction of oil and gas reservoirs by hydrodynamic flushing and water washing. Therefore, a relatively stable, stagnant hydrodynamic environment is one of the important conditions for the preservation of oil and gas reservoirs. Typically, $CaCl_2$-type oil field water with closed characteristics is more favorable for the preservation of oil and gas reservoirs than Na_2SO_4-type oil field water with open characteristics.

(4) Biodegradation

In areas where oil and gas reservoirs are shallowly buried, there are relatively more oxygen and microorganisms in groundwater. Microorganisms selectively consume certain hydrocarbon components, causing changes in the composition of crude oil, which is known as biodegradation. Under the action of microorganisms, light components in oil and gas reservoirs are preferentially consumed, causing the crude oil to become viscous and heavy. Like water washing, this effect generally occurs in shallow layers or areas connected to the surface where atmospheric water intrudes.

5.4.2 *Reformation of Oil and Gas Reservoirs*

(1) The Concept of Primary and Secondary Oil and Gas Reservoirs

The formation of oil and gas reservoirs is a dynamic balance process of hydrocarbon migration and accumulation. Oil and gas reservoirs that have already formed are in a relatively balanced state in the crust. Post-formation tectonic movements can disrupt this balance, causing hydrocarbons to redistribute and reach a new relative balance. After the modification of oil and gas reservoirs, some may be destroyed with all hydrocarbons dissipated. Others may only be partially destroyed, leaving a portion in the original trap. In addition to the component destruction of oil and gas reservoirs caused by biodegradation and water washing, most other types of destruction are accompanied by the outward migration of hydrocarbons from the original oil and gas reservoirs, either vertically or laterally. The hydrocarbons migrated from the original oil and gas reservoirs may re-accumulate in part, and the rest may be lost in the strata or on the surface, becoming hydrocarbon shows. When the cap rock of an oil and gas reservoir is eroded, it can become tar sand or tar lakes. Therefore, the destruction of oil and gas reservoirs may result in various outcomes (Fig. 5.34).

Thus, we can refer to the oil and gas reservoirs formed for the first time in traps through primary migration and secondary migration from a dispersed state as primary oil and gas reservoirs, and those formed by the re-migration and re-accumulation of hydrocarbons after the destruction of primary oil and gas reservoirs as secondary oil and gas reservoirs. Under normal circumstances, secondary oil and gas reservoirs are distributed above or in the up-dip direction of primary oil and gas reservoirs, formed by the accumulation of hydrocarbons transferred from primary oil and gas reservoirs in relatively shallow suitable traps.

As shown in Fig. 5.35, A represents the original oil and gas reservoir. After being disrupted by faults, natural gas and a small amount of oil migrate along the faults to the surface, forming oil and gas seeps. Some of the oil is transported to the C trap through the fault, accumulating to form a secondary oil reservoir. B represents the residual oil reservoir formed after the original oil and gas reservoir A has been destroyed.

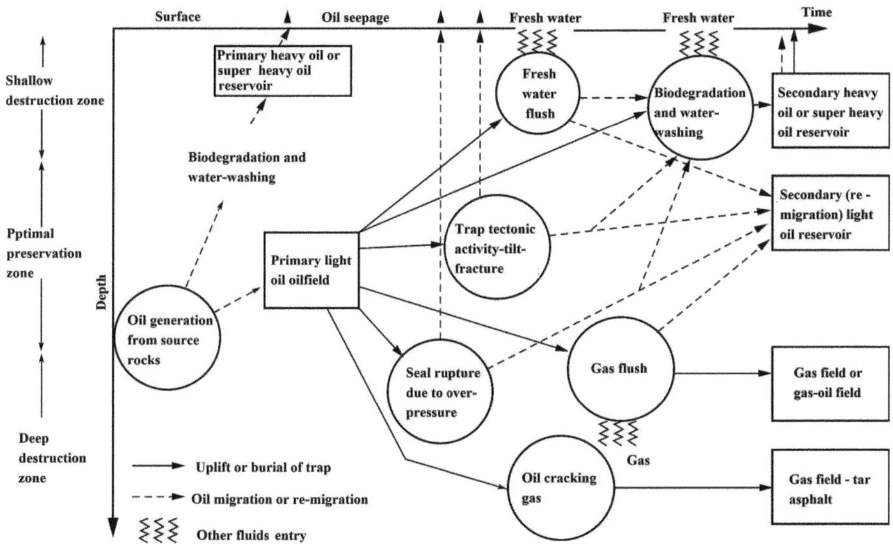

Fig. 5.34 The effects of various destructive reservoir actions and their evolutionary outcomes (according to Macgregor, 1996)

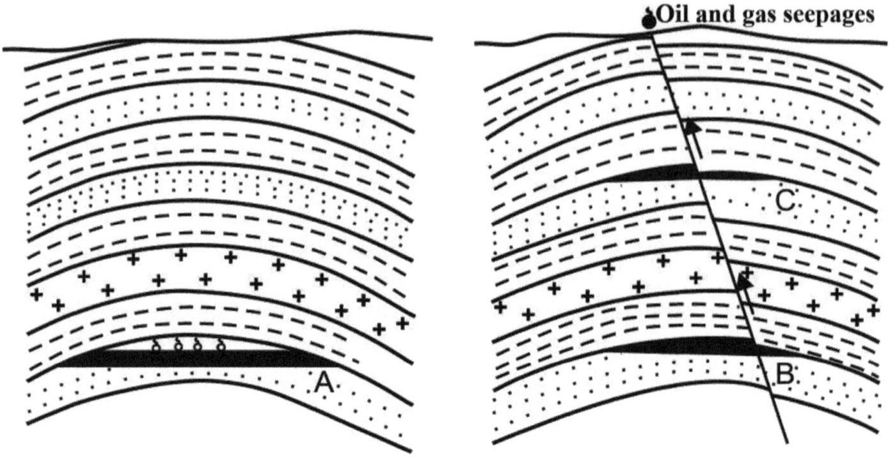

Fig. 5.35 Formation model of primary and secondary oil and gas reservoirs

There is another understanding of the concepts of primary and secondary oil and gas reservoirs. Primary oil and gas reservoirs refer to those found in the source rock series, where hydrocarbons originate from the same stratum and are self-generated and self-stored. Those distributed in non-source rock series are called secondary oil and gas reservoirs, which are formed by the migration and accumulation of dispersed and concentrated oil and gas from the source rock series to traps in non-source rock series. For example, the phrase "formation and preservation conditions of primary oil and gas reservoirs in the Proterozoic or Paleozoic of the North China Basin" refers to oil and gas reservoirs that are self-generated and self-stored within the stratum. The Paleogene in the Bohai Bay Basin has good conditions for oil and gas generation, storage, and preservation, and the oil and gas reservoirs formed there should be primary oil and gas reservoirs. The Neogene does not have oil and gas generation conditions, and the oil and gas reservoirs there are mostly secondary oil and gas reservoirs formed by the migration of oil and gas from the underlying Paleogene primary oil and gas reservoirs to the shallow layers through faults.

Generally, secondary oil and gas reservoirs are the product of re-migration and re-accumulation after the destruction of primary oil and gas reservoirs. In actual work, it is sometimes difficult to strictly distinguish between primary and secondary oil and gas reservoirs. Hydrocarbons in a destroyed primary oil and gas reservoir may be entirely transferred, causing the complete destruction of the primary oil and gas reservoir, or only a part may be transferred, reducing the reserves of the primary oil and gas reservoir. Primary oil and gas reservoirs in the source rock series may also maintain or even increase their scale due to the pressure reduction brought by fault activity and the continued supply of hydrocarbons from surrounding source rocks.

Additionally, hydrocarbons in ancient buried hills mostly come from the overlying new sedimentary series. From the perspective of hydrocarbon migration and accumulation, they are often the first to accumulate and should be called primary oil and gas reservoirs. However, in terms of source-reservoir configuration, hydrocarbons are stored in non-source rock series and can be referred to as secondary oil and gas reservoirs. For example, the Renqiu buried hill oil and gas reservoirs in the Bohai Bay Basin are stored in the Sinian and Lower Paleozoic carbonate reservoirs. Hydrocarbons generated from the overlying Paleogene source rocks enter the buried hills through unconformities and accumulate to form reservoirs, so they can be called both primary and secondary oil and gas reservoirs.

(2) Patterns of Oil and Gas Reservoir Reformation

There are two main patterns for the formation of secondary oil and gas reservoirs from the transformation of primary oil and gas reservoirs:

a. Fault activity disrupts the integrity of the original trap and the equilibrium conditions of the oil and gas reservoir, causing hydrocarbons to migrate upward along the fault and re-accumulate to form new oil and gas reservoirs.

If the primary oil and gas reservoir does not continue to receive hydrocarbons after its formation, that is, there are no new sources of oil and gas later, then after the primary oil and gas reservoir is destroyed by faults, the hydrocarbons in the primary oil and gas reservoir partially or entirely migrate along the faults to form secondary oil and gas reservoirs in the shallower parts, accompanied by partial loss of oil and gas. In this case, the formation of secondary oil and gas reservoirs is predicated on the destruction of primary oil and gas reservoirs, and it can be considered that faults are generally destructive to the formation and preservation of oil and gas reservoirs.

If the primary oil and gas reservoir is in a source rock series with abnormally high pressure, and fault activity coincides with the period of large-scale hydrocarbon migration and reservoir formation, the primary oil and gas reservoirs in the lower source rock series act as a "temporary storage layer" or "transit station" for hydrocarbons. With fault activity, hydrocarbons migrate from the primary oil and gas reservoirs along the faults to the shallower parts. The abnormal pressure in the primary oil and gas reservoirs is released, and the surrounding source rocks still have abnormally high pressure. Driven by high pressure, the hydrocarbons can re-charge the primary oil and gas reservoirs; with further fault activity, hydrocarbons migrate upward again (Fig. 5.36). Due to multiple periods of fault activity, hydrocarbons can aggregate and disperse along the faults multiple times, forming multi-layered oil and gas reservoirs distributed along the faults. In this case, without fault activity, hydrocarbons cannot be transferred on a large scale from the deep to the shallow strata, and faults are generally constructive to the formation of oil and gas reservoirs (Jiang et al., 2014).

This constructive faults in the formation of secondary oil and gas reservoirs is exemplified in the Gulf of Mexico Basin and the Paleogene-Neogene of the Bohai Bay Basin in eastern China. For instance, the largest complex fault block oil and gas field in the Bohai Bay Basin—the Dongxin Oil and Gas Field—has experienced long-term and multiple fault activities, causing multiple aggregations and dispersions of oil and gas along the faults. The original oil and gas reservoirs were destroyed multiple times, and new oil and gas reservoirs were formed multiple times, resulting in multi-layered oil and gas reservoirs distributed comb-like along the faults, with the main controlling fault segments containing oil and gas wells reaching hundreds of

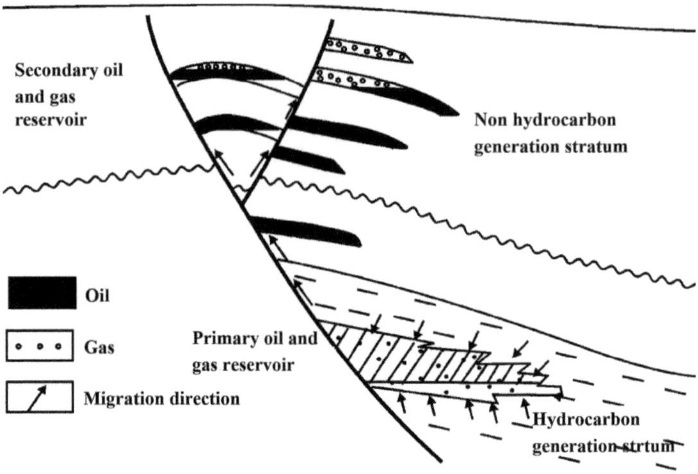

Fig. 5.36 Hydrocarbon migration along faults leading to the formation of secondary oil and gas reservoirs

meters to 2000 m (Jiang & Zhang, 1999). Due to the different activities and sealing properties of faults, complex phenomena such as "alternating oil and water" and "alternating light and heavy" occur vertically. Another example is the Gangdong Oilfield in the Huanghua Depression, where the primary oil and gas reservoirs in the Shahejie Formation of the Paleogene source rock series are transported upward along the faults in the Neogene to form secondary oil and gas reservoirs.

The Neogene oil and gas reservoirs in the Bohai Bay Basin are secondary oil and gas reservoirs, with hydrocarbons originating from the Paleogene. The formation of oil and gas reservoirs is mainly controlled by faults, that is, the formation of Neogene oil and gas reservoirs is closely related to the conductive capacity of faults. The conductive capacity of oil source faults is the result of the combined action of multiple factors. The longer the fault extends, the greater the dip angle, the stronger and longer the activity during the main oil and gas accumulation period, and the stronger the hydrocarbon supply capacity of the cut source rocks, the stronger the fault's ability to conduct oil and gas, which is conducive to the enrichment of oil and gas in the Neogene (Jiang et al., 2022). Through extensive exploration practices in the Bohai Sea area of the Bohai Bay Basin, Xue et al. (2021) proposed a fault-controlled "converging ridge" shallow oil and gas reservoir formation model (Fig. 5.36). The converging ridge refers to the ridge-like deep geological body beneath the shallow structure, which is a low potential area where oil and gas migrate and converge through faults and sand bodies. When the converging ridge communicates the deep and shallow layers through fault activity, oil and gas migrate vertically upward along the faults and accumulate in the shallow sand bodies to form reservoirs. If the storage space of the converging ridge is large enough, the converging ridge itself can form deep oil and gas reservoirs and control the enrichment of shallow oil and gas. If the converging ridge does not have sufficient storage space, it acts as a temporary

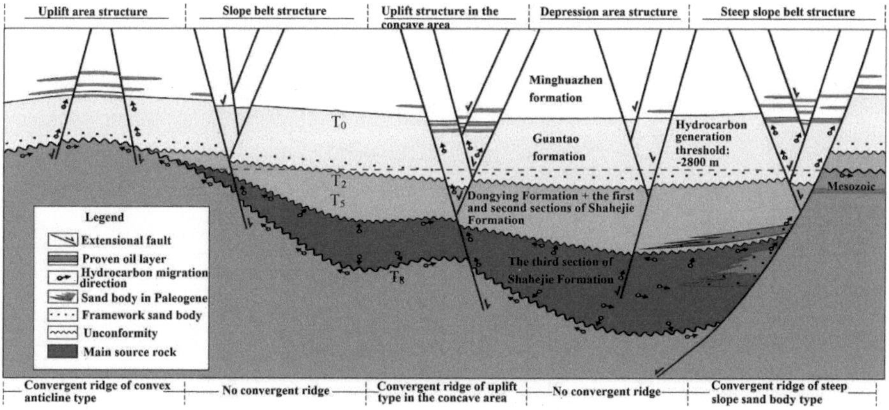

Fig. 5.37 The development of converging ridges in the Bohai Sea area and the enrichment of shallow oil and gas (Xue et al., 2021)

storage area, mainly controlling the migration and accumulation of shallow oil and gas (Fig. 5.37).

b. Crustal movements have altered the original shape of the traps, causing some of the oil and gas to overflow outward or to be entirely transferred, where they accumulate in new traps to form reservoirs.

Subsequent crustal movements generate new traps and raise the spill points of the original traps, while the amplitude of the newly formed traps is larger. The hydrocarbons in the original oil and gas reservoirs will escape from the spill points and re-accumulate in the new traps to form new oil and gas reservoirs. Some of the hydrocarbons in the original oil and gas reservoirs may escape partially, or they may all escape, depending on the degree to which the spill points of the original traps are raised and the strength of the hydrodynamic action.

Subsequent crustal movements can change the inclination direction of large monocline strata, causing the re-distribution and re-accumulation of hydrocarbons within the trap, which is also a reformation of oil and gas reservoirs. For example, the formation of the Sinian gas reservoir in the Weiyuan Gas Field in the Sichuan Basin is believed to be formed after the destruction of the original Ziyang gas reservoir (Fig. 5.38). During the Indosinian period, ancient oil reservoirs were formed in both Weiyuan and Ziyang areas, but due to the lower structural amplitude of Weiyuan and the larger anticline of Ziyang, hydrocarbons were mainly accumulated in the Ziyang area. However, after the Yanshan period, with the gradual disappearance of the Ziyang ancient trap and the increasing size of the Weiyuan anticline during the Himalayan period, Weiyuan structure was able to accumulate natural gas, mainly capturing the natural gas that migrated from the disintegrated ancient gas reservoir in Ziyang (Jiang et al., 2001).

In oil and gas basins with frequent crustal movements, the formation process of oil and gas reservoirs is often very complex. They may go through several cycles of formation-destruction-reformation before reaching the current state of oil and

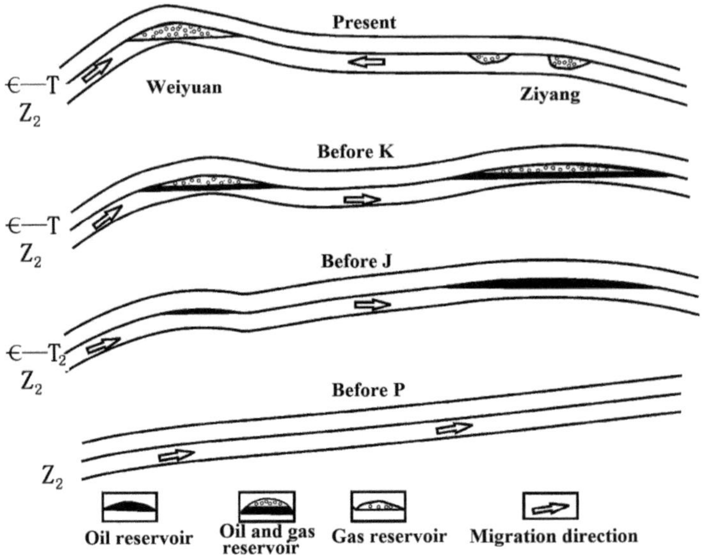

Fig. 5.38 Formation model of Weiyuan gas field in Sichuan Basin

gas reservoirs. Exploration practices have shown that the preservation conditions of oil and gas reservoirs after their formation are a key factor in determining whether they still have value for oil and gas exploration. The long-term lack of significant breakthroughs in oil and gas exploration in the Lower Yangtze region may be directly related to the poor preservation conditions of oil and gas.

5.5 The Period of Oil and Gas Reservoir Formation

The analysis of the formation period of oil and gas reservoirs is one of the important contents in the study of oil and gas reservoir formation. Determining the period and order of oil and gas reservoir formation is not only of great theoretical significance in studying the formation and distribution of oil and gas reservoirs, but also has important practical value in guiding oil and gas field exploration. In a petroliferous area, if we can determine that the oil and gas reservoirs were formed in a certain geological era, then the traps formed before this period are favorable for oil and gas accumulation. On the contrary, traps formed after this period are generally unfavorable for oil and gas accumulation.

Traditional geological analysis methods for determining the formation time of oil and gas reservoirs are mainly based on the formation time of oil and gas sources and traps required for the formation of oil and gas reservoirs, as well as the saturation pressure after the formation of the oil reservoir. The commonly used methods mainly include three types: the main hydrocarbon expulsion period method of source rocks, the trap formation period method, and the oil reservoir saturation pressure method. In the past twenty years, fluid history analysis methods have gradually become the main means of determining the formation period of oil and gas reservoirs, such as reservoir fluid inclusion analysis technology and authigenic mineral isotope dating, which have been widely applied and have achieved good results. Fluid history analysis methods rely on some observable geological records left in the process of oil and gas reservoir formation, with the help of geochemical and organic petrology technical means, to obtain quantitative data of the formation period, and in conjunction with traditional formation period analysis methods, can more accurately determine the formation period of oil and gas reservoirs. Therefore, the combination of fluid history analysis method and traditional geological method has become a universal method for determining the formation period of oil and gas reservoirs in oil and gas bearing basins.

5.5.1 Traditional Geological Analysis Methods

According to the basic geological conditions for the formation of oil and gas reservoirs, only when oil and gas sources and trap conditions are available can oil and gas reservoirs be formed. Therefore, by analyzing the time when oil and gas sources are generated and expelled, and the time when traps are formed, we can roughly judge the earliest time when oil and gas reservoirs may be formed.

(1) Based on the Main Hydrocarbon Expulsion Period of Source Rocks

The formation of oil and gas reservoirs is the result of oil and gas generation, migration, and accumulation. After oil and gas are generated from source rocks, they may accumulate in traps after primary migration and secondary migration. Therefore, the main period when oil and gas are generated and expelled from source rocks represents the earliest possible period for the formation of oil and gas reservoirs, that is, the formation period of oil and gas reservoirs will not be earlier than the main hydrocarbon expulsion period of source rocks.

In some basins where continuous sedimentation occurs without significant uplift and sedimentation interruption geological events, if only one set of source rocks is developed, the maturation and evolution of organic matter in the source rocks are continuous, and there may only be one main hydrocarbon expulsion period, followed by a continuous process of secondary migration and accumulation of oil and gas. The main hydrocarbon expulsion period of source rocks basically represents the main period of oil and gas reservoir formation. If multiple sets of source rocks are developed, the maturation and expulsion periods of different source rocks are different, that is, there are multiple main expulsion periods, resulting in multiple formation periods. Therefore, in petroliferous basins, the formation period of oil sources in different strata may be single or multiple, and different strata may not be the same.

In many basins, the primary hydrocarbon expulsion of source rocks is terminated due to tectonic uplift, followed by secondary hydrocarbon generation due to subsidence, which may result in two formation periods separated by a long time. If there are multiple sets of source rocks, the situation may be more complex. Therefore, understanding the burial history and thermal evolution history of source rocks is key to determining the formation period of oil and gas reservoirs based on the main hydrocarbon expulsion period of source rocks.

Source rocks may reach the main oil generation period at different times under different geological conditions. In areas with large subsidence amplitude and high geothermal gradients, the time for organic matter to reach the main oil generation period is relatively short. In areas with small subsidence amplitude and low geothermal gradients, the time to reach the main oil generation period is longer. According to statistics, rapid oil generation takes about (5–10) $\times$ 10^6 years, while slow oil generation may take 100 $\times$ 10^6 years or even longer (according to Tissot, 1984). The Neogene source rocks in the Los Angeles Basin in the western United States and the Paleogene source rocks in the Bohai Bay Basin in China may both belong to rapid oil generation. The Lower Silurian source rocks in the Hassi Messaoud oil field in North Africa are examples of slow oil generation. From Fig. 5.39, during the initial 300 Ma period (before the Cretaceous), the burial of source rocks was relatively shallow, and during the Permian, due to the uplift of the basin, the burial became even shallower, resulting in the generation of very little oil. It was only from the Cretaceous that the main oil generation period was reached, and the main formation period occurred in the Cretaceous and later geological periods.

(2) Based on the Formation Time of Traps

The formation of oil and gas reservoirs is the result of oil and gas accumulation in traps. Only when traps are formed can oil and gas accumulate in them. Therefore, the formation time of oil and gas reservoirs will not be earlier than the formation time of traps, that is, the formation time of traps limits the earliest time for the formation of oil and gas reservoirs.

In a region, the formation time of different traps may vary greatly. The formation of a trap can be shortly after the formation of the reservoir, or it may be a long time after the formation of the reservoir. It can be formed during a certain tectonic movement curtain, or it may be formed intermittently during a long geological history; and a trap may also undergo multiple modifications (Zhang, 1999).

By analyzing the sequence of strata and the evolution of paleostructures, planar and cross-sectional analysis diagrams of trap formation and evolution can be made, such as developmental evolution profiles and pagoda diagrams, which can effectively analyze the formation history of traps. Figure 5.40 represents the relative time of trap formation. During the deposition of mudstone in era a, the upward pinchout of the underlying sandstone formed a trap, which is the earliest formed trap here. Trap 2 was formed after the occurrence of a fault, that is, in era b. Later, due to weathering and erosion, a secondary pore zone was formed. When mudstone was deposited above the unconformity, that is, in era c, trap 3 was formed. Trap 4 was formed in a lens-shaped sandstone body or sandbar covered by mudstone in era d. Traps 5, 6,

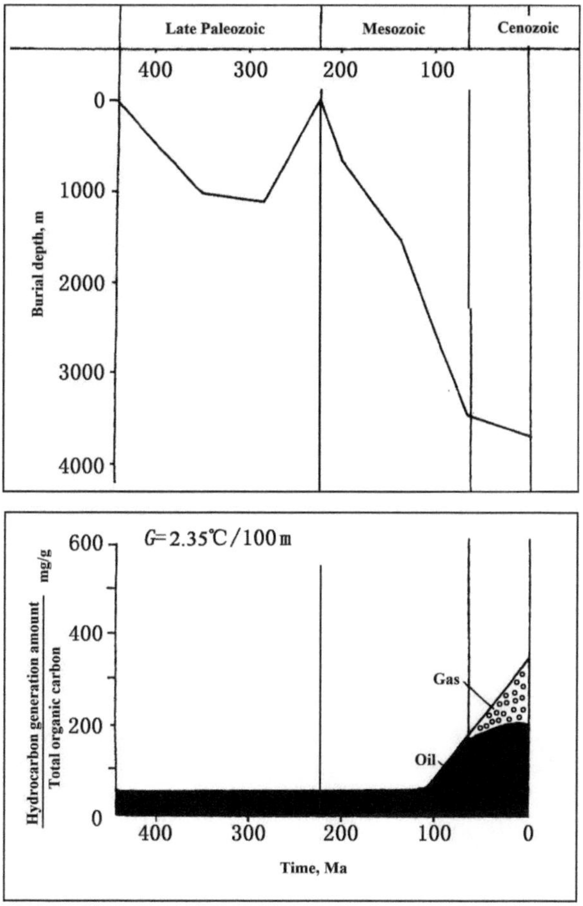

Fig. 5.39 The burial history of Silurian source rocks and the variation of hydrocarbon generation in the Hassi Messaoud oil field area (according to Tissot, 1984)

and 7 were all formed after the deposition of layer e, through folding. Their absolute formation times need to be determined based on the comprehensive research results of paleostructure, lithofacies paleogeography, and absolute age determination.

(3) Based on the Saturation Pressure of Oil Reservoirs

Oil in underground oil reservoirs generally contains a certain amount of natural gas. The pressure at which oil is saturated with natural gas under certain stratum temperatures is called the saturation pressure of the oil reservoir. Since many oil reservoirs are saturated or nearly saturated with natural gas today, according to the fact that oil saturated with natural gas has greater migration capacity, it is believed that oil is saturated with natural gas during its migration and accumulation process. Saturated oil enters traps to form oil reservoirs, and the saturation pressure of the oil reservoir is equal to the stratum pressure at this time, and stratum pressure is related to burial depth. If no free gas enters the oil reservoir after its formation, the

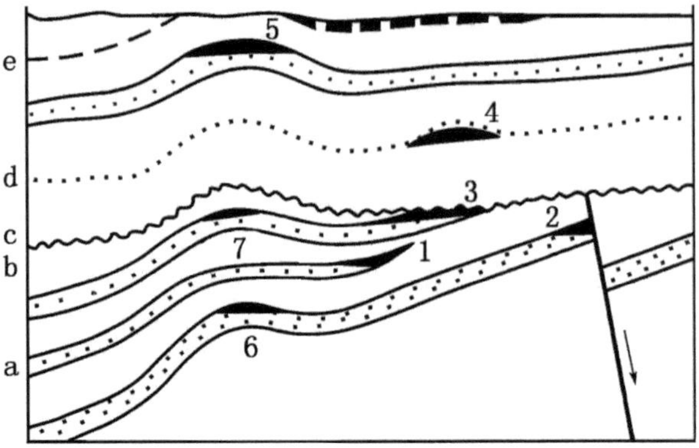

Fig. 5.40 Relative time of trap formation. 1–7: numbers of traps; a–e: stratigraphic era numbers (according to A. I. Levorsen, 1967)

saturation pressure will remain unchanged until today. Therefore, the geological era corresponding to the stratum burial depth equivalent to the saturation pressure is the formation time of the oil reservoir. In other words, from the current oil reservoir, push up to the burial depth equivalent to the saturation pressure, the geological era corresponding to this depth is the formation time of the oil reservoir.

The calculation formula is: $H = p_b/\rho g$.

In the formula, H—the depth of the oil reservoir formation, m; p_b—the saturation pressure of the oil reservoir, Pa; ρ—the density of the stratum water, kg/m^3; g—the acceleration due to gravity, 9.8 m/s^2.

As shown in Fig. 5.41, the saturation pressure of the oil reservoir in the K_1 layer in a certain area is 20 MPa. According to the approximate calculation of hydrostatic pressure, its equivalent stratum burial depth H (assuming the density of water ρ is 1.0×10^3 kg/m^3, and the acceleration due to gravity g is 9.8 m/s^2):

$$H = \frac{p}{\rho g} = \frac{20 \times 10^6}{1 \times 10^3 \times 9.8} = 2041\ (\mathrm{m})$$

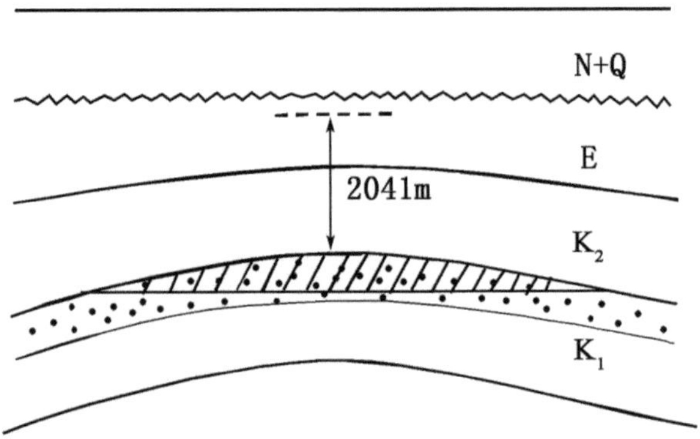

Fig. 5.41 Schematic diagram of using saturation pressure to calculate the formation time of oil reservoirs

Start from the top of the oil reservoir to the upper section of layer E, it can be considered that the oil reservoir in the K1 layer was formed in the late deposition of layer E.

Assuming the oil reservoir was saturated at the time of formation, the stratum pressure at the time of oil reservoir formation is equal to the saturation pressure of the oil reservoir, and the saturation pressure remains unchanged after the formation of the oil reservoir. Therefore, theoretically, the stratum time corresponding to the depth calculated from the saturation pressure of the oil reservoir represents the time of oil accumulation process in the trap, and the saturation pressure method is closer to the true formation time of the oil reservoir than the trap formation time method. However, in practice, the assumptions of this method are strict, some conditions are difficult to meet, and there are certain limitations. The main influencing factors include the following two aspects:

a. The oil reservoir may be oversaturated or undersaturated at the time of formation. If the current oil reservoir was oversaturated at the time of formation, that is, it had a free gas cap oil and gas reservoir, and later the burial depth of the oil and gas reservoir increased, the stratum pressure increased, and the free natural gas was dissolved in oil. At this time, the saturation pressure of the oil reservoir is greater than the original saturation pressure, resulting in an increased depth calculated according to the current saturation pressure of the oil reservoir, that is, later than the actual formation time; if it was undersaturated at the time of formation, the calculated time is earlier than the actual time.
b. After the formation of the oil and gas reservoir, if crustal movement causes the area to be uplifted and the overlying strata are eroded, the temperature and pressure in the oil and gas reservoir will change, thereby changing the magnitude of the saturation pressure. If there is a long-term sedimentation interruption in the overlying stratum profile, the actual oil reservoir may have formed before the sedimentation interruption, but the calculated result may be after the sedimentation interruption.

Therefore, this method is only applicable to oil reservoirs that have not been destroyed and have formed under saturated conditions. It is not suitable for oil and gas reservoirs with gas caps today and oversaturated or undersaturated oil reservoirs in geological history. When using the saturation pressure method to calculate the formation time of oil reservoirs, the impact of various factors must be fully considered, and other methods must be used in conjunction.

5.5.2 *Fluid History Analysis Methods*

Traditional oil and gas reservoir formation period research methods mainly study the effective configuration of various parameters such as source, reservoir, cap, transport, trap, and preservation, and roughly infer or determine the formation period and process based on the structural evolution history, trap formation history, and hydrocarbon source rock generation and expulsion history. There is a lack of research on direct geological records of the oil and gas reservoir formation process.

Oil and gas reservoir formation is a dynamic process in geological history. The interaction between rocks and oil, gas, and water fluids should leave certain fossil records of formation. In the past twenty years, people have been trying to find this more accurate and direct fossil record to determine the oil and gas reservoir formation period, and gradually formed such a consensus: the authigenic minerals in the reservoir and their fluid inclusions directly record the physical and chemical conditions and processes of oil and gas reservoir formation in sedimentary basins. They can be used as fossil records of oil and gas reservoir formation to reconstruct the formation and evolution history of oil and gas reservoirs. On this basis, the experimental testing methods based on geochemistry and petrology have been gradually developed and improved, bringing the study of the formation period into a new stage of quantitative description. There are currently two relatively mature methods: reservoir organic petrology analysis method (reservoir fluid inclusion method) and authigenic mineral isotope geochronology analysis method (authigenic illite dating method).

(1) Reservoir Fluid Inclusion Method

Fluid inclusions are the formation fluids of rock-formation and mineralization captured by minerals during their crystallization process. They are divided into brine inclusions and hydrocarbon inclusions according to composition, and into liquid inclusions, gas inclusions, and liquid–gas inclusions according to phase. Fluid inclusions record the properties, components, physicochemical conditions, and geodynamic conditions of the medium when these authigenic minerals crystallize. Hydrocarbon inclusions are formed by capturing oil and gas in the authigenic minerals in the reservoir during the migration and accumulation process of oil and gas in the reservoir. Their existence indicates that oil and gas charging events have occurred in the reservoir during geological history. Therefore, based on the sequence of oil and gas inclusions in the diagenetic sequence, the relative time of oil and gas charging can be determined. Based on the composition, fluorescence characteristics of hydrocarbon inclusions, and the homogenization temperature of coexisting brine inclusions, combined with the burial and thermal history of the reservoir, the time and period of oil and gas charging can be judged.

During the crystallization process of minerals, the fluids captured in inclusions are mostly in a single liquid phase. After the reservoir samples are brought to the ground, due to the decrease in temperature and pressure, the gases dissolved in the liquid phase are separated to form liquid–gas two-phase inclusions. In the laboratory, when the inclusions are placed on a cold and hot stage and heated until the gas phase disappears, and then restored to a single liquid phase, the temperature at which this occurs is called the homogenization temperature. This temperature represents the temperature at which the inclusions were formed, combined with the burial and thermal history of the reservoir, the temperature that the reservoir experienced at the time of fluid inclusion formation, as well as the corresponding burial depth and geological era, etc., can be determined, thereby judging the time of oil and gas charging.

In theory, only the homogenization temperature of hydrocarbon fluid inclusions can represent the temperature at which hydrocarbons are charged. However, in actual measurement, in addition to directly measuring the homogenization temperature of gas–liquid hydrocarbon or gaseous hydrocarbon inclusions, many fluid inclusions are brine inclusions. By measuring the homogenization temperature of brine inclusions formed simultaneously with hydrocarbon inclusions, the time of hydrocarbon injection can be inferred. Therefore, to use the homogenization temperature of fluid inclusions to infer the formation time of oil and gas reservoirs, the key is to clarify the period of inclusions and their relationship with hydrocarbon inclusions. In addition to measuring the homogenization temperature of fluid inclusions, parameters such as the composition, maturity, and freezing point of inclusions can also be measured to comprehensively judge the oil and gas reservoir formation period and to judge the charging direction and oil and gas source comparison.

The use of fluid inclusion homogenization temperature to judge the oil and gas reservoir formation period has three assumptions: homogeneous system, isochoric system, and closed system. Since oil-phase inclusions are usually not captured under homogeneous phase conditions, their homogenization temperature is usually 10–30 °C lower than that of coexisting brine inclusions. Therefore, brine inclusions coexisting with hydrocarbon inclusions are generally selected to measure their homogenization temperature. Combined with the burial history and thermal evolution history of the inclusion-bearing reservoir, the time of inclusion formation can be determined, and thus the time of oil and gas charging can be determined. The accuracy of this method depends on the division of inclusion formation periods and the precision of homogenization temperature measurement and the accuracy of the reservoir burial and thermal evolution history.

By studying the distribution, phase, type, abundance, fluorescence, and other characteristics of oil and gas inclusions, combined with the formation time sequence and periods of the host minerals of inclusions, the period of oil and gas inclusions can be determined. Micro-thermometric measurements of coexisting brine inclusions of different periods of oil and gas inclusions are carried out, and the oil and gas reservoir formation periods and times are determined by using the homogenization temperature of brine inclusions combined with reservoir burial and thermal history analysis. Liu et al. (2013) used oil and gas inclusion abundance, inclusion petrology, inclusion fluorescence spectroscopy, and micro-thermometric analysis to study the oil and gas charging history of the Ordovician reservoir in the Tazhong area of the Tarim Basin, and believed that there are at least two periods of oil charging, and proposed research procedures and issues to be aware of. That is, first, study the inclusion petrology, that is, divide the inclusion combinations with the help of ultraviolet–visible and fluorescence spectroscopy, then select brine inclusions and oil inclusions within the inclusion combinations to carry out micro-fluorescence spectroscopy and micro-thermometric research, and finally, analyze the oil and gas charging history based on the homogenization temperature, freezing point temperature, and PVT simulation analysis.

When using inclusion analysis, the following issues should be noted: ① Sample selection and processing. ② Use inclusion petrology to comprehensively analyze the oil and gas charging periods and formation history with inclusion combinations and micro-fluorescence spectroscopy. ③ Determine that the used brine inclusions are diagenetic products of the same period as the oil and gas inclusions and avoid using insufficient or unstrictly divided inclusion data to explain the oil and gas charging periods and times. ④ Combine bottom hole temperature, consider the re-equilibrium of inclusions after capture in carbonate rocks and the impact of overpressure capture on fluid inclusion temperature and salinity.

Figure 5.42 is an example of using fluid inclusion analysis combined with hydrocarbon source rock generation history to determine the complex oil and gas reservoir formation process of the Silurian in the Tazhong area (Lu et al., 2012). By comprehensively analyzing the minerals hosting fluid inclusions, their occurrence, color, and phase characteristics, the hydrocarbon inclusions in the study area are divided into three periods. The maturity and source of different periods of inclusions are analyzed by fluid inclusion fluorescence spectroscopy and infrared spectroscopy. It is believed that the maturity of the second period inclusions is low and the density is large, while the maturity of the third period inclusions is higher and the density is smaller. The oil and gas of the first and third periods come from the lower and middle Cambrian hydrocarbon source rocks, and the oil and gas of the second period come from the middle and upper Ordovician source rocks. The current oil and gas are a mixture of multi-period and multi-source charging oil and gas. Finally, by combining the homogenization temperature of fluid inclusions and the burial and thermal history of the study area, it is determined that the Silurian oil and gas in Tazhong has three periods of formation (Fig. 5.42), that is, the first period of formation of oil from the lower and middle Cambrian at the end of Devonian sedimentation (corresponding to black bitumen inclusions and dry bitumen), the destruction at the end of Devonian forming residual bitumen, the second period of formation of oil from the middle Ordovician at the end of Permian sedimentation (corresponding to yellow–brown fluorescent inclusions and brown bitumen), and the third period of formation of oil from the lower and middle Cambrian deep adjustment of oil and gas during the Himalayan period (corresponding to blue-white fluorescent inclusions and oily bitumen), and the current oil and gas are mainly a mixture of the second and third periods of formation oil and gas.

(2) Authigenic Mineral Illite Dating Method

This method uses the isotope age analysis of authigenic illite minerals in the reservoir to determine the time when hydrocarbons entered the reservoir. The basic principle is that authigenic illite in sandstone reservoirs only forms in a flowing potassium-rich water medium environment. After the hydrocarbons enter the reservoir, the formation process of illite will stop. That is, when hydrocarbons are charged into the reservoir, the replacement of formation water in the reservoir by hydrocarbons destroys the formation conditions of authigenic illite, and the formation of authigenic illite is terminated. Therefore, the K–Ar age of authigenic illite in sandstone reservoirs will record the time of hydrocarbon injection events. Since authigenic illite is formed

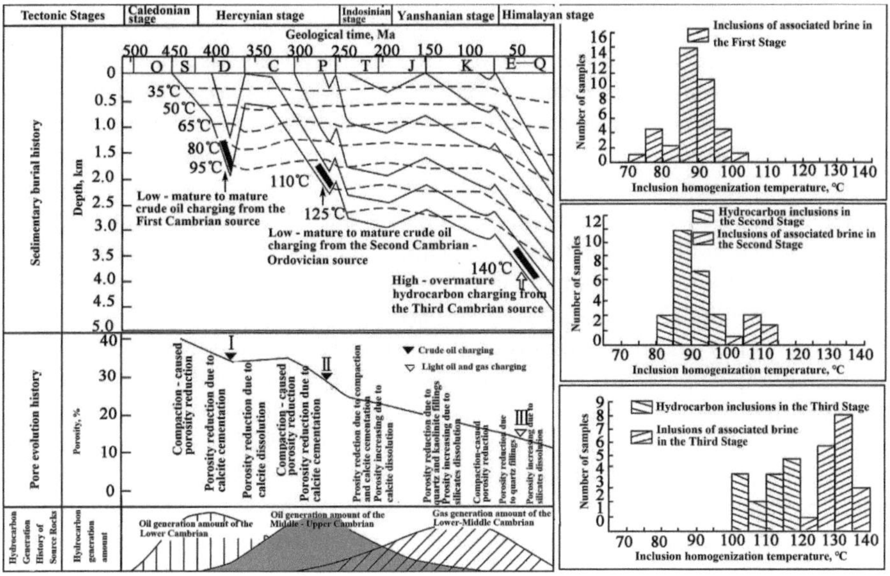

Fig. 5.42 Fluid inclusions and hydrocarbon generation history determine the formation period in the Silurian of Tazhong No. 11 well area (according to Lu, 2012)

before the hydrocarbon injection, and the youngest authigenic illite should be the product of the eve or early stage of hydrocarbon injection, the time of hydrocarbon filling the reservoir should be slightly later than the isotope age of authigenic illite. The speed of formation (fast or slow) and the direction of hydrocarbon migration can be judged based on the distribution of authigenic illite isotope ages on the plane and in the profile.

The prerequisite for using this method to determine the formation time of oil and gas reservoirs is that the studied sandstone reservoirs have authigenic illite, and the diagenetic action of illite must have a causal relationship with the hydrocarbon injection event. Usually, continuous sampling is taken from the oil layer and water layer above and below the oil–water interface of the same oil reservoir. The age of authigenic illite in the oil layer is older, and its youngest age represents the time when illite growth ended, that is, the time of hydrocarbon charging, while the age of authigenic illite in the water layer is much younger than that in the oil layer, which is the result of continued growth in the water-bearing part after the formation of the reservoir. Therefore, there should be a significant change in the age of authigenic illite above and below the oil–water interface.

This method is theoretically feasible, but not all oil-bearing layers have this variation pattern, and fine sandstone and medium sandstone samples usually have better effects. In addition to being affected by sampling, measurement, and other factors, this method is also greatly influenced by the heterogeneity of the reservoir, the distribution of oil and water, and the water medium. Therefore, there are also many variation characteristics of authigenic illite age on the oil and gas reservoir profile.

Hamilton et al. (1989) classified these changes into six types (Fig. 5.43). ① Gradual type, the age of authigenic illite in the oil (gas) layer gradually decreases, indicating continuous and continuous hydrocarbon charging (Fig. 5.43a). ② Constant type, the age of authigenic illite in the oil (gas) layer and the water layer is the same, indicating that the growth of illite is unrelated to hydrocarbon accumulation, possibly caused by structural action (such as uplift, cooling) (Fig. 5.43b). ③ Increasing type, the age of authigenic illite increases with depth, but the age in the water layer is larger than that in the oil (gas) layer, indicating that it is unrelated to hydrocarbon charging, possibly due to changes in the water medium that are not suitable for the growth of illite (Fig. 5.43c). ④ Two-part type, the age of authigenic illite in the oil (gas) layer and the water layer is respectively the same, which may reflect the result of rapid hydrocarbon charging (Fig. 5.43d). ⑤ Two-period type, the age of authigenic illite in the oil (gas) layer has two significantly different values, indicating two distinct periods of hydrocarbon charging (Fig. 5.43e). ⑥ Fluctuating type, the age of authigenic illite in the oil (gas) layer fluctuates with increasing depth, which may reflect the strong heterogeneity of the reservoir or the invasion of external fluids along faults (Fig. 5.43f). Among the six types of authigenic illite, only (a) type belongs to a completely normal change, which is enough to show that this method is significantly affected by external geological factors, changes in oil–water interface and water medium, reservoir heterogeneity, and other aspects. Therefore, when applying it, it is necessary to consider and exclude these interferences, and it is best to use it in conjunction with other methods for comprehensive analysis and judgment to obtain more reliable results (Li, 2013).

In addition to the two main methods of fluid inclusions and authigenic mineral illite dating, there are also methods such as zircon U–Pb, (U–Th)/He dating, apatite fission track dating, and radioactive isotope dating such as U–Pb, Rb–Sr, Sm–Nd, Re-Os in asphalt and crude oil, among which U–Pb dating and Re-Os isotope dating are methods that have been widely used in recent years. The U–Pb dating method is especially applicable to carbonate rock sequences. The 238 U and 235 U in the samples are radioactively decayed to form stable isotopes 206 Pb and 207 Pb, respectively, based on which the content of radiogenic daughter isotopes can be determined to determine their formation age.

Reservoir solid bitumen, as a special "diagenetic mineral," is also used to determine the formation period of oil and gas reservoirs. Reservoir solid bitumen is the product of oil alteration in oil reservoirs, recording the information of oil reservoir modification and destruction. The reflectivity of solid bitumen reflects the thermal history experienced after hydrocarbon fluids are transformed into solid bitumen. By combining the reflectivity of reservoir solid bitumen, the chemical reaction kinetics of bitumen reflectivity, and the quantitative analysis of reservoir burial history and thermal evolution history, the time of oil reservoir destruction can be determined.

However, oil and gas reservoirs are products formed in the geological history process and often undergo complex evolution processes after their formation. Currently, there is no reliable method to determine the formation period, and each method has certain limitations. Therefore, the time and period of oil and gas reservoir formation need to be comprehensively judged by various methods such as

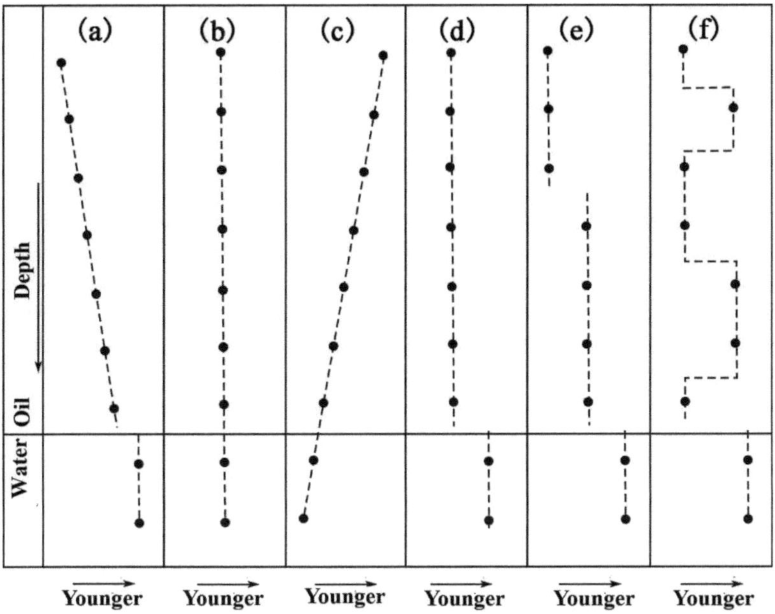

Fig. 5.43 Different types of authigenic illite age profiles on oil reservoir sections (according to Hamilton et al., 1989)

geological research and analysis testing to obtain a more correct conclusion. The widely applied procedural method is: first determine the main hydrocarbon expulsion period of the main hydrocarbon source rocks, and on this basis, make full use of reservoir fluid history analysis methods, mainly diagenetic mineral fluid inclusion testing technology, illite dating technology, and other analysis and testing technologies, combined with basin evolution and reservoir burial history, to comprehensively judge the time and period of oil and gas reservoir formation.

5.6 Formation Mechanisms of Condensate Gas Reservoirs and Natural Gas Hydrates

Condensate gas reservoirs and natural gas hydrates are two special phases of natural gas reservoirs. Their formation, in addition to requiring the basic geological conditions for conventional oil and gas reservoirs, is mainly controlled by stratum pressure and temperature.

5.6.1 Formation Mechanism of Condensate Gas Reservoirs

(1) Concepts

Under certain temperature and pressure conditions, isochoric pressurization of hydrocarbon substances causes condensation, and depressurization leads to evaporation. However, beyond a certain temperature and pressure, inverse evaporation and inverse condensation phenomena occur, that is, isochoric pressurization causes evaporation, while isochoric depressurization causes condensation.

In oil and gas reservoir exploitation, it has been observed that certain hydrocarbons that are in a gaseous state under high temperature and high-pressure conditions deep underground condense into a certain amount of liquid oil (condensate oil) after being brought to the surface due to the reduction in temperature and pressure. Such gas reservoirs are called condensate gas reservoirs.

The main difference between condensate gas reservoirs and conventional oil and gas reservoirs is the existence of a special underground hydrocarbon system, known as the condensate oil system, characterized by a high gas-oil ratio (greater than 600–800 m^3/m^3) and a high enrichment of light hydrocarbon components. Within a certain temperature and pressure range, inverse evaporation and inverse condensation phenomena occur, causing some liquid hydrocarbons to redissolve into the gas phase to form a single gas phase. Its most important feature is that the underground hydrocarbon system is in a gaseous state, while both gas and condensate oil are produced on the surface. Compared with general crude oil, condensate oil has a lower density (<0.78), lower viscosity, lighter color (colorless, light yellow), and a higher proportion of light distillates, with normal alkanes generally greater than 87%, and cycloalkanes + aromatics less than 13%, mainly C_5–C_{10} components.

Unlike conventional oil reservoirs, condensate gas reservoirs are in a single gaseous state underground and can produce some condensate oil and a large amount of natural gas on the surface. Condensate gas reservoirs are also different from general gas reservoirs; gas reservoirs (including wet gas reservoirs) also produce a small amount of condensate oil, but the quantity is far less than that of condensate gas reservoirs. With the decrease of pressure, condensate gas reservoirs will inevitably undergo retrograde condensation, causing a significant loss of condensate oil.

(2) Phase State of Pure Substances

Under certain temperature and pressure conditions, isochoric pressurization causes gaseous substances to liquefy. As the temperature increases, the minimum pressure required for gas liquefaction increases. When a certain temperature is reached, no matter how much pressure is applied, the substance cannot be liquefied. This point is the critical point, and the temperature at the critical point is the highest temperature at which the gaseous substance can maintain a liquid state, known as the critical temperature. The minimum pressure required for the gas to liquefy at the critical temperature is the critical pressure.

The characteristics of any substance in a critical state are that the differences between the coexisting gas and liquid phases disappear. There is no difference in density, viscosity, surface tension, etc., between the two.

Critical temperature and critical pressure are characteristic constants of various substances, and a certain substance has its definite values. For example, the critical temperature of water is 374.2 °C, and that of methane is −82.1 °C, therefore, methane in the underground, except for being dissolved in oil and water, exists in a gaseous state; ethane is also roughly similar.

The relationship diagram between temperature, pressure, and the phase state of a substance is called a phase diagram, also known as a pressure–temperature diagram. From some pVT relationship curves of propane (Fig. 5.44), when the system is pressurized from low to high, the volume of gaseous propane initially decreases with increasing pressure. After passing point A (pressure of 2.8 MPa), the volume continues to decrease, but the pressure remains unchanged. After passing point B, even with a very high pressure, the volume does not change much. The p–V curve at 87.8 °C is similar in nature, the only difference is that the horizontal segment A′B′ gradually shortens as the temperature rises. Finally, at 96.8 °C, it shrinks into a point K, and above this temperature, the horizontal segment disappears entirely (according to Zhang et al., 1999).

The physical significance of the above phenomenon is that at 71.1 °C, propane is compressed to point A and begins to liquefy. As the pressure increases, the gas gradually decreases and the liquid gradually increases. Since the molar volume of the liquid is much smaller than that of the gas, the volume gradually decreases. When reaching point B, the gas has been completely liquefied. At this point, due to the low compressibility of the liquid, even with a very high pressure, the volume does not change much. From point A to B, the pressure does not change, indicating that the liquid has a certain saturated vapor pressure at a certain temperature. During the process from A to B, the liquid phase and the gas phase coexist. As the temperature rises, the saturated vapor pressure of the liquid also increases.

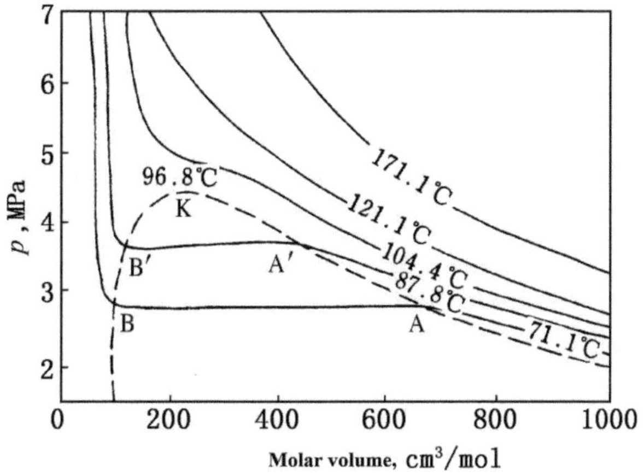

Fig. 5.44 Propane's pVT relationship curves

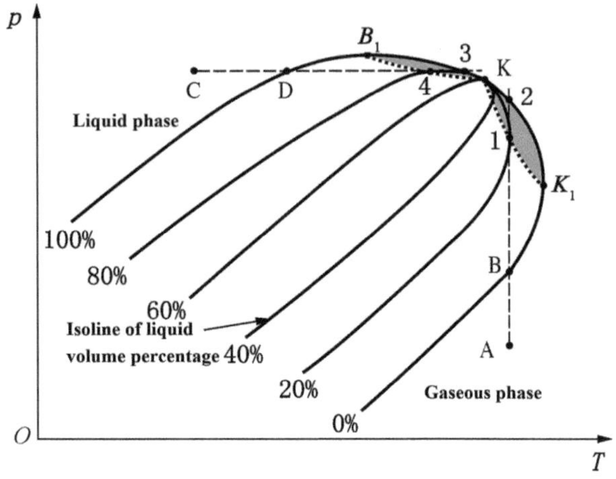

Fig. 5.45 Phase diagram of binary hydrocarbon systems

(3) Phase State of Binary Hydrocarbon Systems

Figure 5.45 is a p–T diagram of a binary hydrocarbon substance. The highest temperature K1 and highest pressure B_1 where the gas and liquid phases coexist in the diagram are called the critical condensation temperature and critical condensation pressure, respectively. The critical point K is the intersection of the bubble point line and the dew point line.

Under isochoric pressurization: A → B → 1 → 2, at point A, the substance is in a gaseous state, and pressurization to point B begins to produce liquid droplets (dew point), and as the pressure continues to increase to point 1, the amount of liquid gradually increases. But from 1 to 2, pressurization causes the liquid to gradually decrease and the gas phase to increase, until at point 2, the substance is completely vaporized. From 1 → 2, isochoric pressurization exhibits vaporization characteristics, known as inverse evaporation. From 2 → 1, isochoric depressurization exhibits liquefaction characteristics, known as inverse condensation.

Under isobaric heating: C → D → 4 → 3, point C is a liquid, and heating to point C begins to produce bubbles (bubble point), and from D → 4, the amount of gas gradually increases. But from 4 → 3, heating instead causes the amount of gas to decrease until it is completely liquefied. From 4 → 3, it is inverse condensation and from 3 → 4, it is inverse evaporation.

The phase transformation of binary hydrocarbon substances (inverse evaporation) is the basic reason for the formation of condensate gas reservoirs.

(4) Phase State of Multi-component Hydrocarbon Systems and the Formation of Condensate Gas Reservoirs

Oil and natural gas are complex multi-component hydrocarbon mixtures. The phase characteristics of multi-component hydrocarbon mixtures are like the above binary situations, but more complex, with a wider coexistence area of gas and liquid. They are usually simplified into two parts: light components and heavy components. If there are more light components, the critical point is located to the left and lower side of the critical condensation point; if there are more heavy components, the critical point is located to the right and lower side of the critical condensation point. Therefore, there can be two retrograde zones or one.

Figure 5.46 shows the physical state of a certain multi-component hydrocarbon system under different temperatures and pressures. K is its critical point, and the critical temperature is 52.8 °C. K_1 is the critical condensation temperature. Curve 4 is the bubble point curve where the gas begins to separate from the liquid, and outside it is the pure liquid phase. Curve 5 is the dew point curve where the liquid begins to condense and separate from the gas, and outside it is the pure gas phase. Within the range enclosed by curves 4 and 5, the mixture is in a two-phase state (liquid and gas), and each isopleth line represents the percentage of liquid in the system.

When the stratum is buried shallowly and the stratum temperature is lower than the critical temperature, the phase state change of the system conforms to the normal concept of condensation and evaporation. For example, at °C, as the pressure increases, the amount of condensed liquid in the system gradually increases, and when the pressure exceeds 18.0 MPa (point C_1), the system is completely condensed into a liquid.

When the stratum is buried deeply and the stratum temperature is between the critical temperature and the critical condensation temperature, the phase state change of the system is more complex, similar to the phase state transformation of binary hydrocarbon substances. In Fig. 5.46, at 82.5 °C, under low pressure, the system is in a two-phase state, but mainly gaseous, with the volume of liquid in the system

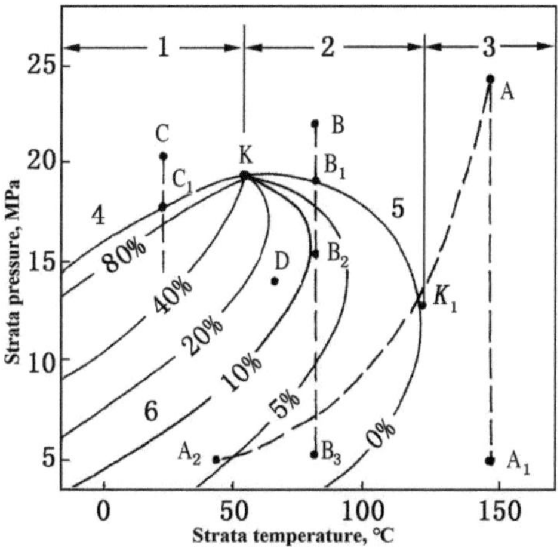

Fig. 5.46 Phase diagram of multi-component hydrocarbon systems. 1—Oil reservoir with pressure exceeding bubble point pressure; 2—condensate gas reservoir with pressure exceeding dew point pressure; 3—single-phase gas reservoir (pure gas reservoir); 4—bubble point curve; 5—dew point curve; 6—volume percentage of liquid in the system; K—critical point (T=); K1—critical condensation temperature; A—pure gas reservoir; B—condensate gas reservoir; C—oil reservoir with dissolved gas; D—oil and gas reservoir

being less than 5–10%. As the pressure increases, the amount of condensed liquid gradually increases. When the pressure increases to 15.5 MPa (point B_2), the amount of condensed liquid is the maximum, accounting for 10% of the total volume of the system. If the pressure continues to increase, the condensed liquid is gasified instead, and the amount of liquid gradually decreases. When the pressure increases to 18.7 MPa (point B_1), the condensate is completely transformed into a gaseous state. Therefore, for this system at 82.5 °C, below 15.5 MPa, it belongs to normal condensation and evaporation, and above 15.5 MPa, it belongs to the category of inverse condensation and inverse evaporation. In other words, when the stratum is buried deeply, the stratum temperature is between the critical temperature and the critical condensation temperature of a certain hydrocarbon system, and the stratum pressure exceeds the dew point pressure (point B in Fig. 5.46), this hydrocarbon can form a condensate gas reservoir (Zhang et al., 1999).

Therefore, the formation of condensate gas reservoirs must have two conditions:

The amount of gas in the hydrocarbon system must surpass the amount of liquid to create conditions for the redissolution of liquids in the gas phase. In the multi-component hydrocarbon system shown in Fig. 5.45, the volume of gas is equivalent to 5–20 times or more of the volume of liquids.

The stratum is buried deeply, the stratum temperature is between the critical temperature and the critical condensation temperature of the hydrocarbon system, and the stratum pressure exceeds the dew point pressure at that temperature, only then can this system undergo significant inverse evaporation.

In oil and gas basins, as the burial depth of the reservoir increases, the stratum pressure and stratum temperature will also increase accordingly. When the stratum temperature reaches the critical temperature of the oil–gas system, the greater the stratum pressure, the easier it is for the oil–gas system to transform into a single-phase gas state, greatly promoting the migration and accumulation of oil and gas in the deep underground reservoirs, forming condensate gas reservoirs.

The special conditions required for the formation of condensate gas reservoirs determine that their distribution on the Earth's crust must have a certain range, as shown in Fig. 5.46, where A, B, C, and D represent four types of oil and gas reservoirs. Type A has a stratum pressure of 24.6 MPa and a temperature of 148.9 °C, exceeding the critical condensation temperature. If exploited isothermally (i.e., pressure decreases along line A-A_1), the system remains in a gaseous state, which is a pure gas reservoir. If exploited to the surface, both temperature and pressure decrease, as shown by the A-A_2 curve, entering the two-phase zone, a small amount of liquid (condensate oil) can be separated in the surface separator. Type C's original stratum pressure is greater than the saturation pressure, but the temperature is lower than the critical temperature, which is a pure oil reservoir with dissolved gas. During isothermal exploitation, as the pressure decreases to C_1, the dissolved gas gradually becomes free, the gas-oil ratio increases, and the energy of the oil reservoir will decrease rapidly. When the original stratum pressure and temperature combination is in the two-phase zone enclosed by the bubble point curve and the dew point curve, such as point D, it has a native free gas cap and is an oil and gas reservoir. The volume percentage of gas and liquid changes greatly, depending on the stratum temperature and pressure. Only type B's stratum temperature is between the critical temperature and the critical condensation temperature. If exploited isothermally, the pressure decreases step by step along B-B_1-B_2-B_3. When the pressure drops below the dew point pressure of 18.7 MPa (point B_1), the system can undergo inverse condensation to form liquids in the stratum, which is condensate oil, different from type A's pure gas reservoir and is a typical condensate gas reservoir.

The stratum temperatures of condensate gas reservoirs and pure gas reservoirs are relatively high, and their burial depths are generally large, mostly distributed at depths of 3000 m or more underground. For example, the Lacq gas field in France has discovered a huge gas reservoir with recoverable reserves of 2000×10^8 m^3 in limestone and dolomite at depths of 3500–4500 m, with a high content of condensate in the gas and no liquid oil has been found. In the United States, the Gulf of Mexico Basin has more than 3000 m deep, mainly dominated by natural gas and condensate gas accumulations, with gas wells accounting for 60–68% and oil wells accounting for 32–40%. In the Permian Basin, which targets the Paleozoic, there are condensate gas reservoirs and pure gas reservoirs at depths exceeding 4500 m, with gas wells accounting for 90–100% and very few oil wells.

In fact, the genesis types of condensate gas reservoirs can take various forms. According to the thermodynamic principles of multi-component hydrocarbon systems, there are two formation mechanisms for gaseous hydrocarbons and liquid hydrocarbons in condensate gas reservoirs to become a single gas phase: retrograde evaporation and normal evaporation.

Retrograde evaporation requires that the amount of gaseous hydrocarbons in the mixed hydrocarbon system be greater than that of liquid hydrocarbons, as described above. The process of forming condensate gas through normal evaporation mainly occurs when the amount of liquid hydrocarbons dominates in the mixed hydrocarbon system. As the stratum temperature and pressure decrease, gaseous hydrocarbons separate from the oil phase through normal evaporation, and at the same time, gaseous hydrocarbons dissolve some liquid hydrocarbons to form condensate gas.

In the process of oil and gas generation, migration, and accumulation, as the composition, temperature, and pressure of the hydrocarbon system change, the hydrocarbon system can form initial and secondary condensate gas reservoirs through retrograde evaporation or normal evaporation in an appropriate environment. The latter can be divided into two types of condensate gas reservoirs: differentiated and enriched. Initial-type condensate gas reservoirs refer to condensate gas phases generated from the hydrocarbon source rock and maintain the initial state of the hydrocarbon system during the migration and accumulation process. In the process of oil and gas migration, accumulation, and formation, the hydrocarbon system differentiates and enriches, with multiple periods of formation and mixing of multiple hydrocarbon systems, and different phase evolution processes can also form condensate gas reservoirs. These condensate gas reservoirs formed during the migration, accumulation, and formation process are called secondary condensate gas reservoirs (Zhou et al., 1996). Differentiated condensate gas reservoirs refer to condensate gas reservoirs formed by the separation of gaseous hydrocarbons from the oil phase system during the oil migration process. Enriched condensate gas reservoirs are mixed condensate gas reservoirs formed by mixing a certain number of liquid hydrocarbons into condensate gas or wet gas and dry gas. Therefore, the natural gas in condensate gas reservoirs is not necessarily condensate gas generated by the hydrocarbon source rock at the condensate gas stage. Under suitable temperature, pressure, and oil and gas generation, migration, and accumulation conditions, different mechanisms and phase genesis condensate gas reservoirs can be formed from the immature to the over-mature stage.

5.6.2 Formation Mechanism of Natural Gas Hydrates

(1) Basic Concepts

Natural gas hydrates (NGH) are ice-like solid compounds formed by natural gas and water (commonly known as "flammable ice"), where methane and other gas molecules are naturally enclosed in the expanded crystal lattice of water molecules under specific low-temperature and high-pressure conditions, forming icy, crystalline cage compounds. It is a special form of natural gas existing in nature, mainly distributed below a certain water depth (usually >300 m) on the seabed and in permafrost zones. The main component of natural gas in natural gas hydrates is methane, so it is also commonly referred to as methane hydrates. Sometimes ethane, propane, isobutane, carbon dioxide, and hydrogen sulfide can also form solid mixed gas hydrates with methane, hence it is also known as solid gas hydrates.

Natural gas hydrates are ice-like solid substances composed of methane and water molecules, with the chemical formula $M{\cdot}nH_2O$, where M is the number of gas molecules mainly methane, and n is the number of water molecules. Essentially, natural gas hydrates are cage-shaped substances where water molecules form a rigid cage lattice, each of which includes a gas molecule primarily methane (Fig. 5.47).

Natural gas hydrates are cage-shaped structures of methane and water, and the methane contained within is controlled by the relationship between methane molecules and water molecules within the structure. Theoretically, in a saturated natural gas hydrate molecular structure, the molar ratio of methane to water is 1:6,

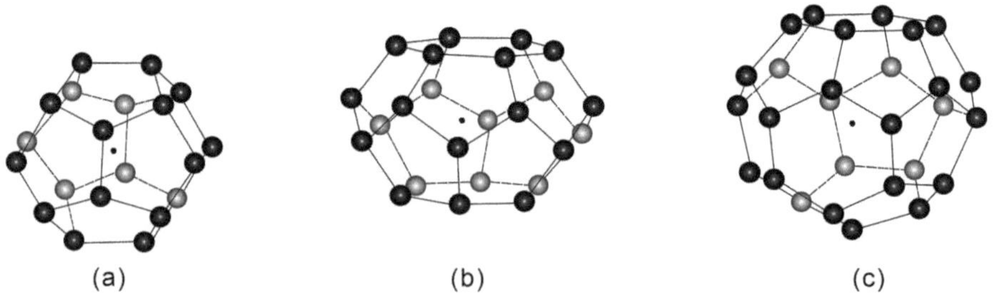

Fig. 5.47 Three types of natural gas hydrate lattice structures

which translates to a volume ratio of 164:1 under standard temperature and pressure conditions. This means that 1 m^3 of gas hydrate can contain 164 m^3 of methane gas and 0.8 m^3 of water. It is evident that the amount of methane gas resources contained is enormous.

(2) Formation and Distribution of Natural Gas Hydrates

The gases in natural gas hydrates mainly come from three sources: ① biogenic gas produced by the degradation of organic matter in sediments by bacterial action, ② thermogenic gas produced by the thermal cracking of deep organic matter or petroleum, and ③ inorganic gas produced by volcanic action. Since natural gas hydrates are currently found mainly in shallow sediments, it is believed that most of the gas in natural gas hydrates is biogenic, especially those formed on the continental margins. However, as natural gas hydrates continue to be discovered on the ocean floor with little terrestrial detritus and low total organic matter, the proportion of thermogenic and inorganic gas in natural gas hydrates may increase.

The formation of natural gas hydrates requires four basic conditions: ① abundant natural gas and water, mainly biogenic gas, and secondarily from thermogenic gas, ② lower temperatures, generally below, ③ higher pressures, generally requiring pressures greater than 10 MPa, ④ favorable reservoir space. The most important are the low-temperature and high-pressure conditions, and temperature and pressure can compensate for each other within a certain range, that is, the lower the temperature at which hydrates form, the lower the required pressure and the shallower the burial depth. The higher the temperature, the greater the required pressure and the deeper the burial depth (Fig. 5.48). However, the formation of hydrates requires pressure to increase logarithmically with linearly increasing temperature, and in most sedimentary basins, the increase in pressure with increasing burial depth is far from meeting the pressure requirements for the formation of natural gas hydrates due to geothermal gradient increases. Hydrates will decompose at temperatures between 21–27 °C, so the geological environment suitable for hydrate formation must be a cold environment that is not too deeply buried, and the lower limit depth for the formation of natural gas hydrates is generally about.

In nature, hydrates can not only form gas hydrate gas fields but also act as a cap layer for free gas below, such as the Messoyakha gas field in Russia, where the hydrate layer and the free gas layer coexist in the reservoir.

Katz (1971) proposed a temperature-depth curve to predict the depth at which gas hydrates generate (Fig. 5.49). The curve represents the hydrate curve of methane and gas with a relative density of 0.6, assuming an increase in static pressure of 0.435 pound/square inch per foot of depth; the temperature data of Simpson Ridge, Prudhoe Bay, and Messoyakha gas fields are plotted on the graph. The shaded area represents the depth and thickness of gas hydrate formation. At Simpson Ridge, where the geothermal gradient is large, the formed gas hydrate belt is very thin, with methane hydrate thickness of only about 100 m, and the hydrate of gas with a relative density of 0.6 is thick. In Prudhoe Bay, methane hydrate depth is between 213–1067 m, and the hydrate of gas with a relative density of 0.6 can extend to

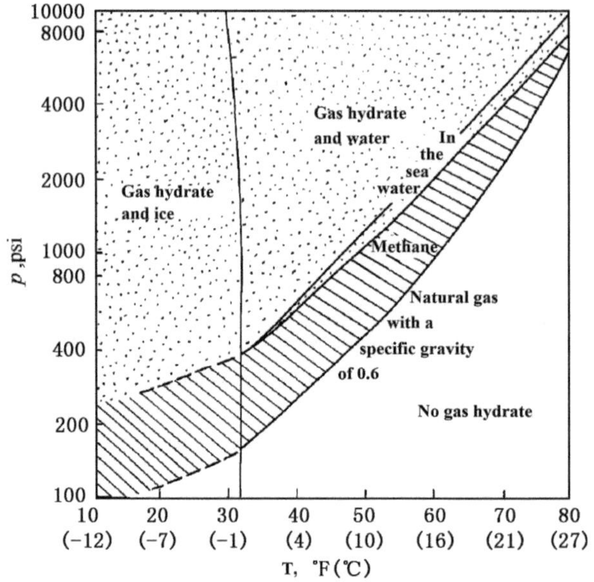

Fig. 5.48 Pressure–temperature diagram of gas hydrates (according to Katz et al., 1959)

1219 m. In the Messoyakha gas field, the measured hydrate depth range is 305–870 m, slightly shallower than the calculated depth.

The high-pressure and low-temperature characteristics of natural gas hydrate formation determine its distribution in geological environments, mainly in polar regions, permafrost zones, and the deep ocean floor. The permafrost zone in the Arctic region is generally 250–600 m thick, with the thickest reaching, and the hydrates in the permafrost zone exist in low-pressure and low-temperature areas.

In the 1960s, natural gas hydrates were first discovered in the Arctic gas fields of the former Soviet Union. It was not until the 1970s, when the huge Messoyakha natural gas hydrate field was discovered in this region, that it attracted people's attention. Later, natural gas hydrates were found in many oil and gas fields in the Arctic. In the early 1980s, the drilling ship of the US Deep Sea Drilling Project even found that there is a wide distribution of natural gas hydrate strata in the Pacific Ocean near Mexico and Central America, and obtained many cores. Over the past forty years, natural gas hydrates have been discovered in many regions around the world. According to statistics, about 27% of the Earth's land can potentially form natural gas hydrates, and about 90% of the world's ocean areas are potential regions, with huge global natural gas hydrate resources, estimated at 1.05×10^{15} m^3 (Pinero et al., 2013).

China's natural gas hydrates are mainly distributed in the South China Sea, the East China Sea, the Qinghai-Tibet Plateau permafrost zone, and the Northeast permafrost zone, with huge resource quantities. Natural gas hydrate samples have been obtained in the Shenhu area of the northern South China Sea and the

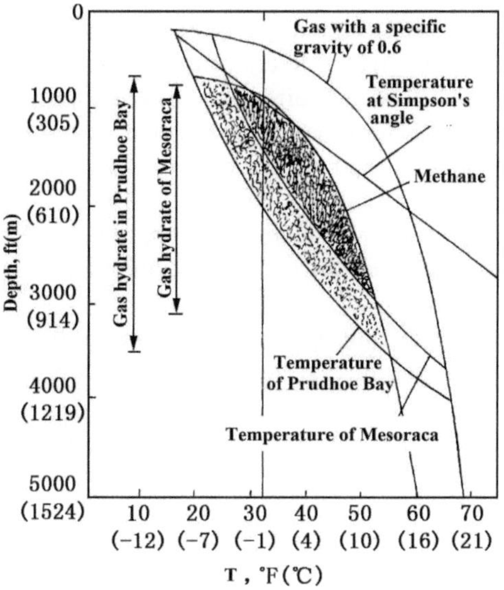

Fig. 5.49 Depth-temperature curve of predicting the depth and thickness of gas hydrates (according to Katz, 1971)

Qilian Mountain permafrost zone in Qinghai Province. In 2013, Chinese marine geological scientists for the first time drilled and obtained high-purity natural gas hydrate samples in the eastern sea area of the Pearl River Mouth Basin, and obtained considerable controlled reserves through drilling.

Natural gas hydrates are a new field in oil and gas exploration, and their development and utilization will inevitably provide more abundant energy for humanity. However, the current exploration and research level of natural gas hydrates is still relatively low, and there are still many issues worth discussing. The exploitation of natural gas hydrates buried in the ocean still faces many new problems and difficulties, such as the potential for hydrate development to cause tsunamis, submarine landslides, and seawater poisoning. For natural gas hydrates to become an important source of energy in the future, a large amount of research work still needs to be carried out.

Questions for Discussion

1. What are traps and oil and gas reservoirs? How to measure traps and oil and gas reservoirs?
2. What geological conditions are required for the formation and preservation of oil and gas reservoirs?

3. What are unconventional oil and gas reservoirs? What are the differences in formation conditions compared to conventional oil and gas reservoirs?
4. Compare the similarities and differences in the formation conditions and formation methods of natural gas reservoirs and oil reservoirs.
5. What are the main actions and mechanisms that cause the destruction of oil and gas reservoirs?
6. What are primary and secondary oil and gas reservoirs? How are they formed?
7. What is the mechanism of oil and gas accumulation in traps? How is the accumulation process?
8. What are the principles and applicable conditions for the differential accumulation of oil and gas in a series of traps of overflow type and seepage type?
9. How to determine the period of oil and gas reservoir formation? What are the principles and applicable conditions of various formation period determination methods?
10. What is a condensate gas reservoir? What are its formation mechanism and conditions?
11. What is a natural gas hydrate? What are its formation mechanism and conditions?

References

Beydoun, Z. R., Clarke, M. W. H., Stoneley, R. (1992.) Petroleum in the zagros basin: a late tertiary foreland basin overprinted onto the outer edge of a vast hydrocarbon-rich paleozoic-mesozoic passive-margin shelf. *Aapg Memoir, 55*, 309–339.

Bogatsky, V., Pankratov, J. (1993). Controls on the distribution of oil, gas, and gas-condensate pools in the timan-pechora basin. *Bulletin, 77*.

Bowering, O. J. W. (1982). Hydrodynamics and hydrocarbon migration—a model for the eromanga basin. *The APPEA Journal, 22*(1), 227.

Carman, G. J., & Hardwick, P. (1983). Geology and regional setting of Kuparuk oil field, Alaska. *AAPG bulletin, 67*(6), 1014–1031.

Chapman, R. E. (1982). Effects of oil and gas accumulation on water movement. *AAPG Bulletin, 66*(3), 368–374.

Chen, Y., Shen, Z., Li, Y., et al. (2002). Characteristics and accumulation mechanisms of the Lunan low-relief condensate gas reservoir in the Tarim Basin. *Journal of Chengdu University of Technology, 29*(5), 481–486.

Cordell, R. J. (1976). How oil migrates in clastic sediments. *Part 2 World Oil, 183*(7), 75–80.

Creaney, S., & Allan, J. (1990). Hydrocarbon generation and migration in the Western Canada Sedimentary Basin. Geological Society, London, *Special Publications, 50*(1), 189–202.

Dahlberg, E. C. (1982). Applied hydrodynamics in petroleum exploration. New York: Springer-Verlag.

Demaison, G. J. (1977). Tar sands and supergiant oilfields. *AAPG Bulletin, 61*, 1950–1961.

Dunnington, H. V. (1958). Generation, migration, accumulation and dissipation of oil in northern Iraq. In: Weeks, L. G. (ed.) Habitat of Oil, AAPG Memoir (no number), 1194–1251.

Earnshaw, J. P., Hogg, A. J. C., Oxtoby, N. H. et al. (1993). Petrographic and fluid inclusion evidence for the timing of diagenesis and petroleum entrapment in the Papuan Basin. In: Carman, G. J.

and Carman, Z. (eds) Petroleum Exploration and Development in New Guinea, Proceedings of the Second PNG Petroleum Convention, Port Moresby. PNG Chamber of Mines and Petroleum, pp. 459–476.

England, W. A., & Mackenzie, A. S. (1989). Some aspects of the organic geochemistry of petroleum fluids. *Geologische Rundschau, 78*(1), 291–303.

Guo, X. (2014). "Binary enrichment" pattern of marine shale gas in South China: Exploration practice of Longmaxi Shale Gas in Sichuan Basin and its margin. *Acta Geologica Sinica, 88*(7), 1209–1218.

Gussow, W. C. (1953). Differential trapping of hydrocarbons. *Bulletin of Canadian Petroleum Geology, 1*(6), 4–5.

Hamilton, P. J., Kelley, S. P., Fallick, A. E. (1989). K-Ar dating of illite in hydrocarbon reservoirs. *Clay Minerals, 24*(2), 215–231.

Hao, F., et al. (2005). *Kinetics of hydrocarbon generation in overpressured basins and mechanisms of oil and gas accumulation.* Science Press.

He, S., Ye, J. R., Xu, S. H., et al. (2010). *Petroleum and natural gas geology*. China University of Geosciences Press.

Hindle, A. D. (1997). Petroleum migration pathways and charge concentration: A three-dimensional model. *AAPG Bulletin, 81*(9), 1451–1481.

Holmquest, H. J. (1965). Deep Pays in Delawarc and Val Verde Basins. In: Fluids in Subsurface Environments, *AAPG Memoir. 4,* 257–279.

Jiang, Y., & Liu, Z. (1997). A pattern of oil and gas migration and accumulation in fault-block oil and gas fields. *Fault-Block Oil and Gas Fields,* (6), 1–4+10.

Jiang, Y., Liu, P., Liu, H., et al. (2014). Differences in hydrocarbon accumulation conditions and accumulation patterns of Neogene in different sub-depressions of the Bohai Bay Basin. *Journal of China University of Petroleum (Natural Science Edition), 38*(1), 14–21.

Jiang, Y., Su, S., Xin, F., et al. (2019). Heterogeneity of oil and gas distribution and controlling factors in oil-rich sub-depressions. *Journal of China University of Petroleum (Natural Science Edition), 43*(5), 34–43.

Jiang, Y., Su, S., & Zhao, K. (2022). Relationship between the conductivity of oil-source faults and hydrocarbon enrichment in non-hydrocarbon-generating strata: A case study of the Chengdao area in the Bohai Bay Basin. *Acta Petrolei Sinica, 43*(08), 1122–1131+1184.

Jiang, Y., & Zhang, Y. (1999). Main geological factors controlling hydrocarbon enrichment in complex fault-block areas: A case study of the Dongxin area in the Bohai Bay Basin. *Petroleum Exploration and Development, 26*(5), 39–42.

Jiang, Y., & Zhang, Y. (2000). Similarities and differences in the formation mechanisms and distribution characteristics of gas and oil reservoirs. *Geological Science and Technology Information, 19*(1), 69–72.

Jiang Y., Zhang Y., Ran L., et al. (2001). Migration and accumulation mechanism of natural gas in the Silurian-Carboniferous petroleum system of east part in Sichuan Basin. *Acta Petrolei Sinica, 22*(1), 25–30.

Jin, Z., & Zhang, J. (2003). Binary mechanism model for natural gas accumulation. *Acta Petrolei Sinica, 24*(4), 13–16.

Katz, D. L., Cornell, D., Robayashi, R. et al. (1959). *Handbook of natural gas engineering*. NewYork: McGraw-Hill.

Katz, D. L. (1971). Depths to which frozen gas fields (gas hydrates) may be expected. *Journal of Petroleum Technology, 23*(4), 419–423.

Levorsen, A. I. (1967). *Geology of petroleum* (2nd ed.). San Francisco: W. H. Freeman and Company.

Li, M. (2013). *Migration of oil and gas* (4th ed.). Petroleum Industry Press.

Liu, K., Julien, B., et al. (2013). Application of fluid inclusions to study oil and gas accumulation: A case study of the Ordovician reservoirs in the Tarim Basin. *Petroleum Exploration and Development, 40*(2), 171–180.

Lu, X., Song, Y., Liu, S., et al. (2012). Application of fluid inclusion microanalysis in the study of oil and gas accumulation in the Silurian of the Tarim Basin. *Journal of China University of Petroleum (Natural Science Edition), 36*(4), 45–50+76.

Macgregor, D. S. (1996). *Factors controlling destruction or preservation of giant light oil field.* Petroleum Geoscience.

Masters, J. A. (1984). Lower Cretaccous oil andgasin western Canada. In: Masters, J. A. (ed.) Elmworth, Case Study of a Deep Basin Gas Field, *AAPG Memoir, 38.*

Magara, K. (1978). Geological model predicting optimum sandstone percent for oil accumulation. *Bulletin of Canadian Petroleum Geology, 26*(3).

Perrodon, A. (1993). *Dynamics of oil and gas accumulations.* Pau France: Elf Aquitaine.

Pinero, E., Marquardt, M., Hensen, C., et al. (2013). Estimation of the global inventory of methane hydrates in marine sediments using transfer functions. *Biogeosciences, 10*(2), 959–975.

Roberts, W. H. (1980). Design and function of oil and gas traps. *AAPG Studies in Geology, 10,* 256–274.

Sales, J. K. (1997). Seal strength VS trap closure: A fundamental control on the distribution of oil and gas. *AAPG Memoir, 67,* 57–83.

Tissot, B. P., & Welte, D. H. (1984). *Petroleum formation and occurrence.* New York: springer-Verlag.

Wilson, R. D., Monaghan, P. H., Osanik, L., et al. (1973). Estimate of Annual Input of Petroleum to the Marine Environment from Natural Seepage. *Transactions of the Gulf Coast Association of Geological Societies 23,* 182–193.

Xue, Y., Niu, C., Wang, D., et al. (2021). *Hydrocarbon migration "convergent ridge" control on reservoir formation and shallow exploration in the Bohai Bay Basin.* Science Press.

Zhai, G., et al. (1996). *Geological atlas of petroleum in China* (Vols. 1, 6, and 10). Petroleum Industry Press.

Zhang, H., Fang, C., Gao, X., et al. (1999). *Petroleum geology* (3rd ed.). Petroleum Industry Press.

Zhang, J., & Wang, Z. (2003). Mechanism of abnormal formation pressure generation in deep basin gas reservoirs. *Petroleum Exploration and Development, 30*(1), 28–31.

Zhou, X., Li, S., & Chen, Y. (1996). Formation of condensate gas in the Tarim Basin. *Petroleum Exploration and Development, 23*(6).

Zou, C., Tao, S., Hou, L., et al. (2014). *Unconventional oil and gas geology.* Geological Publishing House.

Chapter 6
Types and Characteristics of Oil and Gas Reservoirs

Abstract The type of oil and gas reservoir is the foundation for understanding the distribution patterns of oil and gas and is one of the key elements that must be understood in oil and gas exploration. Oil and gas reservoirs are influenced by a variety of factors, including reservoir shape, trap type, reservoir type, seal type, and the phase state of oil and gas, which leads to a highly complex classification scheme for oil and gas reservoirs. Among these, classification based on trap genesis reflects the scientific and practical nature of reservoir classification and is widely used in oil and gas exploration. This chapter mainly classifies oil and gas reservoirs according to trap genesis, dividing them into five major categories: structural, stratigraphic, lithologic, composite, and special types of oil and gas reservoirs, and introduces the characteristics of each type of reservoir.

6.1 Overview

6.1.1 General Introduction to Oil and Gas Reservoir Classification

Oil and gas reservoirs discovered worldwide are numerous and diverse in type. To understand the formation and distribution characteristics of various oil and gas reservoirs and to guide oil and gas exploration work more effectively, petroleum geologists at home and abroad have proposed hundreds of classification schemes from different research and usage perspectives over the years. However, the classifications that are significant to oil and gas exploration mainly include those based on the causes of trap formation, the morphology of oil and gas reservoirs, types of sealing and shielding, types of reservoirs, the magnitude of reserves and production, the phase of hydrocarbons, and the properties of fluids, etc. In addition, classifications can also be made based on the commercial value of oil and gas reservoirs and the types of drive mechanisms.

© China University of Petroleum Press 2025

H. Liu and Y. Jiang, *Petroleum Geology*,
https://doi.org/10.1007/978-981-96-7086-4_6

6.1.1.1 Classification by Trap Formation and Oil and Gas Reservoir Morphology

There are two classifications that have had a significant impact on the petroleum geology and exploration community in China are as follows:

(1) Trap formation classification: Represented by American petroleum geologist A. I. Levorsen, oil and gas reservoirs are divided into structural, stratigraphic, and combined types (Levorsen, 1967).
(2) Classification by oil and gas reservoir morphology: Represented by Soviet scholar I. O. Brod, oil and gas reservoirs are divided into layered, blocky, and irregular types (И.О.Брод and Н.А.Еременко, 1958).

Chinese petroleum geologists, based on the formation and distribution characteristics of oil and gas reservoirs in China's petroliferous basins, have proposed a series of classification schemes, most of which are based on the causes of trap formation. For example, using the cause of trap formation as the classification standard, and using trap morphology, shielding conditions, and reservoir rock types as the basis for dividing subclasses, China's continental basin oil and gas reservoirs are divided into four major types and several subclasses: structural, non-structural, composite, and hydrodynamic types (Hu et al., 1991). According to the dominant sealing factor in the formation of traps, traps are divided into four major types: structural, stratigraphic, hydrodynamic, and composite, each of which can be further divided into several subclasses based on the specific sealing factors of the reservoir's upward direction and combined with reservoir characteristics (He et al., 2010). According to the causes of trap formation, oil and gas reservoirs are divided into five major types: structural, stratigraphic, lithologic, hydrodynamic, and composite, and so on (Jiang & Zhaoming, 2016).

Milton and Bertram (1992) proposed a classification of traps based on the mechanism of sealing oil and gas, dividing traps into one-seal traps and poly-seal traps, which is essentially a classification by trap formation cause. One-seal trap refers to a trap where the bottom of the top sealing bed has a closed contour. Poly-seal trap refers to a trap where the bottom of the top sealing bed does not have a closed contour. In terms of the mechanism of sealing oil and gas, the anticlinal traps, buried hill traps, and sand lens body traps, which are classified by trap formation cause, all have closed structural contour and belong to one-seal traps.

In the process of oil and gas exploration, structural traps such as anticlines and faults usually have obvious characteristics and are easy to identify. Those stratigraphic and lithologic traps that are difficult to discover with conventional technical means often become important exploration targets in the later stages of basin exploration, especially lithologic traps and small-scale stratigraphic traps, which are not easy to identify and are referred to as subtle traps. Halbouty (1982) defined subtle traps as various potential traps that are hidden under unconformities or complex structural zones and are difficult to identify and explore with conventional technical means. It is generally believed that subtle oil and gas reservoirs refer to the general term for oil and gas reservoirs that are difficult to identify with conventional technical means, with the main types being lithologic oil and gas reservoirs and stratigraphic oil and gas reservoirs.

6.1.1.2 Classification by Hydrocarbon Phase

Based on the phase of hydrocarbons, oil and gas reservoirs can be classified into three major categories: oil reservoirs, gas reservoirs, and oil and gas reservoirs, with solid bitumen sand deposits being a special type of oil and gas reservoir. However, in fact, the phase of hydrocarbons underground is diverse, showing many transitional types, and each of these three types of phase oil and gas reservoirs can be further divided into several types. For example, in the gas reservoir category, in addition to dry gas reservoirs, wet gas reservoirs, and condensate gas reservoirs, there are also coalbed methane reservoirs where natural gas is in an adsorbed state and natural gas hydrate deposits where natural gas is in a solid state.

Classifying oil and gas reservoirs based on phase is not only of great significance for formulating development plans for different phase types of oil and gas reservoirs but also helps to understand the formation, evolution, and distribution patterns of oil and gas reservoirs in a region, which is also of great significance for exploration and evaluation. The classification of hydrocarbon phases of oil and gas reservoirs in China is shown in Table 6.1.

6.1.1.3 Classification of Unconventional Oil and Gas Reservoir

The above classification of oil and gas reservoirs is based on conventional oil and gas reservoirs. With the development of unconventional oil and gas geology theory and exploration, oil and gas reservoirs are divided into conventional and unconventional oil and gas reservoirs based on the state of oil and gas occurrence underground and the mechanism of reservoir formation. According to the source-reservoir combination type, unconventional oil and gas (reservoirs) can be divided into three types: source-reservoir integrated, source-reservoir adjacent, and source-reservoir separated. The source-reservoir integrated type includes coalbed methane, shale oil and gas, etc., the source-reservoir adjacent type includes tight sandstone oil and gas, and the source-reservoir separated type mainly includes oil sand oil and natural gas hydrates, etc.

Unconventional oil and gas reservoirs have an essential difference from the traditional single trap oil and gas reservoirs. Most unconventional oil and gas are distributed continuously. Oil and gas are continuously distributed in a large area of tight reservoir rocks. This type of oil and gas accumulation has a large spatial distribution range, but the boundary is blurred and can be called unconventional trap oil and gas reservoirs. Continuous oil and gas reservoirs are mainly distributed in the center and slope of the basin, and are locally enriched.

Table 6.1 Classification of hydrocarbon phases of oil and gas reservoirs in China (according to Sun Zhida, cited from Zhai et al., 1996, simplified)

Phase types of oil and gas reservoir		Original gas-oil ratio m^3/t	Phase		Characteristics of crude oil on the ground			Typical hydrocarbon composition (%)	
Major category	Sub-category		Reservoir	Ground	Color	Relative density	Viscosity (mPa s)	C_1	C_5
Gas reservoir	Dry gas reservoir	No oil	Gas phase	Gas phase	No oil	No oil	No oil	96	0.0
	Wet gas reservoir	>1,500	Gas phase	Gas–liquid	Transparent	0.6	<1.0	91.6	0.94
	Condensate gas reservoir	>1,000	Gas phase	Gas–liquid	Transparent-light yellow	0.6–0.8	<1.0	87	4.6
Critical state oil and gas reservoir	Near-critical condensate gas reservoir	600–1,000	Gas phase	Gas–liquid	Yellow-orange	0.76–0.81	1.0–2.0	70.4	11
	Critical state oil and gas reservoir	526	Gas–liquid	Gas–liquid	Yellow-orange	0.76–0.82	1.0–4.0	59.7	14.5
	Near-critical state oil reservoir (high volatility)	350–650	Liquid phase	Gas–liquid	Orange-yellow-light green	0.76–0.82	1.0–1.0	64	19.7
Oil reservoir	Light oil reservoir	10–350	Liquid phase	Liquid phase	Light green–brown	0.76–0.83	5–10	35	38.3
	Conventional oil reservoir	35–250	Liquid phase	Gas–liquid	Black	0.83–0.87	10–30	49	44.7
	Slightly heavy oil reservoir	<35	Liquid phase	Gas–liquid	Black	>0.87	30–100	20	71
	Moderately heavy oil reservoir	Trace gas	Liquid phase	Gas–liquid	Black	0.9–1.0	100–4000		

(continued)

Table 6.1 (continued)

Phase types of oil and gas reservoir		Original gas-oil ratio m^3/t	Phase		Characteristics of crude oil on the ground			Typical hydrocarbon composition (%)	
Major category	Sub-category		Reservoir	Ground	Color	Relative density	Viscosity (mPa s)	C_1	C_5
	Severely heavy oil reservoir	Virtually no gas	Liquid phase	Liquid phase	Black	>1.0	>400		
Bituminous oil mine		No gas and liquid	Solid phase	Solid phase	Black	>1.0			>90

6.1.2 Basis and Classification Scheme of Oil and Gas Reservoir Classification

6.1.2.1 Basis for Oil and Gas Reservoir Classification

For conventional oil and gas reservoirs, the trap is the basic condition for the formation of oil and gas reservoirs. Under different structural, stratigraphic, and lithological conditions, the causes of traps and the characteristics of oil and gas reservoirs are different, and the types of oil and gas reservoirs are also different. Classifying oil and gas reservoirs based on the causes of traps can fully reflect the formation conditions of various types of oil and gas reservoirs and reflect the differences and connections between various types of oil and gas reservoirs. Therefore, from the perspective of conventional oil and gas exploration, classifying oil and gas reservoirs based on the causes of traps is more conducive to understanding and finding oil and gas reservoirs, scientifically predicting the types of oil and gas reservoirs that may appear in a new area, and using different exploration methods and different exploration and development deployment plans for different types of oil and gas reservoirs.

Like conventional oil and gas reservoirs, unconventional oil and gas reservoirs can be understood as the accumulation of oil and gas in "unconventional traps." However, the mode of occurrence and formation mechanism of oil and gas in such traps are significantly different from those of conventional oil and gas reservoirs, with tight sandstone oil and gas reservoirs and shale oil and gas reservoirs being typical representatives. In addition, considering that most tight sandstone oil and gas reservoirs and shale oil and gas reservoirs have abnormal pressure, it indicates that the geological bodies containing oil and gas (unconventional traps) are relatively closed and should have certain boundaries, which are different from the boundaries of oil and gas distribution in conventional traps. They should be blurred and transitional boundaries. Therefore, oil and gas accumulated in unconventional traps can be classified as a special type—unconventional oil and gas reservoirs.

Classifying oil and gas reservoirs based on the causes of traps should follow two basic principles: scientific and practical. Scientific means that the classification should fully reflect the causes of traps, have a high degree of scientific generalization, and reflect the differences and connections between various types of oil and gas reservoirs. Practical means that the classification should effectively guide the exploration and development work of oil and gas reservoirs and be relatively simple and practical.

6.1.2.2 Oil and Gas Reservoir Classification Scheme

Based on the causes of traps and the scientific and practical principles of oil and gas reservoir classification, this textbook divide oil and gas reservoirs into five major categories: structural, stratigraphic, lithologic, composite, and special types, which are further divided into several types. For the specific classification, names, and typical schematic diagrams of oil and gas reservoirs, see Table 6.2.

Structural oil and gas reservoirs refer to the accumulation of oil and gas in structural traps formed by the deformation or displacement of strata due to crustal movement. Structural movements can form various structural traps, such as anticline traps and fault traps, so the types of oil and gas reservoirs formed are also different, but their common feature is that the cause of the trap is the result of structural action.

Stratigraphic oil and gas reservoirs refer to the accumulation of oil and gas in stratigraphic traps. Here, the concept of stratigraphic traps is narrow, referring to traps formed by the discontinuity of vertical sedimentation continuity of reservoirs, that is, the formation of traps is directly related to stratigraphic unconformities. According to the relationship between stratigraphic unconformities and reservoirs, they can be further divided into subclasses, such as buried hill oil and gas reservoirs, stratigraphic overstep oil and gas reservoirs, etc. It should be noted that the broad sense of stratigraphic traps also includes lithologic traps (He et al., 2010; Levorsen, 1967). This textbook classification adopts the narrow concept.

Lithologic oil and gas reservoirs refer to the accumulation of oil and gas in traps formed by the horizontal changes in the lithology or physical properties of reservoirs. Due to changes in sedimentation conditions or diagenesis, reservoirs gradually become non-permeable rocks in both vertical and horizontal directions, thus forming traps. Such as sandstone updip pinchout, sandstone lens body, etc.

In nature, many phenomena are not simply binary, but there are many transitional types between two extremes or multiple extremes, such as the kerogen types, there are many composite types between Type I and Type III. Oil and gas reservoir types are also like this. In addition to traps formed by single factors such as structure and stratigraphy, there are many composite types. The possibilities of forming traps by combining various geological factors are diverse, which can form traps controlled by single geological factors such as structure, stratigraphy, and lithology, and can also be formed by the combination of two or more factors, forming composite traps, such as structural-lithologic traps, structural-stratigraphic traps.

Special oil and gas reservoirs have different trap characteristics from the three major categories of structural, stratigraphic, and lithologic oil and gas reservoirs, and their modes of oil and gas occurrence and accumulation mechanisms are also significantly different. They mainly include hydrodynamic oil and gas reservoirs, tight sandstone oil and gas reservoirs, shale oil and gas reservoirs, and coalbed methane reservoirs. Hydrodynamic traps are formed by hydrodynamic sealing, that is, sealed by the combination of hydrodynamic forces and non-permeable rocks, making it possible to accumulate oil and gas in places that cannot be accumulated under static water conditions, and this type of oil and gas reservoir is currently very rare. The formation mechanisms of tight sandstone oil and gas reservoirs, shale oil and gas reservoirs, and coalbed methane reservoirs are different from those of conventional oil and gas reservoirs, and they are a type of oil and gas accumulation without buoyancy, which does not require the traditional sense of traps.

Table 6.2 Oil and gas reservoir classification

Major category	Category	Subcategory	Typical pattern
Structural oil and gas reservoir	Anticline oil and gas reservoir	Compressional anticline oil and gas reservoir	
		Basement uplift anticline oil and gas reservoir	
		Diapir uplift anticline oil and gas reservoir	
		Drape anticline oil and gas reservoir	
		Rollover anticline oil and gas reservoir	
	Fault oil and gas reservoir	Faulted nose-like oil and gas reservoir	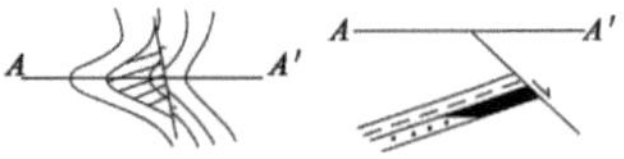
		Fault block oil and gas reservoir	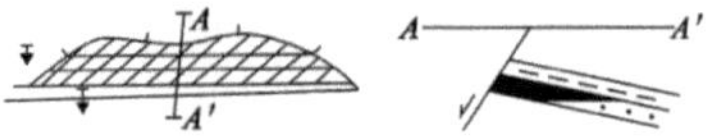
			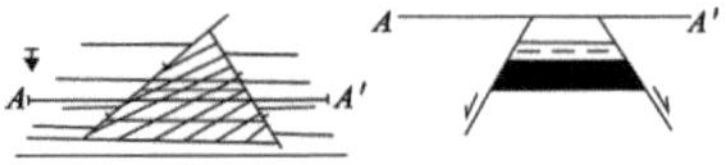

	Rock piercing oil and gas reservoir	Salt piercing oil and gas reservoir	
		Magma piercing oil and gas reservoir	

(continued)

Table 6.2 (continued)

Major category	Category	Subcategory	Typical pattern
		Mud volcano piercing oil and gas reservoir	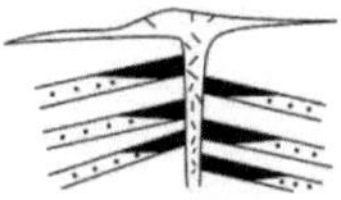
	Fractured oil and gas reservoir		
Stratigraphic oil and gas reservoir	Buried hill oil and gas reservoir		
	Stratigraphic unconformity overlain oil and gas reservoir		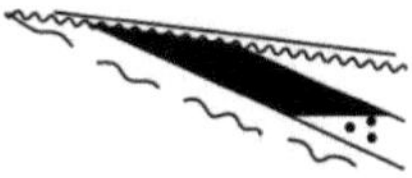
	Stratigraphic overstep oil and gas reservoir		
Lithologic oil and gas reservoir	Lithologic updip pinchout oil and gas reservoir		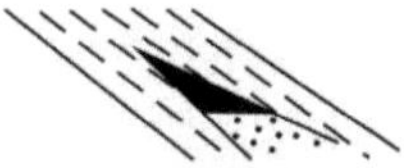
	Sand lens body oil and gas reservoir		
	Physical property sealing oil and gas reservoir		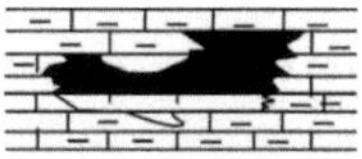
	Biological reef oil and gas reservoir		
Composite oil and gas reservoir	Structural-stratigraphic oil and gas reservoir		
	Structural-lithologic oil and gas reservoir		
Special type oil and gas reservoir	Hydrodynamic oil and gas reservoir		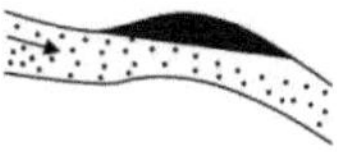

(continued)

Table 6.2 (continued)

Major category	Category	Subcategory	Typical pattern
	Tight sandstone oil and gas reservoir		
	Shale oil and gas reservoir		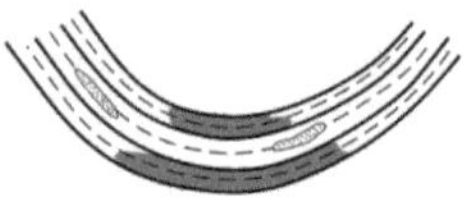
	Coalbed methane reservoir		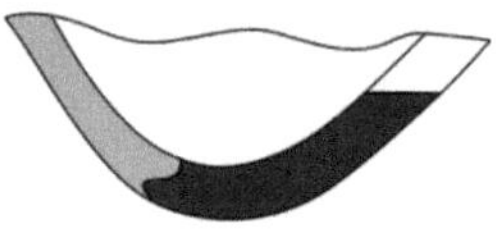

6.2 Structural Oil and Gas Reservoirs

Structural closures formed by the deformation or displacement of strata due to crustal movements are known as structural traps. Oil and gas accumulations within these structural traps are referred to as structural oil and gas reservoirs. Various structural traps can be formed by tectonic movements, leading to a diversity of structural oil and gas reservoirs. Based on the causes of the traps, structural oil and gas reservoirs can be classified into four types: anticlines, faults, piercement structures, and fractured reservoirs.

6.2.1 Anticline Oil and Gas Reservoirs

6.2.1.1 Main Features of Anticline Oil and Gas Reservoirs

Under the influence of tectonic movements, strata undergo bending deformation, forming anticlines that dip towards the surroundings, known as anticline traps. Oil and gas accumulations in anticline traps are called anticline oil and gas reservoirs.

Anticline oil and gas reservoirs have always held the most important position in the history of global oil and gas exploration and are the earliest recognized type of oil and gas reservoir by petroleum geologists. In the mid-to-late nineteenth century, the "anticline theory" proposed by American geologist I. C. White played a significant role in the history of oil and gas exploration. Currently. anticline oil and gas reservoirs still account for most global oil and gas production and reserves. Statistics show that among the more than 200 giant oil fields in the world with ultimate recoverable reserves of more than 7100 $\times$ 10^4 t tons (5 $\times$ 10^8 bbl), anticline oil reservoirs account for more than 75% of the total, and anticline gas reservoirs also dominate in large gas fields (Zhang et al., 1999). Anticline oil and gas reservoirs have always been the most important type for the discovery of large oil and gas fields in oil and gas exploration, especially in the early stages of basin exploration, where anticline oil and gas reservoirs are the main type prioritized by explorers.

6.2.1.2 Types of Anticline Oil and Gas Reservoirs

The existence of anticline traps is the basic condition for the formation of anticline oil and gas reservoirs. Morphologically, there are many types of anticline traps, such as long-axis anticlines, short-axis anticlines, box-shaped anticlines, and recumbent anticlines. In terms of genesis, the anticline traps and anticline oil and gas reservoirs related to oil and gas accumulation mainly have the following five types.

(1) Compressional Anticline Oil and Gas Reservoir

Compressional anticline oil and gas reservoirs refer to oil and gas accumulations in anticline traps formed by folding actions dominated by lateral compressive stress. They are commonly found in folded areas with steeply dipping strata and often asymmetrical. They have a large closure height but a small closure area. Due to the intense deformation of the strata, fractures often accompany the formation of the anticline trap. A typical example is the L-layer oil and gas reservoir in Laojunmiao Oilfield, Jiuquan Basin, China, as shown in Fig. 6.1. It is an asymmetrical anticline trap with a south wing dip of 20°–30° and a north wing dip of 60°–80°. The ratio of the long axis to the short axis is 3:1, and it is cut by thrust faults and cross-faults.

From a regional perspective, these anticlines are distributed within tectonic units such as foredeeps in front of mountains and intermountain foredeeps in folded zones, often appearing in rows or belts. The anticline belt in the southern Qilian Mountain front area of the Jiuquan Basin in China is a typical example. It is an anticline structural belt composed of a series of anticlines, including Qingcao Bay, Ya'er Gorge, Laojun Temple, and Petroleum Gully.

Compressional anticline oil and gas reservoirs are also widely distributed in other petroliferous areas in China that have undergone folding and compression. For instance, the high-steep anticline gas reservoirs in the eastern Sichuan Basin are typical representatives. Figure 6.2 shows the structural plan and cross-section of the Wolonghe Gas Field in Eastern Sichuan (Jia Wu 1), which is a long-axis anticline cut by multiple reverse faults. The two wings are asymmetrical, with the western wing being steep (dip angle of 40°–50°) and the eastern wing being gentle (dip angle of 20°–25°), and the structural axis shifts towards the gentle wing from shallow to deep. The gas field consists of multiple anticline gas reservoirs, with the Jia Wu 1 gas reservoir being the main one [China Petroleum Geology (Volume 10), 1989].

Abroad, many folded regions also widely distribute this type of compressional anticline oil and gas reservoirs. For example, in the Zagros Mountain foredeep of the Persian Gulf Basin, the Appalachian Mountain foredeep in the United States, and

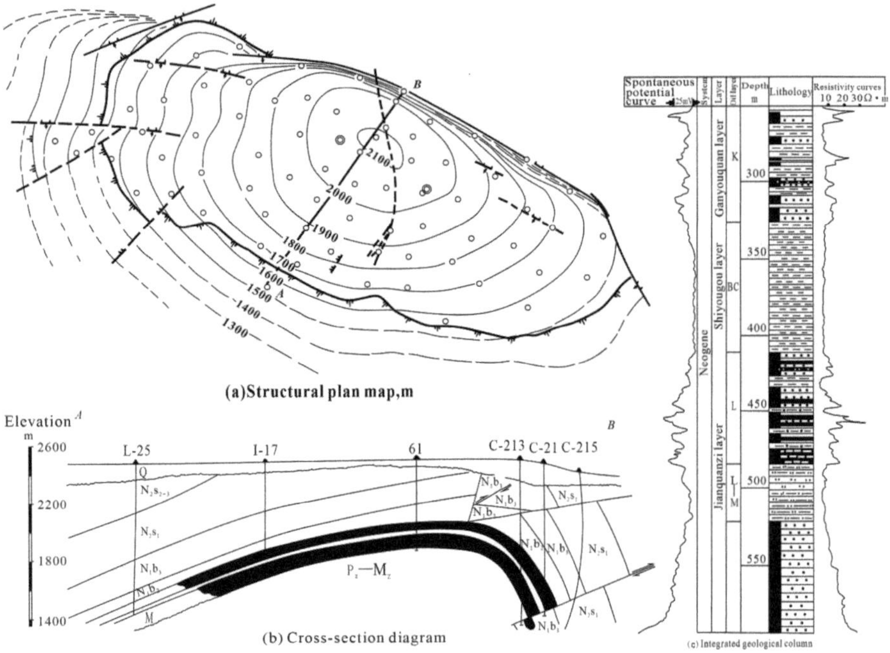

Fig. 6.1 Integrated map of the Laojun Temple anticlinal oil reservoir (according to Yumen Petroleum Administration Bureau)

the Caucasus Mountain foredeep in the former Soviet Union, there are numerous compressional anticline oil and gas reservoirs.

(2) Basement Uplift Anticlinal Oil and Gas Reservoirs

Basement uplift anticlines, formed by the differential subsidence of the basement during sedimentation, are characterized by gentle and large anticline structures. These are commonly found in platform areas where the basement activity primarily forms the anticline traps. The activity of the basement deforms the sedimentary cover, creating anticline traps. The main features are: gentle dips of the wing beds, low closure heights, and large closure areas compared to folded zones. Regionally, these anticline traps often appear as belts in the internal and marginal depressions of platforms, forming long walls or large uplifts. Particularly, the early buried uplift belts in the center of the depression, when they match the formation process of the anticline traps during periods of large-scale oil and gas migration and accumulation, become the best places for oil and gas accumulation, forming a series of this type of oil and gas reservoirs. The oil and gas reservoirs in the Daqing Changyuan and Sartu oilfields in China belong to this type, as shown in Fig. 6.3.

In some platform areas abroad, the distribution of this type of oil and gas reservoir is also widespread, including many famous super-large oil and gas fields. For

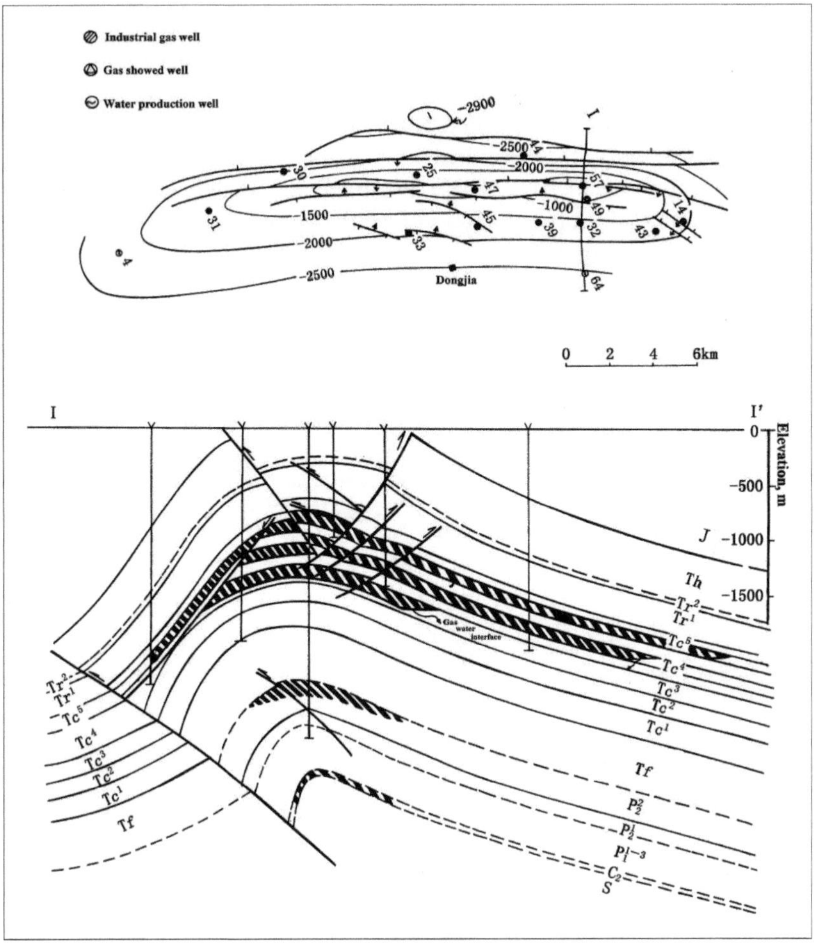

Fig. 6.2 Structural plan and cross-section of Wolonghe Gas Field in Sichuan Basin (according to "China Petroleum Geology," Volume 10, 1989)

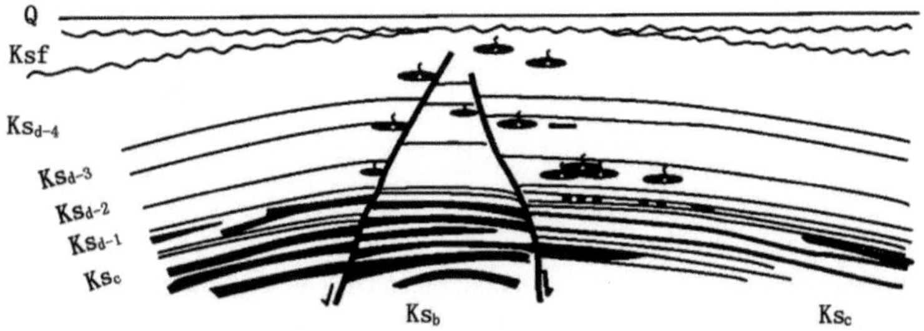

Fig. 6.3 Profile of the Daqing Sartu Oilfield (according to Daqing Petroleum Administration, 2000)

example, the Ghawar oil field in the Persian Gulf Basin, which has the highest production and reserves in the world, the Samotlor oil field in the West Siberian Basin, and the Urengoy gas field, their oil and gas reservoirs are mainly related to basement activity anticline oil and gas reservoirs. The Samotlor oil field is located on the lower Valtov uplift of the West Siberian Basin, with an oil-bearing area of 1575 km^2, original recoverable reserves of 20.6×10^8 t, the main producing layer is the Cretaceous sandstone, buried depth is 1610–2700 m, and the main oil reservoir belongs to the anticline oil reservoir (Fig. 6.4).

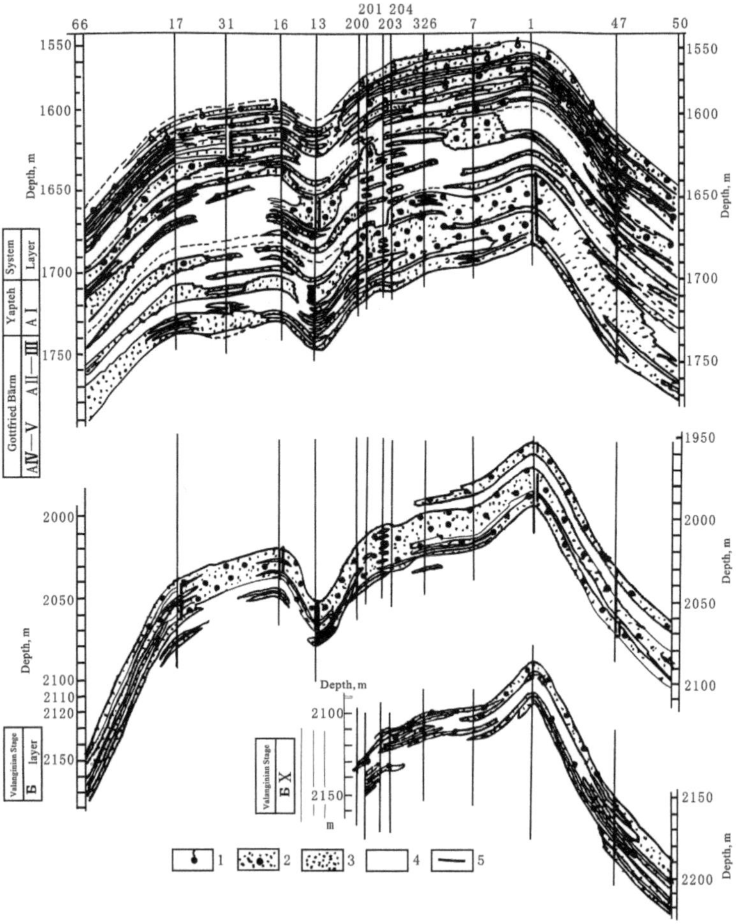

Fig. 6.4 Horizontal cross-section of the Samotlor oil and gas reservoir (according to Zhang et al., 1999). 1. Gas-bearing, 2. oil-bearing, 3. water-bearing, 4. non-reservoir rock, 5. exploration well perforated section

(3) Diapir Uplift Anticlinal Oil and Gas Reservoirs

The formation of these traps is the result of the activity of subsurface plastic materials. Thick layers of plastic sediments such as salt rocks, gypsum, and mudstones within a basin can be deformed by the uneven gravitational load from the overlying rocks and lateral horizontal stress, causing the plastic layers to creep and uplift, deforming the overlying strata and forming diapir uplift anticline traps. Most diapir uplift anticlines associated with oil and gas accumulation are formed from salt rocks or mixed layers of salt rocks, gypsum, and mudstones, with salt domes being the most prominent. These anticlines often develop trench-like or radial fracture systems at their axis, with the crest collapsing and faults complicating their structure. In some cases, while they appear as anticlines macroscopically, the basic units of oil and gas accumulation often no longer have intact anticline traps but are segmented into numerous half-anticlines and fault block traps by faults.

The oil reservoir in the Wangchang Field of the Qianjiang Depression in the Jianghan Basin is a typical representative of this type (Fig. 6.5). The Qianjiang Formation in the Qianjiang Depression is a set of salt lake facies mudstone series rich in salt, with a thickness of over 3500 m. The salt rock layers can reach up to 153 layers, with a cumulative thickness accounting for 50% of the total thickness, especially well-developed in the lower part of the Qian 4 Member. The Wangchang Field is a long-axis anticline with a NW strike, nearly symmetrical wings, and an uplift amplitude of up to 800 m. In the cross-section, the strata dip angle is gentle in the upper part (only 20°) and steep in the lower part (60°–70°), with the underground core being a salt rock uplift. According to seismic data, the structure has completely disappeared at depths of 6000–7000 m.

In the Dongying Depression of the Bohai Bay Basin, there is also a set of plastic strata over 1000 m thick in the lower part of the Paleogene, composed of salt rocks, gypsum, and mudstones. This mixed plastic layer arches in the center of the depression and is the main mechanism for the formation of the central uplift zone. The Dongxin Oilfield on this structural belt has a typical plastic arching anticline as its structural background. The structure consists of the Dongying dome anticline and the Xinzhen long-axis anticline, which are distributed in an east–west direction, and the trench-like fracture system developed at the axis cuts it into trench-like anticlines. The distribution of oil and gas reservoirs is macroscopically controlled by the anticline structure, but most individual oil and gas reservoirs are fault reservoirs (Jiang, 1998).

Many typical instances of this type of oil and gas reservoir have also been found abroad. For example, the Burgan Field, the largest oil field in Kuwait in the Middle East, has the main oil-bearing layers of the Middle Cretaceous Wara sandstone and Burgan sandstone, with the Madud limestone as the interval between the two. The Wara sandstone is a fine-grained to siltstone interbedded with dark claystone, 60 m thick. The Burgan sandstone is a medium to coarse-grained quartz sandstone interbedded with varying thicknesses of dark gray claystone, 335 m thick, and is of deltaic facies deposition. The porosity is 25–30%, and the permeability is (3000–4000) $\times 10^{-3}$ μm^2, with the oil field's recoverable reserves being 90×10^8 t, making

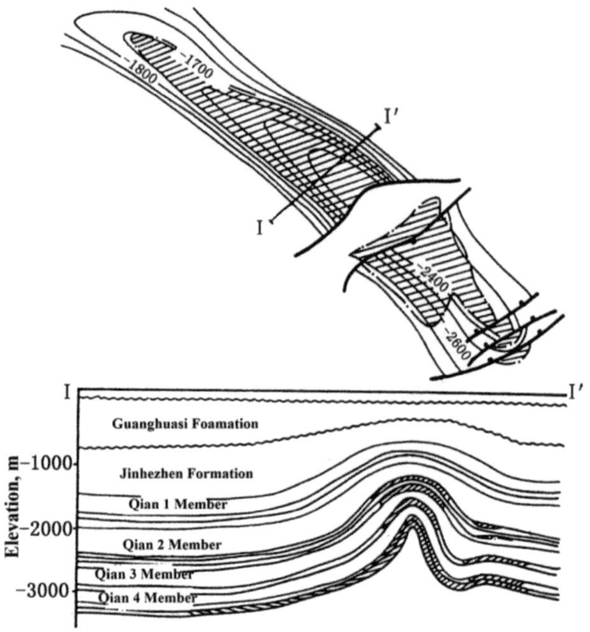

Fig. 6.5 Structural plan and cross-section of Wangchang Field in Jianghan Basin (according to Hu et al., 1991)

it the second-largest oil field in the world (Fig. 6.6). The cause of the Burgan Field's anticline structure trap is the result of long-term activity of the thick Jurassic lagoonal flexible salt layer.

(4) Drape Anticlinal Oil and Gas Reservoirs

The formation of these anticlines is related to topographic highs and differential compaction processes. Various topographic highs often exist on the sedimentary basement, composed of crystalline basement rocks, hard and dense sedimentary rocks, or biological reef blocks. When new sediments are deposited on top, the overlying sediments above these highs are often thinner, while the surrounding sediments are thicker. Consequently, during the lithification process, due to the different thicknesses and rock properties of the sediments, the degree of compaction is uneven. The thicker sediments surrounding the ancient topographic highs are compacted more than the top, resulting in a dome-like uplift of the overlying strata above the topographic high (buried hill), forming an anticline trap. These often present as domal structures with a flat top and slightly steeper wings, with a larger amplitude at the bottom and a smaller one at the top. Anticlines formed in plastic mudstones are more pronounced with steeper dips, while those formed in harder sandstones and limestones are less distinct with more gentle dips. The anticlines on the upper part of a buried hill often reflect the shape of the underlying buried hill, but their closure is always less than the height of the buried hill, decreasing upwards, and the dips also decrease upwards.

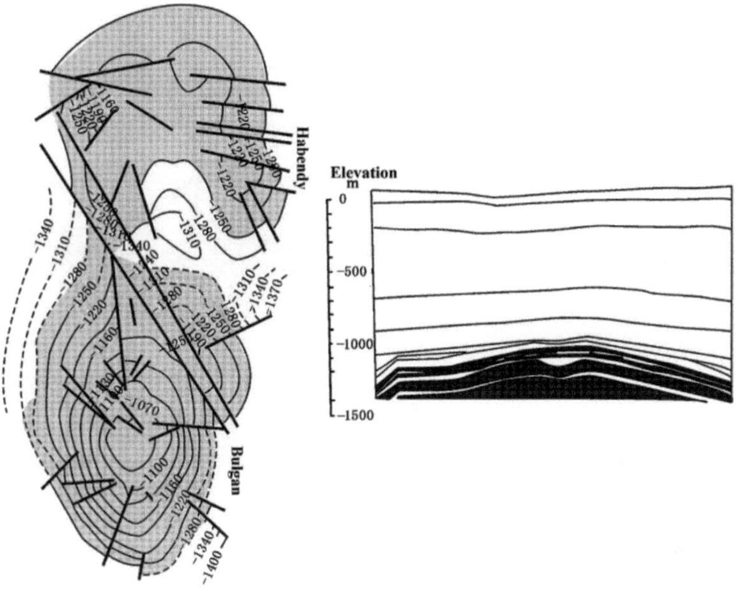

Fig. 6.6 The cause of the Burgan Field's anticline structure trap is the result of long-term activity of the thick Jurassic lagoonal flexible salt layer

These drape anticline structures are also known as draped structures or differential compaction anticlines. For example, in the Jiyang Depression of the Bohai Bay Basin, the Gudao Oilfield and Gudong Oilfield are primarily dominated by drape anticline oil reservoirs. Their "basement" mainly consists of eroded highs (buried hills) composed of Ordovician limestone or dolomite, with overburden deposits of the Paleogene, and the top is covered by the Neogene Guantao Formation and the Minghuazhen Formation, forming a large-scale draped structure. In particular, the Guantao Formation has a typical anticline oil and gas reservoir related to erosion and differential compaction, as shown in Fig. 6.7.

Many oil and gas basins abroad also contain this type of oil and gas reservoir. For instance, in the Pennsylvanian reservoir of the Simaol Oilfield in the Permian Basin of the North American Platform, the underlying coral reef high—the Pennsylvanian anticline—reflects the morphology of the underlying high (Zhang et al., 1999).

(5) Rollover Anticlinal Oil and Gas Reservoirs

In numerous Cenozoic siliciclastic sedimentary basins around the world, many rollover anticline traps and their associated oil and gas reservoirs related to syndepositional faults have been discovered. These oil and gas reservoirs are predominantly located in deltaic areas, characterized by their gentle anticlines, which are primarily the result of syndepositional fault activity during sedimentation. Under the influence of fault block activity and gravity gliding, the sand and mudstone strata accumulated on the downthrown block of the syndepositional fault slide along the fault plane,

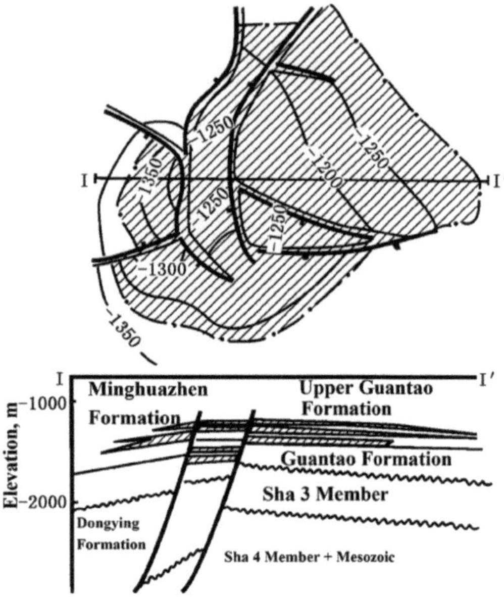

Fig. 6.7 Structural map of the Guantao Formation oil reservoir in Gudong Oilfield (contour interval in meters) and cross-section diagram (according to Shengli Oilfield, 2004)

causing the strata to develop reverse drag, forming this special type of "rollover anticline" trap. These rollover anticlines are situated on the downthrown block of syndepositional faults that dip towards the depression, often forming small, wide, and asymmetric short-axis anticlines with a gentler slope on the side farther from the fault and a steeper slope on the side closer to the fault. The axis is nearly parallel to the fault line, commonly distributed in a beaded or belt-like pattern along the fault. The structural amplitude is larger in the central part and smaller in the deeper and shallower layers. The high point of the anticline is closer to the fault, and the high point gradually shifts towards the deeper part, with the shift trajectory roughly parallel to the fault plane. These rollover anticlines are usually close to the oil source area, facing oil-generating depressions, developed in large deltaic deposits, with thick and high-quality reservoir sands, and forming good source-reservoir-cap combinations. Additionally, as the structure is of syndepositional nature, syndepositional faults can serve as migration pathways for oil and gas. Therefore, these anticlines generally have good conditions for oil and gas accumulation and can often form rich and high-yield oil and gas reservoirs.

A considerable number of such oil and gas reservoirs have been discovered in the Bohai Bay Basin. Some oil fields on the structural belts controlled by syndepositional faults in the Dongying Depression, such as the Shengtuo Oilfield and Yong'an Town Oilfield, belong to this category. The Linpan Oilfield in the Huimin Depression and the Gangdong Oilfield in the Qikou Depression are also rollover anticline structures formed under the control of syndepositional faults. Their main oil-bearing stratum is the Shahejie Formation, which is very rich in oil and gas. Due to the long-term activity of syndepositional faults covering the period of large-scale oil and gas migration and

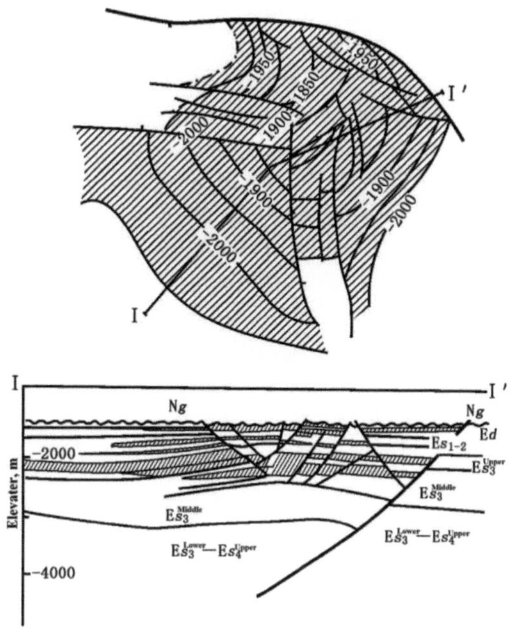

Fig. 6.8 Plan map of the rolling anticlinal oil reservoir in Shengtuo area (contour interval in meters) and cross-section diagram (according to Hu et al., 1991)

accumulation, multiple stratigraphic systems contain oil and gas in the vertical direction. For example, in the Shengtuo area of the Dongying Depression, the rollover anticline structure has a long development time, strong inheritance, and forms simultaneously with the reservoir-cap layers that make up the trap, and is located on the oil and gas migration pathway, possessing extremely favorable conditions for oil accumulation (Fig. 6.8).

There are also many such oil and gas reservoirs abroad, and they are often high-yielding. For example, there are nearly 200 such oil and gas reservoirs in the Niger Delta region of Nigeria. The first offshore oil and gas field in Nigeria, the Okan Field, has an oil and gas reservoir that is a typical rollover anticline type (Fig. 6.9).

The Okan Oil Field is located on the Niger Delta and is a rollover anticline trap. About 3 km to its northeast, there is a major syndepositional fault, which, along with the rollover anticline, was formed syndepositional. The Okan anticline is about 10 km long and 5 km wide. The long axis strikes are in the NW–SE direction. The structure is gentle with three distinct high points. In addition, a considerable number of this type of oil and gas reservoir have also been discovered in areas such as the Gulf of Mexico in the United States.

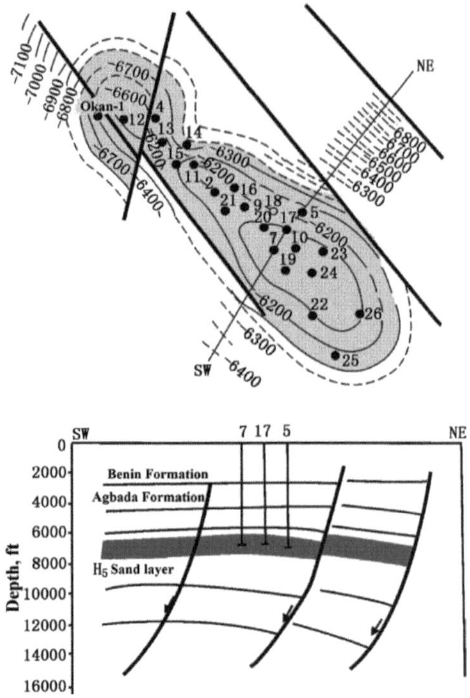

Fig. 6.9 Structural map of the top of the H5 sand layer in the Okan oilfield, Nigeria (contour interval in feet) and cross-section diagram (according to Li, 2005)

6.2.2 *Fault Oil and Gas Reservoirs*

A fault trap refers to the closure formed by the obstruction of a fault along the updip direction of a reservoir. The accumulation of oil and gas within a fault trap is called a fault oil and gas reservoir. This type of reservoir is widely distributed in petroliferous basins around the world. Oil and gas exploration practices in China have shown that fault oil and gas reservoirs are extensively distributed, especially in the eastern regions where faulting has been active since the Mesozoic, leading to the formation of many fault basins. Additionally, many faults have developed in the slope belts and anticline belts of these basins, resulting in a multitude of fault oil and gas reservoirs. For instance, in the Bohai Bay Basin, where faults are highly developed, fault oil and gas reservoirs play a dominant role. Due to the action of faults, multiple stratigraphic systems contain oil and gas, forming numerous complex oil and gas accumulation zones.

6.2.2.1 Mechanism of Fault Trap Formation

The prerequisite for the formation of a fault trap, given the presence of reservoir and seal conditions, is that the fault must be sealed, that is, it must obstruct the migration of oil and gas. At the same time, the reservoir must be blocked by the fault in the updip direction, creating an enclosed space with the fault and the reservoir, allowing oil and gas to accumulate and form a reservoir within this enclosed space.

Over the past two decades, a more comprehensive understanding of fault sealing mechanisms has emerged. Currently, fault sealing mechanisms are primarily divided into the following four types:

(1) Juxtaposition sealing: Reservoir sandstones are juxtaposed against low-permeability mudstones with high displacement pressures, meaning the updip direction of the reservoir sandstone is in contact with mudstones.
(2) Clay smear sealing: Within interbedded sandstone and mudstone strata, plastic mudstone layers sheared by the fault on both sides of the fault are smeared along the fault zone, giving the fault zone itself high displacement pressure.
(3) Cataclasis sealing: Fragmentation reduces the particle size and permeability of the fault zone, such as sand particles breaking to form fine-grained fault gouge. As fault displacement increases, the fragmentation effect is enhanced.
(4) Diagenetic sealing: Changes in subsurface geological conditions cause minerals to precipitate and cement within the fault zone, making the filling material within the fault zone have poorer porosity and permeability, and thus a higher displacement capacity for oil and gas.

In petroliferous basins, juxtaposition sealing and clay smear sealing are the two most important fault sealing mechanisms. In fact, a fault is not a simple plane but a complex geological body composed of a fault core and fractured zones on both sides. The conduction and sealing conditions of the fault core and fractured zones differ, and so do their sealing mechanisms (Fig. 6.10).

Essentially, the sealing capacity of a fault depends on the difference in displacement pressure between the juxtaposed rock layers or the reservoir and the fault rock on both sides of the fault plane. If the displacement pressure on the other side of the fault zone or the reservoir in the updip direction is higher, it can play a certain sealing role and form an oil and gas reservoir under certain closed structural patterns. The most common sealing type is the formation of a fault trap by the juxtaposition of

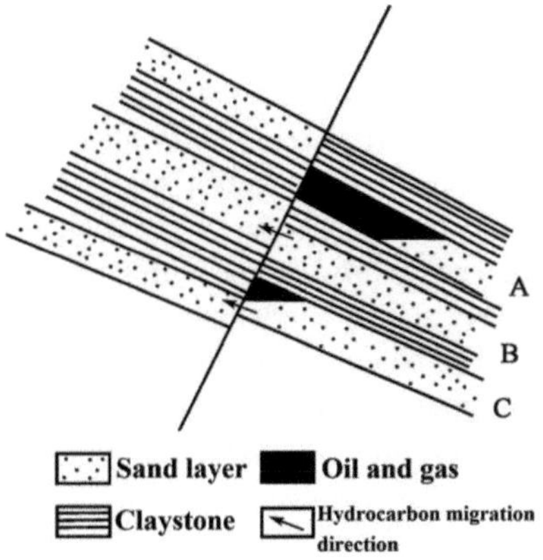

Fig. 6.10 The impact of lithological contact on both sides of the fault on the sealing of the fault trap. Layer A is completely sealed; Layer B is not sealed; Layer C is partially sealed

mudstones and sandstones on the other side of the fault in the updip direction, because the sealing mudstones have a higher displacement pressure, which can prevent the horizontal migration of oil and gas across the fault. The three sealing mechanisms of clay smear sealing, cataclasis sealing, and diagenetic sealing are all about the fault zone forming a seal, with the reservoir being blocked in the updip direction by impermeable fault gouge or other filling materials on the fault zone, creating sealing conditions.

The factors affecting fault sealing are multifaceted. Under the condition that the fault has ceased activity, they can be mainly attributed to the following two aspects:

(1) The lithology on both sides of the fault and its juxtaposition relationship: If the permeable rock layers on both sides of the fault are in direct contact, they often cannot play a sealing role. If the permeable layers and non-permeable layers are juxtaposed on both sides of the fault, the fault sealing is better (Fig. 6.10). The lithology of the faulted strata has a significant impact on the sealing of the fault. In strata with strong plasticity (such as mudstones) where a fault occurs, a dense fault gouge is often formed along the fault plane, which can play a sealing role. The thicker the mudstone in the faulted strata, the better its sealing.
(2) The nature and attitude of the fault: Different forces result in faults of different natures. Faults formed by compressive torque forces show a tight fracture zone, often giving the fault plane a sealing nature. In contrast, the fracture zone of tensile faults is often loose and can act as a channel. However, this does not mean that the sealing of tensile faults is necessarily worse than that of compressive-torsional faults. The orientation of the fault and its relationship with the regional principal stress direction also affect sealing, with faults perpendicular to the regional maximum principal stress having the best sealing, and vice versa. The steepness of the fault plane also has an impact, with steep fault planes having poor sealing and gentle fault planes having good sealing. As the depth of the fault increases and the overburden pressure increases, the sealing improves.

In addition, the precipitation of dissolved substances in groundwater (such as calcium carbonate) can cement the fractured zone together; or during the migration of oil and gas along an open fault, solid asphalt and other substances formed due to the oxidation of oil can block the migration pathways, both playing a sealing role.

When analyzing the sealing of numerous tensile normal faults in the Mesozoic-Cenozoic rift basins of eastern China, the impact of fault fillings on sealing should be considered. Studies have shown that fault fillings is a common geological phenomenon. If the offset rock layers are mainly mudstones, the fillings should also be mainly clayey, with poor permeability, resulting in good sealing. Conversely, if it is mainly sandstone, the sealing ability becomes poor (Lv et al., 2013).

The basic condition for the formation of a fault trap is that the fault must be in the updip direction of the reservoir and be sealed. Therefore, when studying fault sealing, attention must be paid to the combined relationship between the fault plane's dip and the strata's dip to correctly determine whether it is the hanging wall sealing or the footwall sealing. When the strata on both sides of the fault dip in opposite directions, both the upper and lower blocks may form good sealing conditions.

6.2.2.2 The Role of Faults in Oil and Gas Migration and Accumulation and Their Accumulation Patterns

(1) The Role of Faults in Oil and Gas Migration and Accumulation

Faults disrupt the continuity of rock layers. The nature of faults, the degree of fracturing and cementation, and the contact relationships between lithologic combinations on either side of the fault plane are all closely related to the migration, accumulation of oil and gas, and the destruction of oil and gas reservoirs. Sometimes the same fault plays different roles at depth and shallow levels. Throughout geological history, a fault may also play opposite roles at different times, either sealing or destroying. Therefore, the role of faults in the formation of oil and gas reservoirs should be considered from multiple aspects, especially by deeply analyzing the relationship between the fault's evolutionary history and the periods of migration and accumulation, as well as the lithologic combination relationships on either side of the fault and the sealing nature of the fault zone. Only in this way can we correctly understand the role of faults and find the patterns of fault-related oil and gas accumulation. From the perspective of oil and gas migration and accumulation, faults have both positive and negative effects on the formation of oil and gas reservoirs, that is, they have both sealing and conducting roles (Hooper, 1991).

a. Sealing Role

The sealing role refers to the existence of faults that seal oil and gas both vertically and horizontally, preventing their escape and eventually leading to the accumulation of oil and gas reservoirs. The factors affecting the sealing nature of faults have been described above.

b. Conducting and Destructive Role

Due to the high degree of fault activity, the equilibrium state of the original oil and gas reservoirs is disrupted, or oil and gas migrate vertically from the source rock system to the shallower strata to form secondary oil and gas reservoirs, at which point the fault becomes a conduit for oil and gas migration. If a fault is encountered that terminates in an upper stratum and disappears, and there is a good cap rock above, a secondary oil and gas reservoir can be formed. The stratum of such a secondary oil and gas reservoir often coincides with the shallowest stratum where the fault terminates. For example, in the Dagang Oilfield, the highest point where the fault terminates is below the ground at 600–700 m, and the shallow secondary oil and gas reservoirs are also formed below this depth. Another example is the Dongxin Oilfield, where the vertical span of oil and gas-bearing well sections is more than 2000 m, and the shallowest gas-bearing stratum, the Minghuazhen Formation, is also the stratum where the main fault activity ends. This indicates that in these multi-layered oil and gas-rich areas with developed faults, faults are an important conduit connecting deep primary oil and gas reservoirs with shallow secondary oil and gas reservoirs. Under normal circumstances, there are many faults developed in the target reservoir, but not all faults can become conductive pathways for upward oil and gas migration.

Only faults that connect the source rock and the target reservoir and are active during the period of massive oil and gas generation and expulsion from the source rock, i.e., oil-source faults, can become important conductive pathways for oil and gas migration (Jiang & Liu, 2023; Xue et al., 2021).

Some faults extend to the surface, allowing oil and gas to completely escape and destroy the oil and gas reservoirs. For example, in the Qaidam Basin's Youshan Oilfield, which was originally a complete anticline oil reservoir, a large fault with a significant displacement occurred along the vertical structural axis, raising the oil layer on the east side and exposing it to the surface, thus completely destroying the oil reservoir, as shown in Fig. 6.11. The oil layer on the west side dropped and was sealed by the fault, retaining a commercial oil reservoir.

Therefore, the role of faults in the formation of oil and gas reservoirs is dualistic. They can both seal and act as conduits and destructive agents. To determine whether faults within a sedimentary basin play a positive sealing role or a negative destructive role, one should study the relationship between the fault's development history and the periods of sedimentation and oil accumulation. In some fault basins in eastern China, early-developing faults with large displacements often control the boundaries of the sedimentary basin and the depositional range of the source, reservoir, and cap rock combinations. In the depression of the downthrown block of these faults, the deposition is thick, and the source, reservoir, and cap rock combinations are complete. In the upthrown block, the deposition of the source, reservoir, and cap rock layers is often missing, forming a bald topographic prominence.

With the continuous activity of early-developing, large-displacement faults, two sets of faults with strike-slip and oblique intersections often occur at the edges and central uplifts of the basin, complicating the structural belts into fault zones. Some of these faults play a sealing role, some act as separators, and others serve as conduits and destructive agents. However, the main faults are often one of the primary factors contributing to oil and gas enrichment. For example, the Ke-U major fault in the Karamay Oilfield of the Junggar Basin and the Shengbei major fault in the Shengtuo Oilfield of the Dongying Depression have played a very important positive role in the accumulation of oil and gas and the formation of oil and gas reservoirs. In contrast, some secondary faults in local structures, often with smaller displacements, complicate the oil and gas reservoirs, affecting the oil-bearing height, enrichment

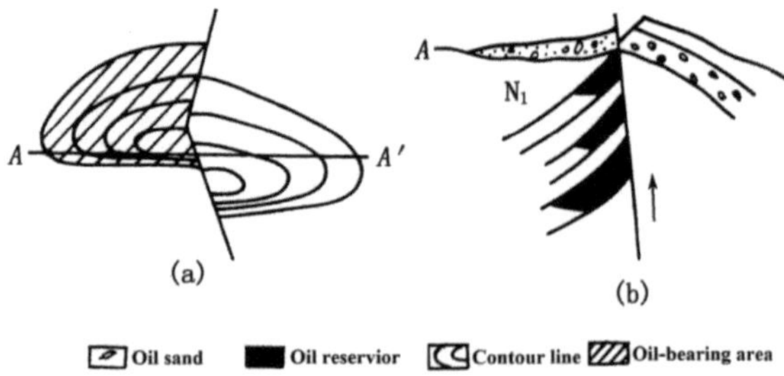

Fig. 6.11 Structural diagram (**a**) and cross-sectional diagram (**b**) of Youshan Oilfield (according to Qinghai Petroleum Exploration Bureau)

degree, control of oil–water interfaces, and the formation of shallow secondary oil and gas reservoirs as conduits.

In summary, regardless of the level of the fault, its changes throughout geological history are very complex, and its roles are diverse. In practical work, one can analyze the role of faults in the formation of oil and gas reservoirs based on the nature of the fault, the height of the faulted strata, the changes in the thickness of the rock layers on either side of the fault, and the activity of the fault. For example, the relationship between fault activity and the oil and gas reservoir formation period often determines whether the fault plays a constructive role or a destructive one. Some faults develop before the oil and gas reservoir formation period and cease activity later. Some faults occur after the formation period, and some faults occur simultaneously with the formation period. In the history of fault development, some faults play a sealing role in the early stages and a conduit or destructive role later. Vertically, the sealing nature of different parts of the fault varies. Some faults are sealed at the top and unsealed at the bottom, or vice versa. In summary, the role of each fault in the formation of oil and gas reservoirs must be analyzed specifically based on the actual circumstances and cannot be subjectively judged from a static perspective. Instead, a comprehensive evaluation should be made based on its developmental history.

(2) Oil and Gas Migration and Accumulation Patterns Along Faults

In petroliferous areas controlled by faults, the structural characteristics of the fault zones and the activity of the faults lead to significant differences in the sealing properties of the faults both spatially and temporally. Consequently, the migration and accumulation of oil and gas along faults are complex and variable. In oil-rich faulted depressions with complex oil and gas accumulation zones, oil and gas migrate from source rocks to shallower reservoirs through a three-dimensional migration process that includes vertical fault movement and lateral fault-sand diversion. Different faults have different conducting and sealing properties, which result in varying patterns of oil and gas migration along the faults. An oil and gas reservoir enriched along a particular fault could have formed by the migration and accumulation of oil and gas along that fault, or it could be due to the conduction of oil and gas from another fault. The alternating periods of active and quiescent phases of fault activity can lead to complex and changeable situations. Therefore, fault activity, the structure of fault zones, and the combination of multiple faults collectively control the differential migration and enrichment of oil and gas in different fault blocks and stratigraphic sequences within oil and gas-bearing areas controlled by faults (Jiang & Liu, 2023) (Fig. 6.12).

6.2.2.3 Main Types of Fault Oil and Gas Reservoirs

The types of fault traps are diverse and can be classified from different perspectives, usually based on the nature of the fault and the combination relationship between the fault and the reservoir.

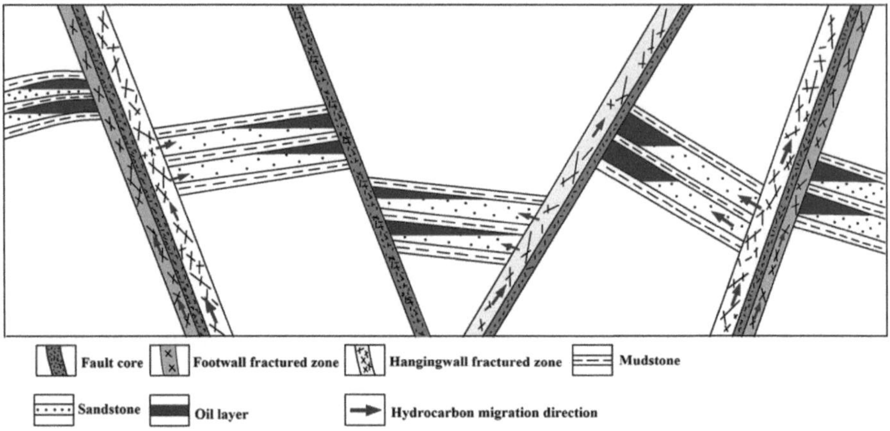

Fig. 6.12 Structural model of fault zones in rich oil faulted depressions and patterns of differential migration and accumulation of oil and gas

(1) Classification Based on Fault Nature

Based on the nature of the fault, fault oil and gas reservoirs can be divided into normal fault blockage reservoirs and reverse fault blockage reservoirs. Almost all oil and gas reservoirs in the Mesozoic-Cenozoic rift basins of eastern China are normal fault blockage reservoirs, whereas the western basins are more developed in reverse fault blockage reservoirs.

a. Normal Fault Blockage Oil and Gas Reservoirs

These types of reservoirs are often found in extensional basins, formed by various shapes of oil and gas-bearing fault blocks combined with multiple normal faults within the basin. Depending on the relationship between the fault dip and the reservoir dip, they can be divided into co-directional normal fault blockage oil and gas reservoirs and counter-directional normal fault blockage oil and gas reservoirs (Fig. 6.13). The former has faults and reservoirs with the same dip, and usually a fault displacement greater than the reservoir thickness is required to form a trap. The latter has faults and reservoirs with opposite dip, forming a gable-like structure, and the resulting oil and gas reservoirs are also known as gable fault blocks. Gable fault block traps are easier to form than co-directional normal fault traps, so most fault blockage oil and gas reservoirs are gable-style. For example, in the Dongxin Oilfield of the Bohai Bay Basin, gable fault block oil reservoirs account for more than 90% (Jiang, 1998).

b. Reverse Fault Blockage Oil and Gas Reservoirs

These types of reservoirs are mainly distributed in the edge areas of compressional basins, formed by various shapes of oil and gas-bearing fault blocks combined with multiple reverse faults or thrust faults at the basin edge. On the hanging wall of the thrust fault, thrust fault blocks oil and gas reservoirs are formed, and on the footwall,

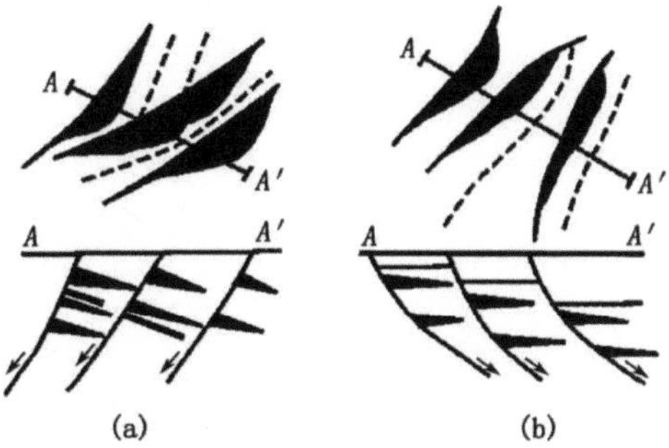

Fig. 6.13 Schematic diagram of oil and gas reservoirs sealed by reverse normal faults (**a**) and oil and gas reservoirs sealed by normal faults in the same direction (**b**)

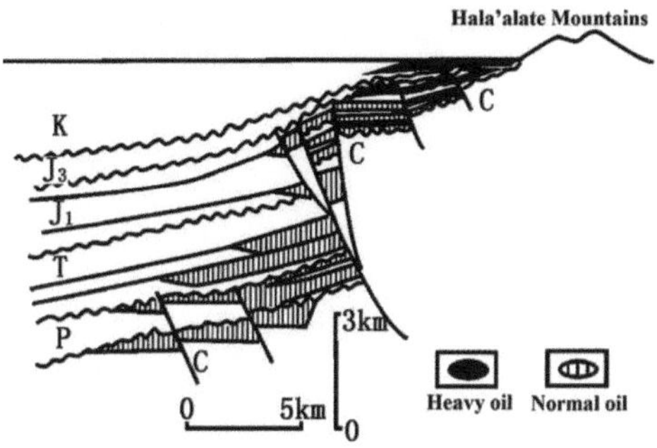

Fig. 6.14 Cross-section diagram of oil reservoirs in the Junggar Basin (according to Hu, 2002)

hidden overthrust fault block oil and gas reservoirs are often formed. For example, oil and gas reservoirs in the northwestern margin of the Junggar Basin (Fig. 6.14).

(2) Classification Based on the Planar Combination Relationship Between Faults and Reservoirs

Based on the combination relationship between the fault and the structural morphology of the reservoir top surface, fault oil and gas reservoirs are divided into faulted nose-like oil and gas reservoirs and fault block oil and gas reservoirs. There is an internal connection in the genesis of various types of fault oil and gas reservoirs, and their most basic common point is that they are all blocked and sealed by faults on the updip direction of the reservoir, forming a closed space with the reservoir and the fault.

a. Faulted Nose-Like Oil and Gas Reservoirs

The trap forming this type of oil and gas reservoir is composed of faults and nose-like structures. Against the background of regional dip, the updip direction of the nose-like structure is blocked by faults, forming a fault trap. When oil and gas accumulate in it, this type of oil and gas reservoir is formed (Fig. 6.15).

The Bohai Bay Basin has a large distribution of this type of oil and gas reservoirs, such as the Yong 12 block of the Shaerxia Sub-member in the Yong'an Town oil and gas field of the Dongying Depression. The reservoir of this oil and gas reservoir is the blocky sandstone of the Shajiazhuang Formation, presenting a nose-like structure that

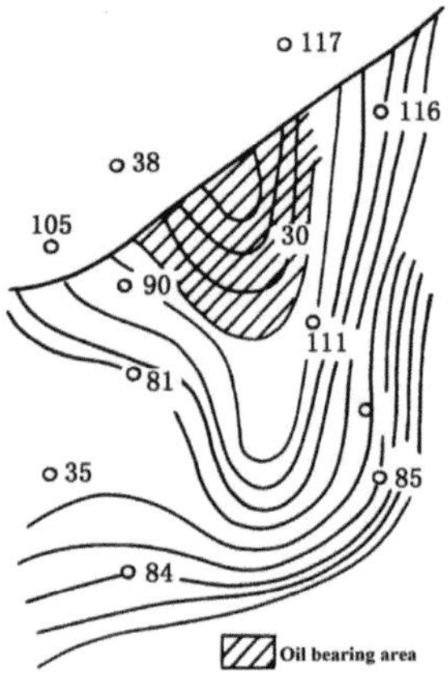

Fig. 6.15 Fault-nose structural trap and oil and gas reservoir

is uplifted to the north, cut by the nearly east–west extending Beiyao Fault, forming a fault nose oil and gas reservoir. Due to the abundance of oil and gas sources, good reservoir properties, and strong fault sealing capacity, the thickness of the oil and gas-bearing layer is quite large, reaching more than 70 m at its thickest (Fig. 6.16).

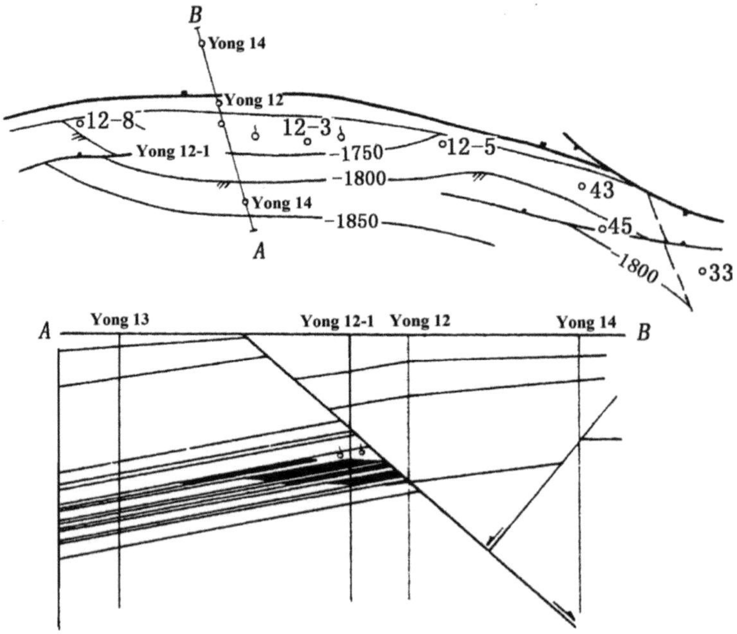

Fig. 6.16 Structure of the Yong 12 fault block in Yong'an Town oilfield (contour interval in meters) and oil reservoir cross-section diagram (according to Wang & Qian, 1992)

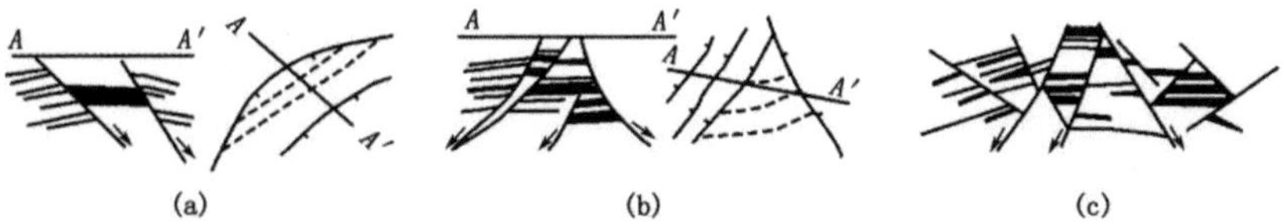

Fig. 6.17 Types of fault block oil and gas reservoir combinations. **a** arc-shaped blocks, **b** cross-shaped blocks, **c** blocks with multiple geometric shapes

b. Fault Block Oil and Gas Reservoirs

Formed by faults and inclined reservoirs that constitute a closed space, the reservoir has no obvious structural form. Due to the different shapes of the sealing faults and the reservoirs being cut by multiple faults, fault block oil and gas reservoirs have various combinations. Common forms include arc-shaped blocks, cross-shaped blocks, and blocks with multiple geometric shapes (Fig. 6.17). Arc-shaped blocks are surrounded by an arc-shaped fault that bulges in the updip direction on the inclined reservoir; on structural maps, they appear as relatively straight structural contour lines intersecting with curved fault lines, forming sealing conditions.

Cross-shaped blocks are surrounded by two intersecting faults on the updip direction of the inclined reservoir, appearing on structural maps as relatively straight structural contour lines intersecting with intersecting faults.

The Bohai Bay Basin has a large distribution of this type of oil and gas reservoirs. In many complex block areas, multiple sets of faults intersect and combine with the stratigraphic attitude to form oil and gas-bearing blocks with various geometric shapes. The sealing faults are often multiple, forming complex block traps, many of which become closed blocks. When the reservoir is cut and sealed in the updip direction and laterally by three or more faults, forming semi-closed or closed blocks, the structural map shows multiple faults and structural contour lines forming a closed area. For example, the Ying 13 block area of the Dongxin Oilfield (Fig. 6.18).

The common point of the above fault trap oil and gas reservoirs is that they must form a closed space. From the structural maps, under the premise that the fault itself is sealing, the necessary condition for forming a fault trap is that the fault line and structural contour lines must be closed. Conversely, if the above conditions are not met, the fault cannot form a trap.

In China's petroliferous regions, especially in the Bohai Bay Basin, faults and reservoirs form various combinations of traps, that is, blocks of different sizes and shapes. Many scholars call this type of fault oil and gas reservoir a block oil and gas reservoir, and the term is widely used in practical work. In different regions, to distinguish different types of block oil and gas reservoirs, they are also classified into fan-shaped, trapezoidal, triangular, rhombic, and other block oil and gas reservoirs based on the shape of the block.

However, most people believe that fault oil and gas reservoirs and block oil and gas reservoirs should not be used as synonyms, and the latter should be a part of the former. Block oil and gas reservoirs generally refer to oil and gas accumulations in traps formed by sealing faults and inclined reservoirs without obvious structural

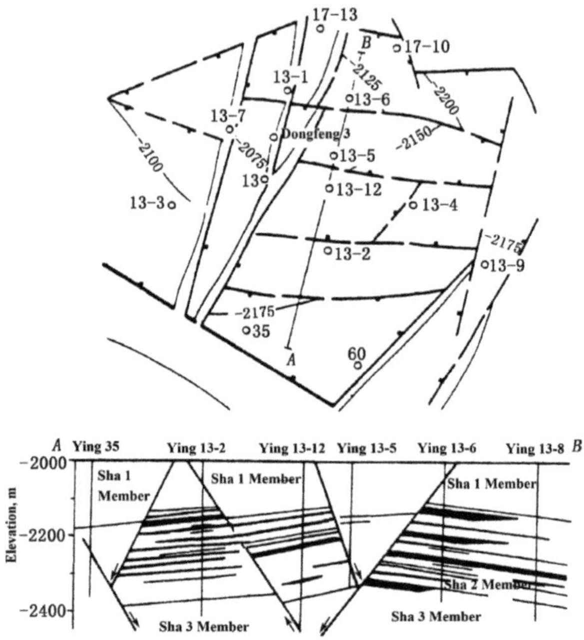

Fig. 6.18 Plan and cross-section diagram of the oil reservoir in the Ying 13 fault block area of Dongxin Oilfield

forms. It is often divided into various shapes of blocks by multiple faults, or formed by a single curved fault and an inclined reservoir, with individual traps being small and fragmented. Fault traps combined with nose-like structures of a certain shape are called faulted nose-like oil and gas reservoirs. These oil and gas reservoirs often have a more regular oil-bearing area, with the reservoir's updip direction blocked by faults, and the oil and gas range often presents a semi-anticline shape.

Fault oil and gas reservoirs have their own characteristics, especially their complexity and diversity, and they change with the nature and strength of tectonic movements in different periods. Therefore, oil and gas geologists must analyze and study the patterns of their changes in complex and changeable situations to make oil and gas exploration work more effective.

6.2.3 Rock Piercing Oil and Gas Reservoirs

6.2.3.1 Concept of Rock Piercing Oil and Gas Reservoirs

A trap formed by the contact and obstruction of rock piercing is called a rock piercing trap. A rock piercing oil and gas reservoir refers to the accumulation of oil and gas within such a trap. Depending on the nature of the piercing rock, they can be classified into salt piercing, mud volcano piercing, and magmatic rock column piercing. Oil and gas reservoirs have been discovered in these three types of rock piercing traps worldwide. However, salt piercing is considered more significant in terms of distribution. For instance, a considerable number of salt piercing oil and gas reservoirs have been found in Romania, Germany, the United States, and Russia. In contrast, oil and gas reservoirs related to mud volcano piercing and magmatic rock column piercing are only found in specific regions.

6.2.3.2 Formation Mechanism and Distribution of Rock Piercing Traps

Subsurface rock masses (including salts, mudstone, soft mud, and various intrusive magmatic rocks) intrude into sedimentary rock layers, causing deformation above the reservoir layer. The updip direction is sealed by the intrusive rock mass, forming a piercing (contact) trap. Deformation and displacement (faulting) of the reservoir layer related to the piercing body can correspondingly form anticline traps and fault traps.

The fundamental characteristic of piercing oil and gas reservoirs is that oil and gas are confined on one side by the piercing rock mass, which serves as the sealing condition of the trap, while the downdip oil (gas) water contact remains parallel to regular contour lines.

Considerable research has been conducted by scholars worldwide on evaporite and mud volcano activity, as well as the formation of related diapir and piercing structures. It is generally believed that gypsum and soft mud often contain a large amount of primary water and have a lower density than other sedimentary rock layers. Under the unequal pressure (static or dynamic) of the overlying denser sedimentary layers, the plastic gypsum or soft mud flow. During the flow process, when they encounter weak zones in the sedimentary layers, such as active syngenetic faults or low-pressure areas with large pressure differences, these plastic salt flows or soft mud flows intrude upward or arch up, causing piercing and diapir structures. Therefore, salt rock and soft mud piercing or diapir structures are often closely related to syngenetic faults.

Based on the above mechanism, the basic conditions for the formation of piercing or diapir structures are: the presence of a considerable thickness of evaporite or soft mud layers deep underground, the greater the thickness, the higher the possibility of forming such structures. The existence of weak zones with significant pressure difference changes in the overlying rock layers, such as syngenetic faults. These two basic conditions control the formation and distribution of piercing contact traps.

6.2.3.3 Examples of Rock Piercing Oil and Gas Reservoirs

(1) Salt Piercing Oil and Gas Reservoirs

Deep underground salt bodies intrude and pierce the overlying sedimentary rock layers, forming salt piercing traps where oil and gas accumulate, thus called salt piercing oil and gas reservoirs. For example, the oil reservoir in the Moleni oilfield in the forecarpath of the Carpathian Mountains in Romania belongs to this type of reservoir. The oilfield is a salt body that has intruded and pierced the sandstone reservoir layer of the Upper Oligocene and Upper Neogene, forming a salt piercing trap and oil and gas reservoir (Fig. 6.19).

Additionally, this type of oil and gas reservoir is widely distributed in regions such as the Gulf of Mexico in the United States, the Emba region in the former Soviet Union, the North German Basin in Germany, the North Sea Basin in Western Europe, and Gabon in West Africa.

(2) Mud Volcano Piercing Oil and Gas Reservoirs

These oil and gas reservoirs are formed by the piercing action of mud volcanoes, which create sealing conditions and accumulate oil and gas. For example, the oil and gas reservoir in the Lokbatan oil and gas field on the Apsheron Peninsula in the former Soviet Union belongs to this category. The oilfield is an anticline structure with the top pierced by a mud volcano. The Upper Oligocene to Upper Neogene reservoir layer along the updip direction contacts the mud volcano piercing body, forming sealing conditions and accumulating oil and gas, thus forming this type of oil and gas reservoir, as shown in Fig. 6.20.

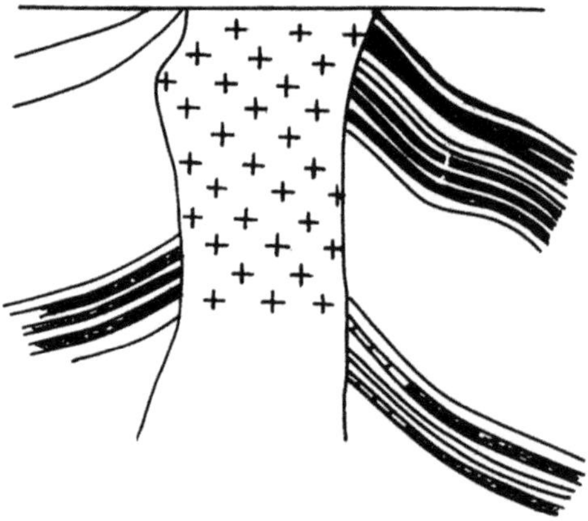

Fig. 6.19 Cross-section diagram of the oil reservoir in the piercing trap formed of Moşni (according to Брод, 1950)

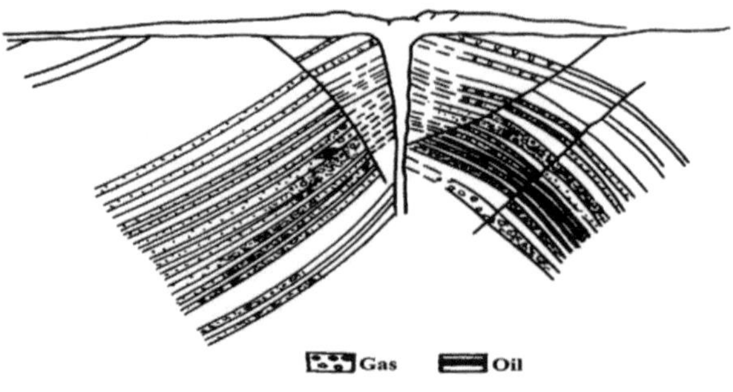

Fig. 6.20 Cross-section diagram of the oil and gas reservoir pierced by the Lokbatan mud volcano (according to Брод, 1950)

Fig. 6.21 Cross-section diagram of the oil reservoir pierced by the magmatic rock body in Mexico (according to И. О. Брод)

(3) Magmatic Piercing Oil and Gas Reservoirs

Deep underground magma intrudes and pierces the overlying sedimentary rock layers, forming a magmatic piercing trap where oil and gas later accumulate, thus forming this type of oil and gas reservoir. For example, an oilfield of this kind was discovered in Mexico, as shown in Fig. 6.21. The oil and gas reservoir there belongs to the magmatic piercing oil and gas reservoir. This type of oil and gas reservoir is relatively rare.

6.2.4 Fractured Oil and Gas Reservoirs

6.2.4.1 Basic Characteristics

Fractured oil and gas reservoirs refer to those in dense rock layers where the storage space and flow pathways for oil and gas are primarily composed of structural fractures and related dissolution pores. In various dense and brittle rock layers, the overall porosity and permeability are very low before tectonic actions occur, making them unsuitable for the storage and flow of oil and gas. However, due to tectonic actions such as folding and faulting, fractures or cavities (pores) are generated under stress. In some easily soluble fracture zones, fluids can also cause dissolution along the fractures (pores), allowing these otherwise impermeable rock layers to develop fractures and closely related dissolution pores at certain depths, thereby creating conditions for fluid storage and flow pathways. When combined with cap rocks and lateral barriers, this can form fractured traps where oil and gas accumulate, thus creating fractured oil and gas reservoirs.

The development of fractures and pores in rock layers is often caused by multiple factors, with tectonic action being the most significant. The formation and development of rock fractures are usually associated with folding and faulting, which is why fractured oil and gas reservoirs are classified as a type of structural oil and gas reservoir. Although fractured oil and gas reservoirs are often closely related to anticlinal and fault oil and gas reservoirs, they possess their own distinct characteristics. Fractured oil and gas reservoirs are primarily controlled by the storage space formed by tectonic and dissolution processes, with oil and gas distributed locally within dense rock layers. Due to the close relationship and significant differences in genesis between fractured oil and gas reservoirs and anticlinal or fault oil and gas reservoirs, they are classified separately as a type of structural oil and gas reservoir.

In some dense, brittle, and insoluble rock layers, such as dense sandstones, faulting leads to the development of fractures along the fault zones, forming what are known as fractured body oil and gas reservoirs. In dense, brittle, and soluble rock layers, such as carbonate rocks, faulting is accompanied by the development of numerous fractures, and subsequent atmospheric precipitation and formation fluids can cause dissolution along the fracture zones, leading to the formation of fault-controlled dissolution bodies. Therefore, in dense sandstone layers, they often manifest as fractured bodies, while in dense carbonate rocks that are prone to dissolution, dissolution bodies can appear. Dissolution bodies are formed based on fractures controlled by faults, so both dissolution bodies and fractured bodies can be classified as fractured oil and gas reservoirs.

Unlike the distribution of oil and gas in classic anticlinal or fault traps, fractured oil and gas reservoirs are controlled by fractures or fault-controlled dissolution cavities. Their oil and gas distribution does not follow the structural contour lines of the reservoir top surface as seen in anticlinal or fault oil and gas reservoirs, but often appears blocky or isolated. In some dense rock layers of anticlinal structures, the development of structural fractures can connect previously isolated pore spaces, cavities, and dissolution voids, forming a unified network of storage space, where the accumulated oil and gas often forms blocky reservoirs. In some fault-controlled fractured bodies or dissolution bodies, or in local fracture development zones of anticlinal structures (such as areas under significant stress), oil and gas are distributed along the fracture development zones, often appearing as bands or isolated accumulations. In fault-controlled fracture zones, the development of dissolution cavities along the faults or fracture zones leads to oil and gas accumulations that are distributed in a beaded or band-like manner along the faults. During drilling, phenomena such as tool emptying and mud loss often occur. Because structural fractures connect various storage spaces within the reservoir, they form a smooth flow system. The measured core permeability often differs significantly from the actual permeability measured during well testing, with the actual measured permeability during well testing often being much higher than the core's measured permeability. Due to the significant variability in porosity and permeability of fractured reservoirs, the storage performance can vary greatly in different parts of the same reservoir, leading to substantial differences in production between different wells. Therefore, when exploring this type of oil and gas reservoir, it is crucial to analyze and understand the distribution patterns of fracture zones, as it is these secondary fracture zones' distribution and development that control the degree of oil and gas enrichment.

6.2.4.2 Examples of Fractured Oil and Gas Reservoirs

(1) Fractured Oil Reservoirs in the Tarim Basin, Well Area Tazhong 45

The Tazhong 45 well area is in a local structural belt at the western end of the Tazhong low uplift in the Tarim Basin, characterized by a typical anticlinal structure with a trap area of 48 km^2 and an amplitude of 60 m (Fig. 6.22).

The reservoir layer is from the Middle and Upper Ordovician Lianglitake Formation, with the primary reservoir space type being structural fractures and dissolution cavities. The rock types mainly include fine crystalline limestone, fully cemented siltstone, and bioclastic limestone, with matrix porosity generally below 2%. Due to the development of fractures and cavities, the average porosity of the reservoir is around 4.2%, indicating relatively good storage conditions. The main factors controlling the reservoir properties are the formation of structural fractures and dissolution processes, with dissolution primarily occurring along structural fractures, reflecting that the main body of the fracture-cavity system is formed by structural fracture dissolution.

The Asmari limestone oil and gas reservoir in the Gachsaran Oilfield in Iran is a typical fractured oil and gas reservoir, characterized by high oil heights, large reserves, and high production rates, and it serves as an important representative of fractured carbonate rock oil and gas reservoirs in the Persian Gulf Basin. Many oil reservoirs in the Yanchang Formation of the Yan'an Oilfield in Shaanxi, China, also belong to fractured sandstone reservoirs. Fractured mudstone and siltstone reservoirs are also distributed in the San Marino Basin in California and the Midland Basin in Texas, USA.

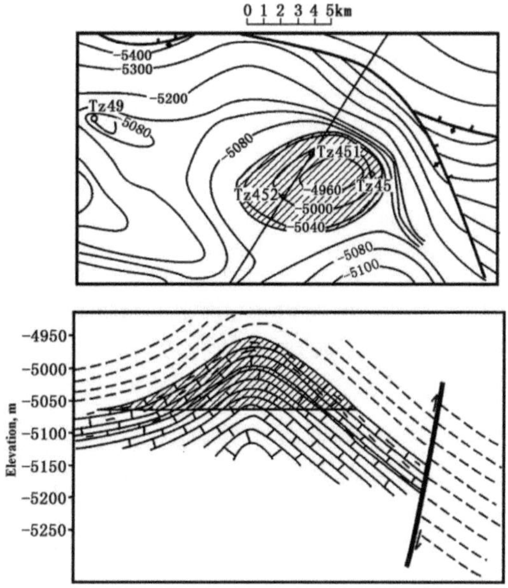

Fig. 6.22 Integrated map of the oil and gas reservoir in the Ta Zhong 45 well area (contour interval in meters) (according to Zhang et al., 2004)

(2) Fault-Controlled Fractured Oil and Gas Reservoirs in the Shunbei Area of the Tarim Basin

A large oil and gas field with reserves of 10×10^8 t has been discovered in the ultra-deep carbonate rocks of the Shunbei area in the Shuntuogole low uplift of the Tarim Basin, with a fault-controlled fractured cavity trap. Under the influence of strike-slip faulting, the Middle to Lower Ordovician carbonate strata in the Shunbei area have developed numerous fractures, cavities, and voids due to structural rupture, forming a carbonate fractured reservoir distributed along the strike-slip fault zone. These reservoirs are sealed under the cap conditions of overlying mudstone regions and dense carbonate rock layers.

The Shunbei area is situated in a structurally lower position, with dense lithology in the Ordovician carbonate strata. The top surface is unconformable, and the karstification and development of karst fracture reservoirs are underdeveloped. However, the area has developed various systems of overlapping strike-slip fault zones, forming fault-controlled fractured reservoirs primarily controlled by the strike-slip fault zones. A set of high-quality source rocks (Yuer Tusi Formation) were developed in the early Cambrian period. During the Cambrian to Middle Ordovician, a carbonate rock layer approximately 3,000 m thick were developed, later modified by strike-slip faulting, forming large-scale fault-controlled reservoirs. In the Late Ordovician, the sediment thickness in the area is about 500–2500 m of shelf mudstone, forming a good source-reservoir-cap combination with the underlying carbonate rock fault-controlled reservoirs. The multiple phases of structural activity from the strike-slip faults primarily generate structural ruptures, with later fluid modifications contributing to the formation of fault-controlled fractured reservoirs, which are sealed by dense carbonate rocks and capped by thick mudstone layers, forming fault-controlled fractured traps. Meanwhile, the strike-slip faults extend downward to the Sinian system, connecting with the hydrocarbon source rocks of the Lower Cambrian Yuer Tusi Formation, allowing oil and gas to migrate upward and accumulate within the fault-controlled fractured traps, forming fault-controlled fractured oil and gas reservoirs. The discovered oil and gas reservoirs are mainly located in the Ordovician Ying Mountain Group and the Yanfang Group, with depths ranging from 7200 to 8800 m, representing deep to ultra-deep oil and gas reservoirs. The oil and gas reservoirs are primarily distributed along the main strike-slip fault zones in the area (Fig. 6.23) (Ma et al., 2022; Zhang et al., 2022).

Exploration and development results indicate that the Ordovician oil and gas in the Shunbei area are primarily enriched within the main strike-slip fault zones and their connected branch fault zones, distributed in a "narrow and long" strip along the fault zones. The internal structure of the reservoirs is complex, with strong heterogeneity in the reservoir in both vertical and horizontal directions, mainly reflected in the number, size, and spatial combinations of cavities, pores, and fractures, leading to significant differences in storage and permeability performance. The internal structure of the reservoirs is mainly influenced by the structural style, activity intensity, and kinematic characteristics of the strike-slip faults. The area has experienced multiple phases of tectonic movement, forming fracture-cavity-type reservoirs dominated by

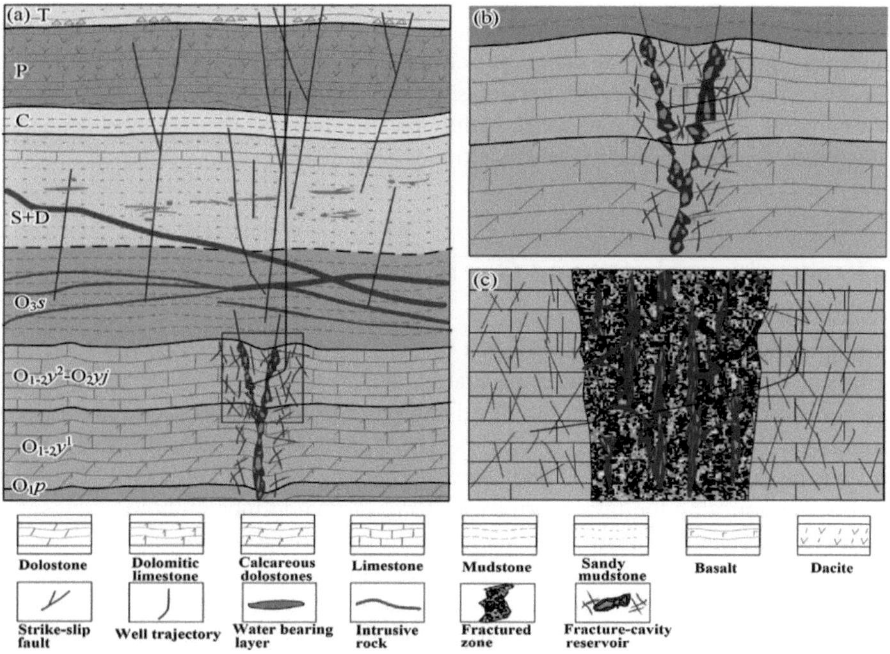

Fig. 6.23 Model of fault-controlled fractured oil and gas reservoirs in the Shunbei area of the Tarim Basin (according to Zhang et al., 2022). In the figure: **a** The main target layer of the Shunbei ultra-deep fault-controlled fractured oil and gas reservoir is the Ordovician Yanfang Group and Ying Mountain Group, with overlying strata developing fault systems and multiple intrusive rocks, **b** an enlarged view of figure (**a**), showing oil and gas enrichment along the fault zone, with multiple faults controlling the development of the reservoir and distribution of oil and gas, **c** an enlarged view of figure (**b**), showing the complex structure of the reservoir within a single fault, exhibiting characteristics of a complex "grid-like" reservoir structure

structural ruptures and modified by multiple causes. The reservoirs are distributed in strips along the strike-slip fault zones, with significant vertical development depth and reservoir development scale controlled by fault activity intensity. The strike-slip fault zones exhibit planar segmentation, with internal reservoirs underdeveloped at segment boundaries, forming lateral barriers along the fault zones. Overlying regional mudstones and dense micritic limestones can form sealing and blocking conditions for vertical loss and lateral migration or adjustment of oil and gas. The formation mechanism, structural style, and planar distribution characteristics of the strike-slip faults determine the morphology of the fault-controlled fracture-cavity traps. Based on the kinematic characteristics and geometric morphology of the strike-slip faults, the fault-controlled fracture-cavity traps are divided into four major subtypes: compressional, tensile, translational, and composite traps (oil and gas reservoirs). Among them, the compressional fault-controlled fracture-cavity traps mainly develop in the overlapping compressional segments of the strike-slip faults. The deep fault surfaces are steep, upright, and narrow, while in the shallower layers, they are controlled by local compressive stress fields, resulting in normal faults and forming multiple high-angle reverse faults, presenting a fanning back-thrust structure. The boundary faults in the overlapping areas are the main slip faults, with high degree of fragmentation of carbonate rocks within the area, forming the fractured zone that constitutes the fracture-cavity system (Yun & Zhu, 2022).

6.3 Stratigraphic Oil and Gas Reservoirs

Stratigraphic traps refer to the closures formed due to the interruption of vertical sedimentary continuity in reservoirs, that is, closures associated with stratigraphic unconformities. The accumulation of oil and gas within these stratigraphic traps is called stratigraphic oil and gas reservoirs. The term "stratigraphic trap" here is used in a narrow sense, referring to the closures formed by the direct contact and sealing of the reservoir with the unconformity surface, excluding those lithologic closures formed due to changes in sedimentation conditions or diagenesis.

Stratigraphic traps differ from the structural traps, which are formed due to deformation, displacement, or fracturing of strata caused by tectonic actions. In contrast, stratigraphic traps are primarily the result of the contact between the reservoir and the unconformity. After the reservoir is subjected to weathering and erosion, it is overlain by non-permeable strata, forming an unconformity seal, or above the eroded surface, permeable layers are overlain along the unconformity surface, with the unconformity acting as the sealing object.

Oil and gas exploration practices have shown that both the upper and lower parts of unconformities can often become favorable areas for oil and gas accumulation. The term "unconformity" here is used in a broad sense, including both angular unconformities and parallel unconformities (disconformities). In an petroliferous area, structural oil and gas reservoirs that are easily identified are always discovered first. However, as the degree of exploration in a region increases, the proportion of non-structural oil and gas reservoirs, including stratigraphic oil and gas reservoirs, will continue to increase. In recent decades, with the continuous advancement of exploration technology, the number of stratigraphic oil and gas reservoirs discovered worldwide has gradually increased. They are not only numerous and widely distributed but often have large reserves and come in a variety of types.

Based on the genesis of the trap and the spatial relationship between the reservoir and the unconformity surface, stratigraphic oil and gas reservoirs can be broadly divided into two categories and three basic types: one category is the stratigraphic unconformity oil and gas reservoirs located below the unconformity surface, and the other is the stratigraphic overlap oil and gas reservoirs located above the unconformity surface, both of which are related to the formation of the unconformity seal. The former can be further divided into buried hill oil and gas reservoirs and stratigraphic unconformity overlain oil and gas reservoirs, with buried hill oil and gas reservoirs occupying an important position. Those reservoirs that are above and below the

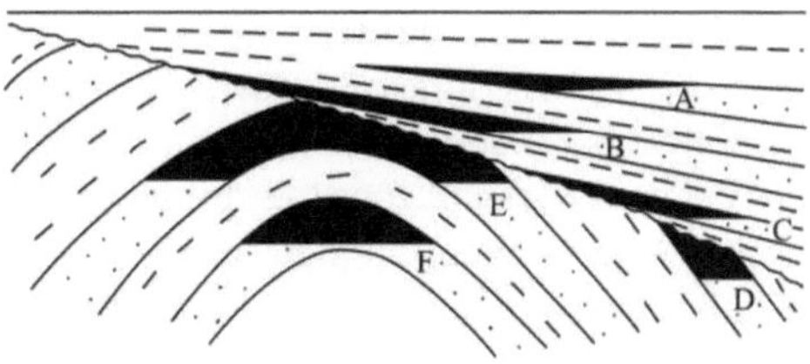

Fig. 6.24 Schematic diagram of the main types of stratigraphic oil and gas reservoirs and their differences from non-stratigraphic oil and gas reservoirs

unconformity surface and do not directly contact the unconformity, and are formed by other factors, do not belong to stratigraphic oil and gas reservoirs.

As shown in Fig. 6.24, B and C are stratigraphic overstep oil and gas reservoirs located above the unconformity surface, while D and E are stratigraphic oil and gas reservoirs below the unconformity surface, which are stratigraphic unconformity overlain oil and gas reservoirs and buried hill oil and gas reservoirs, respectively. A and F are not related to stratigraphic unconformities and are lithologic pinchout oil and gas reservoirs and anticline oil and gas reservoirs, respectively. In China, many oil and gas geologists often collectively refer to oil and gas reservoirs within buried hills that do not have a direct relationship with the unconformity surface as buried hill internal oil and gas reservoirs, most of which are stratiform oil and gas reservoirs. However, from the perspective of trap genesis, most of these internal reservoirs can be classified as fault, anticline, and other types. For example, within the internal strata of a buried hill, stratiform reservoirs with the top and bottom sealed by non-permeable layers and the updip direction blocked by faults form fault oil and gas reservoirs.

6.3.1 Buried Hill Oil and Gas Reservoirs

A buried hill typically refers to the basement rock protrusions of a basin's base that are buried beneath younger sedimentary cover layers due to nonconformity. These include ancient topographic protrusions (remnant hills) and protrusions with certain tectonic shapes formed after the erosion of ancient structures. The formation of a buried hill requires a long period of erosion and is subsequently buried by new sedimentary layers. Relative to its surroundings, it is a local protrusion. Buried hill oil and gas reservoirs refer to the accumulation of oil and gas within these basement rock protrusions, which are covered by overlying impermeable layers to form a trap, leading to the formation of oil and gas reservoirs, also known as "ancient topography" oil and gas reservoirs.

Oil and gas accumulations in the basement rocks of a basin are also referred to as basement oil and gas reservoirs. In a narrow sense, basement oil and gas reservoirs refer to oil and gas accumulations within the crystalline basement rocks of a basin, with reservoir rock types being metamorphic and igneous (intrusive rocks). In a broader sense, basement oil and gas reservoirs refer to oil and gas accumulations formed within the sedimentary basement rocks of a basin, including crystalline basement rocks and sedimentary rocks of different ages before the formation of the basin, meaning that igneous rocks, metamorphic rocks, and sedimentary rocks can all serve as basement rocks. According to the broader concept, basement buried hill oil and gas reservoirs can be divided into two major categories: ancient topography buried hill oil and gas reservoirs and buried hill internal oil and gas reservoirs. The former is primarily characterized by a weathered crust reservoir, nonconformity surface and fault conduction, and overlying tight sedimentary rock layer sealing, while the latter is characterized by tectonic fractures or dissolution pores developed within the buried hill interior, fault or fracture conduction, and internal layer sealing. In terms of sealing conditions, ancient topography buried hill oil and gas reservoirs form traps due to stratigraphic nonconformity and belong to the typical stratigraphic oil and gas reservoir category, but buried hill internal oil and gas reservoirs are mostly unrelated to stratigraphic nonconformity and are sealed by faults or other conditions, thus they can be classified as other types based on the cause of the trap.

Since the basement itself does not have the conditions for generating hydrocarbons, both broad and narrow senses of basement oil and gas reservoirs source their oil and gas from the overlying or lateral new sedimentary rock layers' source rocks. The difference is in the rock types where the oil and gas are stored, with the broader concept including more types of rocks. This book adopts the narrow concept of basement oil and gas reservoirs, meaning that basement oil and gas reservoirs are a type of buried hill oil and gas reservoir.

6.3.1.1 Formation Characteristics of Buried Hill Traps and Oil and Gas Reservoirs

The formation of buried hill traps is related to regional sedimentary hiatuses and erosion processes. At a certain period in geological history, crustal movements caused an area to rise and undergo intense weathering and erosion. Hard and dense rock layers have a strong resistance to weathering and appear as large protrusions on the ancient topography, while rock layers with weaker resistance to weathering form depressions in the ancient topography. Thus, the landscape of ancient mountains, hills, plains, valleys, rivers, and lakes is displayed. Or, the original anticline structure, whose top is eroded, later descends again before the area is eroded into a plain, and at the same time, it is covered and buried by new sediments, forming a series of hidden eroded protrusions or hidden eroded structures on the original ancient topography, also known as "ancient buried hills." These ancient topographic protrusions, due to long-term weathering, erosion, and groundwater circulation, often form fractured zones and dissolution zones, providing good storage spaces. When covered by impermeable layers, they form stratigraphic traps and become favorable sites for oil and gas accumulation (Fig. 6.25).

Scholars have conducted numerous studies on buried hills from different perspectives, leading to a complex and diverse classification and naming system. These

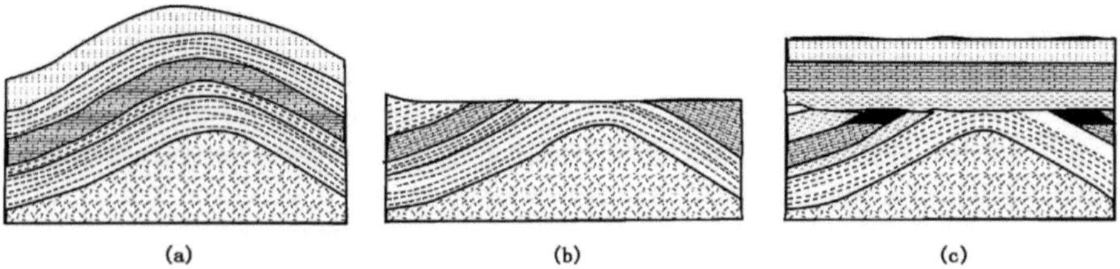

Fig. 6.25 Schematic diagram of the formation process of buried hill oil reservoirs. **a** Stratigraphic uplift forms an anticline. **b** The anticline is eroded, and the top reservoir is eroded away. **c** Sediments cover the angular unconformity, and oil migrates upward to accumulate in the structural wing to form a reservoir

include classifications based on the genesis or formation period of buried hills, reservoir properties and pore types, sealing factors and sealing methods, oil and gas reservoir morphology, and the combination of source, reservoir, and cap rock relationships (Jiang et al., 2021). Based on the morphology and formation characteristics of buried hills, buried hill oil and gas reservoirs can be divided into three major categories: fault block hills, ancient topography hills, and folded hills (Fig. 6.26). Their formation is all influenced by differential weathering factors, with fault block hills and folded hills also controlled by faults and ancient structures.

The rocks that make up buried hills can be crystalline, such as igneous rocks (intrusive rocks) and metamorphic rocks, or sedimentary rocks, such as limestone, dolomite, sandstone, and they can also be volcanic rocks. Their common characteristic is that they are dense and hard, and after long-term weathering, erosion, and groundwater circulation, secondary pores and fractures develop, providing good storage performance.

The types and distribution characteristics of buried hill traps and oil and gas reservoirs are controlled by regional geological structures. The oil and gas in buried hill traps mainly come from the overlying sedimentary source rock layers, so the age of the reservoir in buried hill oil and gas reservoirs is usually older than that of the source rock, which is known as "new-formed ancient storage." However, there are also buried hill oil and gas reservoirs where the reservoir age is the same as the source rock age, or the source rock age is older than the reservoir age, known as "ancient-formed new storage." The main oil and gas migration pathways include two types: oil source faults that connect buried hills and source rocks, and unconformity

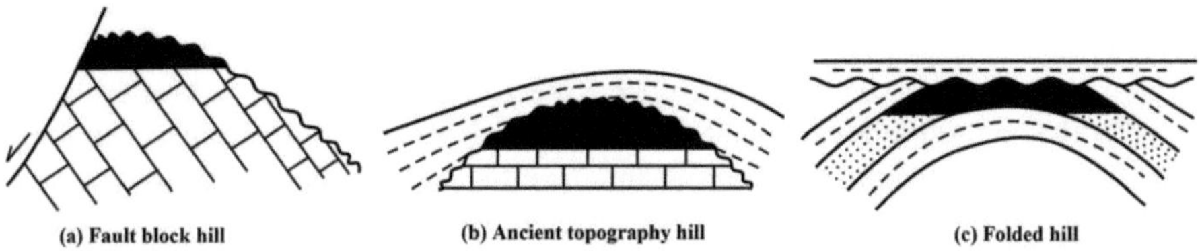

Fig. 6.26 Schematic diagram of the types of subsurface mountain oil and gas reservoirs

surfaces. Oil and gas continuously migrate to buried hill traps along unconformity surfaces and oil source faults to accumulate and form reservoirs.

There is a complex natural fluid migration system between the source rock and the buried hill trap. The formation of buried hill oil and gas reservoirs is mostly derived from other sources and formed elsewhere, and the conduction system plays a bridging role in the formation of buried hill oil and gas reservoirs.

Generally, the geological structure of buried hill oil and gas reservoirs is complex, with severe reservoir heterogeneity and complex elements of oil and gas accumulation. The formation characteristics of buried hill oil and gas reservoirs with different types of reservoirs are not the same.

6.3.1.2 Examples of Buried Hill Oil and Gas Reservoirs

With the continuous development of geophysical exploration methods and the increasing improvement of deep well drilling technology, more and more oil and gas reservoirs related to stratigraphic unconformities have been discovered around the world, many of which are large oil and gas fields.

(1) Panhandle Sandstone and Carbonate Buried Hill Gas Reservoir

The Niobrara buried hill belt, the Virgilian-Amarillo buried hill belt, and the Central Kansas uplift in the western interior basin of the United States are all areas where buried hill oil and gas reservoirs are concentrated. For example, the Panhandle oil and gas field is a super large oil and gas field located on the Virgilian-Amarillo buried hill belt (Fig. 6.27). The Panhandle oil and gas field has an oil and gas-bearing area of 6,000 km^2. The buried hill is composed of a thick block reservoir made up of Precambrian granite, feldspathic sandstone, and Upper Paleozoic carbonate rocks. It is covered by the Permian, especially the Permian salt rock, which serves as a good cap rock, forming a huge block oil and gas reservoir with a unified oil–water interface. The height of the oil and gas-bearing area reaches 400 m, and the oil-bearing part is mainly located on the north side of the buried hill.

(2) Renqiu Carbonate Buried Hill Oil Reservoir

The Renqiu oil field in China is a typical carbonate buried hill oil reservoir (Fig. 6.28). The oil field is one of the high-yielding large oil fields discovered by China in the 1970s in the Ji Zhong depression of the Bohai Bay basin. The buried hill is mainly composed of siliceous dolomite from the Wumishan Formation of the middle and upper Proterozoic era, with the surrounding wings being Cambrian and Ordovician carbonate rock strata. This buried hill has been exposed to the surface from the Late Ordovician to the Paleogene for a long geological period, suffering from long-term weathering, erosion, dissolution, and the effects of various crustal movements, resulting in well-developed fractures, caves, and pores, providing excellent storage performance. Later, it was covered by the thick muddy sediment of the Shahejie Formation of the Paleogene, becoming a good cap rock and forming a trap condition. At the end of the Paleogene and Neogene, the oil generated from the Shahejie

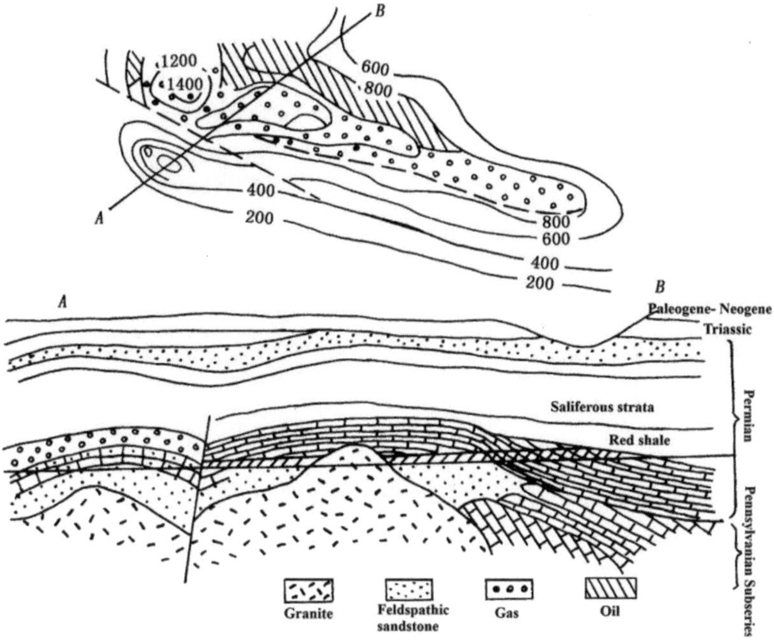

Fig. 6.27 Structural map of the Panhandle oil and gas field in the United States (contour interval in meters) and cross-section diagram (according to Zhang et al., 1999)

Formation's hydrocarbon source rock system entered this trap and accumulated, forming a rich and high-yielding large oil field.

(3) Wangzhuang Metamorphic Buried Hill Oil Reservoir and Bozhong 19-6 Metamorphic Buried Hill Condensate Gas Reservoir

There are also a considerable number of metamorphic buried hill oil and gas reservoirs in the Bohai Bay basin, such as the Wangzhuang metamorphic buried hill oil reservoir in the Jiyang depression, the Xinglongtai metamorphic buried hill oil reservoir in the Liaohe depression, and the Bozhong 19-6 metamorphic buried hill condensate gas reservoir in the Bozhong depression.

The Wangzhuang buried hill oil and gas reservoir is located in the Zhengjia area in the northwestern part of the Dongying depression in the Bohai Bay basin. Based on the background of the south slope of the Chenjiazhuang uplift, due to the development of ancient faults and weathering erosion, a Precambrian metamorphic rock residual hill was formed, which was later covered by the non-permeable mudstone of the Shahejie Formation to form an oil and gas reservoir (Fig. 6.29). Micro fractures are developed in the metamorphic rocks, and there are also a certain number of dissolution holes, caves, and micro pores, as well as mineral fractures and joint fractures. The developed fractures, caves, and pores are both the seepage channels for oil and gas and the storage spaces.

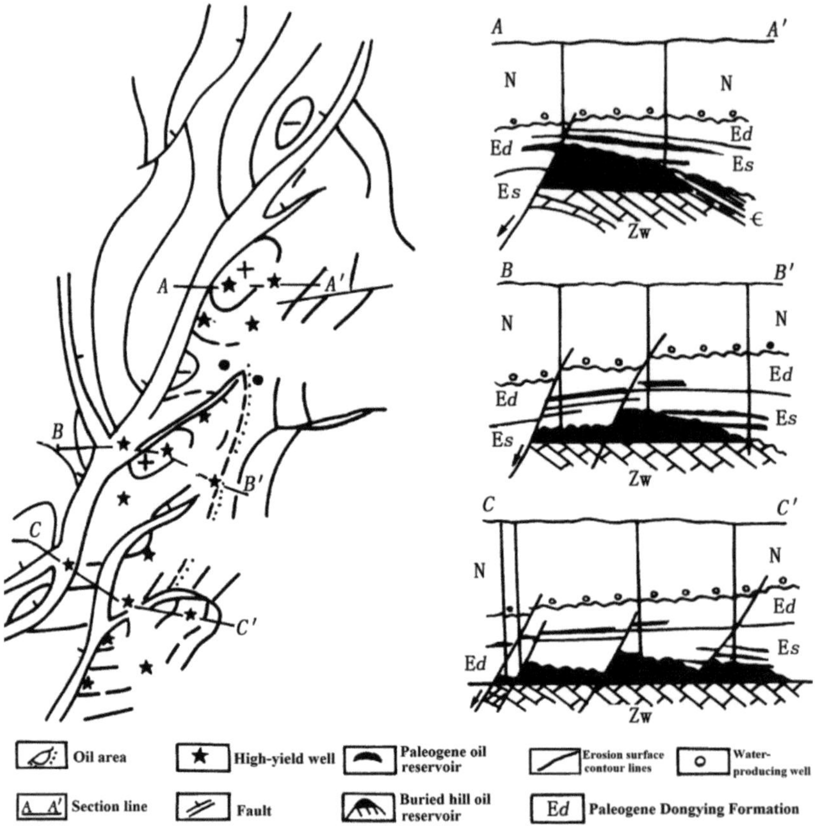

Fig. 6.28 Plan map and cross-section schematic diagram of Renqiu Oilfield (according to North China Oilfield)

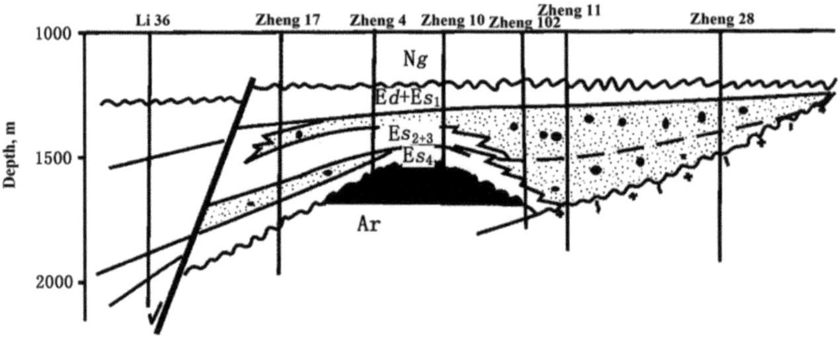

Fig. 6.29 Cross-section diagram of the buried hill oil reservoir in Wangzhuang (according to Shengli Oilfield, 2004)

The Bozhong 19-6 gas field in the Bozhong depression is composed of a large, multi-layered, multi-structured buried hill composite trap group developed on the Precambrian base, with the main gas-bearing layer being the Precambrian metamorphic rock, a typical buried hill gas reservoir. The Bozhong 19-6 structure has experienced multiple periods of tectonic movement, with well-developed fractures in the Precambrian metamorphic rock buried hill, mainly developing three sets of fault bodies. According to the structural characteristics, the Bozhong 19-6 structure can be further divided into the south and north blocks. The gas-bearing reservoirs are mainly metamorphic rocks such as monzonitic gneiss, plagioclase gneiss, and mixed gneiss, with fault breccia rocks visible near the major faults. The Precambrian metamorphic rock reservoir has been continuously sheared by the Tan-Lu fault since the Indosinian period, developing multiple periods of fractures and cataclastic rocks, etc., forming a huge fracture-type high-quality reservoir within the buried hill. The reservoir has obvious vertical zoning, with fracture-type reservoirs dominated by fractures, and also visible along the fracture dissolution pores. The reservoir storage space type is mainly fracture-type, with weathering and crushing zones developed in the high structural parts, such as pore-type and pore-fracture type reservoirs. Bozhong 19-6 is an exposed buried hill, and since 2016, multiple wells have been drilled in the south and north blocks of the Bozhong 19-6 structure, all of which have obtained good production capacities. Among them, the south block tested with a daily oil production of 168 m^3 and a daily gas production of 18.4×10^4 m^3, and the north block tested with a daily oil production of 305 m^3 and a daily gas production of 31.2×10^4 m^3. The fluid is in the condensate gas phase under reservoir conditions, with proven natural gas reserves exceeding 1000×10^8 m^3, making it the largest natural gas field discovered so far in the Bohai Bay basin. The formation of the Bozhong 19-6 buried hill oil and gas reservoir is mainly controlled by various reservoir elements: the deep layer of the Bozhong depression has a huge gas generation potential, which is the material basis for the formation of large condensate gas fields. The regionally stable distribution of the Dongying Formation and the Shahejie Formation's thick overpressure mudstone cap provides good sealing conditions for the preservation of large condensate gas fields. Multiple periods of dynamic crushing make the interior of the buried hill develop a large-scale fracture system and dynamic crushing zone, which is the key to the formation of metamorphic rock high-quality reservoirs (Fig. 6.30) (Xue & Li, 2018; Xu et al., 2019).

(3) Hasi-Messaoud Eroded Structural Type Sandstone Buried Hill Oil Reservoir

The Hasi-Messaoud oil field in North Africa, Algeria, is a famous example of a folded buried hill oil and gas reservoir (Fig. 6.31). The oil field is located in the eastern part of the Algerian Sahara Desert, 560 km from the Mediterranean Sea. Oil and gas are accumulated in the large anticline at the top, which has been eroded, and it is a typical eroded structural trap. The oil-bearing layer is Cambrian sandstone, about 3300 m deep, with an oil-bearing area of 1300 km^2 and an oil reservoir height of 270 m. The petroleum geological reserves are 34.7×10^8 T, making it a super large and high-yielding oil field.

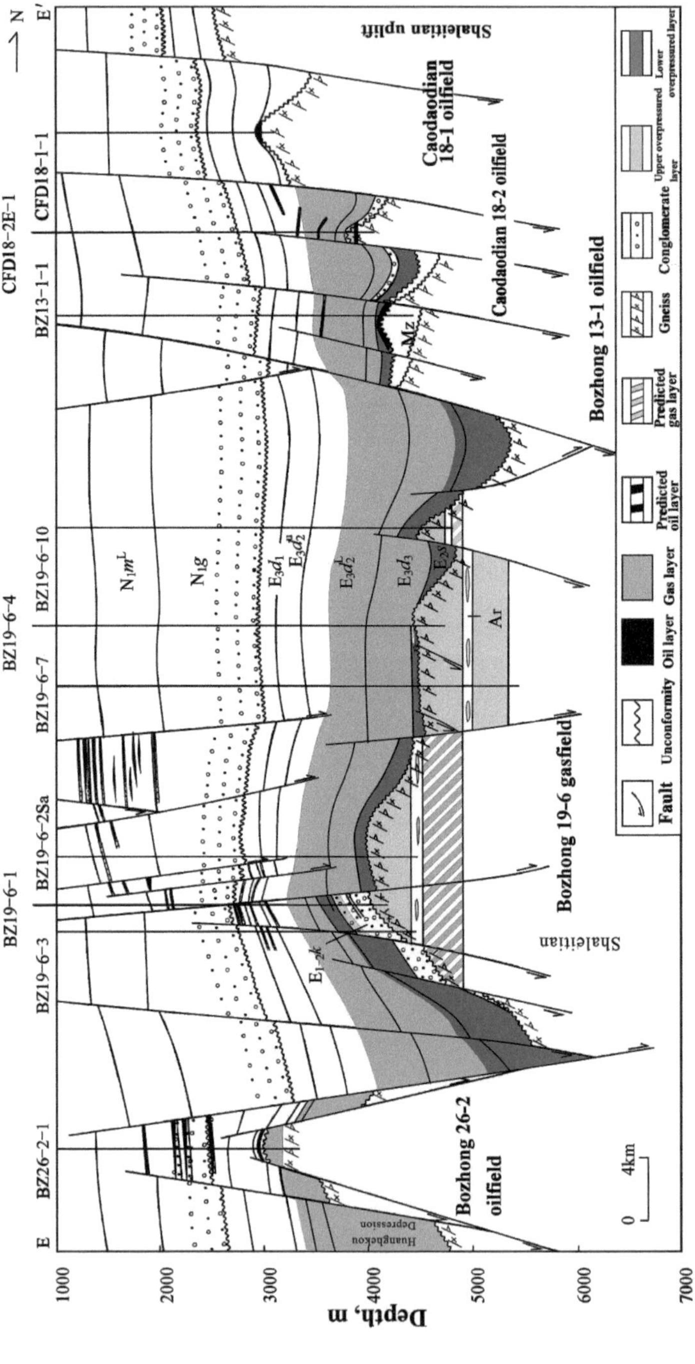

Fig. 6.30 Reservoir formation model of Bozhong 19-6 gas field (Xu et al., 2019). In the figure: Ar—Archean, Mz—Mesozoic, E1-2k—Kongdian formation, E2s—Shahejie formation, E3d3—Dongying third member, E3d2L—Dongying second lower sub-member, E3d2U—Dongying second upper sub-member, E3d1—Dongying first member, N1g—Guantao formation, N1mL—Minghua Zhen lower sub-member

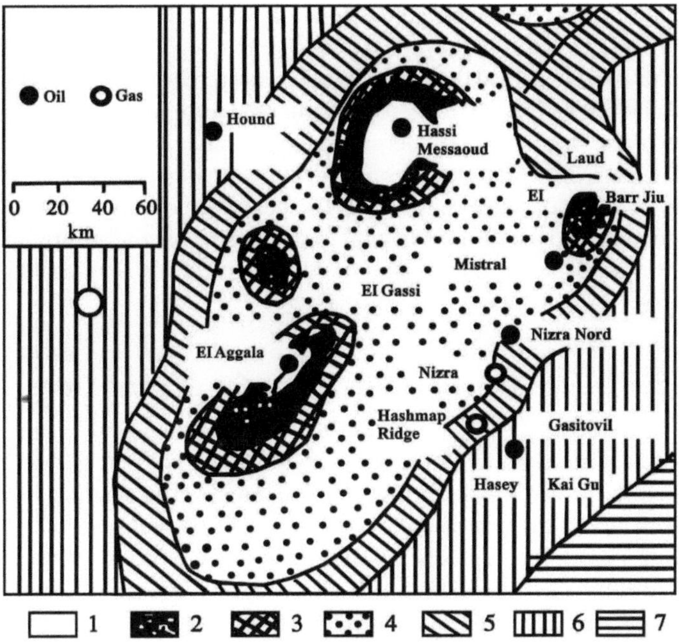

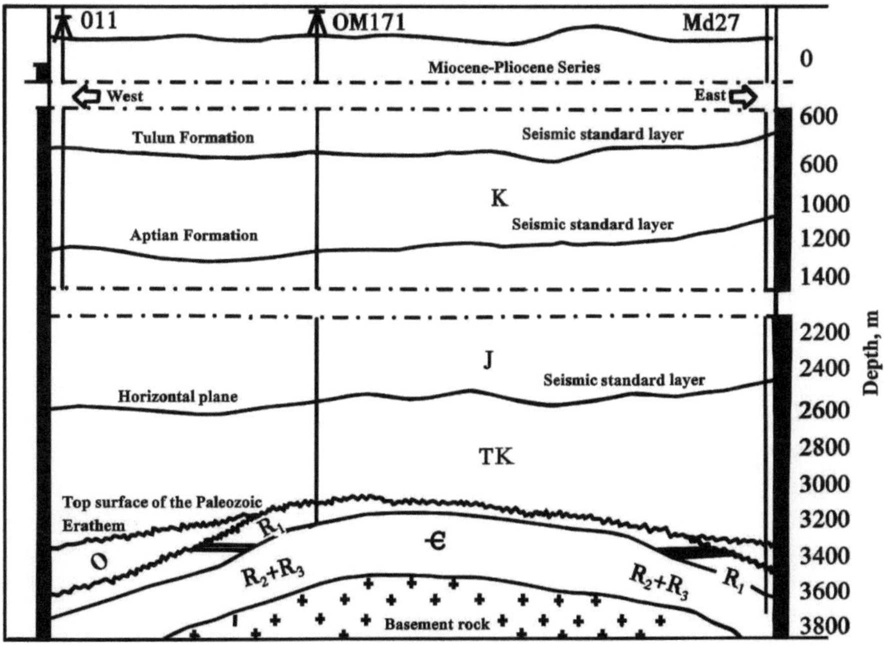

Fig. 6.31 Plan and cross-section of Hasi-Messaoud oil field (according to Zhang et al., 1999). 1—Cambrian, 2—El Gassi Sandstone, 3—El Gassi Clay, Namur Quartzite, 5—Upper Ordovician, 6—Silurian

The anticline structure of the oil field rose during the Caledonian period and was eroded for a long time, with the top of the uplift exposing Cambrian sandstone. It was not until the Triassic period that it began to be covered by salt rock and red shale, forming good conditions for a hidden eroded structural trap. At the same time, due to the deposition of the Triassic, the underground temperature and pressure increased, and the Silurian black shale in the depression 40 km northwest of Hasi-Messaoud had the conditions for secondary oil generation. The generated oil migrated to the Hasi-Messaoud hidden eroded anticline structural trap along the unconformity surface and accumulated, forming the current large oil field.

6.3.2 *Stratigraphic Disconformity Traps for Oil and Gas*

The broad sense of stratigraphic disconformity traps for oil and gas refers to oil and gas reservoirs formed by disconformity traps located beneath the disconformity surface, including the aforementioned buried hill oil and gas reservoirs. The stratigraphic disconformity traps discussed here are in the narrow sense, referring to those primarily found on the edges of basins or on ancient uplifts, where, under certain tectonic settings, the reservoir layers are eroded in the up-dip direction and later covered by newly deposited non-permeable rock layers, forming stratigraphic disconformity closures beneath the disconformity, where oil and gas accumulate to form stratigraphic disconformity traps. Unlike buried hill closures, the disconformity surface of this type of closure generally does not have a significant topographic prominence.

6.3.2.1 Characteristics of Stratigraphic Disconformity Traps and Oil and Gas Reservoir Formation

Stratigraphic disconformity traps for oil and gas are mainly distributed in coarse clastic rocks on the edges of basins. This may be due to the frequent depositional hiatuses between sedimentary rock systems on the edges of basins, where clastic reservoirs in the up-dip areas are prone to erosion. When they are covered by overlying non-permeable rock layers, they form excellent closure conditions. In folded sedimentary basins, folding and faulting are significant, and disconformity phenomena are common, also leading to the development of this type of closure condition.

Like other types of oil and gas reservoirs, the formation of stratigraphic disconformity traps also requires basic conditions for reservoir formation such as source, reservoir, cap, trap, migration, and preservation. The sealing elements of stratigraphic disconformity traps mainly come in two forms: one is the non-permeable argillaceous rock layers overlying the basin disconformity, and the other is the heavy oil seal formed by the oxidation of crude oil. Oil reservoirs formed by heavy oil seals generally suffer from oxidation, resulting in heavier oil quality, while the oil quality gradually becomes lighter away from the heavy oil seal belt.

Stratigraphic disconformity traps and buried hill oil and gas reservoirs are both located beneath the disconformity surface, but their formation characteristics differ significantly. Stratigraphic disconformity traps are more developed on the edges of basins with sedimentary reservoirs. Unlike the formation of buried hill oil and gas reservoirs, which is mainly "new generation and ancient storage," the oil and gas accumulated in stratigraphic disconformity traps mainly come from the source rock developed in the same period in the down-dip direction, and can also come from ancient oil and gas reservoirs that have been reformed and re-accumulated. These reservoirs are mainly sealed by non-permeable argillaceous rock layers on the edges of the basin and heavy oil seals formed by the oxidation of crude oil, with reservoirs often being sand and gravel rocks with primary porosity. Buried hill oil and gas reservoirs, on the other hand, are a special type of "basement" oil and gas reservoir, with the oil and gas layers located beneath the regional disconformity surface, belonging to the basement rock system of the basin, and have a certain structural or paleotopographic prominence. Their oil and gas mainly come from the overlying and lateral younger source rock systems. The reservoirs of buried hill oil and gas reservoirs can be sedimentary rocks, or they can be igneous or metamorphic rocks. Apart from a few with primary porosity in clastic rocks, most are dense and hard rocks with poor primary porosity. Due to various epigenetic modifications, they develop secondary fractures and pores, thus acquiring reservoir capabilities.

The stratigraphic disconformity surface is not a simple plane but a "geological body" that includes multiple structural layers. An ideal disconformity often develops a three-layer structure, namely the rocks above the disconformity surface, the weathered clay layer and semi-weathered rocks below. Due to various factors such as erosion time, rock properties, topography, and climate, most disconformities lack a weathered clay layer. The presence or absence of the disconformity weathered clay layer, the rock property configuration and lithology situation of the upper and lower rock layers of the disconformity, and the development and combination form of the semi-weathered rocks' permeability have a significant impact on the migration and accumulation of oil and gas near the disconformity.

If a stable weathered clay layer or semi-weathered rocks develop above a stable non-permeable layer (such as thick mudstone) as a regional cap layer, the disconformity mainly acts as a seal, and oil and gas will accumulate in suitable positions beneath the disconformity surface, forming stratigraphic disconformity traps for oil and gas. In this case, if the rocks beneath the disconformity surface are easily weathered carbonate rocks, oil and gas can migrate long distances along the semi-weathered rock layers beneath the disconformity surface. If the rocks beneath the disconformity surface are interbedded sandstone and mudstone that are not easily weathered, and the sandstone and mudstone layers cannot form a continuous conductive layer beneath the disconformity surface, oil and gas find it difficult to migrate through the layers. However, due to the good sealing effect of the mudstone layer on the disconformity surface for oil and gas migration, oil and gas can form stratigraphic disconformity traps in various sand layers beneath the disconformity surface (Fig. 6.32).

If the semi-weathered rocks or permeable layers beneath the disconformity surface directly contact with permeable layers (such as sandstone layers), that is, when the

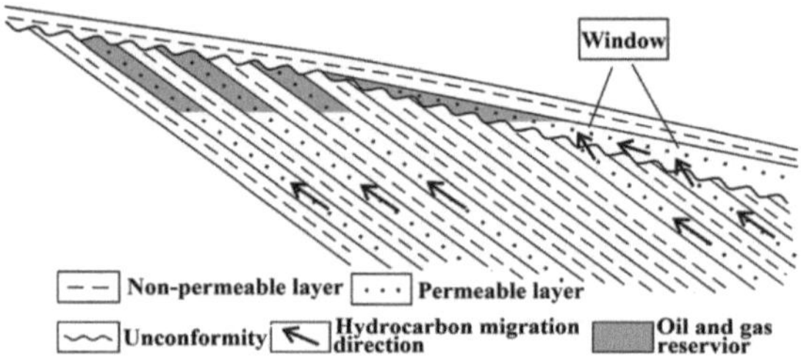

Fig. 6.32 Schematic diagram of oil and gas migration and accumulation along unconformities in strata

upper and lower layers of the disconformity surface are permeable layers, the permeable layers beneath the disconformity surface are prone to forming "windows" due to the lack of a cap layer. Oil and gas migrate through the "windows" across the disconformity surface into the permeable layers and move in the up-dip direction, forming stratigraphic superimposed oil and gas reservoirs (Fig. 6.32). Therefore, in the interbedded sandstone and mudstone systems, due to the poor continuity of the permeable layers beneath the disconformity surface and the non-permeable layers at the top in the horizontal direction, it is difficult for oil and gas to migrate long distances along the disconformity.

6.3.2.2 Example of Stratigraphic Disconformity Traps for Oil and Gas

The Jinkang Oilfield in the Shahejie Formation within the Dongying Sag's southwest slope area of the Bohai Bay Basin is a typical representative of stratigraphic disconformity traps for oil and gas. The oilfield is in the southern gentle slope zone of the Dongying Sag, adjacent to the Luxi Uplift in the south, with the Paleogene strata dipping from south to north. Tectonic movements at the end of the Oligocene formed disconformities between the first, second, and third members of the Shahejie Formation and between the Guantao Formation and the underlying strata. The base of the Guantao Formation is developed with mudstones 10–50 m thick, which, in conjunction with the nose-like structural background, has formed a series of stratigraphic disconformity oil and gas reservoirs. Due to the relatively shallow burial depth of the oil and gas reservoirs (800–1200 m), the crude oil has become viscous due to biodegradation and oxidation (Fig. 6.33).

6.3.3 Stratigraphic Overstep Oil and Gas Reservoirs

Crustal uplift and subsidence movements and their differences often cause the advance and retreat of seawater or lake water. The result of these water movements is manifested in geological sections as "transgression" and "regression" phenomena, as shown in Fig. 6.34.

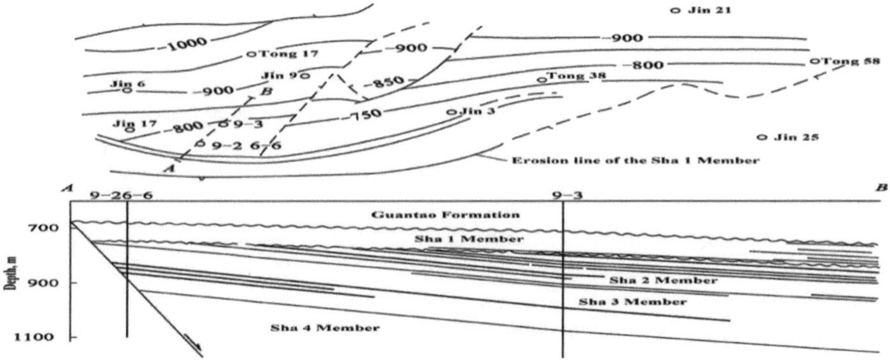

Fig. 6.33 Structure of Jinjia Oilfield (contour interval in meters) and oil reservoir cross-section diagram (according to Wang & Qian, 1992)

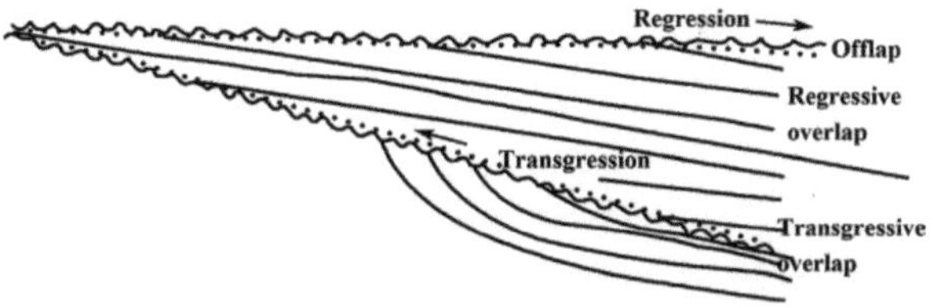

Fig. 6.34 Schematic diagram of transgression and regression (according to Zhang et al., 1999)

Stratigraphic transgression refers to the gradual expansion of the depositional range as the water body advances, with newer sedimentary layers covering older ones and extending towards the land, making an unconformable contact with the eroded surface of the older strata. From a profile view, transgression is characterized by each stratum in the overlying sequence extending beyond the edges of the underlying older strata, and within the same columnar section, the sediments become finer from bottom to top. Regression occurs as the water body recedes, with the range of the newer sedimentary layers becoming smaller. In actual geological environments, pure water advance rock system strata migration and pure water retreat rock system strata migration are rare. What is more commonly seen is the alternation of water advance and retreat, which is manifested in the section as the intersection of transgression unconformities and regression erosion surfaces (Fig. 6.34). In terms of rock structure, the particles change from coarse to fine and then back to coarse from bottom to top, forming a complete sedimentary cycle. Due to the constant changes in the direction, speed, and magnitude of crustal movement, the advance and retreat of seawater or lake water are also varied, and the alternation of transgression and regression reflected in the geological sections is diverse. All these changes can form various types of stratigraphic traps. Therefore, in sedimentary basins, a detailed analysis of the historical changes in water and land and the paleogeographic conditions of each geological period is of great significance for the search for stratigraphic transgression oil and gas reservoirs.

6.3.3.1 Characteristics of Stratigraphic Transgression Traps and Oil and Gas Reservoir Formation

When the water body advances, the water basin gradually expands, and porous sandstones are deposited along the erosion surfaces at the edges of the sedimentary depressions, which are well sorted and have good reservoir properties. As the water basin continues to expand and the water deepens, impermeable mudstones are deposited on top of the sandstone layers, resulting in stratigraphic transgression traps, and the accumulation of oil and gas within them forms stratigraphic transgression oil and gas reservoirs.

These types of stratigraphic transgression traps are formed in the zones where water and land alternate, especially during the water advance stage, where the basin bottom is mainly characterized by stable subsidence with minor oscillations, often associated with the reducing environment of the shallow marine continental shelf or large and deep lake basins. Therefore, the sand layers above and below, as well as the deeper lateral transitions into muddy deposits, are often rich in organic matter, making them good source rock layers and also good cap layers. This forms a cyclic and lateral variation in the generation, storage, and sealing combination. After the oil and gas are generated, they migrate to the stratigraphic transgression traps and accumulate, forming stratigraphic transgression oil and gas reservoirs. These types of oil and gas reservoirs are concentrated in the zones where water and land alternated historically, and can be found in the coastal areas of marine sedimentary basins and the shallow lake areas of large and deep lacustrine sedimentary basins.

Stratigraphic transgression oil and gas reservoirs are generally distributed in the marginal areas of basins, and their special structural position determines that they have different reservoir control elements from lithological oil and gas reservoirs, such as large-scale unconformities that form the basis for stratigraphic traps. Sufficient oil sources, nose-like structures, oil and gas migration dynamics, and a composite conductive system composed of high-permeability sand bodies, faults, and unconformities are necessary conditions for the long-distance migration and accumulation of oil and gas. The action of shallow atmospheric water makes the original heavy oil more viscous.

6.3.3.2 Examples of Stratigraphic Transgression Oil and Gas Reservoirs

Many such oil and gas reservoirs have been discovered worldwide, among which the oil and gas reservoirs of the East Texas oil field are quite famous, as shown in Fig. 6.35.

The East Texas oil field is located on the west side of the Sabine uplift in the western Gulf of Mexico Basin, where Upper Cretaceous Woodbine Formation sandstones are deposited on the Lower Cretaceous unconformity surface, and are overlain by the Upper Cretaceous Austin Group, which is also in unconformable contact. The top and bottom of the sandstone, which are two unconformities, intersect in the updip

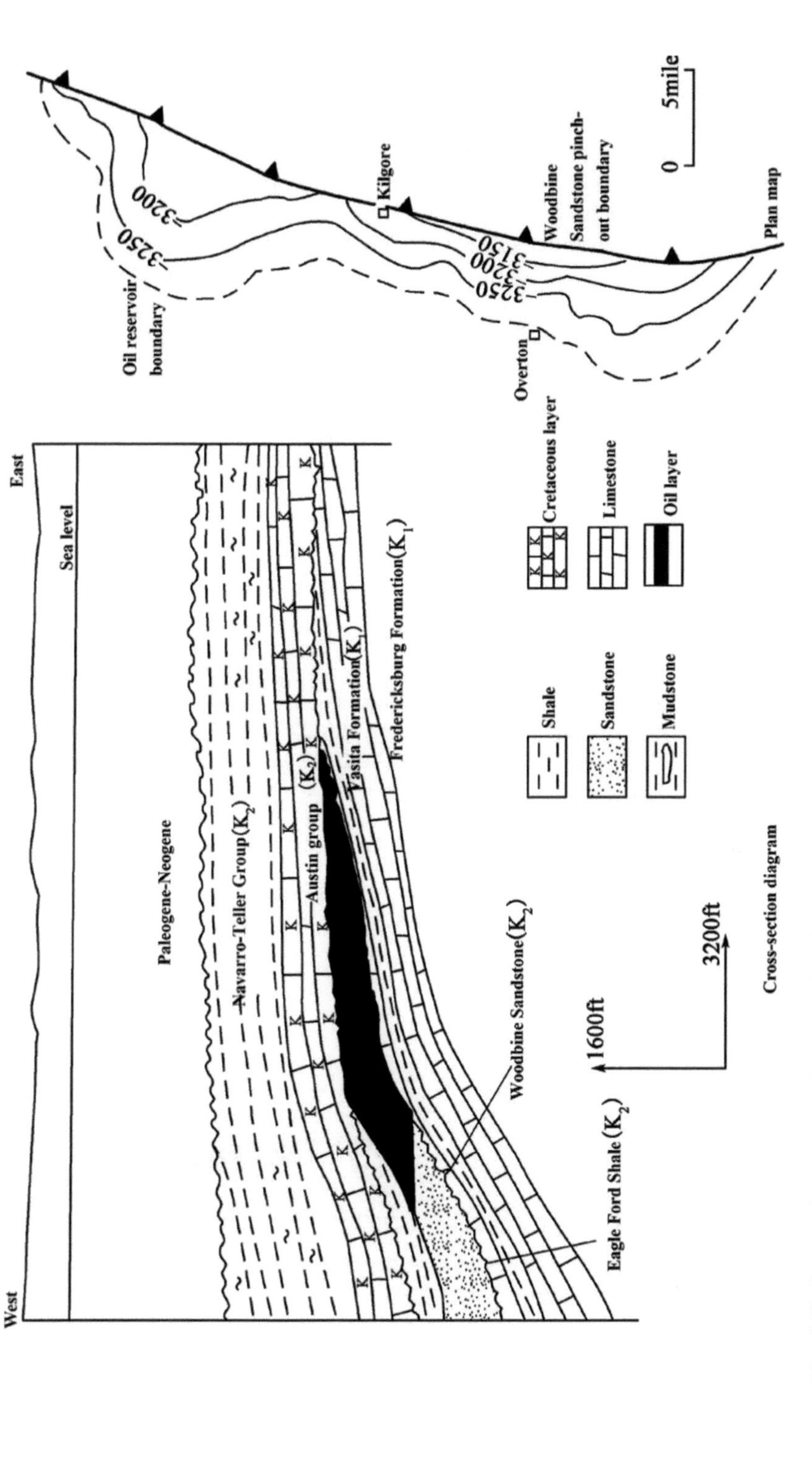

Fig. 6.35 Structural contour map (contour unit in feet) and cross-sectional view of the top of the Woodbine (Cretaceous) oil-producing layer in the East Texas oil field (A. I. Levorson)

direction, and oil and gas accumulate within them, forming stratigraphic transgression oil and gas reservoirs. The total recoverable reserves of this oil field are 7.3×10^8 t, making it one of the largest oil fields in the United States.

China has also discovered many stratigraphic transgression oil and gas reservoirs, among which the Shanjiasi oil field in the Dongying Depression is a typical representative, as shown in Fig. 6.36. The oil field is in the western Bin County uplift edge area of the Dongying Depression, where Paleogene Shahejie Formation sandstones are deposited on the unconformity surface, and oil and gas accumulate within them, forming stratigraphic transgression oil and gas reservoirs. In addition, stratigraphic transgression oil and gas reservoirs have also been found on the marginal slopes of many sedimentary basins in the eastern part of China, as well as on the slopes of large uplifts, but they are not large in scale.

The oil reservoir of the Curaqual oil field in eastern Venezuela is a typical large-scale stratigraphic transgression example (Fig. 6.37), and the oil field is one of the large oil fields in South America. The Pliocene–Pleistocene sandstones are deposited on the underlying unconformity surface and are overlain by impermeable layers, forming stratigraphic transgression trap conditions, and oil and gas accumulate within them, forming a giant stratigraphic transgression oil reservoir.

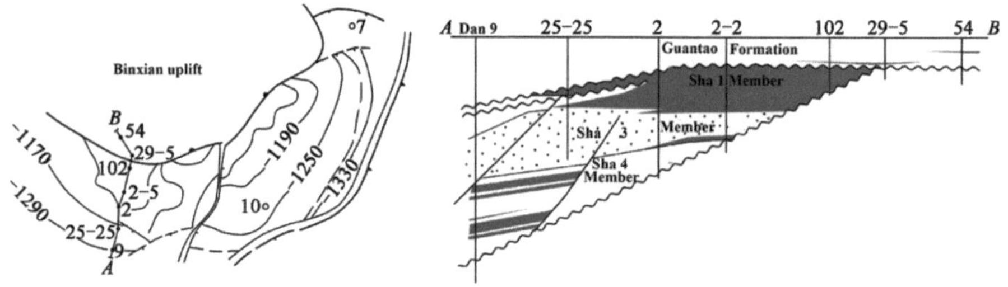

Fig. 6.36 Stratigraphic overstep hydrocarbon reservoir in the Danjia Temple Oilfield of the Dongying Depression (according to Wang & Qian, 1992)

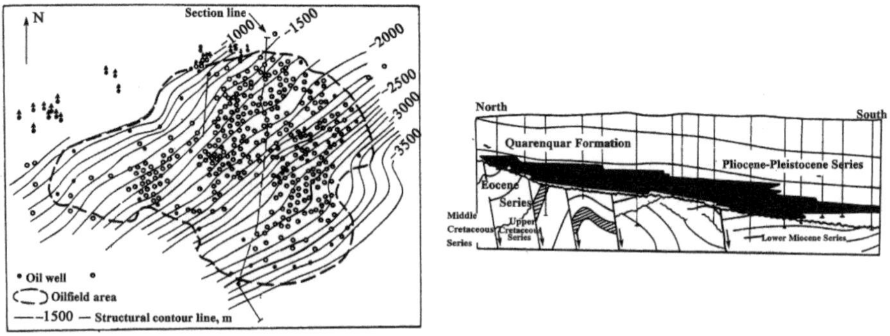

Fig. 6.37 Plan and cross-section diagram of the Quarenquar oilfield in eastern Venezuela (according to A. I. Levorsen)

6.4 Lithologic Oil and Gas Reservoirs

6.4.1 *Formation Mechanism and Characteristics of Lithologic Traps and Oil and Gas Reservoirs*

6.4.1.1 Formation Mechanism of Lithologic Traps

Lithologic traps refer to the closures formed by changes in the lithology or physical properties of the reservoir rocks, within which oil and gas are accumulated, and are called lithologic oil and gas reservoirs. The vertical and horizontal changes in reservoir lithology can be formed during the sedimentation process or during the diagenesis process. However, most lithologic traps are direct products of the sedimentary environment. Due to different sedimentary environments and diagenetic actions, the lithology or physical properties of the sediments change, forming lithologic updip pinchouts, lens bodies, and physical property sealing traps, etc.

In sandstone and mudstone sedimentary profiles with significant lithologic changes, many thin layers of sandstone are commonly interbedded and interlaced. Some stratified sandstone bodies are bounded by impermeable mudstones at the top and bottom and gradually transition into impermeable mudstones laterally, with the sandstone body wedge-shaped pinching out within the mudstones, which is known as the sandstone updip pinchout trap, as shown in Fig. 6.38a. Some sandstone bodies are lenticular and surrounded by impermeable layers, then they are sandstone lenticular body traps, as shown in Fig. 6.38b. These two types of sandstone bodies (or conglomerate bodies) often occur in the same profile because their genesis is similar, occurring within the same basin where different sedimentary environments lead to the simultaneous deposition of materials of different properties, resulting in lateral lithologic changes in the sediments. Either sandstone grades into mudstone, or mudstone grades into sandstone, or uneven permeability changes in sandstone. At the pinchout end of the sandstone pinchout body and at both ends of the lenticular body, the content of mud increases and permeability becomes worse. While towards the main body of the sandstone, the mud content decreases and permeability improve, forming lenticular, lithologic pinchout closures, or physical property filling traps, etc. In addition to sandstone facies changes forming lithologic traps, carbonate rocks (such as clastic limestone) can also form lithologic traps due to lithologic changes, such as bio-reef oil and gas reservoirs.

During diagenesis and epigenetic actions, secondary actions can alter the original lithologic traps and make parts of the reservoir into non-permeable rock layers, or make parts of non-permeable rock layers into permeable rock layers, forming lithologic traps. For example, in thick sandstone layers, due to uneven permeability, there can also be local high-permeability zones within low-permeability sandstones, as shown in Fig. 6.38c. In carbonate rock areas, due to the ease of dissolution and secondary actions, it is easy to form lithologic traps during the diagenetic stage.

Near the ancient coastlines, such as coastal sandbars, ancient river channels, and ancient delta channel sand layers, as well as along the strike-valley sand layers

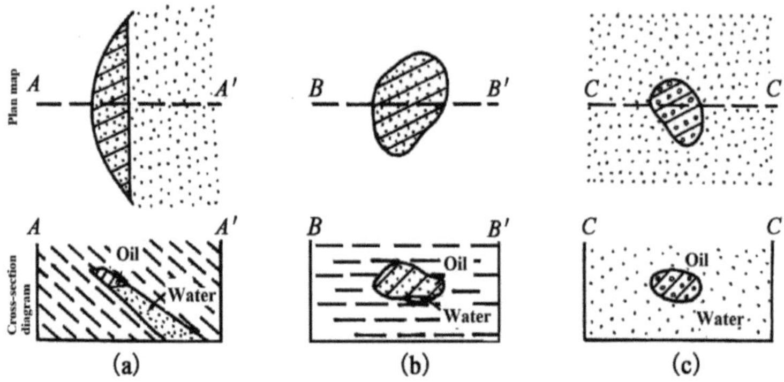

Fig. 6.38 Basic types of sandstone lithological traps and oil and gas reservoirs. **a** Lithologic updip pinchout oil and gas reservoir, **b** sand lens body oil and gas reservoir, **c** Physical property sealing oil and gas reservoir

distributed along the steep cliffs or fault scarps of ancient topography, when they are overlain by impermeable mudstones, they can also form sandstone body lithologic traps. They are lenticular in cross-section and extend in an irregular strip-like pattern in plan view.

In continental sedimentary basins, lithologic and lithofacies changes are frequent, and there are many types of reservoir rock bodies. Different types of reservoir rock bodies are superimposed on each other, which is conducive to the formation of various types of lithologic traps. Lithologic traps are mainly divided into reservoir (sandstone and carbonate) up-dip pinchout oil and gas reservoirs and lenticular lithologic oil and gas reservoirs. Their common features are: (1) Reservoir bodies often interfinger and pinch out within the source rock bodies, not only having abundant oil and gas sources but also having good reservoir-seal combination conditions, (2) The trap formation period is early, and oil and gas primary migration directly enters the reservoir, which is conducive to the accumulation and formation of oil and gas reservoirs. (3) The distribution of lithologic oil and gas reservoirs is mostly related to river–lake sedimentary systems and ancient topography.

Lithologic traps are the result of lithologic changes and are related to changes in sedimentary environments in terms of genesis. Therefore, they often appear in groups and sets, forming larger multi-layered lithologic traps. In actual exploration work, if a sandstone pinchout body or lenticular body oil and gas reservoir is found, it is likely to find more similar oil and gas reservoirs nearby.

6.4.1.2 Types and Formation Mechanisms of Lithologic Oil and Gas Reservoirs

Based on the types of reservoirs, lithologic oil and gas reservoirs can be divided into four categories: sandstone, mudstone, carbonate, and igneous rock lithologic oil and gas reservoirs, with the sandstone type being the primary one. According to the causes of the traps, lithologic oil and gas reservoirs can be classified into sandstone updip pinchout oil and gas reservoirs, sandstone lens body oil and gas reservoirs, physical property sealing lithologic oil and gas reservoirs, and biological reef oil and gas reservoirs. The main types are sandstone updip pinchout traps and sandstone lens body traps, which are primarily controlled by sedimentation conditions, especially in continental oil and gas basins.

Based on the relationship between effective source rocks and reservoir bodies, lithologic oil and gas reservoirs can be divided into two categories: those that are in contact with source rocks and those that are not in contact with source rocks. The former is surrounded by or partially in contact with source rocks, and the oil and gas generated by the source rocks can directly enter the reservoir bodies through bedding, fractures, and sand layers within the source rocks. The latter has a mudstone layer tens or even hundreds of meters thick between the source rock and the reservoir body, and oil and gas can only potentially accumulate through conduits such as faults and fractures.

Regarding the formation mechanism of sandstone lens body oil and gas reservoirs, England et al. (1987) suggested that during the burial process, due to differential compaction and the pressure increase from the generation of hydrocarbons in mudstones, the surrounding mudstones have higher abnormal pressures compared to the lens sandstones. There is a significant residual pressure difference between the sandstone lens body and the mudstones. This pressure difference points towards the lower overpressure amplitude of the sandstone. Since the sandstone lens has high permeability, it is easy to achieve pressure equilibrium within it, ultimately leading to the residual pressure in the lower half of the sandstone lens being lower than the surrounding rock pressure (the water potential Φ_w contour line is concave downwards), while the residual pressure in the upper half is higher than the surrounding rock pressure (the water potential Φ_w contour line is convex upwards). This results in the formation water and oil and gas from the lower half of the sandstone lens entering the lens due to the residual pressure difference. Within the sandstone lens, oil and gas can reach the top of the lens under the influence of buoyancy. Oil and gas are retained at the top of the lens due to the difference in capillary forces between the reservoir and the cap, while water is not affected by capillary forces and is expelled from the upper part of the lens under the drive of the residual pressure difference (Fig. 6.39).

Therefore, whether a sandstone lens can be charged and form a reservoir depends on whether the charging dynamics and resistances between the surrounding rocks (source rocks) and the sandstone body can achieve a balance. The charging dynamics of oil and gas come from the capillary pressure difference and the expulsion pressure of the source rock, while the charging resistance comes from the capillary resistance experienced by the displacement of pore fluids in the sandstone lens. As the burial depth increases, the source rocks surrounding the lens have a strong expulsion capability, and the oil and gas charging dynamics are greater than the resistance, thus meeting the conditions for the accumulation of hydrocarbons in the sandstone lens. When the burial depth increases to a certain extent, the sandstone lens gradually

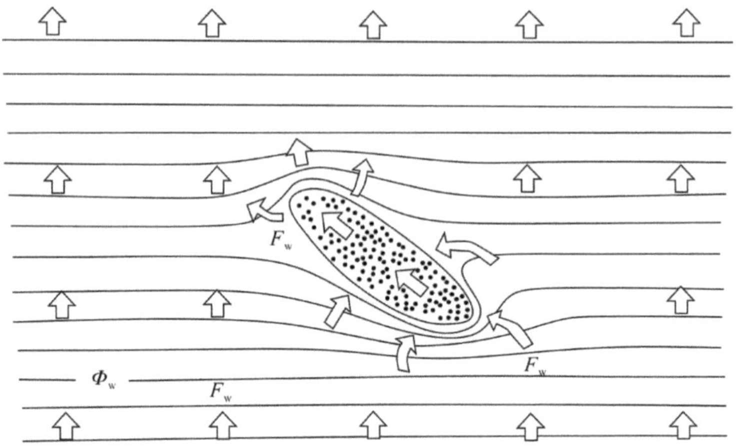

Fig. 6.39 Analysis diagram of the water potential field and force direction of a sandstone lens oil reservoir (England et al., 1987)

becomes compact, and the charging dynamics are no longer sufficient to provide the conditions for expulsion into the lens, thus ending the charging of the lens.

Based on the formation, evolution, and accumulation and release processes of the oil and gas migration and accumulation driving forces, the oil and gas charging, migration, and accumulation modes in the source rock system should be mainly episodic, and they play a "transit station" role in the entire process of oil and gas migration and accumulation in the oil and gas basin, directly providing oil and gas for large and medium-sized structural oil and gas reservoirs and stratigraphic oil and gas reservoirs.

Exploration practices have shown that there is a significant difference in oil content in lithologic oil and gas reservoirs, which is due to the influence of the spatial and temporal organic configuration of the main controlling factors of oil and gas reservoir formation, such as source rocks, formation dynamics, conduit types, and the various elements during the formation period.

6.4.2 Lithologic Pinchout Oil and Gas Reservoirs

These oil and gas reservoirs are formed due to the pinchout of the reservoir along the up-dip direction, creating closure conditions where oil and gas accumulate. In continental lacustrine basins, the front edge zones of various types of sandstone bodies, in conjunction with large uplifts or local structural closures, cause the up-dip pinchout line of the sandstone to intersect with the top surface contour line of the reservoir, forming an up-dip pinchout closure. The distribution and size of these oil and gas reservoirs depend on the relationship between different parts of the sandstone body and different levels of structural configuration. Composite sandstone bodies formed by multiple rhythmic layers, when combined with the slope zone of a depression or a large uplift zone, cause the updip pinchout lines of multiple sand layer groups to be tangent to the structural contour lines, forming large to medium-sized lithologic up-dip pinchout oil reservoirs. These reservoirs are characterized by a large oil-bearing area, multiple oil-bearing layer groups, and a high degree of oil and gas enrichment.

For example, the front pinchout zone of the Shahejie Formation lake bottom fan sand body in the Biyang Depression is situated on the slope zone background, and each lobe of the lake bottom fan sand body correspondingly forms an updip pinchout oil and gas reservoir (Fig. 6.40).

In the Fularji region on the western slope of the Songliao Basin, the sandstone reservoir of the Cretaceous Yaojia Formation becomes less permeable along the up-dip direction, creating closure conditions and forming a lithologic pinchout oil and gas reservoir.

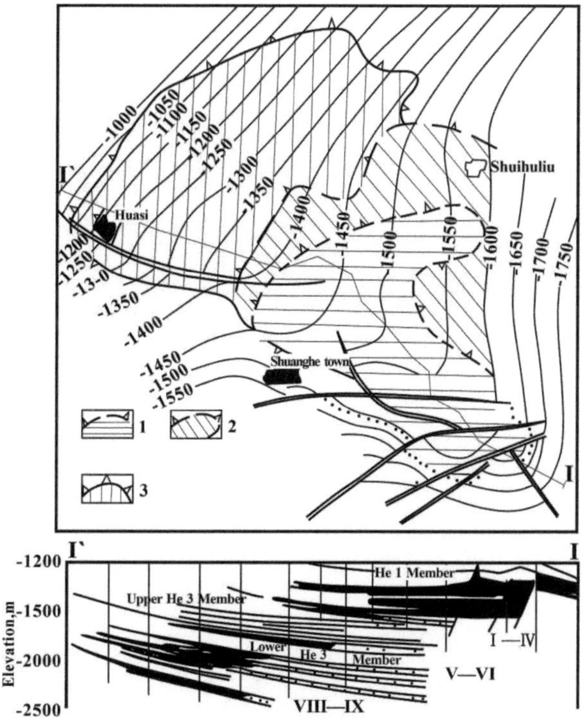

Fig. 6.40 Plan of the Shuanghe Sandstone updip pinch-out oil and gas reservoir in the Biyang Depression (contour interval in meters) and cross-section diagram (according to Hu et al., 1991). 1—Oil-bearing range of I–IV oil groups, 2—oil-bearing range of V–VI oil groups, 3—oil-bearing range of VII–IX oil groups

6.4.3 Sandstone Lens Oil and Gas Reservoirs

These oil and gas reservoirs are formed by lenticular or other irregular reservoirs that are enclosed by impermeable strata, creating lenticular trap conditions, and when oil and gas accumulate within them, they form lens oil and gas reservoirs. The most common type is sandstone lenses surrounded by mudstone. The scale of lenticular oil and gas reservoirs is usually not large.

In the Cenozoic Shahejie Formation of the Bohai Bay Basin, there are many sandstone lens oil and gas reservoirs developed within the large set of mudstones. For example, in the Ying 11 area in the southwestern part of the Dongying depression structural belt, there is the largest sandstone lenticular lithologic oil and gas reservoir in the eastern basins of China. The oil-bearing stratum is a turbidite sandstone body in the middle of the Shahejie Formation, and the oil reservoir is located within the Shahejie Formation source rock. The original formation pressure is high, and the oil quality is light, reflecting the characteristics of a primary oil reservoir (Fig. 6.41).

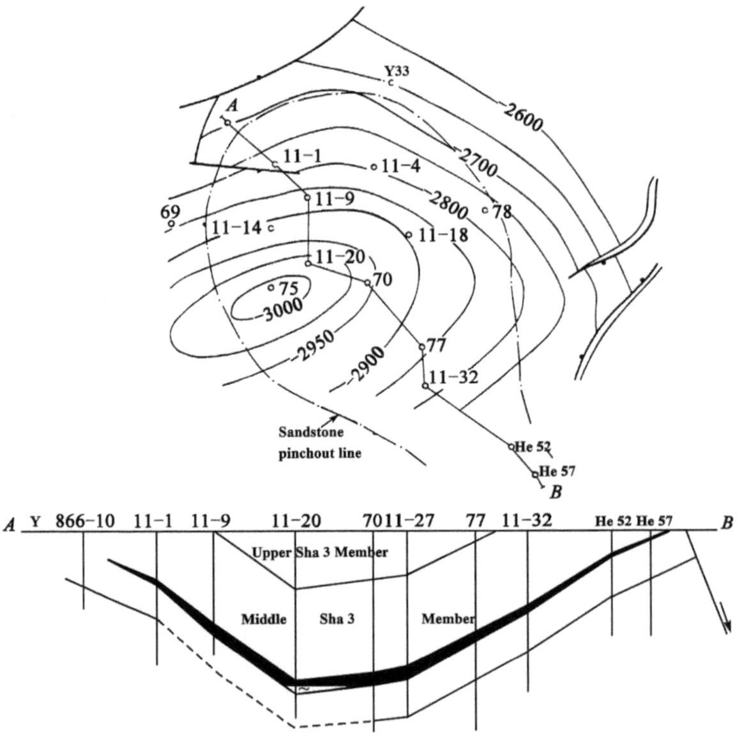

Fig. 6.41 Plan of the Yingshi 11 sandstone lens oil reservoir in the Dongying Depression (contour interval in meters) and cross-section diagram

6.4.4 Physical Property Sealing Oil and Gas Reservoirs

Physical property sealing traps, also known as diagenetic traps, refer to the types of closures formed due to various secondary diagenetic actions that change the porosity of originally deposited rock layers. There are mainly two situations: First, due to cementation, the porosity and permeability of the reservoir in the updip direction are reduced, and because the reservoir's physical properties deteriorate in the updip direction, it forms a shielding condition, thus creating a physical property sealing traps. Second, due to secondary changes such as dolomitization and dissolution, the porosity and permeability of parts of rock layers that were originally non-permeable are increased, forming high porosity and high permeability sections within low permeability layers, thus creating a physical property sealing traps. The accumulation of oil and gas in these closures formed by changes in physical properties is known as physical property sealing traps lithologic oil and gas reservoirs.

Physical property sealing oil and gas reservoirs are widely developed in various sand and gravel fan bodies. For example, subaqueous fans form a shield in the updip direction due to the dense physical properties at the fan root. The Yong 921 sand and gravel fan body in the Dongying Depression is a typical physical property sealing oil and gas reservoir (Fig. 6.42). This sand and gravel body is located in front of the ancient alluvial fan deposit between the Yanjia and Yong'an Town nose-like structures, with the fan top presenting a nose-like structure that lifts to the north. The fan body can be further divided into three subfacies: fan root, fan middle, and fan end. The fan root has very poor physical properties due to large gravel particles, mixed composition, and extremely poor sorting, becoming an irregular shield, forming a physical property sealing traps and oil and gas reservoir.

In addition, in low permeability rock layers, there often exist high permeability belt sand body oil and gas reservoirs. The permeability of the reservoir changes greatly, with oil and gas accumulating in the parts with good permeability, known as the "sweet spot," while the poorly permeable parts are filled with water, also belonging to physical property sealing lithologic oil and gas reservoirs. The shape and distribution of these oil and gas reservoirs are very irregular. The oil and gas reservoirs in the "hundred-foot sandstone" of the Lower Carboniferous in the Appalachian oil and gas basin in the United States can serve as a typical example.

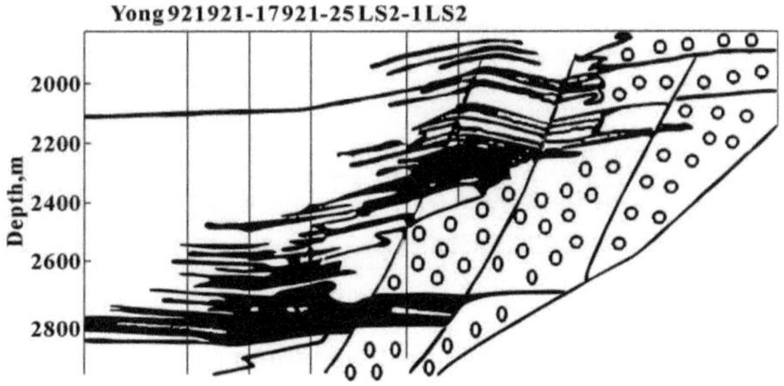

Fig. 6.42 Cross-section diagram of the property-sealed oil and gas reservoir in the Yong 921 sand and gravel fan body (according to Shengli Oilfield, 2004)

In some oil and gas basins in China, it is also common to see high permeability belt oil and gas reservoirs in low permeability rock layers. For example, the Triassic and Jurassic in the Shaan-Gan-Ning Basin have this type of oil and gas reservoir.

6.4.5 *Biological Reef Oil and Gas Reservoirs*

6.4.5.1 Characteristics of Formation of Biological Reef Oil and Gas Reservoirs

A biological reef trap refers to a biological reef reservoir rock with good porosity and permeability that is sealed by overlying and surrounding impermeable rock layers, and the oil and gas accumulation formed within it is called a biological reef oil and gas reservoir. The shape of the biological reef trap and the oil and gas reservoir is related to the shape of the reservoir body in the reef complex.

Biological reefs are carbonate rock constructions composed of reef-building organisms such as corals, sponges, bryozoans, algae, and archaeocyaths, which are buried in situ. In addition to the reef-building organisms, biological reefs also contain reef-loving organisms such as crinoids and foraminifera. Different geological eras have different reef-building organisms.

Ancient biological reefs are similar in origin to modern biological reefs, and the parts of the biological reefs and their lithofacies distribution can be compared with modern biological reefs. Figure 6.43 shows the parts of ancient biological reefs and their lithofacies characteristics (Levorsen, 1967): Lithofacies A of the lagoonal sediments behind the biological reef, including interlayers of dolomite, limestone, sandstone, red shale, and anhydrite, etc., collectively referred to as the back-reef facies; transitioning from the back-reef facies to the main body of the biological reef, B; the lithofacies immediately in front of the biological reef, facing the sea, is limestone and sandstone with biological reef debris, called the fore-reef facies C; further towards the sea, it transitions to lithofacies including gray to black shale and limestone, called the basin facies D.

In some areas, within a thick rock series at different heights and different stratigraphic positions, ancient biological reefs are often found simultaneously, forming a

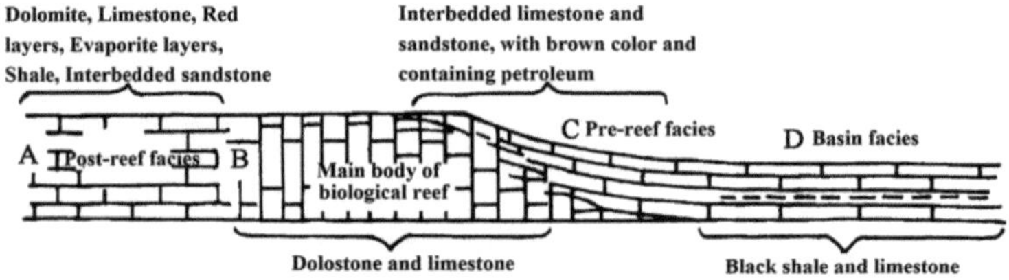

Fig. 6.43 Schematic diagram of the various parts of ancient biological reefs and their lithofacies distribution characteristics (according to A. I. Levorsen)

composite biological reef body. This situation is due to the alternation of transgression and regression in these areas suitable for reef-building.

If the reef-building organisms develop, biological reefs can be formed under both transgressive and regressive conditions. However, during regression, as the seawater recedes, the suitable conditions for reef-building shift towards the center of the basin, and the biological reefs develop in the direction of the basin center; during transgression, as the seawater deepens, the suitable conditions for reef-building shift towards the coast, and the biological reef blocks develop towards the coastal direction.

From the perspective of oil and gas reservoir formation conditions, the main body of the biological reef block and the fore-reef facies are the most favorable. Firstly, these two zones have abundant sources of oil and gas. In addition to their own good oil-generating conditions, a large amount of oil and gas can be transported from the adjacent basin facies. Secondly, these two zones have good reservoir conditions. The biological reef itself has well-developed primary pores and secondary solution caves, and the fore-reef facies also has this condition. Exploration practice has also proven that oil and gas are mainly concentrated in these two lithofacies zones.

6.4.5.2 Examples of Biological Reef Oil and Gas Reservoirs

Abundant oil and gas resources have been discovered in biological reefs of different geological eras around the world. According to existing data, biological reef oil and gas reservoirs have been found from the Silurian period of the Paleozoic era to the Miocene period of the Cenozoic era, with those from the Silurian, Devonian, Permian, Cretaceous, and Paleogene-Neogene periods being more significant. In terms of distribution, important regions for biological reef oil and gas reservoirs include the Western Canada Sedimentary Basin, the Permian Basin in the United States, the Ural Foredeep in the former Soviet Union, the Gulf of Mexico Basin (including both Mexican and American parts, with the Mexican part being more important), the Persian Gulf Basin in the Middle East, the Sirte Basin in Libya, and the Salawati Basin in Indonesia. In these basins, biological reef oil and gas reservoirs are often distributed in belts, forming rich oil and gas-producing areas.

Biological reef oil and gas reservoirs hold an important position in the world's oil reserves, especially in major oil-producing countries such as Mexico and Canada. Approximately 60% of Canada's oil and gas production comes from biological reef oil and gas reservoirs, and about 70% of Mexico's national oil production comes from these reservoirs.

(1) The Golden Lane Atoll Field Cluster

Located in the Tampico Bay of Mexico, the atoll belt is divided into three parts: the area north of San Isidro is called Old Golden Lane, the southeast onshore part is called New Golden Lane, and the offshore part is called Offshore Golden Lane, as shown in Fig. 6.44.

The entire Golden Lane Atoll Belt is elliptical, with a NW–SE oriented long axis of about 150 km in length and about 70 km in width; the onshore branch protrudes

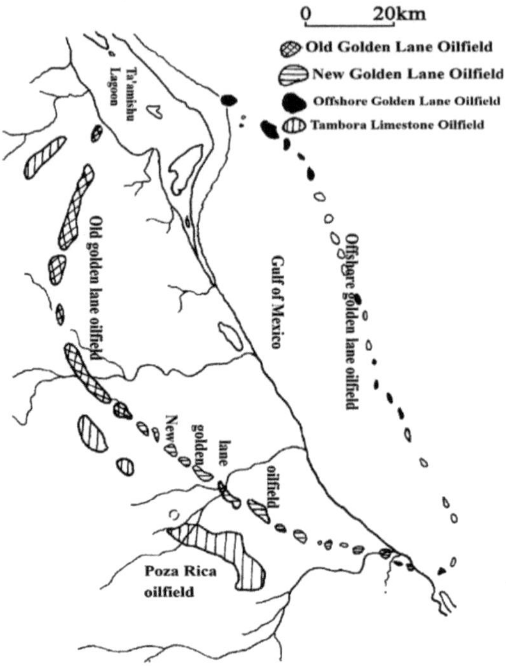

Fig. 6.44 Plan map of the Huangjin Lane Oilfield and Pozo-Rica Oilfield (according to "Geology of Giant Petroleum Fields")

westward in a bow shape, about 180 km in length, and the width of the reef is generally 2 km. This oil field is famous for having three high-yield oil wells, one of which initially produced 37,000 tons per day, making it the highest single-well daily oil production well in the world. From the mid-1950s to the late 1960s, more than 50 biological reef oil fields were discovered onshore, and more than 20 oil and gas fields were discovered offshore.

(2) Liuhua 11-1 Biological Reef Oil Reservoir

The Liuhua 11-1 oil field has an oil reserve of more than 100 million tons and is the first large oil field discovered at sea in China, as well as the largest biological reef oil field in the country. The oil field is located in the southwestern part of the Dongsha Uplift in the Pearl River Mouth Basin, a large anticline structure surrounded by concave areas on three sides and raised to the northeast. The reservoir is composed of Neogene biological reef limestone from the Miocene Series. The reef body shows the characteristics of interdistribution between the reef and the beach. The reef limestone mainly consists of coral-algal boundstone and peloidal limestone. The reef limestone has a strong amplitude reflection at the top boundary on the seismic profile, slightly protruding like a hill, which is very distinctive (Fig. 6.45). The main part of the structure is relatively complete and gentle, with a dip angle of 1.5°–2°, and the wing part becomes steeper, with a dip angle of 6°–7°. Small faults developed in the wing part do not control the distribution of oil and gas.

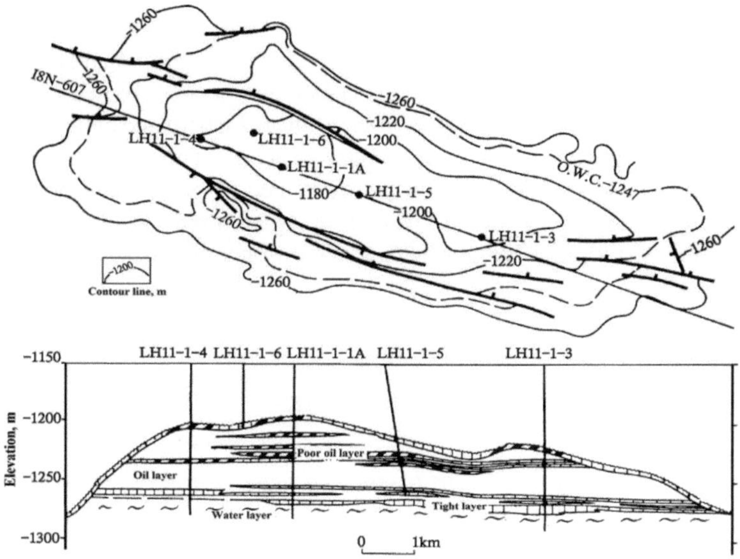

Fig. 6.45 Plan and cross-section of the Liuhua 11-1 biological reef oil reservoir (from "The Petroleum Geology of China," Volume 16, 1992)

The area has three reef-forming periods and has undergone multiple complex diagenetic and epigenetic processes involving uplift and exposure, leading to the interdistribution of large segments with well-developed pores and relatively dense segments within the reef blocks, resulting in various types of reservoirs mainly dominated by dissolution cavities and pores. Based on the vertical changes in lithology, electrical properties, and physical properties, the oil-bearing stratum can be divided into three high-permeability segments and three low-permeability segments. According to physical properties, the reservoirs can be classified into three types: Type I is loose, well-developed reef limestone with a porosity greater than 20% and a permeability greater than 3000×10^{-3} μm^2. Type II is relatively dense reef and beach limestone with a porosity of 15–20% and a permeability of $(50–300) \times 10^{-3}$ μm^2. Type III is dense reef and beach limestone with a porosity less than 15% and a permeability less than 50×10^{-3} μm^2.

The Liuhua 11-1 biological reef oil reservoir is relatively shallow (1,200–1,300 m) and is a block-shaped biological reef oil reservoir with underlying water, having a roughly uniform oil–water interface. The oil column height is 74.5 m, the structural trap amplitude is 75 m, and the reef body is essentially filled with oil. The crude oil is relatively viscous and heavy, characterized by high density, high viscosity, low sulfur content, low wax content, and low pour point. The ground crude oil density is 0.9182–0.9587 g/cm^3, and the viscosity is 50–270 mPa s.

It should be noted that the discovery of the first biological reef in an area is often accidental because biological reefs are not as easily identifiable as structural traps. Although biological reefs can cause structural anomalies, some of which can be detected by geophysical methods, these anomalies are often small and cannot fully indicate the presence of a biological reef trap. Therefore, in an area where no biological reefs have been discovered, it is usually not possible to make an accurate judgment based on such anomalies. However, in an area where one biological reef has been found, more biological reefs can usually be found. This is because biological reefs are rarely isolated; they are always distributed in groups and belts, and are usually related to ancient coastlines. Therefore, when a biological reef is found in an area, further exploration should be conducted in nearby areas to discover more biological reefs.

6.5 Composite Oil and Gas Reservoirs

6.5.1 Basic Concept of Composite Oil and Gas Reservoirs

Oil and gas traps are often controlled by a variety of factors. When a single factor plays a dominant role, oil and gas reservoirs can be classified based on that single factor. However, when multiple factors act together to have roughly the same effect, they become composite traps. That is, if the trap is formed by the joint sealing of two or more factors such as structure, stratigraphy, and lithology above and in the updip direction of the reservoir, it can be called a composite trap. The oil and gas reservoirs formed within these are known as composite oil and gas reservoirs.

There are both traps and oil and gas reservoirs formed by single factors and many composite traps and oil and gas reservoirs formed by factors such as structure, stratigraphy, and lithology. Their causes and methods of oil and gas exploration are different. The characteristics of composite traps and reservoirs differ from those formed by single factors. Therefore, the identification of composite oil and gas reservoirs and the classification of composite oil and gas reservoirs as a separate major category have practical value for oil and gas exploration.

6.5.2 Main Types of Composite Oil and Gas Reservoirs

According to the combinations formed by the three main factors of oil and gas reservoir classification—structure, stratigraphy, and lithology—various types of composite oil and gas reservoirs can be formed. However, from the perspective of exploration practice, the majority that appear are composite oil and gas reservoirs such as structure-stratigraphy and structure-lithology.

(1) Structure-Stratigraphy Oil and Gas Reservoirs

Any oil and gas reservoir formed by the joint sealing of any structural and stratigraphic factors above and in the updip direction of the reservoir is called a structure-stratigraphy oil and gas reservoir. The most common types include anticline-stratigraphic unconformity oil and gas reservoirs and stratigraphic unconformity-fault oil and gas reservoirs. The Kaltsegi gas field in Texas, USA, the Chiren gas field in Bulgaria (Fig. 6.46), and the Rhodesha oil field in Louisiana, USA, are all typical examples of this type of oil and gas field.

For example, the Ya 13-1 gas field in the Qiongdongnan Basin has a gas-bearing layer of LingShui Formation deltaic sandstone, a semi-anticlinal structure, and is covered by marine mudstone unconformity, forming a structure-stratigraphy composite trap. Natural gas from the overpressured gas source areas of both terrestrial and marine origins in the Ya Nan depression migrates bidirectionally to the normal pressure Ya 13-1 structure, forming a large gas field (Fig. 6.47).

(2) Structure-Lithology Oil and Gas Reservoirs

A trap formed by the dual control of structure and lithology is called a structure-lithology trap, and the oil and gas gathered within it is known as a structure-lithology oil and gas reservoir. Common types include anticline-lithology oil and gas reservoirs and fault-lithology oil and gas reservoirs, such as the structure-lithology oil reservoir of the Shahejie Formation, third member, in the Liangjialou oil field of the Jiyang depression. The turbidite sandstone of the Shahejie Formation, third member, is cut by faults, forming a series of fault-lithology traps (Fig. 6.48) (Wang & Qian, 1992).

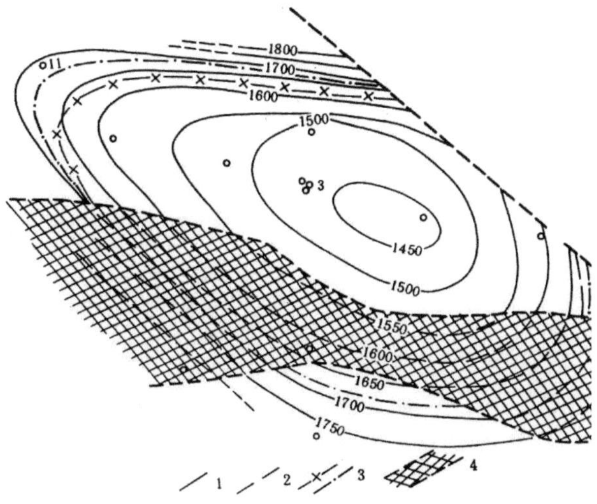

Fig. 6.46 Structural map of the top of the J11 sandstone and gas reservoir distribution map in the Chiren Gas Field, Bulgaria. 1. Structural contour lines (m), 2. faults, 3. gas–water contact, 4. pinch-out line of J11 sandstone

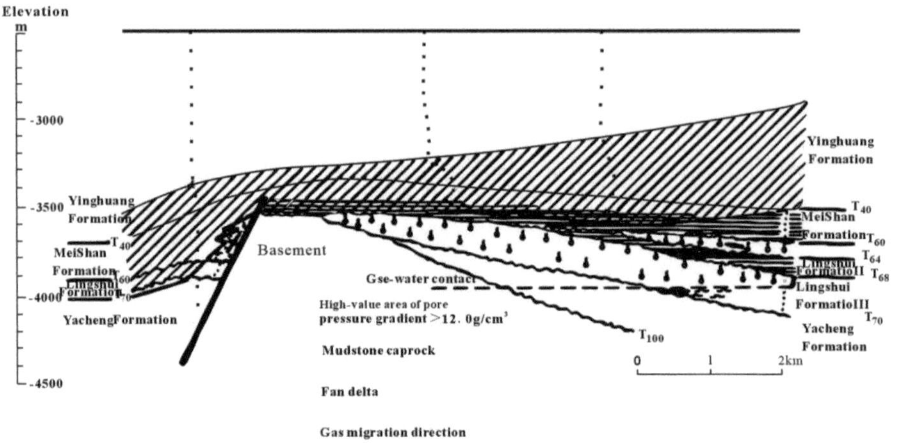

Fig. 6.47 Cross-section diagram of the Yacheng 13-1 Gas Field

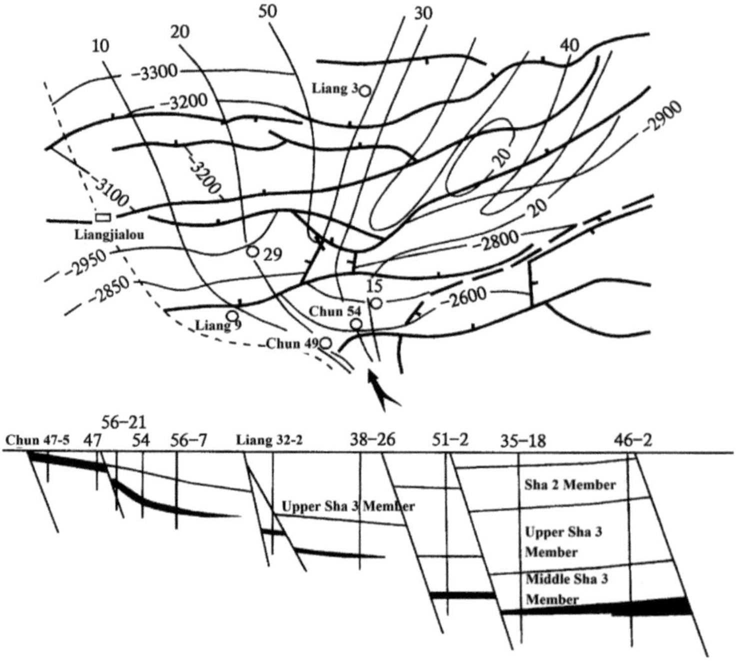

Fig. 6.48 Plan and cross-section of the oil reservoir in the third member of the Shahejie Formation at Liangjialou, Jiyang depression (according to Wang & Qian, 1992)

6.6 Special Types of Oil and Gas Reservoirs

According to the causes of trap formation, in addition to the four major categories of oil and gas reservoirs—structural, stratigraphic, lithologic, and composite—there are numerous special types of oil and gas reservoirs in nature, such as hydrodynamic oil and gas reservoirs, tight sandstone oil and gas reservoirs, shale oil and gas reservoirs, and coalbed gas reservoirs. The formation mechanisms and storage spaces of oil and gas in these reservoirs, as well as the modes of occurrence, differ from those of the aforementioned four major categories. The distribution of oil and gas is not controlled by traditional traps. However, as geological bodies that store oil and gas, these special oil and gas reservoirs still require certain conditions for storage and sealing, which can also be referred to as special trap conditions. They have specific conditions for reservoir formation and characteristics of oil and gas reservoirs, and thus can be classified as special types of oil and gas reservoirs.

6.6.1 Hydrodynamic Oil and Gas Reservoirs

(1) Basic Concept and Formation Mechanism of Hydrodynamic Oil and Gas Reservoirs

A trap formed by the combined sealing of hydrodynamic forces and impermeable rock layers, which creates a hydrocarbon accumulation trap where oil and gas would not typically accumulate under static water conditions, is called a hydrodynamic trap. When commercial-scale oil and gas accumulate in such a trap, it is referred to as a hydrodynamic oil and gas reservoir. These types of oil and gas reservoirs are likely to form in areas where the stratigraphic attitude undergoes slight changes, such as structural noses and flexural zones, areas with non-uniform lithology and thickness variations in monocline reservoirs, and near stratigraphic unconformities. In these areas, when the dynamic water pressure of the seeping groundwater is in the opposite direction and roughly equal in magnitude to the buoyancy force of oil and gas migration, it can block and accumulate oil and gas, forming hydrodynamic oil and gas reservoirs.

Under hydrodynamic action, the force field intensity of oil and gas should be the resultant force of net buoyancy and hydrodynamic force. Therefore, the direction of the oil and gas potential surfaces (which are perpendicular to the force field intensity of oil and gas) also changes accordingly, tilting towards the direction of the water's force field intensity (i.e., the oil–water interface tilts towards the direction of the water's force field intensity), and the oil and gas potential surfaces are no longer parallel to the structural contour lines of the reservoir top. In this case, inclined or curved oil and gas potential surfaces can form hydrocarbon accumulation traps in areas where there are no traps under static water conditions. The closure range of the trap can be defined by the closed oil and gas potential surfaces (or closed h_o, h_g lines) (Chen, 1994).

The influence of hydrodynamic action and fluid properties on the effectiveness of traps has been discussed in Sect. 5.3 of Chap. 5. Due to the different inclinations of the oil–water interface and the gas–water interface, the positions of hydrodynamic traps for oil and gas are also different under the same hydraulic gradient. If the trap accumulates oil, it shifts more towards the direction of water pressure drop, and this shift increases with the increase in the water pressure gradient. However, this shift is limited. When the inclination angle of the oil–water interface is greater than the inclination angle of the reservoir on the side of the anticline that is down the hydraulic gradient, the anticline cannot effectively trap oil, but it can still be a trap for accumulating natural gas. If the inclination angle of the gas–water interface is greater than the inclination angle of the wing of the anticline in the direction of water flow, then even natural gas cannot be trapped. In this case, both oil and gas will be driven out of the anticline and can only be re-accumulated in new traps at appropriate locations in their migration direction (Fig. 5.32).

(2) Basic Types of Hydrodynamic Oil and Gas Reservoirs

The most important characteristic of hydrodynamic oil and gas reservoirs, when viewed in cross-section, is that the oil–water (or gas–water) interface is inclined or curved, forming a suspended configuration. The oil–water boundary intersects with the structural contour lines in the plane, creating low oil and gas potential areas.

Based on the characteristics of hydrodynamic sealing and current exploration results, the main types of hydrodynamic oil and gas reservoirs that develop are structural nose-type and monocline-type. Sometimes, hydrodynamic factors combine with other factors such as stratigraphy, lithology, and faults to form composite traps.

This structure does not close under static water conditions and cannot form a trap. However, under the action of water flowing towards the down-dip direction of the reservoir, the oil–water (or gas–water) interface tilts or curves in the direction of water flow, and when it satisfies the condition $\alpha_1 < \theta_{Hc/w} < \alpha_2$ (or $\alpha_1 < \theta_{g/w} < \alpha_2$), a closed low potential area for oil and gas will form at the change in inclination of the structural nose or terrace (Fig. 6.49). Here, $\theta_{o/w}$ is the inclination angle of the oil–water interface, $\theta_{g/w}$ is the inclination angle of the gas–water interface, α_1 is the low inclination angle, and α_2 is the high inclination angle. The gas reservoir in the Abilene sandstone of the lower Cretaceous in the Sokolov gas field and the West Texas Whittier oil field (Fig. 6.50) can serve as examples of this type of oil and gas reservoir.

The top contour map of the Abilene gas layer in the Sokolov gas field shows a NNE-trending nose-shaped structure, with the direction of water pressure drop nearly SN, dropping from south to north. A hydrodynamic trap is formed on the north side of the nose-shaped structure axis. The direction of water head drop in this gas reservoir is not consistent with the down-dip direction of the reservoir, and there is a large angle between them, yet a closed area can still be formed. If the two were consistent, a larger trap and gas reservoir might be formed.

From the characteristics of hydrodynamic oil and gas reservoirs, when groundwater flows towards the down-dip direction of the reservoir, causing the oil and gas

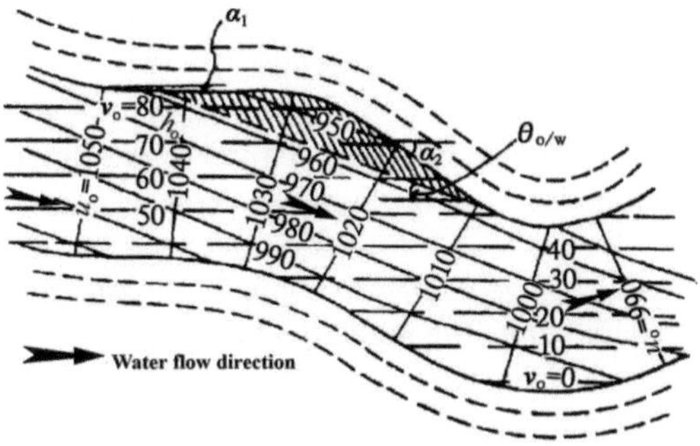

Fig. 6.49 Schematic diagram of the formation mechanism of hydrodynamic traps in nose-shaped structures (according to Hubbert, 1953)

$$h_o = u_o - v_o \; ; \; u_o = \frac{\rho_w}{\rho_o} h_w \; ; \; v_o = \frac{\rho_w - \rho_o}{\rho_w} \cdot Z$$

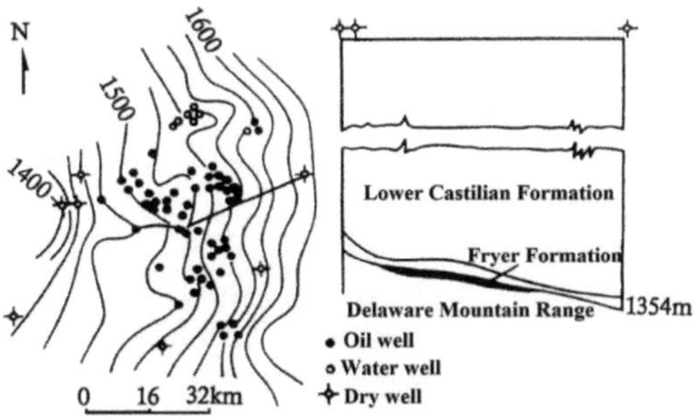

Fig. 6.50 Structural map of the Wett Oilfield in Texas (contour interval in meters) and cross-section diagram

potential surfaces to tilt or curve is the main reason for the formation of hydrodynamic traps. However, in different types of oil and gas reservoirs, their roles and specific methods vary. Hydrodynamic traps do not have a fixed location. The specific location of the trap depends on the changes in the water head gradient.

6.6.2 Tight Sandstone Oil and Gas Reservoirs

Tight sandstone oil and gas reservoirs and shale oil and gas reservoirs are the most important types of unconventional oil and gas reservoirs. Unconventional oil and gas accumulations are not controlled by obvious or fixed boundaries of traps and are distributed in a large area as continuous or quasi-continuous types.

6.6.2.1 Concept and Basic Characteristics of Tight Sandstone Oil and Gas Reservoirs

Tight sandstone oil and gas reservoirs refer to the oil and gas accumulations that are contained within tight sandstones. The oil and gas have not undergone large-scale long-distance migration, the reservoir air permeability is less than 1 mD, and single wells generally do not have natural production capacity or natural production capacity below the industrial oil and gas flow limit. It is necessary to transform the reservoir through technical measures to obtain industrial production capacity.

Oil and gas contained in tight rocks mainly include tight sandstone oil and gas and tight carbonate rock oil and gas. Oil and gas in tight carbonate rocks are mostly controlled by fractures and solution pores, and their storage space and accumulation mode are different from those in tight sandstones. Here, we mainly discuss tight sandstone oil and gas reservoirs.

Tight sandstone oil and gas reservoirs are divided into two basic types: tight sandstone oil reservoirs and tight sandstone gas reservoirs. Before the widespread use of the term "tight sandstone gas reservoirs," many scholars at home and abroad proposed many related concepts to describe the natural gas accumulation in tight sandstone reservoirs, such as deep basin gas reservoirs, basin-centered gas reservoirs, and continuous gas reservoirs (Law & Curtis, 2002; Masters, 1979; Zhang & Jin, 2005). In the 1950s–1970s, several giant natural gas accumulations lacking edge-bottom water were discovered in the San Juan and Alberta basins in the United States and Canada, mainly distributed in the central or deep parts of the basin structure. In 1979, Masters proposed the concept of deep basin gas traps. It is generally believed that deep basin gas traps are natural gas accumulations with inverted gas–water relationships in tight sandstone reservoirs developed in the downdip direction or central area of a basin. Because these types of natural gas were mainly discovered in the central or downdip deep parts of the basin in the early stage, they are named deep basin gas or basin-centered gas.

The basic characteristics of tight sandstone oil and gas reservoirs are: (1) Large area source-reservoir coexistence, source-reservoir symbiosis, mostly near-source or closely contacted source-reservoir. (2) No obvious trap boundaries, oil and gas distribution is not controlled by structural zones, slope zones and depression areas can all become favorable areas, with a wide distribution range and local enrichment. (3) Due to the tightness of the reservoir, the buoyancy effect is limited, non-Darcy seepage is dominant, non-buoyancy accumulation, hydrodynamic effects are not obvious, oil and gas short-distance migration is the main, and there is a starting pressure gradient. (4) Oil and gas resource abundance is low, generally no natural production or economic production, appropriate technical measures are needed to form industrial production. (5) The pressure coefficient of tight sandstone gas reservoirs varies greatly, but mainly abnormally high pressure, some are normal pressure or low-pressure tight sandstone gas reservoirs, which have abnormally high pressure during formation. (6) The distribution of oil and water in tight oil reservoirs is complex, and the oil quality is lighter.

6.6.2.2 Formation Characteristics of Tight Sandstone Oil and Gas Reservoirs

Compared with conventional gas reservoirs, the formation of tight sandstone oil and gas reservoirs does not require traditional traps, but still requires sufficient oil and gas sources and good post-preservation conditions. The formation characteristics of tight sandstone oil reservoirs and tight sandstone gas reservoirs have both commonalities and differences. For tight sandstone oil and gas reservoirs where the reservoir is tight first and then charged, their reservoir formation characteristics can be summarized as follows:

(1) High-quality hydrocarbon source rocks with high organic matter abundance, good quality, and moderate to high thermal evolution are the prerequisites for the formation of tight oil and gas reservoirs. Generally, the total organic carbon (TOC) of effective hydrocarbon source rocks is greater than 2%. For tight oil reservoirs, the maturity is moderate (R_o is 0.7–1.3%), but for tight gas reservoirs, the maturity is higher (R_o is greater than 1.3%, most are greater than 2.0%), and there are differences in hydrocarbon source rock standards in different basins.

(2) Different from conventional reservoirs, low porosity, low permeability, and large area development of reservoirs are important conditions for the formation of tight oil and gas reservoirs. The reservoir is tight but still requires a certain porosity. Tight oil reservoirs require higher porosity than tight gas reservoirs. The reservoir distribution is stable and has a large distribution area, ensuring a large resource scale. Under the condition of high reservoir diagenesis and poor reservoir physical properties, hydrocarbons expelled from hydrocarbon source rocks can be expelled in a large area and as a whole to displace pore water in tight reservoirs, forming tight sandstone oil and gas reservoirs. The existence of "sweet spots" in the reservoir is also one of the necessary conditions for the formation of tight oil reservoirs. The matching relationship and distribution range of high-quality hydrocarbon source rocks and reservoir sweet spots control the scale of tight oil and gas reservoirs. The "sweet spots" of tight oil and gas can be understood as areas with relatively good porosity and permeability within the tight reservoir that are conducive to oil and gas accumulation.

(3) Source-reservoir coexistence and near-source accumulation are favorable conditions for the large-scale accumulation of tight oil and gas. The source rocks and reservoirs are in proximity or in close contact over a large area. Vertically, there are three types of contact relationships between the source rocks and reservoirs, namely interbedding, being overlying or underlying. In the process of oil and gas reservoir formation, due to the tightness of the reservoir, there are large changes in lithology and physical properties in the horizontal direction. Oil and gas have poor seepage ability in tight reservoir bodies, and the buoyancy effect is limited. It is difficult to undergo large-scale, long-distance migration and form enriched areas, and can only be short-distance, near-source reservoir formation.

Therefore, oil and gas are mainly transported and enriched within the hydrocarbon source rock layer and near-source reservoir bodies, and generally are low in abundant, large area in distribution.

(4) Large, gentle structural backgrounds and favorable matching of gas sources, reservoirs, sealing layers, and structural conditions. The main body of tight sandstone gas reservoirs is located in the central and slope areas of oil and gas basins, with stable structures and gentle slopes, and limited control by structural zones. The original deposition was gentle and had a small slope, and the current strata are generally gentle. Large, gentle structural backgrounds are conducive to the large-area enrichment of natural gas.

In negative structures, the combination of hydrocarbon source rocks located below tight reservoirs and the close contact between the two is the most favorable source-reservoir combination for the formation of tight sandstone gas reservoirs. At the same time, good top and bottom sealing layers can reduce the loss of natural gas and are conducive to the accumulation of natural gas reservoirs.

6.6.3 Shale Oil and Gas Reservoirs

6.6.3.1 Concept and Basic Characteristics of Shale Oil and Gas Reservoirs

There are two understandings of shale oil and gas in China: the narrow sense and the broad sense. The narrow sense refers to oil and gas that are contained within organic-rich shales and their interbedded siltstones, fine sandstones, carbonate rocks, and other thin interlayers (with individual layer thickness generally less than 3 m). The broad sense refers to oil and gas contained within organic-rich shale sequences, including both narrow-sense shale oil and gas and tight oil and gas (Zhao et al., 2020). According to the current national standards, shale oil and gas refer to oil and gas that are contained within organic-rich shale sequences. Organic-rich shale sequences have siltstones, fine sandstones, carbonate rocks, and other single layers with a thickness not greater than 5 m, and the cumulative thickness accounts for less than 30% of the total thickness of the shale sequence. They have no natural production capacity or are below the lower limit of industrial oil production, and special technological measures are required to achieve industrial oil production.

Obviously, shale oil and gas reservoirs mainly include two types of reservoirs: one is tight sandstone or carbonate interlayer oil and gas reservoirs within the source rock sequence, and the other is oil and gas reservoirs within pure shale, which is a type of in-source oil and gas accumulation. Based on the latest progress in the research and exploration and development of terrestrial shale oil in China, it can be divided into interlayer type, fracture type, and pure shale type according to the type of shale oil reservoir rock and the space where it is contained (Jin et al., 2023). Within the overall oil-containing terrestrial hydrocarbon source rock sequence, the favorable reservoirs that are relatively richer in oil, have better physical properties, are easier to transform, and have commercial development value are referred to as "sweet spots," which mainly include three types of "sweet spots": interlayer type, mixed accumulation type, and shale type (Fig. 6.51) (Jiao et al., 2020).

According to the phase state of oil and gas, shale oil and gas reservoirs are divided into shale oil reservoirs and shale gas reservoirs. The oil in shale oil reservoirs has basically not undergone secondary migration and is the result of in-situ retention, existing in free and adsorbed states, and generally has lighter oil quality and lower viscosity. The majority of commercial shale oil that has been discovered comes from tight sandstones or fractured mud shales within the shale sequence, which is similar to the characteristics of tight oil and gas. This section focuses on introducing shale gas reservoirs.

Shale gas reservoirs are unconventional natural gas accumulations that are contained within organic-rich mud shale and its interlayers, mainly in adsorbed and free states. The main body is self-generated and self-stored, with a large area of continuous natural gas accumulation, which is the result of natural gas being retained in situ on a large scale within the hydrocarbon source rock, including natural gas contained within pure mud shale and thin interlayers of siltstones, carbonate rocks, etc.

As a shale reservoir collective with a certain scale and commercial exploitation value, shale gas reservoirs mainly have the following characteristics:

(1) Source and reservoir integration, in-situ retention. Rich organic matter dark mud shale is both a hydrocarbon source rock and a reservoir, self-generated and self-stored, forming a closed reservoir collective, with natural gas having no or extremely short-distance migration, basically retained and accumulated in situ, which is a typical source and reservoir integration, continuous accumulation, and continuous enrichment of natural gas.
(2) Reservoir tightness and micro-fracture development. Shale gas reservoirs are mainly rich in organic matter black shale, which generally has low porosity and ultra-low permeability characteristics, with nano-scale pores forming the main pores of shale. Micro-fractures can provide ample storage space for shale gas and also serve as migration channels.
(3) Adsorption and free state are the main methods of occurrence. There are two main methods for shale gas to occur: adsorbed state and free state, mainly in natural fractures and pores in a free manner or on the surface of kerogen and mineral particles in an adsorbed state. Factors affecting the content of adsorbed gas and free gas in shale reservoirs mainly include rock mineral content, organic matter content, stratigraphic pressure, and the degree of fracture development.
(4) Natural gas is widely and continuously distributed without obvious trap boundaries. The distribution of shale gas is not controlled by structures and has no trap boundaries. The range of gas-bearing areas is controlled by the area of hydrocarbon source rocks and good sealing layers. Rich organic matter dark shale is the source rock of shale gas, and the distribution of effective gas source rocks

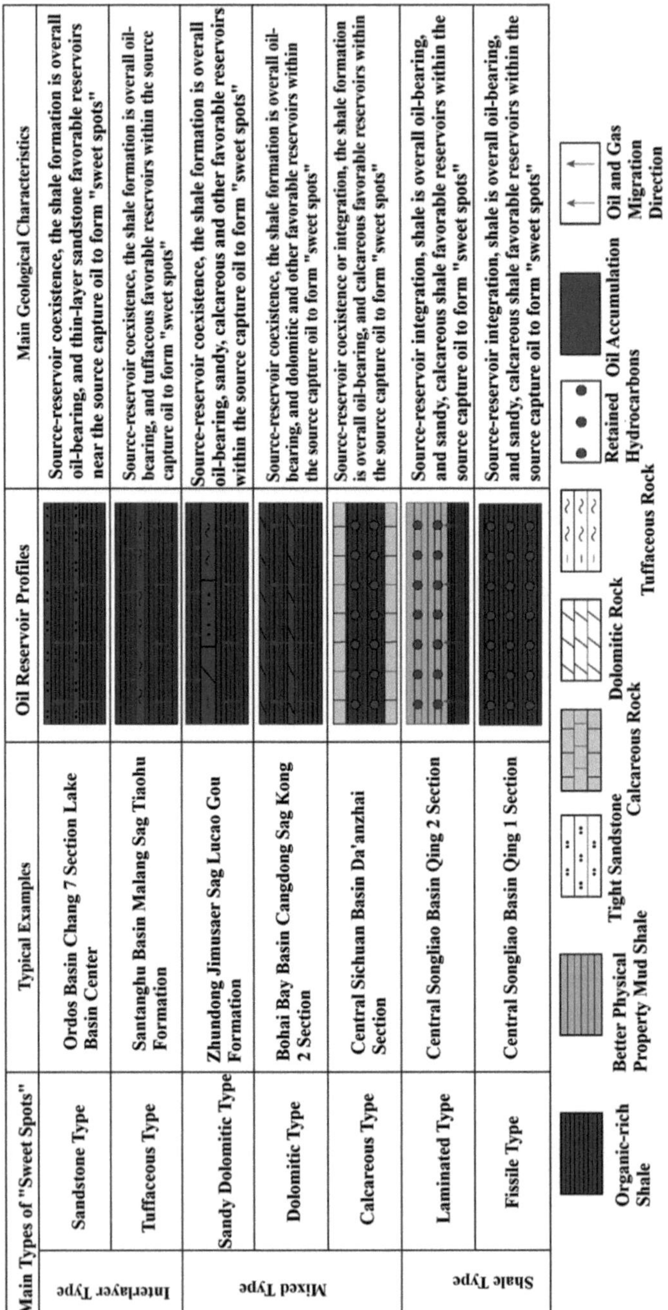

Fig. 6.51 Main types and geological characteristics of terrestrial in-source oil accumulation "sweet spots" (Jiao et al., 2020)

is usually the range of favorable exploration areas for shale gas, often widely and continuously distributed in the central or slope areas of the basin.

(5) High stratigraphic pressure, easy to flow and exploit. Shale gas is widely enriched in mature dark organic matter mud shale layers, with sufficient stratigraphic energy, and the pressure coefficient is generally greater than 1.2. The greater the abnormal pressure of the stratum, the stronger the adsorption capacity of the mud shale reservoir for oil and gas, which is conducive to exploitation.

(6) No natural production capacity, large-scale fracturing is required for exploitation. Shale gas reservoirs generally have no natural production capacity in single wells and require certain technological measures to obtain industrial gas flow. The production of natural gas is mainly non-Darcy flow, with the transformation of adsorption, diffusion, and seepage in terms of phase and flow mechanisms. Shale gas production is mainly free gas in the early stage, similar to the development of conventional oil and gas, and later production is similar to coalbed gas, mainly adsorbed gas desorption and diffusion.

6.6.3.2 Formation Characteristics of Shale Gas Reservoirs

The formation of shale gas reservoirs is like that of tight sandstone gas reservoirs, with sufficient gas sources and good preservation conditions being key. Chinese scholars, based on the enrichment patterns of marine shale gas in the south and the exploration and development practices of the Fuling shale gas field, have proposed the "dual enrichment" theory of shale gas. That is, the development of high-quality shale in deepwater continental shelves is the basis for "hydrocarbon generation controlling reservoirs" in shale gas, and good preservation conditions are the key to "trap formation controlling production". The foundation for the enrichment of shale gas lies in generation and storage, with high hydrocarbon generation intensity, well-developed organic matter pores, which are conducive to reservoir modification, and good shale top and bottom boards that can effectively prevent the vertical loss of hydrocarbons and retain their accumulation from the beginning of shale hydrocarbon generation. The intensity and duration of later tectonic actions determine the preservation conditions of shale gas, and good preservation conditions are the key geological factors in "trap formation controlling production" of shale gas (Guo, 2014).

Since the composition of shale has a decisive impact on the formation of fractures in reservoir technical modification, the mineral composition of shale is also an important condition for the formation of commercial shale gas reservoirs. The characteristics of shale gas reservoir formation can be summarized as follows:

(1) Hydrocarbon source characteristics: Shales with a certain thickness and wide distribution of organic-rich mud shale are the foundation for the generation and occurrence of shale gas. The generation of natural gas in organic-rich shale mainly depends on the following three factors: the abundance of original organic matter in the rock, the type of organic matter and its original gas generation capacity, and the degree of transformation of organic matter into natural gas. According to statistics from rich shale gas basins in the United States, shale has

a high organic matter abundance, and the lower limit of average organic carbon content in gas-producing shale is about 2–3% (Bowker, 2007), which is significantly higher than the lower limit of organic matter abundance in conventional oil and gas hydrocarbon source rocks. The type of kerogen is mainly Types I and II, with less Type III. High-yield shale gas hydrocarbon source rocks have a maturity of $R_o > 1.4\%$, especially the part with $R_o > 2.0\%$ as the main part of shale gas production. It can be seen that abundant organic matter and a high degree of evolution are necessary conditions for the formation of shale gas reservoirs.

(2) Reservoir characteristics: Shale layers are both source rocks and reservoirs, so shale gas has a typical "self-generation and self-storage" reservoir formation characteristic, which is the result of natural gas gathering within or near the source rock after its generation. Due to special storage conditions, natural gas exists in various phases such as free, adsorbed, and dissolved states, mainly in free and adsorbed states. Sufficient burial depth and thickness are the prerequisites for ensuring shale gas storage. Shale has low porosity and permeability, but the presence of natural fractures can improve the storage performance of shale gas reservoirs.

(3) Trap and sealing preservation characteristics: Shale gas reservoirs are a type of continuous natural gas accumulation, mainly forming sealing conditions through lithological changes and hydraulic sealing. Due to the low porosity and permeability of tight shale, the shale body can form a sealed and non-leaky reservoir collective to seal shale gas within the shale layers, equivalent to traps in conventional oil and gas reservoirs. Shale gas reservoirs do not have a unified gas–water interface and do not have obvious trap boundaries. They contain less water and are widely layered with gas, but to form high-yield enrichment, regional sealing conditions of the cap rock and the top and bottom boards are essential. Shale gas is a dense fine-grained sedimentary rock, and the shale layer itself can act as a cap rock, but the dense rocks above and below the shale gas are crucial for the formation and preservation of shale gas reservoirs.

(4) Mineral content: The content of brittle minerals is an important factor affecting the development of shale matrix pores and micro-fractures, gas content, and fracturing modification methods, and it is also an important factor affecting the commercial development of shale gas reservoirs. The mineral composition of rocks is crucial for the later development of shale gas, generally requiring brittle mineral content to be greater than 40% and clay mineral content to be less than 30%. The lower the content of clay minerals in shale, and the higher the content of brittle minerals such as quartz, feldspar, and calcite, the stronger the brittleness of the rock. Under the action of artificial fracturing forces, it is easier to form natural fractures and induced fractures, which is beneficial for shale gas extraction. In contrast, shale with high clay mineral content is plastic and energy-absorbing, which is not conducive to shale reservoir modification.

6.6.4 Coalbed Methane Reservoirs

6.6.4.1 Basic Concept and Types of Coalbed Methane (CBM)

(1) Basic Concept of Coalbed Methane

Coalbed methane is a type of naturally occurring, self-generating, and self-storing natural gas that is generated during the coalification process and is primarily stored in coal seams in the form of adsorption. Since the main component of this type of natural gas is methane, it is also known as coalbed methane. In China's coal industry, it is referred to as coal mine gas.

Coalbed methane is a part of coal-type gas, which is the coal-derived gas that remains in the coal seams. Coal itself is a resource and also a type of hydrocarbon source rock that can generate natural gas. If it contains a certain grade of coalbed methane, it also becomes a natural gas reservoir. Coalbed methane is different from "coal-type gas" (also known as coal-derived gas or coal-series gas), which is the natural gas generated and migrated to non-coal reservoirs after the thermal evolution of coal and coal series reaches maturity, and it is conventional natural gas in a free state in non-coal layers.

The occurrence of coalbed methane does not depend on the existence of traps, which is fundamentally different from the storage mechanism of natural gas in conventional reservoirs. Coalbed methane, mainly methane, is considered dry gas. Under geological conditions, coalbed methane can be stored in coal seams in the form of adsorbed, free, and dissolved states. Generally, the adsorbed state can account for 70–95%, the free state about 10–20%, and the dissolved state a very small proportion. Coalbed methane is self-generating and self-storing, accumulating in situ without significant migration. The coal rock is both the source rock and the reservoir.

The migration of coalbed methane is mostly related to changes in the pressure of the coal seam. The main modes of migration are molecular diffusion and seepage. Therefore, the extraction of coalbed methane requires a reduction in pressure to transform a large amount of adsorbed natural gas into a free state and initiate migration. When the pressure of the coal seam decreases, coalbed methane is desorbed from the internal surface of the matrix, enters the cleats, fractures, and larger pores (>100 nm) through molecular diffusion in micropores (<100 nm) due to concentration differences, becoming free gas, and is extracted or lost through single-phase or gas–water two-phase seepage under the influence of potential differences. Thus, the migration process of coalbed methane is mainly desorption diffusion in the matrix micropores and seepage discharge in the cleats and fractures (Fig. 6.52).

(2) Types of Coalbed Methane Reservoirs

A coalbed methane reservoir refers to a coal that is controlled by similar geological factors, contains a certain scale of coalbed methane resources mainly in the form of adsorption, and has a relatively independent fluid system. This concept emphasizes, first, that the coal seam contains a certain scale of coalbed methane resources, rather than referring to a general gas-bearing coal, and second, that the coal has a relatively

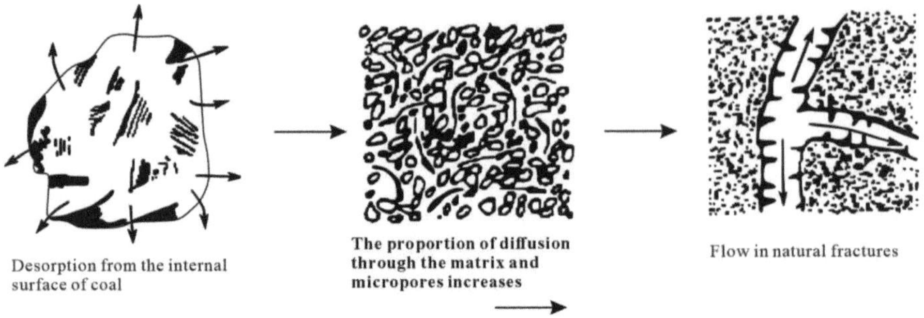

Fig. 6.52 The migration process of coalbed methane desorption-diffusion-seepage (according to Luo et al., 1997)

independent fluid system, that is, a basic fluid unit that has undergone the same evolution process and similar geological actions.

Unlike tight gas reservoirs and shale gas reservoirs, coalbed methane reservoirs usually have clear boundaries that separate them from surrounding geological bodies. Coalbed methane reservoirs mainly have six types of geological boundaries: economic boundaries, hydrodynamic boundaries, weathering zone boundaries, physical property boundaries, fault boundaries, and lithological boundaries. These boundary types are important basis for the division of coalbed methane reservoirs.

Coalbed methane, which is mainly in the adsorbed state, is controlled by temperature, pressure, and the properties of coal. Among them, the spatial variation laws of temperature are obvious. The properties of coal are the result of various factors acting together over geological history, and its current state is basically stable. The only thing that is constantly changing is pressure, which varies with the recharge, migration, and discharge of groundwater. Therefore, taking pressure as the main line, combined with boundary types and the structural characteristics of coalbed methane reservoirs themselves, Song et al. (2010) proposed a classification scheme for coalbed methane reservoirs.

This classification scheme first divides coalbed methane reservoirs into two major categories based on the pressure formation mechanism of coalbed methane reservoirs: hydrodynamically sealed (Fig. 6.53a–h) and self-sealed coalbed methane reservoirs (Fig. 6.53i–k). The hydrodynamically sealed type can be further distinguished into two subclasses: hydrodynamically blocked and hydrodynamically driven coalbed methane reservoirs (Fig. 6.53f–h). Combining boundary types, the hydrodynamically blocked coalbed methane reservoirs are further divided into 5 types, the hydrodynamically driven type is divided into 3 types based on structural characteristics, and the self-sealed type can be further divided into 3 types (Fig. 6.53).

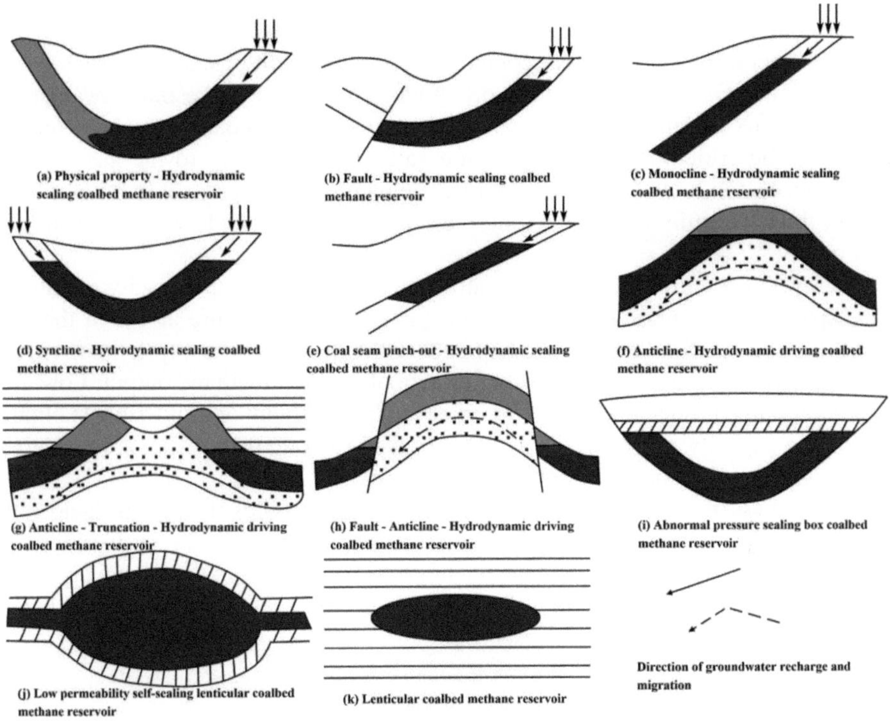

Fig. 6.53 Schematic diagram of coalbed methane reservoir types (Song et al., 2010)

6.6.4.2 Formation Conditions of Coalbed Methane Reservoirs

The formation conditions of coalbed methane reservoirs mainly include coal seam thickness, gas content, permeability, preservation conditions, and hydrogeological conditions. The organic combination of these conditions is essential for the formation of coalbed methane reservoirs.

(1) Coal Seam Thickness, Distribution, and Gas Content

As a relatively gas-rich coal geological body with commercial mining value, a coalbed methane reservoir requires good conditions for the generation and occurrence of coalbed methane, and a certain thickness of coal seam is the material basis for the formation of coalbed methane reservoirs. It provides both the gas source and the storage space. Therefore, the thicker the coal seam, the more favorable it is for the formation of the gas reservoir. Under current technical conditions, coal seams with a thickness of 1–30 m and a burial depth of 45–2730 m have commercial mining value for coalbed methane, and the coal rank can range from lignite to anthracite. According to domestic and international experience, a single coal seam thickness greater than 3 m is possible for high production. In terms of resource potential, a total thickness greater than 10 m is considered to have significant resource potential.

Gas content is one of the most important comprehensive conditions for coalbed methane reservoirs, including gas content, desorption conditions, and reservoir pressure gradients. Gas content is an important basic parameter for resource evaluation and a necessary parameter for the evaluability assessment of coal reservoirs. Considering that the critical desorption pressure value corresponding to the gas content should not be too low, the lower limit of gas content is generally set to greater than 8 m^3/t, thus determining that the minimum thickness of a single coal seam should be 4 m.

Desorption conditions affect the coalbed methane production capacity mainly in two aspects: gas saturation and Langmuir pressure (the pressure at which the gas adsorption on the coal surface reaches half of the maximum adsorption capacity). They control the difficulty and production rate of coalbed methane output. Low gas saturation (large ground pressure difference) makes it difficult or impossible to obtain production capacity; generally, a Langmuir pressure greater than 3 MPa is considered to have high production conditions for coal seams.

The richness of coalbed methane resources in a region is positively correlated with the distribution, thickness, and gas content of the coal-bearing strata. Continuous distribution of coal-bearing strata with large area, great thickness, and high gas content, under good sealing conditions, is conducive to the formation of enriched and high-yield coalbed methane reservoirs.

(2) Coal Seam Permeability

Coal seam permeability is a key parameter controlling the size of production capacity. High coal seam permeability indicates strong flow capacity of the fracture system. If the gas migration passage is affected by mineral filling or increased ground stress, causing the passage to be closed, the coal seam permeability will be greatly reduced, and the gas flow will also be significantly reduced. Evaluation of reservoir permeability should be based on permeability data obtained from well testing or historical fitting. There is a significant difference in permeability between coal reservoirs in China and the United States. The permeability of coal reservoirs in the United States is generally good, ranging from 0.1 to 100 mD, while the permeability of coalbed methane reservoirs in China's coal-bearing areas is generally low, mostly less than $10 \times 10^{-3}\ \mu m^2$.

Coal reservoir permeability is controlled by a variety of complex geological factors, among which the development characteristics of natural fractures have a crucial impact on permeability. Therefore, in the absence of coal seam permeability data, the permeability of coal seams can be predicted using the tectonic stress field and the principal stress difference method.

(3) Sealing and Preservation Conditions

The geological factors that restrict the preservation of coalbed methane mainly include the effective thickness of the overlying strata, the characteristics of the top and bottom plates of the coal seam, tectonic conditions, and hydrogeological conditions.

Tectonic uplift and subsidence movements can change the temperature and pressure conditions of the coal seam, breaking the original adsorption equilibrium state of coalbed methane, thus affecting its preservation. Due to tectonic uplift, the overlying strata of the coal seam are eroded and thinned, causing the loss of coalbed methane.

Good top and bottom plates of the coal seam can reduce the seepage migration and diffusion loss of coalbed methane, maintain the formation pressure and the adsorption amount of coalbed methane, prevent the alternation of formation water, maintain the balance among the three, keep the gas in an adsorbed state, and thus allow its preservation and enrichment in the coal seam.

(4) Hydrogeological Conditions

In addition to good sealing conditions, the enrichment and preservation of coalbed methane also require favorable hydrogeological conditions, namely hydrodynamic sealing and formation water overpressure. Alternating hydrodynamic conditions will break the balance between adsorption and dissolution and free gas, causing the adsorbed gas to gradually decrease, affecting the preservation of coalbed methane. At the same time, flowing groundwater significantly affects the content and geochemical characteristics of coalbed methane. In areas with strong hydrodynamic conditions, the content of coalbed methane is low, and methane carbon isotopes are light. Hydrogeological conditions in a confined area help prevent the escape of coalbed methane, with high gas content and easy production of coalbed methane.

Questions for Discussion

1. Based on the causes of trap formation, the morphology of oil and gas reservoirs, and the phase state of hydrocarbons, what types of oil and gas reservoirs can be classified? What are their main characteristics?
2. What are structural traps and oil and gas reservoirs? What specific types of structural oil and gas reservoirs are included?
3. What are the genetic types and main characteristics of anticline traps and oil and gas reservoirs?
4. What are the conditions for the formation of fault traps? What is the role of faults in the formation of oil and gas reservoirs?
5. What are piercing oil and gas reservoirs and fractured oil and gas reservoirs? How are they formed?
6. What are stratigraphic traps and oil and gas reservoirs? What specific types of stratigraphic oil and gas reservoirs are included, and what are their main characteristics?
7. What role does stratigraphic unconformity play in the migration and accumulation of oil and gas? What are the main characteristics of buried hill oil and gas reservoirs?
8. What are the common points and differences in the formation characteristics between buried hill oil and gas reservoirs and unconformity-sealed oil and gas reservoirs?

9. What are lithologic traps and oil and gas reservoirs? What specific types of lithologic oil and gas reservoirs are included, and what are their main characteristics?
10. What are composite traps and oil and gas reservoirs? What are their main characteristics? What is the mechanism of trap formation?
11. What is a hydrodynamic oil and gas reservoir? How is it formed?
12. What are tight sandstone oil and gas reservoirs, shale oil and gas reservoirs, and coalbed methane reservoirs? What are their main geological characteristics?

References

Брод, И. О., Еременко, Н. А. (1958). Principles of petroleum and natural gas geology (M). Beijing: Petroleum Industry Press.

Bowker, K. A. (2007). Barnett shale gas production, Fort Worth Basin: Issues and Discussion. *AAPG Bulletin, 91*(4): 523–533.

Chen, R. (1994). *Petroleum and natural gas geology*. China University of Geosciences Press.

England, W. A., Mackenzie, A. S., Mann, D. M, et al., (1987). The movement and entrapment of petroleum fluids in the subsurface. *Journal of the Geological Society, 144*(2): 327–347.

Guo, X. (2014). "Dual-enrichment" law of southern marine shale gas: Exploration practice of Longmaxi shale gas in the Sichuan Basin and its margins. *Acta Geologica Sinica, 88*(7), 1209–1218.

He, S., Ye, J., Xu, S., et al. (2010). *Petroleum and natural gas geology*. China University of Geosciences Press.

Hooper, E. C. D. (1991). Fluid migration along growth faults in compacting sediments. *Journal of Petroleum Geology, 14*(2), 161–180.

Hu, J., Huang, D., et al. (1991). *Theoretical basis of continental petroleum geology in China*. Petroleum Industry Press.

Hubbert, M. K. (1953). Entrapment of petroleum under hydrodynamic conditions. *AAPG Bulletin, 37*(6), 1954–2026.

Jiang, Y. (1998). A discussion on the relationship between the thickness of reservoir seals and the height of the hydrocarbon columns they seal. *Natural Gas Industry, 18*(2), 20–23.

Jiang, Y., & Liu, H. (2023). *Differences in oil and gas enrichment and dominant controlling factors in the Bohai Bay Basin*. China University of Petroleum Press.

Jiang, Y., Liu, J., & Su, S. (2021). *Formation mechanism of submerged hill oil and gas reservoirs in fault-controlled basins*. China University of Petroleum Press.

Jiang, Y., & Zhaoming. (2016). *Petroleum and natural gas geology and exploration* (2nd ed.). Petroleum Industry Press.

Jiao, F., Zou, C., & Yang, Z. (2020). Geological theory of intrabasin oil accumulation and exploration and development practice. *Petroleum Exploration and Development, 47*(6), 1067–1078.

Jin, Z., Zhang, Q., Zhu, R., et al. (2023). Classification of continental shale oil in China and its significance. *Petroleum and Natural Gas Geology, 44*(4), 801–819.

Law, B. E., & Curtis, J. B. (2002). Introduction to unconventional petroleum systems. *AAPG Bulletin, 86*(11), 1851–1852.

Levorsen, A. I. (1967). *Geology of Petroleum*. San Francisco: W. H. Freeman and Company.

Li, G. (2005). *World Atlas of Oil and Gas Basins*. Beijing: Petroleum Industry Press.

Luo, Z. (1997). *Research on the kinetic model of coal seam methane migration*. Beijing: China Coal Research Institute.

Lv, Y., Fu, G., Fu, X., et al. (2013). *Faults' conduction and sealing of oil and gas*. Petroleum Industry Press.

Ma, Y., Cai, X., Yun, L., et al. (2022). Exploration and development practice and theoretical and technical progress of super-deep carbonate oil and gas fields in the Shunbei area of the Tarim Basin. *Petroleum Exploration and Development, 49*(1), 1–17.

Masters, J. A. (1979). Deep basin gas trap, Western Canada. *AAPG Bulletin, 63*(2), 152–181.

Milton, N. J., & Bertram, G. T. (1992). Trap styles: A new classification based on sealing surfaces. *AAPG Bulletin, 76*, 986–999.

Song, Y., Liu, H., et al. (2010). *Geological conditions of coalbed gas accumulation in China*. Science Press.

Wang, B., & Qian, K. (1992). *Geological research and exploration practice in the Shengli Oilfield*. Petroleum University Press.

Xu C., Yu H., Wang J., et al. (2019). Formation conditions and accumulation characteristics of Bozhong 19-6 large condensate gas field in offshore Bohai Bay Basin. *Petroleum Exploration and Development, 46*(1), 25–38.

Xue, Y., & Li, H. (2018). Discovery of a large-scale condensate gas field in the deep Archaean metamorphic rock submerged hills in the Bohai Sea and its geological significance. *China Offshore Oil and Gas, 30*(3), 1–9.

Xue, Y., Niu, C., Wang, D., et al. (2021). *Oil and gas migration "convergence ridge" control of reservoirs and shallow exploration in the Bohai Bay Basin*. Science Press.

Yun, L., & Zhu, X. (2022). A new type of trap: Fault-controlled fracture-cave trap. *Petroleum and Natural Gas Geology, 43*(1), 34–42.

Zhang, H., Fang, Z., Gao, X., et al. (1999). *Petroleum geology* (3rd ed.). Petroleum Industry Press.

Zhang, J., & Jin, Z. (2005). *Mechanism of deep-basin gas accumulation and distribution prediction*. Petroleum Industry Press.

Zhang, S., et al. (2004). *The formation of marine oil and gas in the Tarim Basin*. Petroleum Industry Press.

Zhang, Y., Li, H., Chen, X., et al. (2022). Geological-engineering integration practice and achievements of super-deep fault-controlled fracture-cave oil and gas reservoirs in the Shunbei area of the Tarim Basin. *Petroleum and Natural Gas Geology, 43*(6), 1466–1480.

Zhai, G., et al. (1989–1996). *Petroleum geology of China* (Vols. 1, 6, and 10). Petroleum Industry Press.

Zhao, W., Jia, A., Wei, Y., et al. (2020). Progress and development prospects of shale gas exploration and development in China. *China Petroleum Exploration, 25*(1), 31–44.

Chapter 7
Petroleum and Natural Gas Accumulation Units and Distribution Patterns

Abstract The distribution of oil and gas in the Earth's crust is controlled by regional geological structures and geological conditions such as lithology and facies, and appears in clusters, belts, and zones. Dividing oil and gas accumulation units at different scales can effectively reveal the distribution and enrichment patterns of oil and gas, providing guidance for oil and gas exploration. This chapter introduces oil and gas reservoirs, oil and gas fields, oil and gas accumulation belts, petroleum systems, and petroleum basins as different levels of oil and gas accumulation units. It compares and analyzes the petroleum geological characteristics of different types of petroleum basins and introduces the distribution characteristics of global and Chinese oil and gas resources and the main geological factors affecting the distribution and enrichment of oil and gas.

Oil and gas reservoirs are the basic units of oil and gas accumulation in the Earth's crust. The distribution of oil and gas in the Earth's crust often occurs in groups, belts, and areas, and is controlled by regional geological structures as well as lithological and facies conditions, with distinct patterns that can be identified. Within the same area controlled by a single local structure, several oil and gas reservoirs can form an oil and gas field. Oil and gas fields do not exist in isolation either, but are often restricted by certain geological conditions and appear in groups and belts, forming oil and gas accumulation belts. Some oil and gas accumulation belts often have the same source of oil and gas, are located within the same oil and gas -bearing area, and have undergone the processes of oil and gas generation and accumulation. An area with a unified geological history and one or several oil and gas -bearing areas can constitute an petroliferous basin.

Since the 1990s, the concept of "petroleum system" has been developed, which integrates the static elements and dynamic processes involved in the formation of oil and gas reservoirs. It uses the ideas of systems theory to study the evolutionary history of basins and the formation of oil and gas reservoirs. The research has evolved from studying the static conditions of oil and gas reservoir formation, such as generation, reservoir, cap rock, migration, trap, and preservation, to dynamically analyzing the formation process of oil and gas reservoirs.

© China University of Petroleum Press 2025

H. Liu and Y. Jiang, *Petroleum Geology*,
https://doi.org/10.1007/978-981-96-7086-4_7

7.1 Oil and Gas Fields

7.1.1 Basic Concepts of Oil and Gas Fields

An oil and gas field refers to the sum of oil reservoirs, gas reservoirs, and oil–gas reservoirs within the same area, controlled by structural, stratigraphic, or lithological factors. If there are only oil reservoirs, it is called an oil field. If there are only gas reservoirs, it is called a gas field.

The basic concept of an oil and gas field includes the following implications:

(1) An oil and gas field is the current place where petroleum and natural gas are gathered, regardless of their place of generation.
(2) The controlling factors of an oil and gas field can be a single structural, stratigraphic, or lithological factor, or a combination of multiple geological factors.
(3) An oil and gas field occupies a certain area, which geographically includes a specific range. The size of independent oil and gas fields can vary greatly, with some as small as a few square kilometers and others as large as thousands of square kilometers.
(4) The same area refers to the total area of oil and gas-producing layers that are superimposed and connected at different levels. In an oil and gas field, oil and gas-producing layers at different levels can exist within a single geological body controlled by the same structural or stratigraphic factors, or within a composite geological body controlled by multiple factors. Some oil and gas fields have several individual oil and gas-producing areas that are not directly connected but are close in location, and have similar oil and gas-producing layers, reservoir types and characteristics, as well as similar mechanisms of trap formation, and can often be considered as one oil and gas field.
(5) An oil and gas field can include one or several oil or gas reservoirs.

As shown in Fig. 7.1, this is an oil and gas field mainly controlled by a salt body piercing structure. This oil and gas field contains both oil and gas reservoirs, including multiple series and types of traps in the cross-section, and these reservoirs are superimposed and connected in the plane. The maximum envelope range of these reservoirs constitutes an oil and gas field.

7.1.2 Main Types of Oil and Gas Fields

There are various classification schemes for oil and gas fields. Zhang et al. (1999) classified oil and gas fields based on lithology and structure, first emphasizing the differences in the formation conditions of clastic and carbonate rock oil and gas fields. They divided oil and gas fields into two main categories: sandstone oil and gas fields

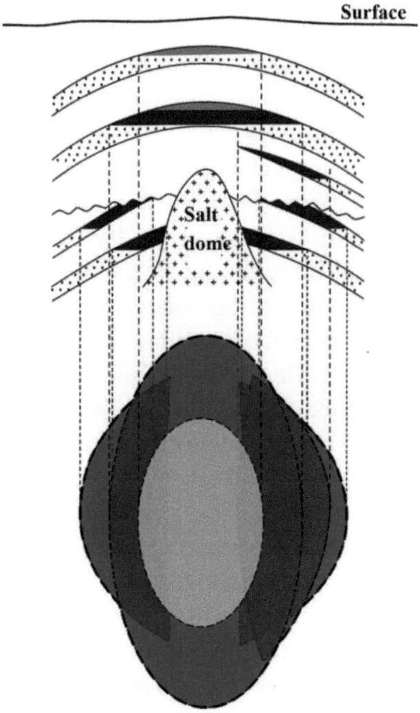

Fig. 7.1 Cross-section and plan schematic diagram of the same oil and gas field

Table 7.1 Classification of oil and gas fields based on lithology and structure

Sandstone oil and gas fields	Anticline sandstone oil and gas fields
	Monocline sandstone oil and gas fields
	Piercing structural sandstone oil and gas fields
	Irregular band sandstone oil and gas fields
	Sandstone buried hill oil and gas fields
Carbonate oil and gas fields	Large anticline carbonate oil and gas fields
	Fracture-type carbonate oil and gas fields
	Biological reef carbonate oil and gas fields
	Carbonate buried hill oil and gas fields

and carbonate oil and gas fields. Secondly, they considered the genetic characteristics of individual local structural units for detailed classification, as shown in Table 7.1.

7.1.2.1 Structural Oil and Gas Fields

Structural oil and gas fields are those where the oil and gas-producing area is controlled by a single structural factor, such as folds or faults. In most cases, folds are accompanied by faults, but the folds are predominant, and these are called anticline oil and gas fields. Sometimes, the fields are mainly controlled by faults, and these are called fault or fault block oil and gas fields.

(1) Anticline Oil and Gas Fields

The geological units controlling the oil and gas-producing area in anticline oil and gas fields are the anticline structures formed by fold deformation. The fold deformation generally extends vertically through thick multi-layered sedimentary rock layers. Within the anticline range, if the reservoir layers are sealed by the cap rock and have good sealing conditions, they can form anticline traps. Therefore, the vertical superposition of multiple oil and gas layers, forming a thick oil and gas-bearing layer group, is often one of the most significant characteristics of anticline oil and gas fields. The thick oil and gas-bearing layer group can compensate for the deficiency in the oil and gas-producing area, making some relatively small anticline oil and gas fields become large oil and gas fields. Many of the world's super large oil and gas fields are anticline oil and gas fields.

It should be noted that the structural morphology of anticlines is not consistent at different burial depths. The position of the anticline high point and the shape of the folds can change with depth. Secondly, the oil-producing area of anticline oil and gas fields is controlled by the closure area of the anticline. Generally, it is controlled by the closure area of a single anticline, but sometimes several adjacent anticline structures that are closely related in genesis, although the oil and gas-producing areas are not completely connected, are also considered as the same oil and gas field, such as the Burgan oil field and the Daqing oil field. They are oil and gas accumulation zones.

The fold morphology of anticline oil and gas fields can be diverse. The rock types of the reservoir layers in anticline oil and gas fields can be clastic rocks or carbonate rocks. Clastic rock reservoirs are mainly medium-fine sandstones, with good porosity and permeability, and are relatively stable laterally. Carbonate rock reservoirs can be pore-type grainy limestone, but most are pore-fracture or fracture-pore type reservoirs. Some oil and gas fields have both carbonate and clastic rock reservoirs, such as the Alran and Kulikovo oil fields in the Volga-Ural oil and gas basin (Fig. 7.2).

In the Daqing structural belt of the Songliao Basin, there are 7 anticline oil and gas fields, including Lamadian, Sartu, and Putuohua, forming the Changyuan anticline oil and gas accumulation zone. The reservoir layers in this accumulation zone are mainly lacustrine deltaic sandstone bodies, with good oil storage properties. Multiple oil-bearing layers are superimposed vertically in the profile, and the oil-producing areas are connected in the plane, as shown in Fig. 7.3.

The types of oil and gas reservoirs in anticline oil and gas fields are usually dominated by anticline oil and gas reservoirs, but they often also include other types of oil and gas reservoirs, such as fault and lithological types. Some oil and gas fields are composed of a single anticline oil and gas reservoir. However, in any case, the oil and gas-producing areas at different levels are often superimposed vertically and connected in the plane.

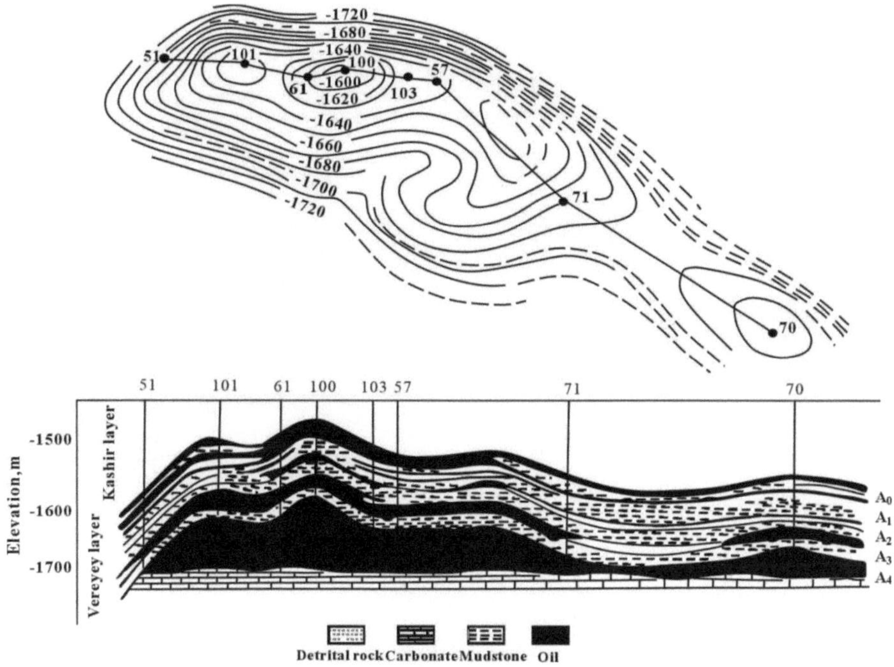

Fig. 7.2 Structural plan map of the Kulikovo Oilfield in the former Soviet Union (contour interval in meters) and cross-section diagram (according to Максимов, 1970)

(2) Fault (Fault Block) Oil and Gas Fields

Fault (fault block) oil and gas fields refer to the sum of oil and gas reservoirs within the same oil-bearing area controlled by multiple faults. The representative classifications are mainly as follows.

Graben or half-graben type rift (or graben) basins, such as the Oman Graben, Rhine Graben, Magdalena Basin, and the fault (fault block) oil and gas fields in the Bohai Bay Basin of China.

Basin slope belts or flexure belts, such as the fault flexures along the Gulf of Mexico coast, or synsedimentary fault belts, such as the synsedimentary fault belt in the Niger Delta. The main faults in these oil and gas fields are often synsedimentary faults, which not only form one side boundary of the oil and gas field but also play an important controlling role in the formation of source rocks, reservoir layers, and oil and gas traps.

Fault (fault block) oil and gas fields are generally medium to small, but some fault block oil and gas fields developed on an anticline background can also form large oil and gas fields, such as the Dongxin Oilfield and Linpan Oilfield in the Bohai Bay Basin. The oil and gas reservoirs are mainly fault oil and gas reservoirs, with a small number of unconformity oil and gas reservoirs and lithological oil and gas reservoirs.

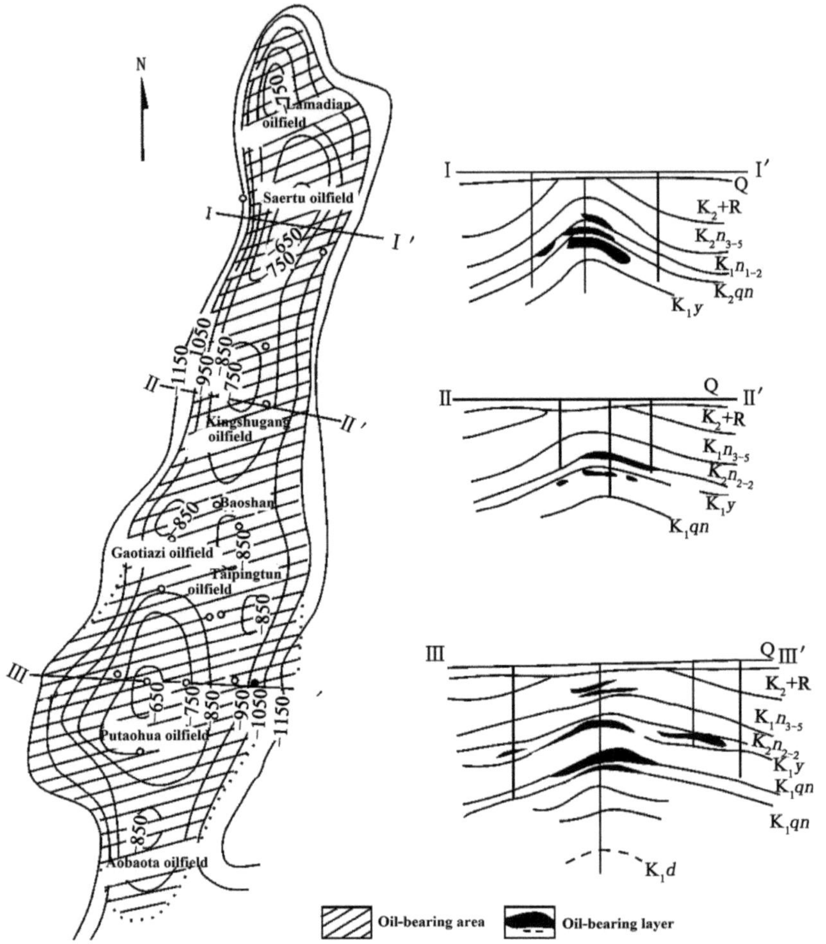

Fig. 7.3 Structural plan of the Daqing Oilfield (contour interval in meters) and cross-section diagram

7.1.2.2 Stratigraphic Oil and Gas Fields

Stratigraphic oil and gas fields refer to the sum of oil and gas reservoirs within the same oil-bearing area controlled by unconformities. These fields are mostly controlled by unconformity surfaces, and their structural background is often regional monoclines, commonly distributed along the edges of basins and large anticline slope areas. Due to frequent transgressions and regressions of water bodies, various oil and gas reservoirs related to unconformity surfaces are formed. The types of oil and gas reservoirs are mainly unconformity oil and gas reservoirs and lithological oil and gas reservoirs, with a small number of fault oil and gas reservoirs. The East Texas Oilfield in the United States is an example of this type of oilfield.

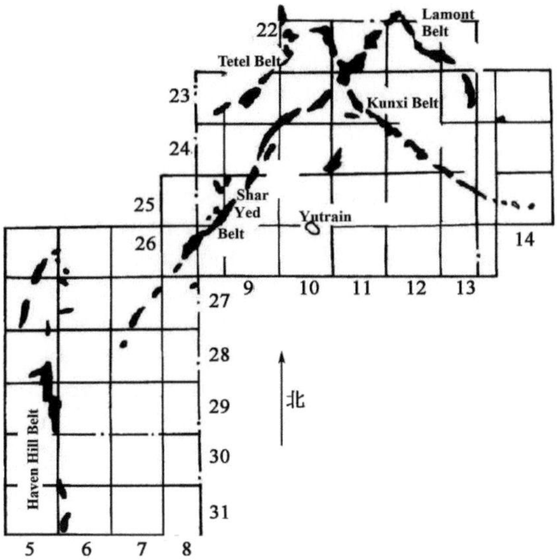

Fig. 7.4 Plan map of the shoestring oil and gas field in Kansas, USA (according to Cadman)

7.1.2.3 Lithological Oil and Gas Fields

Lithological oil and gas fields refer to the sum of lithological oil and gas reservoirs within the same oil-bearing area controlled by sedimentary conditions, including sandstone lens oil and gas fields, lithological pinch-out oil and gas fields, and biological reef oil and gas fields. For example, the shoestring oilfields in Greenwood and Butler counties, Kansas, USA. These fields consist of many offshore sandbars, forming many elongated lenses. Each lens is 50–100 ft thick, 2–6 miles long, and up to 1.5 miles wide, discontinuously arranged in a band stretching 25–45 miles (Fig. 7.4).

7.1.2.4 Composite Oil and Gas Fields

Composite oil and gas fields refer to oil and gas fields where oil and gas reservoirs at different levels and depths within the oil and gas field are controlled by two or more factors such as structure, stratigraphy, and lithology. The main types are as follows:

(1) Salt (Mud) Dome Type Composite Oil and Gas Fields

In piercing salt (mud) dome oil and gas fields, the salt core pierces the oil and gas reservoir layers, forming not only salt core, salt cap barriers, and lens oil and gas reservoirs within the salt cap but also causing reservoir fracturing, pinch-out, and even erosion, forming fault, unconformity, and lithological oil and gas reservoirs. This is a typical composite oil and gas field. Hidden piercing can also form anticlines above the salt dome. Whether it is piercing or hidden piercing salt (mud) dome structures, if there are composite oil and gas reservoirs formed by structural and stratigraphic factors, they are called salt (mud) dome type composite oil and gas fields. If multiple oil and gas reservoirs are formed under the control of a single structural factor, they cannot be called composite oil and gas fields.

(2) Biological Reef Type Composite Oil and Gas Fields

In biological oil and gas fields, some are deep biological oil and gas reservoirs, while the shallow parts (above the biological reef) are synsedimentary anticlines or compaction anticlines formed during the growth of the biological reef, or anticline oil and gas reservoirs formed in strongly folded anticlines. Oil and gas fields formed by the development of multiple genetic oil and gas reservoirs in different series under the background of biological reef are called biological reef type composite oil and gas fields. The Snyder-Skelly biological reef type composite oil and gas field in Scurry County, Permian Basin, USA, is a typical example. If there are only biological reef oil and gas reservoirs and no structural oil and gas reservoirs above the biological reef, it belongs to the lithological type biological reef oil and gas field.

(3) Buried Hill Type Composite Oil and Gas Fields

Burial hill type composite oil and gas fields have a burial hill oil and gas reservoir in the deep part, while the overlying strata may form anticline oil and gas reservoirs and fault oil and gas reservoirs due to draping and compaction. Above the unconformity surface, there may also be lithological or stratigraphic oil and gas reservoirs that pinch out towards the subsurface mountain tip or overstep.

The geological structure and types of oil and gas reservoirs beneath the unconformity surface and above the burial hill can vary significantly. If there is only a superimposed geological body without the superimposition of different types of oil and gas reservoirs, it cannot be called a burial hill type composite oil and gas field. Some oil and gas fields have a superimposition of different types of oil and gas reservoirs (such as stratigraphic and structural) above the burial hill (i.e., above the unconformity surface), but if there are no oil and gas reservoirs within the burial hill, they also cannot be called burial hill type composite oil and gas fields. The schematic cross-section of a burial hill type composite oil and gas field is shown in Fig. 7.5.

(4) Structural-Stratigraphic Overlapping Composite Oil and Gas Fields

These types of oil and gas fields refer to those where the oil and gas reservoirs, which are primarily structural in some layers and primarily stratigraphic in others, are not vertically superimposed but laterally adjacent, or where the oil and gas-bearing areas have some degree of overlap, forming a unified oil and gas field.

The Midway-Sunset Oilfield in California is a typical example of this type of oil and gas field (Fig. 7.6). The southwestern part of the field has the Seraes Anticline oil and gas reservoir (N_{11}) and the unconformity-trapped oil and gas reservoir (N_{12}, minor) beneath the unconformity surface, while above the unconformity surface, there are stratigraphic onlap oil and gas reservoirs. Although the oil-bearing areas of the two are not superimposed and connected, they are closely adjacent and exist within a unified geological body, and can be defined as a structural-stratigraphic overlapping composite oil and gas field.

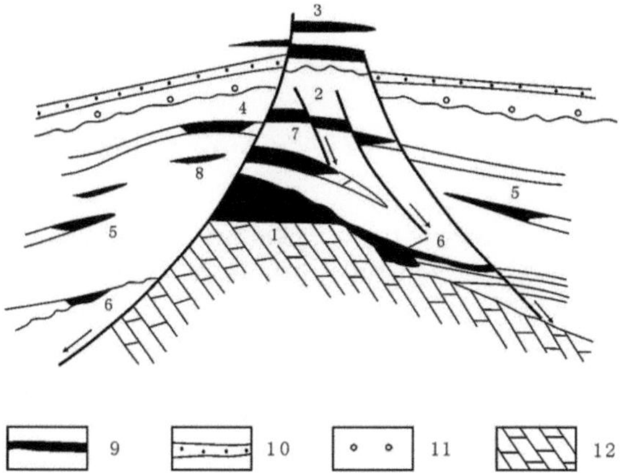

Fig. 7.5 Schematic cross-section diagram of oilfield patterns associated with buried hill (according to North China Oilfield). 1—Buried hill oil reservoir, 2—compacted anticlinal oil and gas reservoirs on buried hills cut by faults, 3—shallow anticlinal and fault oil and gas reservoirs, 4—step fault or reverse drag anticlinal oil and gas reservoirs, 5—lithological oil and gas reservoirs, 6—stratigraphic oil and gas reservoirs, 7—compacted anticlinal oil and gas reservoirs above buried hills, 8—lithological oil and gas reservoirs, 9—oil reservoir, 10—sandstone, 11—conglomerate, 12—limestone

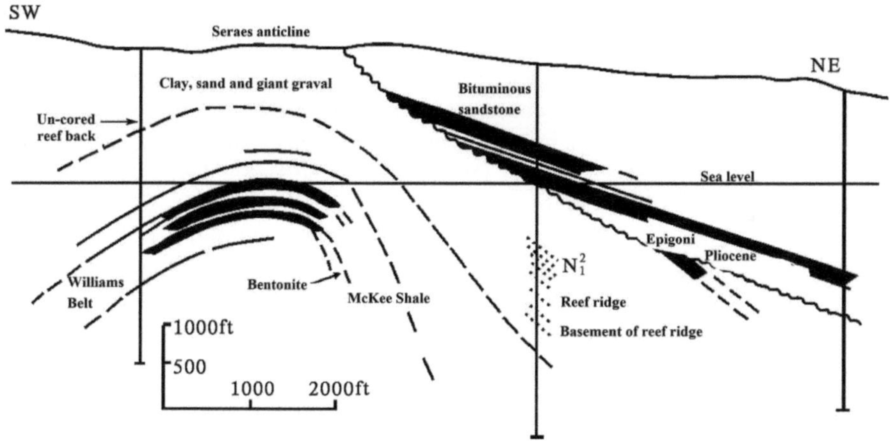

Fig. 7.6 Structural cross-section diagram of the Midway-Sunset Oilfield in California, USA, including the Williams and Twenty-Five Mile Hills areas (according to Levorsen, 1967)

7.2 Oil and Gas Accumulation Zones and Petroliferous Areas

7.2.1 The Concept of Oil and Gas Accumulation Zones

An oil and gas accumulation zone refers to the sum of a series of oil and gas fields that have causal connections and similar conditions for oil and gas accumulation within the same secondary structural zone or lithofacies change zone. Oil and gas exploration practices have proven that oil and gas fields do not exist in isolation in the Earth's crust. After discovering an oil and gas field, new oil and gas fields are often found in adjacent structures, or oil and gas shows are encountered during drilling. This phenomenon fully illustrates that the migration of oil and gas is regional and is often controlled by secondary structural zones or lithofacies change zones. When these secondary structural zones or lithofacies change zones are well connected to or close to the oil source areas, with the continuous supply of oil and gas, a series of traps throughout the entire secondary structural zone or lithofacies change zone may form oil and gas reservoirs, resulting in the clustered and belt-like appearance of oil and gas fields, thus forming oil and gas accumulation zones.

The formation of oil and gas accumulation zones is the result of the organic cooperation between secondary structural zones and oil source areas and reservoir facies zones. The petroleum and natural gas generated in the oil source areas within sedimentary basins first migrate to the upper, lower, and surrounding adjacent reservoir facies zones. Therefore, the secondary structural zones here are often the main direction of oil and gas migration and become favorable oil and gas accumulation zones. The oil and gas generated in the lower oil-generating area of the Shahejie Formation in the Dongying Sag and Huanghua Depression of the Bohai Bay Basin migrate nearby and preferentially accumulate in the secondary structural zones near the oil source areas. The Tuozhuang-Shenglicun and Dongxin structural zones concentrate 63% of the geological reserves of the Dongying Sag. The geological reserves of the Dagang Oilfield account for 60% of the entire Huanghua Depression (Fig. 7.7).

If in the depression area of a sedimentary basin, the source rock layer and the reservoir layer are interbedded and interlaced, the oil source area is the reservoir area. Here, the secondary structural zones have the advantage of receiving the oil and gas, becoming the most favorable oil and gas accumulation zones, and it is easy to form super-large oil fields. For example, the Daqing Changyuan Oil and Gas Accumulation Zone concentrates more than 80% of the geological reserves of the Songliao Basin (Fig. 7.8).

In sedimentary basins of different large-scale tectonic units in the Earth's crust, due to the differences in regional geological tectonic conditions, various secondary structural zones can be formed. Therefore, oil and gas accumulation zones also present various types according to the different large-scale tectonic positions they are in.

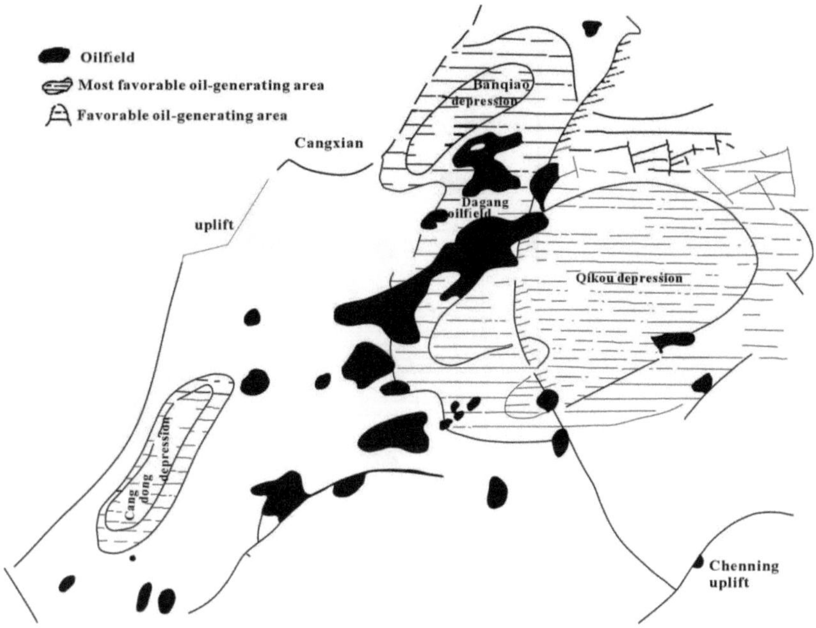

Fig. 7.7 Relationship between the distribution of oil and gas fields and the oil-generating areas in the Huanghua Depression (according to the Petroleum Exploration and Development Research Institute, 1977)

7.2.2 Types of Oil and Gas Accumulation Zones

Based on the main geological factors controlling oil and gas accumulation zones, they can be classified into the several types.

(1) Anticline Oil and Gas Accumulation Zones

Anticline oil and gas accumulation zones refer to a series of anticlinal oil and gas fields controlled by anticlinal belts. Some are anticlinal oil and gas accumulation zones related to basement uplifts, such as the Daqing Changyuan Oil and Gas Accumulation Zone and the Saudi Arabian Ghawar Oil and Gas Accumulation Zone. These types of oil and gas accumulation zones generally have gentle folds, small dip angles on both wings, large oil-bearing areas, and abundant oil and gas. There are also anticlinal oil and gas accumulation zones related to folding actions, with intense folding, steep dip angles on both wings, large closure heights, and individual anticlinal oil and gas fields often arranged in linear or en-echelon patterns, extending up to hundreds of kilometers. A series of such oil and gas accumulation zones are distributed in the Zagros Basin of the Persian Gulf Basin, and the oil and gas reserves are also quite abundant.

(2) Fault-Controlled Oil and Gas Accumulation Zones

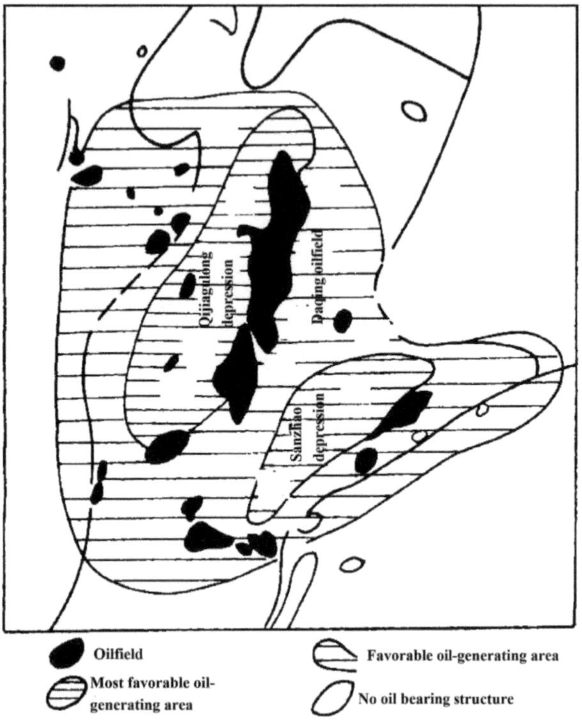

Fig. 7.8 Relationship between the distribution of oil and gas fields and the oil-generating areas in the Songliao Basin (according to the Petroleum Exploration and Development Research Institute, 1977)

Fault-controlled oil and gas accumulation zones refer to the sum of a series of oil and gas fields controlled by faults. According to the relationship between faults and oil and gas accumulation, fault-controlled oil and gas accumulation zones can be divided into the following three types.

Fault-block oil and gas accumulation zones consist of several fault-block oil and gas fields or anticlinal oil and gas fields complicated by faults, distributed in a belt-like pattern. Fault activity controls the formation of traps, the migration, and accumulation of oil and gas. For example, the Dongxin fault-block oil and gas accumulation zone in the Jiyang Depression of the Bohai Bay Basin.

Growth fault-controlled oil and gas accumulation zones, in which growth faults control the development of traps and the accumulation of oil and gas. A series of reverse-drag anticlinal oil and gas fields and fault-block oil and gas fields are distributed in the accumulation zone. For example, in Nigeria's Niger Delta, growth fault oil and gas accumulation zones are well-developed, with a series of reverse-drag anticlinal oil and gas fields developed on this zone.

Thrust fault-controlled oil and gas accumulation zones consist of anticlinal oil and gas fields and fault oil and gas fields developed on thrust fault zones. The geological structure of this zone is very complex, which brings certain difficulties to the exploration and development of oil and gas fields. For example, the thrust fault oil and gas accumulation zone on the eastern side of the Rocky Mountains in the United States.

(3) Monocline-Type Oil and Gas Accumulation Zones

At the edge of sedimentary basins, it is often possible to see large monocline zones where the strata dip in a basically consistent direction, the structural contour lines are basically parallel, and they dip towards the center of the basin. Local structures are not well-developed on the monocline zones. However, due to the location at the edge of the basin, phenomena such as unconformities and lithological pinchouts are very common. Coupled with the effects of nose-like structures and faults, this leads to the development of unconformity oil and gas fields and lithological oil and gas fields on the monocline zones, which are distributed in clusters and in belts, forming monocline-type oil and gas accumulation zones. For example, the Bolivar Monocline Oil and Gas Accumulation Zone in the eastern part of the Maracaibo Basin in Venezuela.

(4) Biological Reef-Type Oil and Gas Accumulation Zones

In certain sedimentary environments, biological reefs that are distributed in groups and belts control the distribution of biological reef-type oil and gas fields, forming biological reef-type oil and gas accumulation zones in the form of atoll-like, horseshoe-like, and linear reefs. For example, the Mexican Golden Lane Atoll-Type Biological Reef Oil and Gas Accumulation Zone.

In addition, there are also oil and gas accumulation zones of the buried hill type, salt dome and mud volcano type, and sandstone lens type, etc. The distribution of oil and gas fields is controlled by oil and gas accumulation zones. Studying the distribution patterns and characteristics of oil and gas accumulation zones is of great significance for oil and gas exploration. From the perspective of geological development, favorable oil and gas accumulation zones should be:

In sedimentary basins, there are long-term inherited anticlinal oil and gas accumulation zones in the oil-source areas or their vicinity. These zones are close to the oil-source areas, have well-developed reservoir facies zones, and form structural traps early. During the uplift process, the already generated oil and gas can gather nearby.

In the process of geological history development, oil and gas accumulation zones that generally form earlier are more favorable for containing oil and gas.

The large monocline zones at the edges of sedimentary basins are often areas where favorable reservoir facies zones develop and where various stratigraphic and fault traps are easily formed. They are favorable target areas for regional oil and gas migration and are conducive to the formation of large monocline oil and gas accumulation zones.

Areas where biological reefs, salt domes, ancient buried hills, and coastal sandbars develop can form various corresponding types of oil and gas accumulation zones.

7.2.3 Petroliferous Areas

Favorable oil and gas accumulation zones are mostly located in the long-term subsiding and sediment-receiving low-lying areas within sedimentary basins, which are conducive to the generation and accumulation of petroleum and natural gas. These low-lying areas are mostly distributed in sedimentary depressions, where the geological development history and the characteristics of sedimentary rock series development have unity, and the processes of oil and gas generation and accumulation also have common regularity.

In petroleum geology, sedimentary depressions that belong to the same large-scale tectonic unit, have a unified geological development history, and have conditions for oil and gas generation and accumulation are called petroliferous areas or oil and gas-bearing areas.

7.3 Petroliferous Basins

7.3.1 The Concept and Structural Elements of Petroliferous Basins

7.3.1.1 The Concept of Petroliferous Basins

During a certain geological period, the unified subsidence areas on the Earth's crust that once stably sank and received thick sedimentary deposits are called sedimentary basins. If oil and gas fields of industrial value are discovered within a sedimentary basin, such a basin can be regarded as a petroliferous basin. That is, any sedimentary basin on the Earth's crust that has a unified geological development history, features good source, reservoir, and seal combinations as well as trapping conditions, and has discovered oil and gas fields is called a petroliferous basin. It is evident that a petroleum basin must first be a sedimentary basin. During the long geological history, it has continuously subsided to receive sedimentation, has favorable conditions for the generation and accumulation of oil and gas, and contains oil and gas fields. The understanding of a petroliferous basin includes the following aspects:

It has a unified geological development history.

It was a long-term developing low-lying area on the Earth's crust, with thick sedimentary rock layers. It maintained a state of continuous subsidence in a certain underwater environment for a considerable geological period to form thick sedimentary rocks and an environment favorable for oil generation.

It has undergone a certain degree of tectonic movement, which is conducive to oil and gas migration and trap formation. Of course, tectonic movements should not be too intense, otherwise they will destroy the generation and accumulation of oil and gas.

The size of petroleum basins varies greatly, with areas ranging from tens of square kilometers to tens of thousands of square kilometers. For example, the world's largest petroliferous basin, the Persian Gulf Basin, has an area of 305×10^4 km^2. The largest petroliferous basin in China, the Tarim Basin, covers an area of 55.7×10^4 km^2. The Minhe Basin in western China has an area of only 9,000 km^2. The Los Angeles Basin in the western United States has an area of only 3,900 km^2.

A petroliferous basin can be further divided into different tectonic units internally. On the basis of the commonality of the whole basin, there may be certain differences in the sedimentation process and oil and gas generation and accumulation conditions of each unit. For example, in the Bohai Bay Cenozoic Petroleum Basin, there are significant differences in the degree of oil and gas enrichment among different depressions.

7.3.1.2 The Structural Elements of Petroliferous Basins

The basement, periphery, and sedimentary cover are the three major elements that make up the basin structure.

Basement: It is the foundation on which the petroliferous basin exists, composed of the rock series before the formation of the basin. It can be composed of ancient crystalline or metamorphic rocks, sedimentary rock series, or a mixture of both.

Periphery: It refers to the boundary around the basin. The contact relationship between the basin periphery and its boundary geological bodies mainly includes onlap contact and fault contact. The nature of the basin periphery determines the basic type of the basin. According to the nature of the basin periphery, basins are divided into depression basins and faulted basins (which can be further divided into two types: single-faulted and double-faulted).

Sedimentary Cover: It refers to the sedimentary rock layer developed above the basement.

The geological characteristics of the basement and periphery of petroliferous basins have an important controlling effect on the shape of the basin, the development of sedimentary rock series, and geological structures.

The oldest basement of the basin is the pre-Sinian Series (mostly pre-Cambrian Series abroad), which belongs to the crystalline metamorphic basement, with hard lithology. This type of basin is generally large in area. The depression-type basins are often nearly circular-elliptical, while the faulted-type basins are nearly rectangular or rhombic, with the periphery controlled by large faults. The sedimentary rock series in the basin are mainly Paleozoic, and sometimes Mesozoic and Cenozoic are also developed. The thickness is generally small, about 2000–4000 m, and the thickest can reach tens of thousands of meters. The tectonic activity of the basin is generally small, and the cover structure is mostly controlled by the basement activity, with gentle folding and the development of small-scale normal faults. Another type of basin basement belongs to the young basement, including Caledonian, Hercynian or Mesozoic basement, mostly in the shape of a strip. The sedimentary and structural characteristics in the depression are mostly controlled by the adjacent fold belt. The thickness of the sedimentary rock series is large, generally 6000–7000 m, and the thickest exceeds 10,000 m; folding and various fractures are all intense.

7.3.2 *Types of Petroliferous Basins and Tectonic Units*

(1) Types of Petroliferous Basins

Different types of basins have different probabilities of containing oil and gas, different abundances of oil and gas, and different possibilities of finding large oil and gas fields. Therefore, when conducting a prospective evaluation of the oil and gas potential of a basin, it is essential to first clarify its type. Since the 1950s, various classification schemes for sedimentary basins have emerged both domestically and internationally, with the theoretical basis for division mainly including tectonics and geodynamics.

Classification based on mobilism (plate tectonics theory) is represented by figures such as Halbouty, Dickinson, Klemme, Bally, Chen Fajing, and Kingston. For example, Dickinson (1976), from a plate tectonics perspective, classified basins into two major categories and 16 types of basins (Table 7.2), where rift-type basins are primarily characterized by divergent plate motion and tensional structures, resulting from crustal thinning and subsidence; while orogenic-type basins are mainly characterized by convergent plate motion and compressional structures, caused by plate subduction leading to crustal subsidence, or possibly due to increased sedimentary load prompting crustal descent.

Classification based on geodynamics is represented by Liu Hepu, Lu Kezheng, Allen, and others. Many scholars, according to the various stress actions suffered by petroliferous basins during different geological historical periods, have divided petroliferous basins into four geodynamic environments: rifting environment, compressional environment, gravitational environment, and shearing environment. The corresponding four major types of basins are generally referred to as rift basins, foreland basins, cratonic basins, and strike-slip basins (Table 7.3). Liu (1986) believed that from the perspective of the dynamic system of basin formation, there are mainly three stress environments: (a) rift basin, where the maximum principal compressive stress axis is vertical, (b) compressional basin, where the maximum principal compressive stress axis is horizontal, (c) strike-slip basin, where both the maximum and minimum principal compressive stress axes are horizontal (Fig. 7.9). This classification is consistent with the three basic types of plate boundaries and the controlling faults of basin boundaries. The geodynamic environment of basin formation not only controls the pattern of the basin, the characteristics of sediments within the basin, the characteristics of the source-reservoir-seal assemblages and their combinations, but more importantly, it controls the tectonic pattern of the basin, the migration direction of oil and gas within the basin, and the favorable accumulation positions. Therefore, the classification scheme based on geodynamics is widely applied in current petroliferous basin analysis.

Table 7.2 Basin classification based on mobilism (plate tectonics theory) (according to Dickinson, 1976)

Major category	Sub-classification of basin types	Basic characteristics of basins
I. Rift environment basins	I1. Intracratonic basin	Rift basin within the continental interior, with thinned basin basement
	I2. Marginal trough	An aborted rift extending inward from the continental margin, with a basement of oceanic or transitional crust
	I3. Proto-oceanic rift	A nascent narrow oceanic crust forming between two continental blocks, with sedimentation still influenced by the adjacent continents
	I4. Passive margin sedimentary prism	A sedimentary complex developed on the continental shelf, slope, and rise along the transitional zone between continent and ocean
	I5. Continental rise	Sediments formed along the outer edge of a rifting continental margin, gradually advancing towards the ocean
	I6. New oceanic basin	A newly formed basin between the mid-ocean ridge and continental block, due to the growth and subsidence of the oceanic lithosphere
	I7. Transtensional basin	A tensional or wedge-shaped graben basin developed in locally thinned crust along complex transform fault systems
	I8. Arc-intertidal basin	A small oceanic basin formed by the subsidence of oceanic crust between an inactive residual arc and an active fore-arc due to the splitting of a magmatic arc
II. Orogenic environment basins	II1. Trench	A deep-sea trench formed as a subduction zone of plate convergence
	II2. Slope basin	A graben basin between the axis of the trench and the inflection point of the trench slope
	II3. Fore-arc basin	A basin in the gap between the inflection point of the trench slope and the magmatic island arc

(continued)

Table 7.2 (continued)

Major category	Sub-classification of basin types	Basic characteristics of basins
	II4. Peripheral foreland basin	A basin formed at the junction of the continental block periphery and the collisional orogenic suture zone
	II5. Back-arc foreland basin	A foreland basin behind the continental block margin magmatic arc, adjacent to the island arc orogenic belt
	II6. Rifting foreland basin	A foreland basin of the orogenic belt, with peripheral or back-arc structural depressions formed due to basement deformation and block faulting
	II7. Transpressional basin	A basin formed by wrench folding and fault-bounded depressions along complex transform fault systems
	II8. Remnant oceanic basin	A contracting oceanic basin produced along one side of the island arc-trench system due to the subduction of old lithosphere

Overall, basin evolution is a series of continuous processes that occur, develop, mature, reverse, and perish during the relative motion of plates, which includes the transformation and combination of basin dynamic properties. Basins generated under a single geodynamic system or during a single tectonic cycle are considered prototype basins or single-cycle basins, while those formed under multiple geodynamic systems or during re-cyclic tectonics are classified as composite basins or re-cyclic basins, also known as superimposed basins, which in fact include the horizontal combination and vertical superposition of basins. The phased developmental characteristics exhibited by sedimentary basins in their evolutionary history also manifest in petroliferous basins as different types in their sedimentary development history, such as single-age monophase oil-generating layer combinations and multi-age polyphase oil-generating layer combinations.

(2) Tectonic Units of Petroliferous Basins

A petroliferous basin is a unified subsidence area. However, internally, neither the basement nor the sedimentary cover is a simple smooth concave surface or plane. The basement has undulations, and the sedimentary cover often undergoes various deformations. Due to the different properties of the basement and cover, the tectonic characteristics of petroliferous basins are also complex. Therefore, the interior can be further divided into several secondary tectonic units, and there are differences in

Table 7.3 Classification of sedimentary basins based on geodynamics (according to Liu, 1993, modified)

Major category	Basin type	Example
I. Rift basins	1. Continental rift basin	North Sea basin
	2. Intercontinental sea basin	Red Sea basin
	3. Tensional continental margin basin	Atlantic offshore basin
	4. Marginal sea-back-arc basin	Andaman Sea basin
	5. Trough basin	Southern Oklahoma basin
II. Compressional basins	1. Deep-sea trench basin	Peru–Chile trench
	2. Fore-arc basin	Dagu basin
	3. Remnant basin	Black Sea basin
	4. Forland basin	Alberta basin
	5. Intermontane basin	Fergana basin
III. Strike-slip basins	1. Strike-slip-Pull-apart basin	Death Valley, USA; Yilan-Yitong basin, China
	2. Strike-slip flexural basin	Baise basin, China; Tarim southwest depression, China, Qaidam northwest depression, China
IV. Cratonic basins	1. Intra-cratonic basin	West Siberian basin, Williston basin, Ordos basin
	2. Cratonic margin basin	Permian basin

the division of tectonic units for different types of petroliferous basins, as shown in Table 7.4.

Based on the basic characteristics such as basement undulations, fracture structures, sedimentary rock thickness, and the morphology of the main exploration target layers, the first-level, second-level, and third-level tectonic units of petroliferous basins are divided as follows:

First-level tectonic units: These refer to the areas where the relative uplift and subsidence movements have significant differences during the developmental history of the basin. The uplifts and depressions formed by basement undulations are the most important first-level tectonic units. Uplifts are characterized by relative uplift dominance, thinner and often incomplete sedimentary cover, more sedimentary discontinuities, and are prone to stratal onlap and lithological pinchout belts on the flanks adjacent to depressions, which are favorable for oil and gas accumulation. Depressions are the areas within the basin where the basement is buried deepest, with complete development of the sedimentary cover, large thickness, and stable lithofacies, making them regions conducive to oil and gas generation and thus the source areas of petroliferous basins.

Second-level tectonic units: These refer to the belt-like distributed tectonic unit combinations with obvious geological structural differences within the first-level

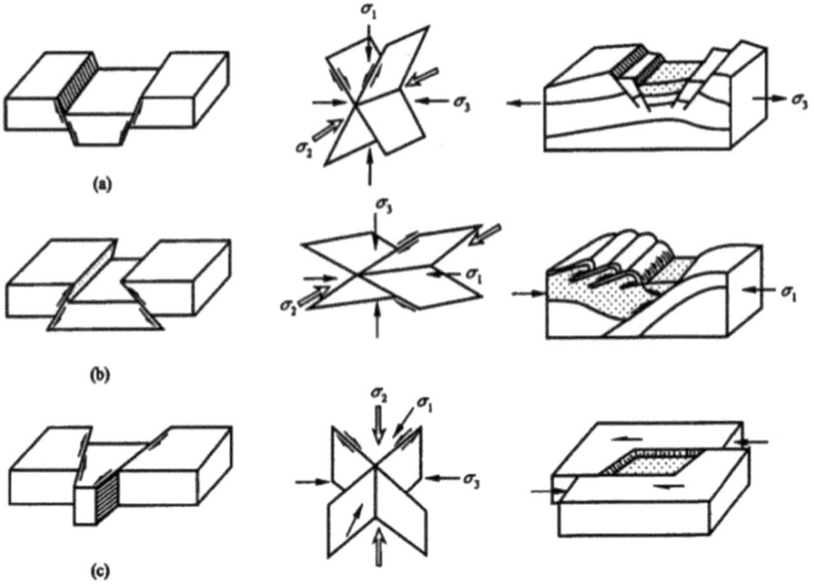

Fig. 7.9 Formation of sedimentary basins and the orientation of three principal stress systems (Liu, 1986). **a** Normal faults and rift basins, **b** thrust fault systems and compressional basins, **c** strike-slip faults and strike-slip basins

tectonic units. Inside the first-level tectonic units such as depressions, the belt-like undulations caused by basement undulations or differential uplift and subsidence of the two sides of the main faults lead to the formation of second-level tectonic units, mainly consisting of bulge, sag, low bulge. Among them, sags are the areas with the thickest sedimentary cover and are also the main oil-generating areas within the depressions. Bulges or low bulges are positive tectonic features and represent favorable directions for oil and gas migration and accumulation.

Sub-second-level tectonic units: These refer to local tectonic groups that are adjacent and have similar genesis and morphology within the first-or second-level tectonic units. Based on the tectonic differences of the second-level tectonic units, they can be further divided into multiple sub-second-level tectonic units, including anticlinal belt (sub-bulge, low bulge), synclinal belt (sub-sag), slope belt (steep slope belt, gentle slope belt, slope break belt, monocline belt), fracture belt (fault-block belt, fault-nose belt), which are the most direct oil and gas generation and migration areas. Positive tectonic features control the formation of oil and gas accumulation belts.

Third-level tectonic units: These refer to individual local tectonic structures or traps that constitute the second-level tectonic units, also known as third-level tectonics, including anticline (fault-related anticline), syncline, nose-like structure (fault related nose), fault block, buried hill, which are the tectonic units for the formation of oil and gas fields.

Table 7.4 Classification of tectonic units and oil and gas units within basins (modified from the Petroleum and Natural Gas Industry Standard of the People's Republic of China, 2016)

Basic tectonic units	First-level tectonic units	Second-level tectonic units	Sub-second-level tectonic units	Third-level tectonic units
Faulted basin	Uplift Depression	Bulge Sag Low bulge	Anticlinal belt (sub-bulge, low bulge) Synclinal belt (sub-sag) Slope belt (steep slope belt, gentle slope belt, slope break belt, monocline belt) Fracture belt (fault-block belt, fault-nose belt) Transfer belt (Hinge belt)	Anticline (fault-related anticline) Syncline Nose-like structure (fault related nose) Fault block Buried hill
Depression basin	Uplift Depression Slope	Bulge Sag Low bulge	Anticlinal belt (placanticline) Synclinal belt Slope belt (terrace, steep slope belt, gentle slope belt, slope break belt, monocline belt) Fracture belt (fault-block belt, fault-nose belt)	Anticline (fault-related anticline Syncline) Nose-like structure (fault related nose) Fault block
Foreland basin	Thrust belt Depression Uplift	Wedge-top belt Foreland depression Forland slope Forland uplift	Fractured belt (reverse fault belt, fault-block belt) Monocline belt Anticlinal belt Synclinal belt	Anticline Syncline Monocline Nose-like structure
Cratonic basin	Uplift Depression	Bulge Sag Low bulge	Anticlinal belt Synclinal belt Slope belt Fractured belt	Anticline Syncline Nose-like structure Fault block
Petroliferous basin	Petroliferous basin area		Oil and gas accumulation belt (sub-second level tectonic belt)	Oil and gas field

7.3.3 *Petroleum Geologic Characteristics of Major Types of Petroliferous Basins*

Different types of petroliferous basins have distinct regional tectonic and sedimentary filling characteristics, which determine the differences in the petroleum geologic characteristics of different types of basins. Among the types of petroliferous basins, foreland basins, cratonic basins, and rift basins are the most significant. Purely strike-slip basins are relatively few, and the strike-slip phenomena in petroliferous basins often occur within the three types of basins. For composite or superimposed basins, they are formed by the superposition and combination of basins of different types and periods, possessing the composite petroleum geologic characteristics of multiple types of basins. Therefore, this section primarily discusses the petroleum geologic characteristics of foreland basins, cratonic basins, and rift basins.

7.3.3.1 Foreland Basins

(1) The Concept and Characteristics of Foreland Basins

The classic foreland basin refers to an asymmetric thrust-flexural basin that develops between a contractional orogenic belt and an adjacent craton, extending in a narrow, elongated band parallel to the orogenic belt. It is formed by flexure and subsidence caused by the thrust load of the orogenic belt. Concepts such as compressional basins, pre-mountain depressions, pre-mountain depressions-craton margin depressions, and pre-mountain depressions-craton slopes all fall within the category of foreland basins. Foreland basins generally have the following characteristics: (a) The tectonic load of the adjacent fold-thrust belt causes the basin to flex and subside, (b) During the basin evolution, the side adjacent to the orogenic belt experiences stronger deformation, (c) The side adjacent to the craton is wider and more gentle, gradually merging with the craton sequence, (d) The basin extends in a narrow, elongated band parallel to the orogenic belt, roughly equivalent in length to the thrust-fold belt at the front of the adjacent orogenic belt, (e) The cross-sectional structure of the basin is distinctly asymmetrical, with most having a four-part structural characteristic from the orogenic belt towards the craton direction, namely thrust-fold belt → deep foreland sag → slope belt and foreland uplift.

(2) Petroleum Geologic Characteristics of Foreland Basins

Foreland basins are one of the most oil- and gas-rich types of basins in the world. Many famous petroliferous basins, such as the Persian Gulf Basin, Volga-Ural Basin, Western Canada Basin, Rocky Mountain Basin, Eastern Venezuela Basin, Maracaibo Basin, North Slope of Alaska Basin, and Appalachian Basin, are all foreland basins. In China, foreland basins are mainly developed in the western regions, including: the Tarim North Kuche Carboniferous-Triassic and Cenozoic Basin, Sichuan West

Mesozoic Basin, Tarim Southwest Silurian-Devonian and Cenozoic Basin, Zhungarian South and Northwest Permian and Cenozoic Basin, Qaidam North Cenozoic Basin, Jiuquan West Cenozoic Basin, Hubei West Mesozoic Basin, etc.

Affected by multi-stage sedimentary filling before and during the evolution of foreland basins, foreland basins generally develop two sets of sedimentary systems: passive continental margin depositional type and foreland depression type. Corresponding to the sedimentary systems, this type of basin mainly develops two sets of source rock series. The former is the marine carbonate source rock series, with lithologies mainly consisting of mud shale, marl, and limestone. The latter is the clastic rock source rock series, primarily composed of mud shale. The mature oil-generating centers of the source rocks are close to the side of the deep sag, and under the influence of compression during the orogenic period and the effect of stratigraphic load, the oil and gas in the deep sag areas migrate upward or towards the craton side along faults, unconformities, or permeable rock layers.

This type of basin also develops two sets of reservoir rock series, including the reservoir system mainly composed of platform-phase carbonates and the reservoir system mainly composed of terrestrial clastic rocks. In the two major sedimentary systems, there also exist cap rock systems mainly composed of platform-phase carbonates and terrestrial clastic rocks, with lithologies mainly consisting of evaporites, mud shale, and lignite, forming multiple favorable reservoir-cap combinations corresponding to the reservoir systems.

The most common and important trap types in foreland basins are anticlinal traps, fault traps, and stratigraphic traps. Anticlinal traps are mainly some folds related to thrust faults, distributed on the side close to the basin's thrust-fold belt, often extending in a belt-like pattern parallel to the thrust zone. Fault traps include both early-formed fault block traps and traps formed by later orogenic thrust faulting, with relatively well-developed early sedimentary series in the thrust-fold belt and other structural belts. The deep foreland depressions and slope belt have weaker structural deformation, and since the basin strata always onlap gradually towards the craton direction, stratigraphic and lithologic oil and gas reservoirs are the main types of oil and gas reservoirs, but low-amplitude structural oil and gas reservoirs also exist. In structurally complex foreland basins or in the foreland basin's front-depression areas strongly affected by late thrusting, in addition to lithologic traps, structural oil and gas reservoirs and composite structural-lithologic oil and gas reservoirs are also relatively common. The front-range uplift is an important part of the foreland basin, consisting of arch-shaped anticlines and fault block traps, as well as stratigraphic oil and gas reservoirs related to unconformities.

Foreland basins are a type of basin where fluid abnormal pressure is universally present, with both overpressure and underpressure, predominantly overpressure. For example, in the United States, the Powder River Basin and Green River Basin in the Rocky Mountain oil and gas area are both overpressured basins. The Cretaceous series is the main oil- and gas-producing layer in both basins and the series where abnormal pressure occurs. Overpressure mainly occurs at depths of 2700–3500 m and 2800–6100 m, with maximum pressure gradients reaching 17.55 kPa/m and $\pm$18.48 kPa/m, respectively. In China, abnormal pressure has been observed in foreland basins such as the Sichuan Basin's western part, the Kuche area of the Tarim Basin, the southwest of the Tarim Basin, the southern margin of the Zhungarian Basin, the northern margin of the Qaidam Basin, and the Turpan-Hami Basin. Overpressure is a key condition for the formation of some gas reservoirs in foreland basins. Natural gas generated from shale-dominated source rocks can directly charge into adjacent reservoirs under the drive of overpressure or can be injected through faults into upper reservoirs not in direct contact with the source rocks, especially for tight sandstone reservoirs, the role of abnormal high pressure is more important.

The distribution of oil and gas in foreland basins is mainly controlled by the distribution characteristics of traps (Fig. 7.10). On the side close to the thrust belt or within the thrust belt, anticlinal and fault oil and gas reservoirs are mainly found. For example, the Laojunmiao structural belt in the Jiuquan West Basin is a fault-fold belt controlled by thrust faults, where anticlinal oil fields such as Laojunmiao and Yaxia have been discovered. On the front-slope belt and front-range uplift belt close to the craton side, sandstone bodies with updip pinchout or stratigraphic onlap oil and gas reservoirs, as well as fault block oil and gas reservoirs related to tensional or trans-tensional faults, are mainly distributed. On the slope belt and foreland depression belt close to the foreland depression, oil and gas reservoirs related to lithology and stratigraphy are mainly found. In the plane, oil and gas in foreland basins are distributed in strip-like patterns around the oil-generating centers on the structural belts parallel to the orogenic belt.

Due to the activity of the orogenic belt and the continuous compression of the thrust belt, oil and gas reservoirs within the basin are constantly adjusted, transformed, and redistributed by tectonic movements. Therefore, foreland basins are also a type of basin where oil and gas reservoirs are relatively severely destroyed. For example, the Western Canada Basin and the Eastern Venezuela Basin are typical heavy oil and asphaltic sand basins in the world.

(3) Typical Case—The Persian Gulf Basin

The Persian Gulf Basin is located within the Arabian Plate, between latitudes 13°–38° N and longitudes 35°–60° E, covering areas including Yemen, Oman, Saudi Arabia,

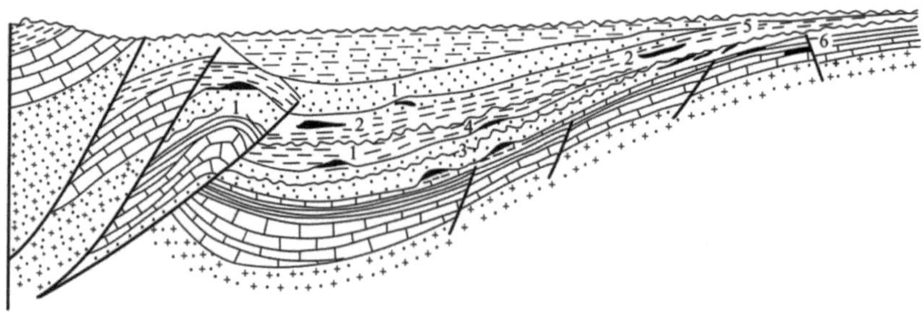

Fig. 7.10 Distribution pattern of oil and gas reservoirs in foreland basins. 1—Compressional anticlines, 2—lithological, 3—biological reef, 4—draped anticlines, 5—stratigraphic, 6—fault blocks

the United Arab Emirates, Qatar, Bahrain, Jordan, Israel, Palestine, Lebanon, Syria, Iraq, Kuwait, southeastern Turkey, and southwestern Iran, with a total area of 305 $\times$ 10^4 km^2. The Persian Gulf Basin is extremely rich in oil and gas resources. As of 2017, the proven reserves of oil and gas account for over 40% of the global reserves, with a total of 1692 oil and gas fields discovered.

The Persian Gulf Basin is a superimposed basin of a Paleozoic–Mesozoic passive continental margin basin and a Cenozoic large-scale foreland basin, composed of the eastern marginal slope of the Arabian Platform and the pre-mountain depression belt at the southwest foot of the Zagros Orogenic Belt. The basin structure is asymmetrical, with the basin axis trending northwest-southeast, roughly corresponding to the present-day Euphrates and Tigris River valleys, extending to the Persian Gulf. Above the Precambrian crystalline basement, the Persian Gulf Basin has deposited a thick sequence of Paleozoic to Cenozoic cover layers, thin in the west and thick in the east. The sedimentary thickness is less than 1500 m around the Arabian Shield and can reach up to 13,000 m at the Zagros foreland, as shown in Fig. 7.11. The Paleozoic sedimentary cover is mainly composed of clastic rocks, while the Mesozoic and Cenozoic are predominantly carbonate rocks.

In the Persian Gulf Basin, source rocks have developed in the Precambrian, Paleozoic, Mesozoic, and Cenozoic. Due to the low maturity of Cenozoic source rocks, the main source rocks in the basin are the Precambrian-Cambrian, Lower Silurian of the Paleozoic, and Middle Jurassic-Lower Cretaceous source rocks of the Mesozoic, with dozens of source rock series accumulated (Fig. 7.12). During the passive continental margin period, the basin environment was generally transgressive, with the development of deep-sea/shallow-sea carbonates, depositing extensive source rock series. The Precambrian-Cambrian source rocks are distributed in the salt basins within Oman, with developed strata being the Nafun Formation and Ara Formation of the Huqf Group. The Silurian hot shale is widely but discontinuously distributed in the basin, with strata including the Qusaiba Member of the Qalibah Formation of the Lower Silurian and its time-equivalent formations. Mesozoic source rocks are distributed in the Jurassic-Cretaceous shelf interior basins near the Persian Gulf, with developed strata including the Sargelu Formation of the Middle Jurassic, the Tuwaiq Mountain/Hanifa Formation of the Middle-Upper Jurassic, and the Diyab Formation of the Upper Jurassic, as well as the Garau Formation and Kazhdumi Formation of the Lower Cretaceous. Horizontally, the distribution of source rocks migrates from west to east with the progression of time. Vertically, the organic matter type shifts from sapropelic to humic from bottom to top. The TOC of the three main source rock series is generally greater than 2%, and the distribution of mature source rocks (R_o > 0.7%) is extensive, forming the material basis for the rich oil and gas accumulation in the Persian Gulf Basin.

The reservoir layers in the basin also have developed in more than ten sets, stacked vertically (Fig. 7.12). Typical representatives of carbonate reservoirs include the Lower Cretaceous Shuaiba Formation and the Upper Cretaceous Mishrif Formation. Typical representatives of sandstone reservoirs include the Lower Cretaceous Zubair Formation and the Upper Cretaceous Nahr Umr and Burgan Formations. Whether it is carbonate or clastic reservoirs, the physical properties of the reservoirs are good.

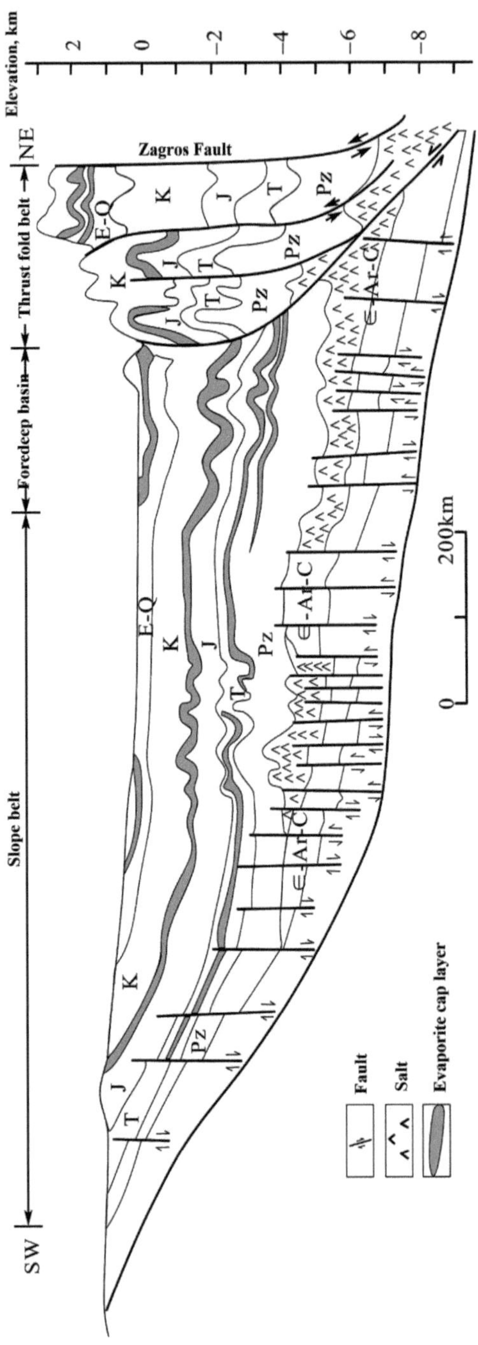

Fig. 7.11 Cross-section diagram of the Persian Gulf Basin (according to Zhang et al., 2013)

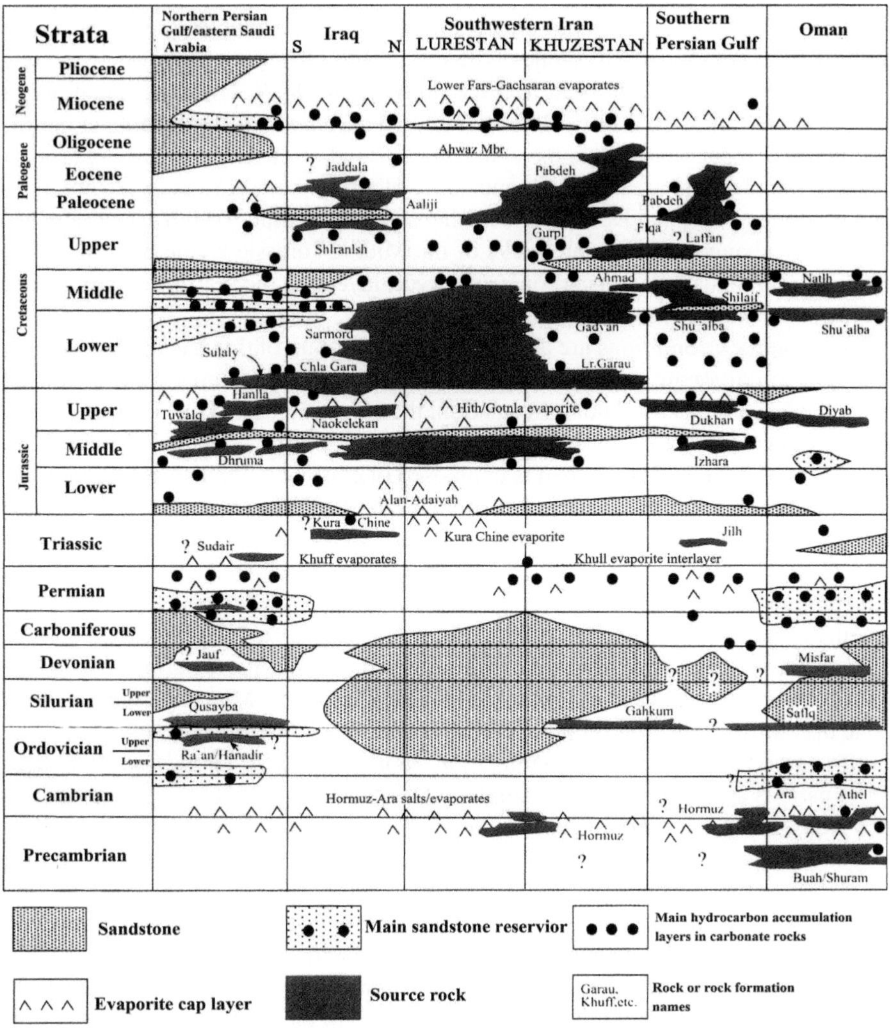

Fig. 7.12 Main source, reservoir, and cap rock systems in the oil and gas-rich areas of the Persian Gulf (Beydoun et al., 1992)

In the pre-mountain depression belt, the main reservoirs are the Oligocene–Miocene fractured limestones of the Tertiary, whose original fabric and reservoir characteristics have been strongly modified by tectonic actions. The reservoir capacity and production capacity have basically no connection with the original sedimentary facies and depend on the fracturing associated with folding. For example, the Asmari Limestone in southwestern Iran has an average porosity of less than 7%, but it has very well-developed structural fractures, with oil-yielding fracture widths ranging from 0.5 to 5 cm and connected lengths reaching 32–100 km. In the Kirkuk oil field in northern Iraq and the Gachsaran oil field in southwestern Iran, the reservoir layers are biological reef block limestones. The pre-reef limestones are recrystallized into a honeycomb structure, with an average porosity of 18–36% and permeability as high as (5–1,000) $\times 10^{-3}$ μm^2. Porous-fractured carbonate oil-bearing layers of the Middle and Upper Cretaceous have also been found at the edge of the depression. Towards the platform area, porous sandstones of the Lower Cretaceous are predominant in southern Iraq and Kuwait, and further southeast to Saudi Arabia, Bahrain, and Qatar, they change to carbonate rocks of the Middle Jurassic as the main reservoir layers. These sandstone and carbonate reservoir layers on the platform margin slope mainly store oil in pores.

Good caprock conditions are also an important feature of this basin. Multiple sets of evaporite, carbonate, and shale caprocks have developed within the basin. Among them, the Lower Triassic Sudair Formation, the Upper Jurassic Hith Formation, and the Miocene Gachsaran Formation are three typical regional evaporite caprocks. The pre-mountain depression belt is composed of extremely thick evaporites, marls, and limestones, forming a regional caprock up to 200–1500 m thick. The platform margin area also has gypsum layers and shale layers dozens of meters thick as caprocks. Spatially, regional and local caprocks are distributed in a superimposed and continuous manner, jointly sealing the oil and gas in the Persian Gulf Basin, with excellent sealing performance.

In different areas and stratigraphic levels of the basin, the combinations of the source, reservoir, and caprock layers vary. The Oligocene–Miocene Asmari Limestone in southwestern Iran serves as both a source and a reservoir layer, representing a self-sourced and self-stored combination. In northern Iraq, lateral transformation combinations are common. For example, in the Ain-Zala oil field, oil and gas generated from the radiolarian marl and shale of the Middle and Lower Cretaceous have migrated laterally into the limestone reservoirs of the Middle Cretaceous. At the edge of the Arabian Platform, the oil-bearing limestone of the Upper Jurassic Arab Formation and its underlying Zubair source rock are in a cyclical combination.

The Persian Gulf Basin has developed multiple hydrocarbon migration systems, including faults and fractures. Oil and gas generated from the lower source rocks migrate laterally and vertically over short distances to the upper reservoir layers and are sealed by regional caprocks. Oil and gas from the pre-passive continental margin source rocks migrate through faults to the Upper Permian-Lower Triassic reservoir layers for accumulation. Oil and gas from the passive continental margin source rocks gather in the Oligocene–Miocene through the fracture network formed during the foreland period and are sealed by the overlying layers.

The discovered oil and gas reservoirs in the Persian Gulf Basin are mainly structural oil and gas reservoirs, with large oil fields mainly distributed in the slope belt and the foreland depression belt. For example, in the slope belt and foreland depression belt at the edge of the Arabian Platform, close to the Persian Gulf in the Hasha structural stage area, many of the world's largest oil fields have been discovered. They are huge domes and short-axis anticlines with a near NS trend and gentle dips, rarely exceeding 6°–7°. The uplift height ranges from 300 to 700 m, and some are locally disrupted by faults. The largest oil fields in Saudi Arabia and Kuwait belong to these uplift belts, such as the An-Nara oil-bearing structural belt in Saudi Arabia (the Ghawar giant anticline oil field is composed of the high points of Haradh, Hawiyah, Imam, Shedgum, and Ain Dar in this structural belt) and the Abu Qaiq-Qatif oil-bearing structural belt. These uplift structures may be related to basement faults and salt diapirs. In the Zagros foreland depression belt and the thrust-fold belt, oil and gas fields are distributed in a strip-like pattern along the NS direction. The axes of the oil and gas fields are mostly consistent with the structural trend, mainly featuring anticlines and fault traps formed by tectonic compression. For example, the famous Gachsaran oil field was formed in a structural trap caused by lateral compression. The oil field appears on the surface as a distinctly asymmetric plunging anticlinal structure.

In summary, the Persian Gulf Basin is a superimposed basin of a Paleozoic–Mesozoic passive continental margin basin and a Cenozoic large-scale foreland basin, with extremely abundant oil sources, good reservoir conditions, thick and dense regional distribution of caprocks, and numerous huge oil-bearing structures. It is these favorable conditions that coexist, forming an extremely rich and unparalleled giant petroliferous basin.

7.3.3.2 Cratonic Basins

(1) The Concept and Characteristics of Cratonic Basins

In the concept of plate tectonics, a craton mainly refers to the part of the continental plate that can be approximated as a rigid block, representing a stable continental mass. A sedimentary sequence characterized by a wide area, irregular shape, relatively slow subsidence rate, and mainly composed of sags, formed based on a craton, is called a cratonic basin.

Cratonic basins can be divided into two major categories based on their position in the large-scale tectonic setting: intra-cratonic basins and marginal cratonic basins. They can also be classified into single-cycle cratonic basins and multi-cycle cratonic basins according to their development. The former is mainly characterized by Paleozoic marine deposits, with a lack of Mesozoic and Cenozoic cover, such as in the areas of western Hubei, Yunnan-Guizhou-Guangxi, etc. The latter has the Paleozoic marine deposits as the first cycle and the Mesozoic and Cenozoic as the second cycle, such as the Ordos, Sichuan, and Tarim basins. The Paleozoic prototype basins are mostly subjected to later deformation and modification, and what is preserved now are mostly residual basins. Large cratonic basins are often multi-periodic, with each period developing a composite of different types of prototype basins.

(2) Petroleum Geologic Characteristics of Cratonic Basins

Approximately 25% of the world's oil and gas is distributed in cratonic basins, and the proportion of natural gas reserves is much higher than that of oil reserves. Cratonic basins where large-scale oil fields have been discovered include the West Siberian Basin in Russia, the Illinois Basin, Williston Basin, and Michigan Basin in North

America, the Ilizi Basin and Goudermis Basin in Africa, the Northwest German Basin in Europe, and the Cooper Basin in the Asia–Pacific region. The main parts of the Sichuan, Ordos, and Tarim basins in China belong to cratonic basins.

The geological conditions of cratonic basins determine that they have a complex history of oil and gas accumulation. Multiple sets of source rocks, multiple reservoirs, various types of traps, and multiple oil and gas combinations are important characteristics of the oil and gas areas in cratonic basins.

Source rocks are distributed from the Cambrian to the Cretaceous in these basins, with lithologies mainly consisting of mudstone, shale, and carbonate rocks, etc. The thickness of the source rocks varies greatly, generally ranging from 20 to 1000 m. The distribution of organic matter in different basins differs significantly.

In cratonic basins without underlying rifts, sedimentation and subsidence remain synchronous. During the development of the basin, a faster subsidence rate forms a starved intra-cratonic basin, surrounded by carbonate banks and deltaic margins. Rapid subsidence leads to the distribution of reservoirs along the basin margins and the formation of typical deltaic and coastal sandstones at the basin edges, as well as carbonate banks and platforms associated with biological reefs. In cratonic basins with underlying rifts, rift action forms grabens and tilted blocks, both of which are distributed with reservoirs. In addition, fracture reservoirs in the basin are also important, with increased fracturing enhancing the porosity or permeability of the reservoirs.

The rapid subsidence periods of cratonic basins often coincide with the deposition of caprock strata. Vertically, caprocks and reservoirs form various reservoir-cap combinations. Laterally, reservoirs can transform into impermeable rock layers, creating lateral reservoir-cap combinations.

Oil and gas traps in cratonic basins are mainly stratigraphic-structural composite traps, as well as basement-related buried hill traps, (new) structural traps above basement uplifts, and lithologic traps. Under conditions where fracture systems exist and sea-level changes relatively rapidly, many laterally discontinuous reservoirs often develop, causing the oil and gas in cratonic basins to be relatively dispersed and separated.

Oil and gas in cratonic basins often undergo long-distance migration and exhibit various migration patterns, such as radial, vertical, and long-distance lateral migration. The three factors of sedimentary compaction within the basin, topographic relief, and the carrier beds for long-distance migration jointly promote the long-distance lateral migration of oil and gas.

The oil and gas accumulation patterns in cratonic basins are complex, with the issue of optimal accumulation timing. After oil and gas accumulation, relatively stable tectonic conditions, regional caprocks, and sealed hydrogeological conditions are the basic conditions for their effective preservation.

(3) Typical Case—The West Siberian Basin

The West Siberian Basin is located between the Ural Mountains and the Yenisei River in northwestern Asia and is one of the world's most petroliferous basin. It is also the largest known cratonic petroliferous basin, with an area of about 350×10^4 km^2. The

West Siberian Basin is currently Russia's main oil and gas production base, with oil fields mainly distributed in the central and southern areas, while gas fields are mainly located in the northern and western areas. The basin's oil production accounts for about 70% of Russia's total oil production, and its natural gas production accounts for about 90% of Russia's total natural gas production. Among the discovered gas fields in the basin, there are 8 giant gas fields with reserves exceeding one trillion cubic meters, including the Urengoy gas field with reserves of 10.2×10^{12} m^3, which is also the world's second-largest gas field, as well as the Yangpo, Medvezhye, and Bovanenkovskoye giant gas fields.

In terms of tectonics, the West Siberian Basin is bordered by the Ural and Novaya Zemlya uplifts to the west, the Siberian Craton and Taimyr uplift to the east, the Kazakh and Altai-Sayan Ridge uplifts to the south, and the Kara Sea to the north. It is an overall north-dipping platform-like monocline. The basin's basement is composed of Proterozoic-Paleozoic strata from the Baikal to the Hercynian periods, with peripheral folded mountain ranges extending into the basin from different directions, forming the basement of different periods. Based on the basement, the basin has experienced stable platform-type sedimentation during the Mesozoic-Cenozoic era (Fig. 7.13). The main productive layers of the basin are mainly from the Cretaceous system, with a wide range of production depths. The depth range of 1500–3000 m concentrates more than 98% of the recoverable oil reserves of the large oil fields in the basin.

The West Siberian Basin mainly develops multiple mature argillaceous source rocks, including the Lower-Middle Jurassic, Upper Jurassic, and Upper Cretaceous. The Jurassic-Cretaceous source rocks belong to the marginal marine basin deposits. Influenced by the sedimentary environment and sediment source, the organic matter abundance and type of the source rocks gradually improve from the basin margin to the center. The source rocks in the northern part of the basin have a higher maturity,

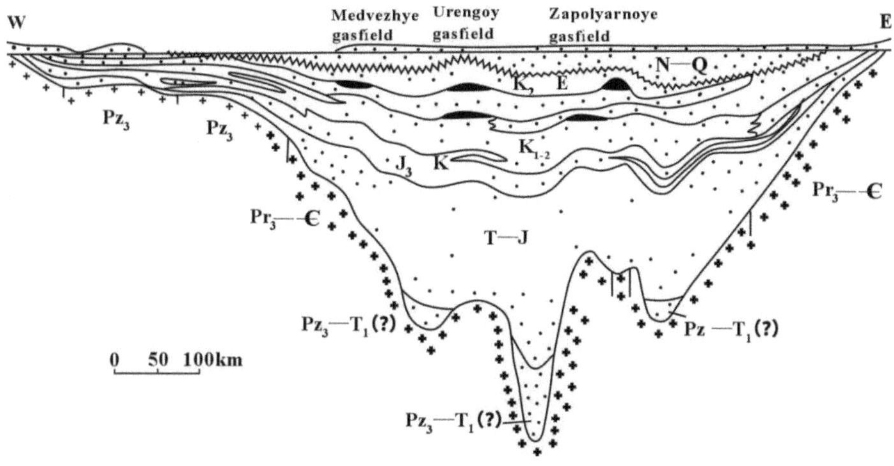

Fig. 7.13 Northern Cross-Section of the West Siberian Basin

reaching the gas-generating stage; the source rocks in the southern part of the basin have high organic matter content, moderate maturity, and great thickness, which is an important reason for the enrichment of oil.

Affected by the sedimentary environment and diagenetic secondary action, the basin forms various types of reservoir rock bodies, such as large-scale deltaic sandstone bodies and coastal sandbars. They are buried at a moderate depth, with good physical properties of the reservoir bodies, which are conducive to the formation of high-yield super-large oil and gas fields. The marine carbonate rocks of the Devonian-Lower Carboniferous, after the secondary weathering leaching in the Jurassic, form karst-fracture-type carbonate reservoir bodies with good reservoir conditions. The Lower-Middle Jurassic is a set of interbedded continental sandstone and mudstone, with a formation thickness of 0–800 m and great lithological changes, belonging to medium-porosity and low-permeability sandstone and siltstone layers. The Cretaceous is a set of interbedded sandstone and mudstone, with a formation thickness of 300–600 m and distributed throughout the area.

During the Mesozoic-Cenozoic era, the basin developed multiple sets of shale caprocks, among which the shale of the Upper Jurassic and the shale of the Upper Cretaceous-Paleogene form the regional caprocks of the basin.

The favorable structures within the basin include large-scale uplift areas, placanticline anticlinal belts, and local structural areas. The uplift area in the central part of the basin and the placanticline structural belt has the highest degree of oil and gas enrichment and are the main places for the formation of giant oil and gas fields. In the contemporaneous structures, some reservoirs are missing at the top or on the flanks of the structures, forming structural-lithologic composite traps. Representative giant gas fields such as Urengoy, Bovanenkovskoye, and Yangpo belong to this category. Vertically, the basin mainly develops four oil and gas combinations of the Lower-Middle Jurassic, Upper Jurassic, Lower Cretaceous, and Upper Cretaceous. The first two oil and gas combinations are mainly oil reservoirs in the central part of the basin and gas and condensate gas reservoirs in the marginal zones. The Lower Cretaceous is the most important oil and gas combination in the basin, with oil production mainly in the south and gas production mainly in the north; the Upper Cretaceous is the main gas-bearing combination in the northern part of the basin.

7.3.3.3 Rift Basins

(1) The Concept and Characteristics of Rift Basins

A rift basin refers to the valley with a symmetric or asymmetric central deep depression that develops in the crust when the lithospheric plates move horizontally apart or the mantle upwells. It is formed due to mantle upwelling, crustal thinning, horizontal tension, and strike-slip action. During the process of plate movement, certain specific areas within the continental lithospheric plates may be subjected to extension, leading to the thinning of the continental lithosphere and the formation of rift-depression basins or rift basins. If the extension causes the entire continental lithosphere to split, a continental rift is formed. As the continental rift further expands and stretches, and a new oceanic crust forms at the expansion center, the continental rift will evolve towards an intercontinental rift, passive continental margin, and oceanic rift. In a triple rift system, one branch may be restricted in its extension by the expansion and stretching of the other two rifts during the process, and this branch of the continental rift will regress and evolve into a trough.

Under the influence of different levels of fracture action, rift basins generally exhibit a tectonic pattern of alternating depressions and uplifts, with adjacent concave and convex structures, and have all experienced a transition from rapid subsidence to stable subsidence. Typically, the rift phase of rapid subsidence is filled with a large amount of sediment. Under favorable conditions of source, reservoir, and caprock, the generated oil and gas enter anticlinal, fault block, unconformity, and lithologic traps to form oil and gas reservoirs.

(2) Petroleum Geologic Characteristics of Rift Basins

Rift basins are extremely important petroliferous basins, and significant progress has been made in oil and gas exploration in such basins both domestically and internationally, with abundant oil and gas resources proven. The Gulf of Mexico Basin, North Sea Basin, Sirte Basin, Dnieper-Donets Basin, Kutei Basin, and Gulf of Suez Basin are all rift basins rich in oil and gas resources. In the eastern part of China, Mesozoic and Paleogene-Neogene rift basins are well-developed, such as the Songliao Basin and Erlian Basin, which are Mesozoic fault-bound basins, and the Bohai Bay Basin, Jianghan Basin, as well as the East China Sea Basin, Pearl River Mouth Basin, Qiongdongnan Basin, Yinggehai Basin, and Beibu Gulf Basin off the coast of China, which are mainly Paleogene fault-bound basins.

The oil and gas potential of rift basins depends on the favorable combination of conditions such as the development of source rocks, reservoir-cap combinations, overburden sufficient to mature the source rocks, traps, and the preservation of oil and gas reservoirs. In the world's major rift basins, source rocks are distributed from the Cambrian to the Paleogene, with lithologies mainly consisting of mudstone, shale, and carbonate rocks. Source rocks containing a large amount of aquatic-derived organic matter mainly formed during the main period of rift basin development, characterized by great thickness, high abundance, wide distribution, and diverse types of source rocks. Due to the high geothermal background, the time required for organic matter to evolve into hydrocarbons is relatively short, and the oil-generating threshold depth is relatively shallow (generally less than 2700 m).

The sedimentary characteristics of rift basins are controlled by the tectonic evolution and development degree of the basin. In depression-type rifts, reservoirs develop on a large scale, are laterally stable, and have high maturity under stable sedimentary conditions. In fault-bound basins, reservoirs develop on a small scale, have large lateral changes, and have multiple genetic types of reservoirs under the action of block faulting. The main types of caprock in rift basins include mudstone, shale, salt rock, gypsum, and dense carbonate rocks. The source-reservoir-cap combinations differ greatly at different stages of rift basin development. In the pre-rift stage, the combination of new-born and ancient-stored is mainly present. In the rift fault-bound stage, the combination of self-born and self-stored is predominant. In the post-rift stage, the combination of ancient-born and new-stored is mainly present.

The overall oil and gas migration in rift basins is mainly vertical with lateral migration as a supplement. Positive structural belts often have vertical migration as the main feature, while sags and slope areas have lateral migration as the main feature. In fault-bound rift basins, the fracture system is well-developed, and vertical oil and gas migration is very active, with characteristics of multi-phase migration and accumulation, redistribution, and multi-phase reservoir formation. Oil and gas often migrate upward along fractures and accumulate on both sides of the fractures. The oil-and gas-bearing intervals in the vertical direction are long, generally ranging from tens of meters to hundreds of meters, and even exceeding 2000 m, thus forming complex oil and gas accumulations in positive structural belts.

Rift basins have a variety of oil and gas reservoir types, including anticlinal oil and gas reservoirs, fault block oil and gas reservoirs, lithologic oil and gas reservoirs, stratigraphic unconformity oil and gas reservoirs, and stratigraphic onlap oil and gas reservoirs. The differential subsidence and multi-cyclical sedimentation determine the widespread development of stratigraphic and lithologic traps in the sags and slope areas of rift basins. In the sags and slope areas of oil-rich basins, multi-series, large-area stratigraphic and lithologic oil and gas reservoirs are often produced in complex, continuous, and superimposed patterns. Together with the complex oil and gas accumulation belts in the positive areas, they constitute the oil and gas distribution pattern of full-sag oil-bearing and complex-superimposed reservoir formation.

(3) Typical Cases

a. Depression-Type Rift Basin Example—Songliao Basin

The basement of depression-type rift basins is relatively stable, often being areas of long-term developed depressions and stable subsidence. Shallow layers commonly develop Mesozoic source rock series, while deep layers often have Paleozoic or marine source rock series of the middle and upper Proterozoic, which mostly show gradual onlap contact with the exposed crystalline basement around the basin. The basin contains large-scale secondary structural belts such as anticlines, placanticline, uplifts, domes, and slopes. Due to the influence of the surroundings, steep anticlines are generally located at the basin margins, gentle domes and linear uplifts are in the central part of the basin, and slopes and flexures are mostly distributed in the large monoclines. Huge oil and gas fields can be formed in the central part or marginal slopes of the basin.

The Songliao Basin is the largest petroliferous basin in Northeast China, extending in an NNE direction with an area of about 226×10^4 km^2. The basin basement mainly consists of Paleozoic metamorphic sandstone, marble, slate, and phyllite, with large-scale Indosinian-early Yanshanian, Hercynian, and Caledonian granite intrusions. The sedimentary cover is mainly composed of Mesozoic-Cenozoic sedimentary rocks, with a cumulative maximum thickness of ten thousand meters. The sedimentary cover from bottom to top is the Upper Jurassic, Cretaceous, Paleogene, and Neogene.

The Songliao Basin is a rift basin mainly characterized by depression-type subsidence. During the late Jurassic fault-bound development stage, a series of graben-like fault-bound belts were formed, developing a set of coal-bearing constructions of fluvial-swamp facies, which were in a high-temperature evolution stage and had certain hydrocarbon-generating capabilities. In the early Cretaceous, the basin entered a large-scale depression-type development stage, and the deposition period of the Nenjiang Formation was the peak period of lake basin development. The basin mainly had three sedimentary systems in the early Cretaceous, with scales reaching thousands or even tens of thousands of square kilometers. The sedimentary systems were mainly composed of fluvial-deltaic-lacustrine systems, with reservoirs mainly consisting of fluvial sand bodies and delta front sand bodies. Large deltas surrounded the lake basin, extending towards the center, providing widely distributed reservoir conditions. Multi-cyclical sedimentation led to the development of multiple source-reservoir-cap combinations in the basin, centered around the main hydrocarbon-generating source rocks of the first section of the Lower Cretaceous Qingshankou Formation, the first section of the Nenjiang Formation, and the second section of the Nenjiang Formation. The basin experienced multiple tectonic movements during the rift development period, forming various structural belts or local structures. The anticlinal traps had a wide and gentle shape and a large area, providing large-scale trap conditions for oil and gas accumulation. In addition, there were various types of nose-like structural traps, stratigraphic traps, and lithologic traps.

Multiple oil-generating sags have developed in the Songliao Basin, and various types of oil and gas reservoirs are mostly distributed around the sags in a ring shape. The Daqing Placanticline Uplift is located between two sags and is the best place for oil and gas accumulation. In each sag, oil and gas reservoirs are regularly distributed from the center to the edge: the center of the sag is dominated by lithologic oil and gas reservoirs, moving outward to fault-nose structural oil and gas reservoirs and fault-lithologic composite oil and gas reservoirs, and the edge of the sag is dominated by anticlinal, fault block oil and gas reservoirs or gas reservoirs (Fig. 7.14). Fault oil and gas reservoirs are mainly controlled by complex fracture belts and are distributed in a belt-like pattern.

b. Fault-Bound Rift Basin-Bohai Bay Basin

The Bohai Bay Basin is in the Bohai Sea area and its coastal regions in the eastern part of China, spanning provinces and cities such as Liaoning, Hebei, Henan, Shandong, and Beijing and Tianjin. It extends north to Shenyang, south near Kaifeng, west to the Beijing-Shijiazhuang line, and east to the Weifang-Yingkou line. It is surrounded by the Yanshan, Taihang, western Shandong, and Jiaoliao mountains. The basin is 2,600 km long from north to south and 1,200 km wide from east to west, with a total area of about 20×10^4 km^2 (Fig. 7.15). It includes the Jizhong, Huanghua, Linqing-Dongpu, Liaohe, Bozhong, and Jiyang depressions.

The Bohai Bay Basin is a typical fault-bound rift basin. During the fault-bound development stage, the basin has a strong segmentation, with a tectonic pattern of multiple domes, multiple sags, multiple uplifts, and multiple depressions. The sags are mostly separated by uplifts, and each sag is an independent oil-generating unit.

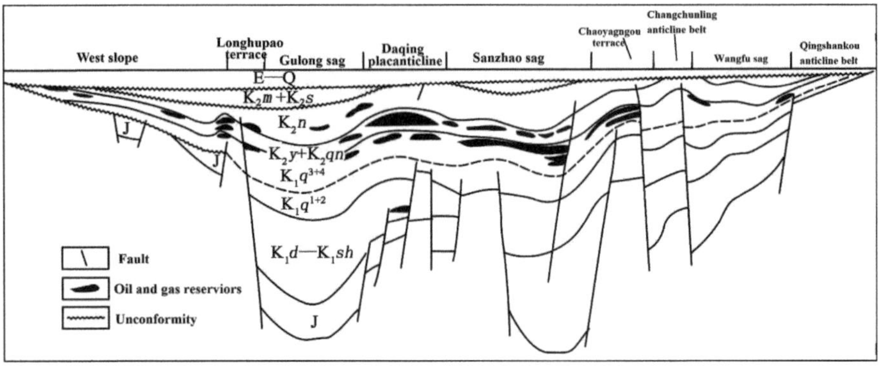

Fig. 7.14 Distribution pattern of oil and gas reservoirs in the Songliao Basin (according to Zhao et al., 2004)

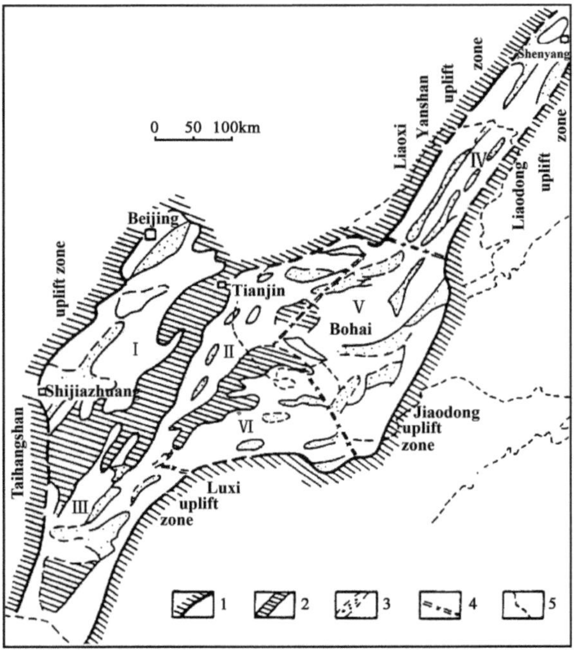

Fig. 7.15 Structural division map of the Bohai Bay Basin (according to Qi et al., 1995). 1—Basin boundary, 2—uplift, 3—convexities within the depression, 4—approximate boundary between structural units, 5—coastline, I—Jizhong depression, II—Huanghua depression, III—Linqing-Dongpu depression, IV—Liaohe depression, V—Bozhong depression, VI—Jiyang depression

The fault activity within the sags is very intense, often with one side of the main fracture belt strongly active, controlling the development of the sag and forming a graben-like sag. The interior is further cut by secondary fractures, presenting many basement-tilted fault blocks. They are numerous and have large undulations, which are conducive to the formation of stratigraphic onlap, unconformity, and buried hill traps.

Small sags and large subsidence amplitude are one of the important characteristics of the Bohai Bay Basin. During the long-term development and evolution, the derivative fractures within the sags control the formation of sedimentation, secondary structural belts, and local structures, such as draping anticlines, rollover anticlines, and salt (mud) arch anticlines. Moreover, multiple phases and groups of normal faults with different attitudes have transformed the existing anticlinal traps into fault-related anticlines or fault block clusters. Therefore, fault block traps are the most widely developed basic type.

Due to the strong segmentation of the basin, with sags as units, multiple types of sedimentary systems have developed. Each system is small in scale, only tens to hundreds of square kilometers, with small sand bodies that often have large lateral changes and continuous superposition in the vertical direction. The main sedimentary facies types are alluvial fans, fan deltas, deltas, beach bars, lacustrine fans, and turbidite facies. Multiple phases of block faulting and frequent transgression and regression of the lake basin have led to the extensive development of multiple sets of oil- and gas-generating and reservoir rock series and the formation of favorable source-reservoir-cap combinations. The multi-source, near-source, and rapid accumulation of various sedimentary systems extend from the margins to the lake basin center, inserting into the hydrocarbon-generating area, forming large-area, multilayer superimposed reservoir distribution characteristics and presenting various types of lithologic traps. The mutual configuration of these geological factors allows the abundant organic matter to form hydrocarbons and, through short-distance migration, accumulate into reservoirs, resulting in the widespread distribution of various types of oil and gas reservoirs. The overall Paleogene sag sedimentation also plays a good role in preserving the Cenozoic oil and gas reservoirs, making the oil and gas reservoirs in the Bohai Bay Basin not only diverse and widely distributed but also very rich.

The characteristics of oil and gas reservoir formation in the steep slope belt, sag belt, and gentle slope belt of the basin are different (Fig. 7.16). The steep slope belt is the abrupt transition zone between the sag belt and the uplift, close to the source area, with well-developed underwater fans and alluvial fans, and stratigraphic onlap is common, with developed faults. The inner side has a rolling anticline belt and lithologic updip pinchout belt distributed on the downthrown block of the coeval fault, while the outer side has developed fault block traps and stratigraphic traps. The oil source conditions are good, which is conducive to the formation of various types of trap oil and gas reservoirs. The marginal zone is distributed with stratigraphic onlap oil and gas reservoirs, buried hill oil and gas reservoirs, and fault-lithologic oil and gas reservoirs.

The sag belt is the area with the most favorable oil source conditions inside the fault-bound basin. The positive structural belts developed inside not only have structural traps and vertical fault migration channels conducive to oil and gas accumulation but also have abundant oil and gas sources from surrounding multiple sags, which is very conducive to the formation of oil and gas reservoirs. The types of oil and gas reservoirs mainly include various anticlinal oil and gas reservoirs, fault oil and gas reservoirs, fault-lithologic oil and gas reservoirs, and buried hill oil and gas

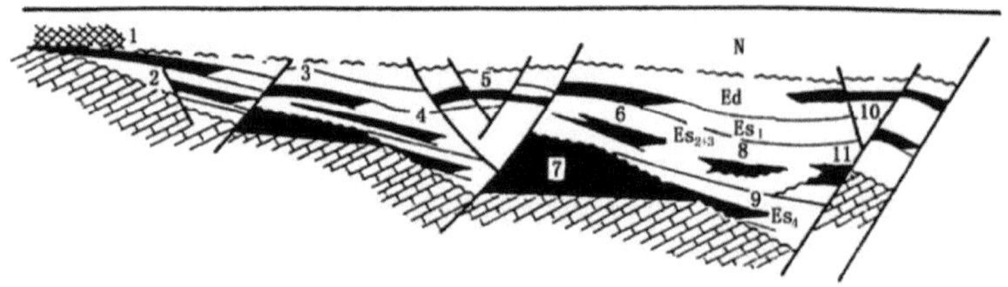

Fig. 7.16 Distribution pattern of oil and gas reservoirs in faulted basins (according to Hu Jianyi). 1—Stratigraphic unconformity (or asphalt seal) oil and gas reservoirs, 2—fault block lithological oil and gas reservoirs, 3—drape anticlinal oil and gas reservoirs, 4—clastic limestone lithological oil and gas reservoirs, 5—compressional anticlinal oil and gas reservoirs, 6—lithological updip pinch-out oil and gas reservoirs, 7—buried hill oil and gas reservoirs, 8—lens-shaped sandstone lithological oil and gas reservoirs; 9—stratigraphic onlap oil and gas reservoirs, 10—rolling anticlinal oil and gas reservoirs. 11—Fault-lithological oil and gas reservoirs

reservoirs. The sag belt has the best oil source conditions, and hidden traps such as sandstone lenses and fault-lithologic are more developed, which is conducive to the formation of oil and gas reservoirs with high oil and gas abundance.

The gentle slope belt is the part of the fault-bound basin where the basement is shallowly buried and the sedimentary cover is thin. The basement fractures and coeval fractures often coexist, and buried hill oil and gas reservoirs and draping anticline oil and gas reservoirs are commonly seen in the deep part. Stratigraphic unconformity oil and gas reservoirs or asphalt-sealed unconformity oil and gas reservoirs are developed on the outer side of the coeval fractures, and the inner side downthrown block often develops rolling anticline oil and gas reservoirs, followed by fault-lithologic oil and gas reservoirs and stratigraphic onlap oil and gas reservoirs.

7.4 Petroleum Systems

The concept of petroleum systems has become an essential way of thinking and a tool for effectively predicting and discovering oil and gas resources. Petroleum systems study the evolution history of basins and the formation of oil and gas reservoirs using systems theory, evolving from the static conditions of oil and gas reservoir formation such as source, reservoir, seal, migration, trap, and preservation, to dynamic analysis of the formation process of oil and gas reservoirs. It has shown significant importance in reducing risks, lowering costs, and improving exploration efficiency.

7.4.1 The Concept and Connotation of Petroleum Systems

(1) The Concept of Petroleum Systems

The concept of oil and gas bearing systems was first introduced by W. D. Dow in 1972 at the AAPG Annual Convention, referred to as "petroleum systems." W. D. Dow, based on oil-source correlation, proposed the concept of a source-reservoir system, suggesting that each petroleum system comprises a set of source rocks and a group of reservoir rocks, sealed by cap rocks and isolated from other petroleum systems. In 1987, L. B. Magoon introduced the concept of petroleum systems, and for the first time, used the term "elements," explaining that these elements "must have the appropriate spatial and temporal configuration to enable oil accumulation." In 1994, the book "Petroleum Systems—From Source Rock to Trap," edited by Dow and Magoon, was published, systematically introducing the concept, classification, research content, research methods, and applications of petroleum systems.

As early as the mid-1960s, Chinese scholars proposed the concept of "oil-generating systems," pointing out that an "oil-generating system" is a whole composed of oil sources, reservoir layers, cap rocks, traps, and other oil-generating elements connected by unified oil and gas migration and accumulation processes in various periods (Hu & Liao, 1996), encompassing the basic connotations of foreign oil systems.

A petroleum system refers to mature source rocks and all oil and gas reservoirs formed in relation to these source rocks, including all necessary geological elements and processes for the formation of oil and gas reservoirs. The term "oil and gas" includes all highly concentrated hydrocarbon substances in any state; "system" refers to the interdependent geological elements and processes in the oil-generating process. "Geological elements" include static elements such as oil and gas source rocks, reservoir rocks, cap rocks, and overlying strata. "Geological processes" refer to the formation of traps and the processes of hydrocarbon generation, migration, and accumulation. These basic elements and processes cooperate with each other in time and space to enable the organic matter in the source rocks to be transformed into oil and gas accumulations. Only with the simultaneous presence of these geological elements and processes can a petroleum system be constituted.

(2) The Connotation of Petroleum Systems

The essential attribute of a petroleum system is that it is an integrated natural fluid movement system. In general, the connotation of petroleum systems includes the following aspects:

Emphasizing the importance of the quality and effectiveness of hydrocarbon generation in source rocks, which is the basis for the formation of petroleum systems.

Emphasizing the division of petroleum systems centered on hydrocarbon generation. That is, an effective set of source rock layers forms a petroleum system.

Considering petroleum systems as an oil and gas geological unit between petroliferous basin and oil and gas-bearing zones (Fig. 7.17). It has a certain spatial distribution range and a certain period of evolution. This geological unit is centered on

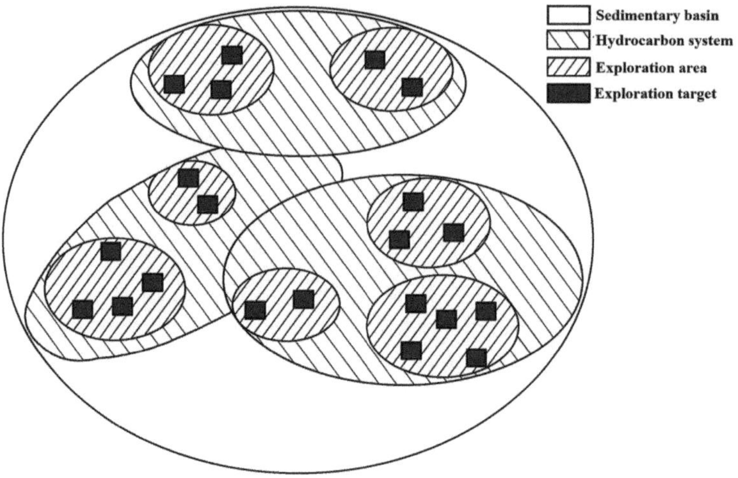

Fig. 7.17 Four levels of oil and gas survey (according to Jiang et al., 2002)

effective source rocks, and the boundary of the unit is the maximum outer boundary of hydrocarbon migration generated by the source rocks (Jiang et al., 2002).

Emphasizing that the purpose of dividing petroleum systems is to determine the basic units for oil and gas resource evaluation, to improve the reliability of oil and gas geological evaluation, and the research results can serve as the basis for strategic selection of oil and gas exploration areas.

7.4.2 Research Content of Petroleum Systems

Petroleum systems are objectively existing hydrocarbon fluid systems in nature, occurring and existing in specific geological bodies, and they have concepts of space and time. Petroleum systems have their specific regions, stratigraphic distribution, and time ranges, and their research content can be summarized as the description of static geological elements, the determination of geological processes, and the establishment of their interrelationships.

(1) Description of Static Geological Elements

The static geological elements of petroleum systems mainly include source rocks, reservoir layers, cap rocks, conduits, and overlying strata, which are the basic geological elements for oil and gas accumulation. Source rocks, reservoir layers, and cap rocks are the most fundamental elements for the existence of petroleum systems. Overlying strata refer to all rock layers deposited above effective source rocks before the critical moment, which not only serve as necessary conditions for the thermal maturation of source rocks but also have a certain influence on the geometry of migration pathways and traps in the underlying strata.

(2) Description of Geological Processes

Basic elements are the material basis of petroleum systems, and establishing the combination relationship of static elements in space and time is the description of the geological processes and actions of petroleum systems. The geological processes of petroleum systems refer to the oil and gas accumulation process "from source rocks to traps," including the analysis of the generation and expulsion process of oil and gas, the migration and accumulation process of oil and gas, the formation and evolution process of traps, and the modification process after the formation of oil and gas reservoirs. These processes can be studied through methods such as basin simulation, organic geochemistry, and fluid history analysis.

The analysis between oil and gas geological elements and geological processes should choose a time for reference, that is, the critical moment of the petroleum system. The critical moment refers to the time when most of the oil and gas in the petroleum system are generated-migrated-accumulated, which is the formation period of the main oil and gas reservoirs. It is based on the burial history curve of strata, and the time–temperature index (TTI value) can show the generation time of most hydrocarbons. From a geological perspective, the migration and accumulation of oil and gas occur in a short period of time, which usually happens slightly later than the maximum burial depth of source rocks, that is, the critical moment.

(3) Key Diagrams for Petroleum System Research

The key diagrams for petroleum system research include: the burial history curve of the petroleum system, the plan and section distribution diagrams of the petroleum system at the critical moment, and the event diagram of the petroleum system.

a. The Burial History Curve of Petroleum Systems

The burial history curve can be used to illustrate the critical moment, era, and basic elements at a given location. The era refers to the geological period of the petroleum system; the basic elements refer to source rocks, reservoir rocks, cap rocks, and overlying strata.

Figure 7.18 shows the burial history curve of the Deer-Boar (*) petroleum system. All rock layers in the figure are fictional. The Deer shale is the source rock, the Boar sandstone is the reservoir rock, the George shale is the cap rock, and the rocks above the Deer shale are the overlying strata. The figure shows that the source rock entered the oil-generating window at 260×10^6 years ago in the Permian period, reached its maximum burial depth at 255×10^6 years ago, and the critical moment was at 250×10^6 years ago. The time for oil and gas generation, migration, and accumulation was from $(260–240) \times 10^6$ years, which is also the time of the petroleum system.

b. Plan and Section Distribution Diagrams of Petroleum Systems

The regional distribution range of the petroleum system at the critical moment is defined by the boundary of active source rocks and all conventional and unconventional oil and gas reservoirs and shows derived from these source rocks.

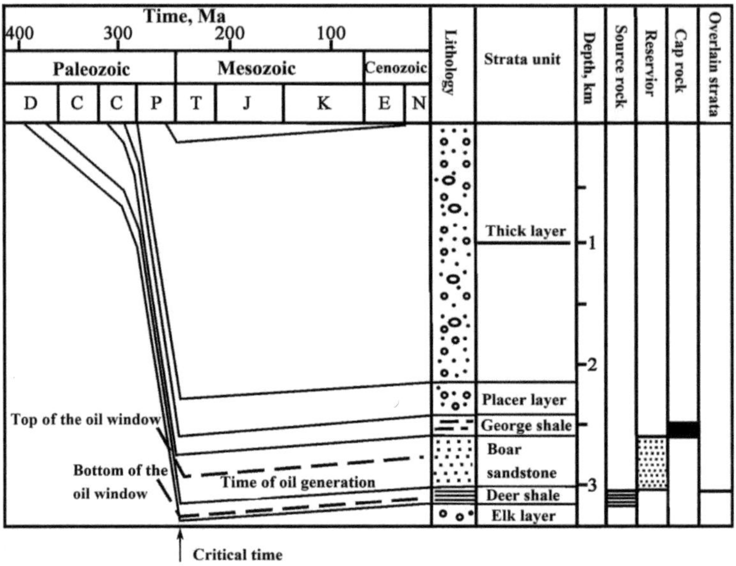

Fig. 7.18 Burial history curve of the Deer-Boar (*) petroleum system (according to Magoon & Dow, 1994)

Figure 7.19 shows the plan view of the Deer-Boar (*) petroleum system at the end of the Paleozoic era at the critical moment. The area within the oil- and gas-generating window is the effective source rock, and the outside is the immature source rock. The regional distribution range of the petroleum system is defined by a line that encloses the effective source rocks and all relevant discovered oil and gas shows.

Figure 7.20 shows the section view across the regional distribution range of the petroleum system. The figure illustrates the stratigraphic distribution and the spatial relationships of the basic elements such as source rocks, reservoir rocks, cap rocks, and overlying strata at the critical moment of the Deer-Boar (*) petroleum system. The source rocks below the top of the oil-generating window are active source rocks, and those above are immature source rocks. Oil and gas are stored in the reservoir rocks, which are sealed by cap rocks.

c. Event Diagram of Petroleum Systems

Figure 7.21 shows the event diagram of the Deer-Boar (*) petroleum system. The figure displays eight different events: the first four events record the sedimentation times of several basic elements, and the next two events record the formation process of the petroleum system. The formation time of traps is determined based on geophysical data, various geological data, and structural geological analysis. The generation-migration-accumulation of oil and gas, or the time of the petroleum system, is determined based on stratigraphic and oil and gas geochemical research and burial history diagrams.

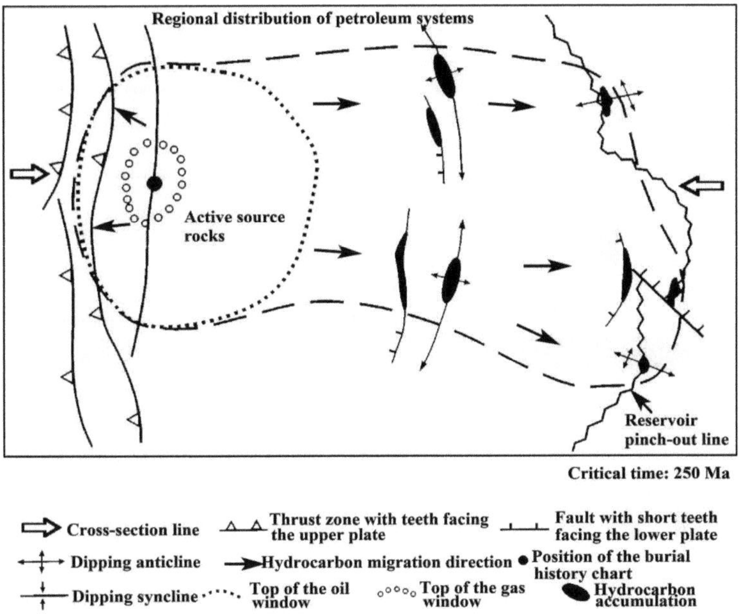

Fig. 7.19 Regional distribution of the Deer-Boar (*) petroleum system at the critical moment (according to Magoon & Dow, 1994)

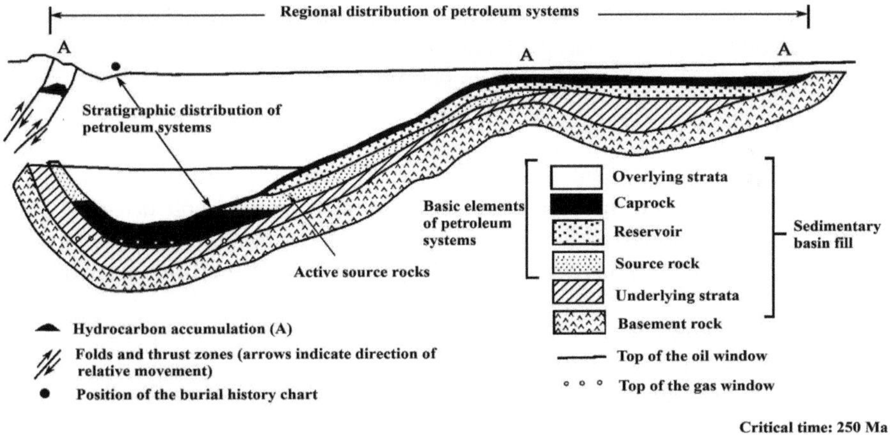

Fig. 7.20 Stratigraphic distribution of the Deer-Boar (*) petroleum system at the critical moment (according to Magoon & Dow, 1994)

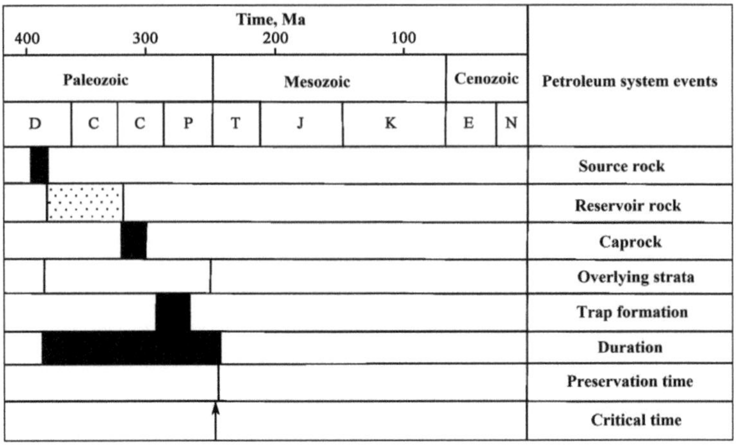

Fig. 7.21 Event chart of the Deer-Boar (*) petroleum system (according to Magoon & Dow, 1994)

(4) Division and Evaluation of Petroleum Systems

a. Division of Petroleum Systems

According to the classic method of dividing petroleum systems, the division of petroleum systems in a basin mainly depends on the distribution of effective source rock bodies, with the maximum range that the oil and gas provided by this set of source rock bodies can reach as the boundary. Each source rock layer and its related reservoir layer constitute an independent petroleum system.

b. Reliability Grades of Petroleum Systems

The reliability grade is essentially an issue of oil source reliability, indicating the degree of reliability that the oil and gas in an oil and gas reservoir originate from a certain mature source rock. Magoon divided petroleum systems into three grades based on the reliability of oil generation and accumulation: known, hypothetical, and speculative.

Known petroleum systems: There is a good geochemical match between the oil and gas reservoir and the source rock.

Hypothetical petroleum systems: The existence of source rocks can be determined using geochemical data, but there is a lack of comparison basis between the oil and gas reservoir and the source rock.

Speculative petroleum systems: Inferred only based on geological and geophysical data.

Known is represented by (!), hypothetical by (*) or (·), and speculative by (?).

c. Naming of Petroleum Systems

The name of a petroleum system includes the name of the source rock, followed by the name of the main reservoir layer, connected by a hyphen, and finally the symbol indicating its reliability grade.

For example, the Kuche Sag Jurassic-Paleogene-Neogene (!) petroleum system in the Tarim Basin, and the Dongying Sag Sha III-Sha II, Sha III (!) petroleum system in the Bohai Bay Basin, etc.

The division and evaluation of petroleum systems have become an important method for effectively predicting the potential and distribution of oil and gas resources. It emphasizes the source-reservoir relationship with the source at the core, connecting the basic elements of oil and gas accumulation through the processes of oil and gas migration and accumulation into a whole, and studying the combination of oil and gas geological elements and geological actions in time and space. In an oil and gas-bearing basin or area, there can be several overlapping petroleum systems. On the plane, different petroleum systems can be presented in one or several oil and gas accumulation zones.

(5) Petroleum Accumulation Systems

a. The Concept and Connotation of Petroleum Accumulation Systems

Although the study of petroleum systems is valued by oil and gas geologists and explorers for its rigorous logic and scientific methodology, this method is more suitable for the early and mid-stages of oil and gas exploration. In areas with a high degree of exploration, the genetic relationship between source rocks and known oil and gas reservoirs is relatively clear, and there is a considerable understanding of the basic geological laws of oil and gas reservoir formation and distribution. It is difficult to simply apply petroleum systems to guide oil and gas exploration in such areas. In addition, when an oil and gas reservoir is supplied with oil and gas by two or more effective source rock bodies, the reservoir belongs to two or more petroleum systems, causing confusion in the division of petroleum systems. Therefore, the division method of petroleum systems should be improved and developed in practical applications to better meet the requirements of exploration evaluation.

In order to apply the petroleum system research method more effectively to high-maturity exploration areas and avoid the superposition of different petroleum systems, some scholars have proposed the concept of accumulation systems, believing that petroleum accumulation systems are relatively independent oil and gas migration and accumulation units within or between petroleum systems. They include effective source rocks within the same accumulation system and their related oil and gas reservoirs, as well as all geological elements and actions required for the formation of oil and gas reservoirs. The same source rock can supply oil and gas to one or more petroleum accumulation systems, and an accumulation system can be supplied with oil and gas by one or more sets of source rocks. Therefore, in space, an accumulation system includes the oil- and gas-generating area involved in one or a group of related oil and gas reservoirs.

Petroleum accumulation systems are centered on the "reservoir," or rather, on the structural units that control the direction of oil and gas migration. Vertically, they are separated by regionally stable sealing layers that divide the source-reservoir configuration. Their boundaries on the profile are the vertical migration and accumulation boundaries of oil and gas, that is, the regionally distributed source rock layers and cap layers. Horizontally, their boundary lines are the "watersheds" or termination boundaries of oil and gas migration, that is, the inherited development of syncline axes or fluid high-potential boundaries and other sealing surfaces, such as closed large faults, lithologic pinch-out lines, unconformity lines, and basin boundaries. Accumulation systems can also be understood as including in space the oil- and gas-generating areas and the series involved in the migration and accumulation of oil and gas to form one or a group of related oil and gas reservoirs. Each accumulation system has similar oil and gas migration and accumulation characteristics.

b. Division Principles of Petroleum Accumulation Systems

The concept of accumulation systems is very similar to that of oil and gas migration and accumulation units, and can also be called oil and gas migration and accumulation systems. They are divided and evaluated based on the study of petroleum systems. In order to better evaluate the oil and gas prospects of an area and make exploration targets clearer, several accumulation systems can be further divided based on the study of petroleum systems, and then the oil and gas potential of the accumulation systems can be quantitatively evaluated.

Figure 7.22 illustrates the difference between the concepts of petroleum systems and petroleum accumulation systems. According to the definition and division method of petroleum systems, the two sets of source—reservoir-cap combinations separated by the sealing layer in the figure can be divided into 4 petroleum systems. They are superimposed on the plane, but do not completely overlap. In each set of source—reservoir-cap combinations, 3 petroleum accumulation systems can be further divided, so there are 6 accumulation systems in total. It can be clearly seen from the figure that petroleum systems emphasize the genetic relationship between "source" and "reservoir", centering on the "source" and including a certain effective source rock body and its related oil and gas reservoirs in the same system. An oil and gas reservoir that comes from oil and gas of two effective source rock bodies A and B can be attributed to both system A and system B. Such division is obviously confusing. Petroleum accumulation systems emphasize the core of "reservoir", and spatially include the oil- and gas-generating areas related to "reservoir". Accumulation systems can be understood as relatively independent units of oil and gas generation, migration, and accumulation.

Petroleum accumulation systems are mainly divided according to the migration and accumulation characteristics of oil and gas in the vertical and horizontal directions. A petroleum accumulation system is a relatively independent unit of oil and gas migration and accumulation. Vertically, the petroleum accumulation system is sealed by the cap layer that is regionally distributed; horizontally, it is bounded by the separation troug of fluid migration and the outer boundary of oil and gas migration that can be reached. In complex oil and gas accumulation areas with multiple sources, the sources of oil and gas in different series or regions are not the same. In this case, the positive structural belt can be taken as the core, and the boundary of the oil and gas reservoir involved in the belt can be used to divide and evaluate the petroleum accumulation system.

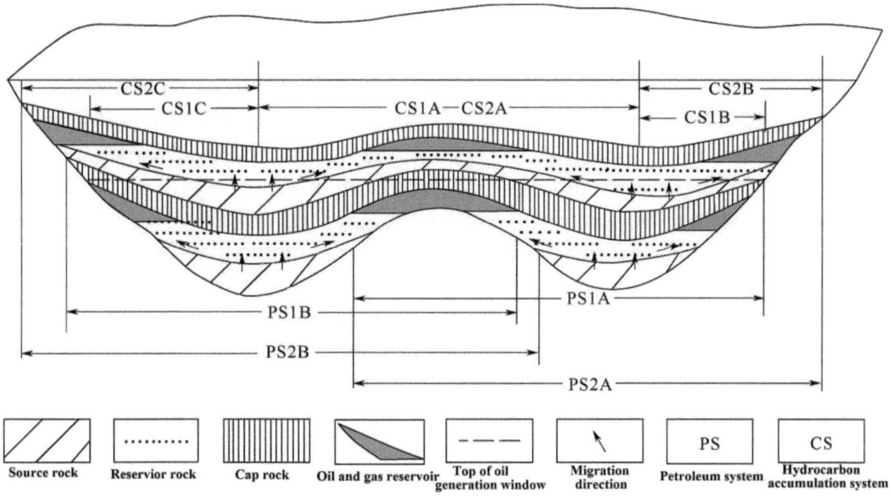

Fig. 7.22 The relationship between petroleum systems and hydrocarbon accumulation systems (Jiang et al., 2002)

7.4.3 Whole Petroleum Systems

(1) The Concept and Connotation of Whole Petroleum Systems

The development of unconventional oil and gas has shifted the focus of oil and gas exploration from "traps" back to "source rocks," recognizing that conventional and unconventional oil and gas form a unified hydrocarbon accumulation system. This has endowed theories such as petroleum systems with new connotations and led to the concept of Whole Petroleum Systems (WPS).

The Whole Petroleum System (WPS) is a natural system that encompasses all oil and gas, reservoirs, and resources within an petroliferous basin, including their formation conditions, evolutionary processes, and distribution characteristics. It is based on and developed from various concepts such as petroleum systems and accumulation systems (Jia et al., 2023). In addition to emphasizing oil and gas resources that have been accumulated through the processes of generation, expulsion, migration, accumulation, and preservation, the WPS also highlights hydrocarbons that are retained within source rocks and charged into near-source tight layers. This includes both conventional and unconventional hydrocarbon resources such as long-distance migrated hydrocarbons, short-distance migrated hydrocarbons, and retained hydrocarbons. Research on WPS covers the entire process of hydrocarbon generation and evolution, the migration and accumulation of conventional oil and gas as well as their adjustment and destruction, the in-situ accumulation or migration and accumulation of unconventional oil and gas and their adjustments, and the unified distribution and enrichment patterns of conventional and unconventional oil and gas. It is evident that the WPS is the sum of all accumulation elements, actions, processes, products, and their spatiotemporal domains.

Compared to "petroleum systems," the WPS has the following main features: (a) The reservoirs not only include the sweet spot reservoirs at the terminal of accumulation but also those in the source rock series and along the migration pathways, focusing on all reservoir spaces within the hydrocarbon-supplying range, (b) Oil and gas resources are not limited to conventional trapped oil and gas and some unconventional natural gases but also cover all types of conventional and unconventional hydrocarbon resources, including shale gas, shale oil, tight oil, etc., (c) It is not confined to the perspective of "from source rocks to traps" but rather adopts a new perspective of "source-reservoir coupling and orderly accumulation" for the analysis of the entire process of oil and gas generation, migration, and accumulation, the discovery of all types of oil and gas resources, and the integration of exploration, development, engineering, and transportation across all sectors. This leads to a holistic research and evaluation, three-dimensional exploration and development, and ultimately the maximum economic recovery of both conventional and unconventional oil and gas within the entire oil and gas-bearing unit.

(2) Oil and Gas Distribution Patterns Controlled by Whole Petroleum Systems

The WPS is applicable not only for studying the formation and distribution of conventional oil and gas reservoirs but also for investigating the formation and distribution of various unconventional oil and gas reservoirs, providing a new theoretical framework and methodological guidance for oil and gas exploration and development under complex geological conditions. The basic pattern of orderly distribution of conventional and unconventional oil and gas reservoirs controlled by the WPS (Fig. 7.23) is as follows:

Shale-type oil and gas resources (S) are mainly formed and distributed within the source rock layers at the bottom of the system. Above them, tight-type oil and gas resources (T) are formed, which are in close contact with or interbedded with the source rock layers. Further up, conventional oil and gas resources (C) are formed, separated from the source rock layers. Near the surface, there may also be heavy oil asphalt and natural gas hydrates, which are oil and gas resources with abnormal characteristics due to adjustment and transformation (A). The distribution characteristics from bottom to top can be represented as S\T-C-A. Under complex geological conditions, the orderly distribution of oil and gas (reservoir) resources within the WPS may vary, presenting three special modes of orderly distribution: (a) Shale oil and gas are separated from tight oil and gas, showing a distribution characteristic of "S-T-C-A," as seen in the Junggar Basin, (b) Oil and gas are orderly distributed in two directions, upwards and downwards from the source rock layer, showing a distribution characteristic of "C-T\S\T-C," as seen in the Ordos Basin, (c) Lateral orderly distribution of oil and gas, presenting a distribution characteristic of "S\T-C," as seen in the Songliao Basin.

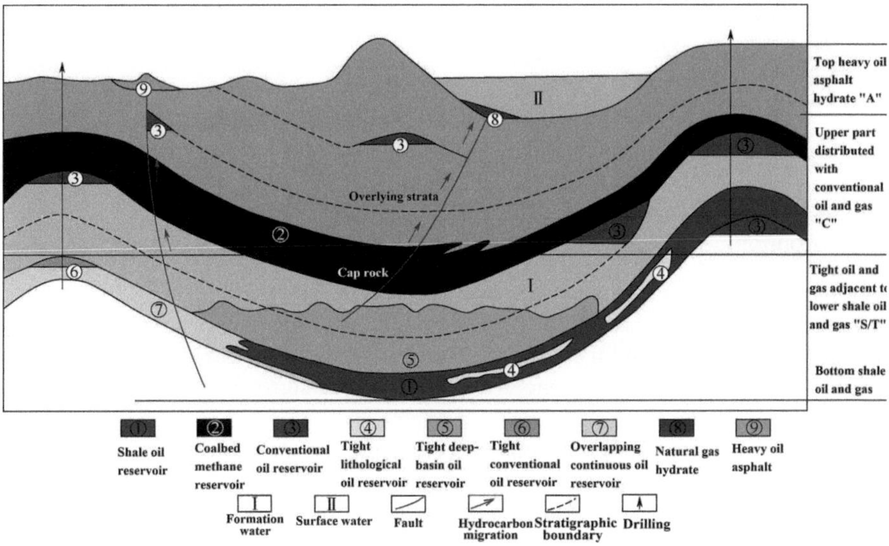

Fig. 7.23 Distribution pattern of hydrocarbon reservoirs in the entire petroleum system (Jia et al., 2023)

(3) The Significance of Whole Petroleum System Research

The future of petroleum geology should be based on a new theoretical model of the whole petroleum system. The significance of WPS research is mainly reflected in the following aspects: (a) Starting from the interconnections between various oil and gas reservoirs or resources, studying the formation and evolution of the WPS helps to deeply analyze the relationships between conventional and unconventional oil and gas reservoirs, establish a unified genetic model and classification scheme, and improve and develop oil and gas geological theories, providing new theoretical guidance for oil and gas exploration under complex geological conditions, (b) Starting from the differences between oil and gas reservoirs or resources, studying the formation and evolution of the WPS helps to reveal the genetic characteristics, dynamic mechanisms, controlling factors, and distribution patterns of different types of oil and gas reservoirs or resources, and determine the boundary conditions and domain ranges for their formation and distribution, providing new technical support for the prediction and evaluation of oil and gas resources under complex geological conditions, (c) Starting from the principle of aggregation and dispersion balance between oil and gas reservoirs or resources, studying the formation and evolution of the WPS helps to predict and evaluate the total generation, loss, and accumulation of oil and gas, as well as the remaining resource potential under different technological conditions, indicating the direction for future oil and gas exploration and development.

7.5 Distribution Characteristics of Oil and Gas Resources in China and the World

7.5.1 Distribution Characteristics of Oil and Gas Resources in China

(1) Distribution of Oil and Gas Resources in Different Tectonic Units in China

The distribution of oil and gas resources in China is directly controlled by regional tectonic characteristics. Due to the long-term effects of the northward subduction of the Indian Ocean Plate and the northwestward subduction of the Pacific Plate, there is a significant east–west difference in the crustal structure and tectonic properties of the Chinese Plate.

The western tectonic lines are mainly oriented in the NWW direction, with high tectonic activity. The arrangement of tectonic lines becomes denser towards the south, and both tectonic and magmatic activities show a trend of increasing intensity towards the south.

The eastern tectonic lines were mainly oriented in the EW direction during the Paleozoic era and in the NNE direction during the Mesozoic and Cenozoic eras. The tectonic activity was weak during the Paleozoic era, and the sedimentation was relatively stable. Since the Mesozoic and Cenozoic eras, the entire eastern region has been affected by the Yanshanian and Himalayan movements, with NE- and NNE-oriented tectonic lines developing, presenting five subsidence zones and forming sedimentary basins with distinct block-faulting and varying sizes, which received uneven thicknesses of Mesozoic and Cenozoic continental-dominated sediments.

According to tectonic characteristics, since the Mesozoic and Cenozoic eras, the Chinese Plate can be divided into three different tectonic patterns: the western convergent zone, the eastern extensional zone, and the central transitional zone. Correspondingly, the petroliferous basins in China can be classified into three major categories, belonging to three major oil and gas-bearing regions (Fig. 7.24).

(a) Western Orogenic Belt Compressive Basins (some with trans-pressure type), belonging to the western oil and gas-bearing region. Due to the northward thrust of the Indian Ocean Plate, the compressive convergence is evident, leading to crustal thickening and the alternating arrangement of a series of NWW-oriented compressive orogenic belts and large basins. At the front edge of the orogenic belt, foreland basins and intermediate blocks or continental blocks form large composite basins, which are rich in oil and gas resources, such as the Junggar, Tarim, Qaidam, and northern Tibet Qiangtang basins; within the orogenic belt, inter-mountain basins are formed, such as the Turpan-Hami Basin and the Hexi Corridor Basin group.

(b) Eastern Rift Belt Extensional Basins (some with tension-twist type), belonging to the eastern oil and gas-bearing region. Due to the westward subduction of the Pacific Plate and the uplift of the Chinese mainland, the crust thins, the

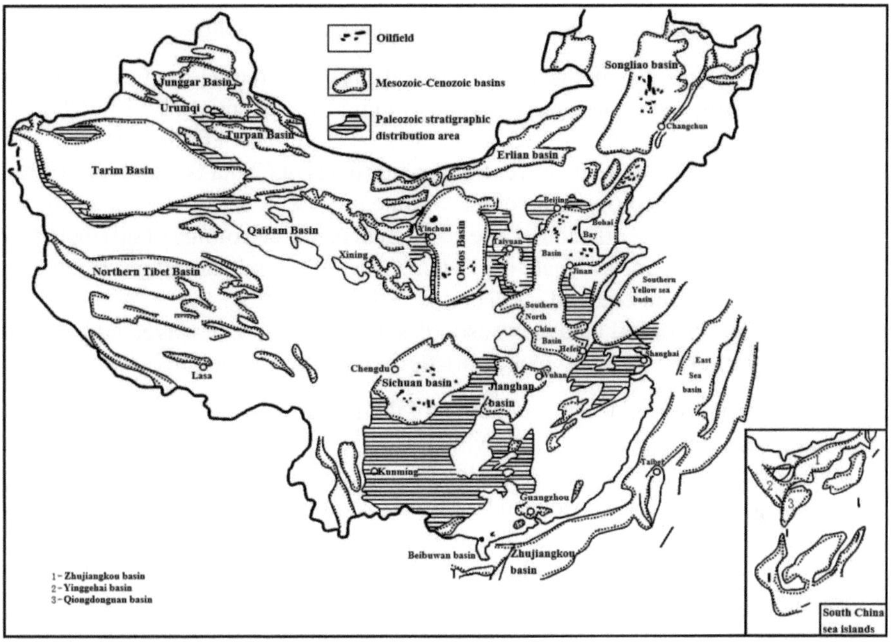

Fig. 7.24 Schematic diagram of the distribution of oil and gas basins in China (Li & Tang, 2002)

mantle arches upward, and the thermal tectonic action is evident. It is characterized by continental rift-style or continental margin rift-style, fault-sag type, with frequent basic magmatic activity, forming a series of NNE- or NE-oriented magmatic arcs as the main expansion uplift zones and expansion subsidence zones. In these subsidence zones, oil and gas-bearing basins such as the Songliao, Bohai Bay, and Jianghan basins have developed, as well as large-scale sedimentary basins along the southeast coast, such as the South Yellow Sea-Northern Jiangsu, Beibu Gulf, Yinggehai, Qiongdongnan, Pearl River Mouth, East China Sea, western Taiwan, central South China Sea, and the Taiping-Liyue Shoal.

(c) Central Cratonic Transitional Basins, belonging to the central oil and gas-bearing region, located between the first two major regions. From north to south, there are large basins such as Erlian, Ordos, and Sichuan. Due to the northward thrust of the Indian Ocean Plate, a series of nearly SN-oriented compressive thrust fault zones have formed on the western edges of these basins, such as the Helan Mountains, Zibai Mountain, Longmenshan, and Ailao Mountain. To the east, there is a rapid transition from the mountain-front zone to the stable cratonic large basins, with both compressive and extensional (or tension-twist) mechanisms.

(2) Distribution Characteristics of Oil and Gas Resources in China

According to the results of the fourth oil and gas resource assessment in China published in 2019, the geological resource volume of conventional oil in China is 1,080.31 × 10^8 t, of which 792.16 × 10^8 t is onshore, accounting for 73%. The geological resource volume of conventional natural gas in China is 78.44 × 10^{12} m^3, of which the onshore geological resource volume is 41.00 × 10^{12} m^3, accounting for 52%. For unconventional oil and gas, the geological resource volume of oil (including tight oil, oil sands oil, and oil shale oil) is 672.08 × 10^8 t, and natural gas (including tight gas, shale gas, and coalbed methane) is 284.95 × 10^{12} m^3.

The distribution of oil and gas resources in China is extensive, and can be summarized with the following characteristics:

(a) Due to the influence of the subduction of the Indian Ocean Plate and the Pacific Plate, many petroliferous basins with a wide variety of types have been formed in China. Under the influence of the Yanshanian-Himalayan movements, they have received thick Mesozoic and Cenozoic continental deposits. The Mesozoic and Cenozoic strata have become some of the most important oil and gas-bearing areas in China at present, with recoverable oil resources accounting for 85.9% of the national recoverable oil resources and recoverable natural gas resources accounting for 66% of the national recoverable natural gas resources.

(b) Both marine and continental source and reservoir oil and gas series are well-developed in China. Below the Mesozoic and Cenozoic continental deposits, there are also Paleozoic and Mesoproterozoic-Neoproterozoic marine strata, forming a multi-layer structure with overlapping multi-era source and reservoir oil and gas series. Therefore, the oil and gas-producing strata span a long geological time, from the Mesoproterozoic-Neoproterozoic to the Quaternary, almost all of which have relatively rich oil and gas resources.

(c) The geographical distribution of oil resources in China is mainly in the east, west, offshore, and central regions, while the geographical distribution of natural gas resources is mainly in the west, central, and offshore areas (Table 7.5).

From the analysis of different geographical conditions, oil and gas are distributed in plains + grasslands, loess plateaus, hills + mountains, deserts + Gobi, swamps + tidal flats, plateaus, and shallow and deep seas of the offshore areas. Among them, the conventional oil geological resource volume in the onshore plains + grasslands is 390.9 × 10^8 t, accounting for 45.1%. Other complex terrain areas account for 475.9 × 10^8 t, or 54.9%. The conventional oil geological resource volumes in the shallow and deep seas of the offshore areas are 143.7 × 10^8 t and 64.5 × 10^8 t, respectively, accounting for 69.0 and 31.0%. The geological resource volume of conventional natural gas in the onshore plains + grasslands area is 5.2 × 10^{12} m^3, accounting for 11.7%, while other complex terrain areas account for 38.8 × 10^{12} m^3, or 88.3%, among which hills + mountains reach 21.6 × 10^{12} m^3, accounting for 48.8%; the geological resource volumes of conventional natural gas in the shallow and deep seas of the offshore areas are 1.2 × 10^{12} m^3 and 37.2 × 10^{12} m^3, respectively, mainly distributed in the deep-water areas of the offshore.

Table 7.5 Geographical distribution of oil and gas resources in China (according to the results of the fourth oil and gas resource assessment in China, 2019)

Basin (oil and gas area)	Basin (oil and gas area) area (km^2)	Number of discovered oil and gas fields	Oil geological resource ($\times 10^8$ t)	Natural gas geological resources ($\times 10^{12}$ m^3)	Proven reserves total resources	Proven reservestotal resources
Northeast oil and gas area	552,860	Oilfield:108 Gasfield:29	82.101804	149.547482	0.439568	3.411198
North China oil and gas area	648,043	Oilfield:329 Gasfield:13	166.819495	351.075928	0.958151	5.229594
Northwest oil and gas area	1,115,037	Oilfield:107 Gasfield:39	60.57243	214.572743	2.306153	17.708252
South China oil and gas area	408,605	Oilfield:55 Gasfield:123	4.861615	12.00351	2.157588	13.171671
Qinghai-Tibet oil and gas area	428,972	Oilfield:0 Gasfield:0	0	64.96	0	1.4839
Onshore total	3,153,517		314.355344	792.159663	5.86146	41.004615
Bohai Bay (Offshore)	61,800	Oilfield:66 Gasfield:2	33.144084	110.2915	0.06795	1.2977
Nroth Yellow Sea	24,000		0	4.24	0	0
South Yellow Sea	145,000		0	2.9801	0	0.1847
East China Sea	250,000	Oilfield:6 Gasfield:10	0.27095	7.2304	0.315487	3.6361
Northern South China Sea	566,439	Oilfield:79 Gasfield:27	12.346275	47.6144	1.005594	9.7343
Southern South China Sea	550,313		47.36	115.797327	7.262749	22.5848
Offshore total	1,597,552		93.121309	288.153727	8.65178	37.4376
National total	4,751,069		407.476653	1,080.31339	14.51324	78.442215

(d) From the resource depth distribution, 49.57% of China's oil resources are distributed in the depth range of 2,000–3,500 m; the second largest proportion is in the depth range of 3,500–4,500 m, accounting for 23.26%. The rest are mainly distributed in the depth range shallower than 2000 m. Relatively speaking, the oil resources in the eastern basins are shallower than those in the western basins, but with the strengthening of exploration, the reserves discovered in the deep parts will gradually increase.

In the past decade or so, the exploration of oil and gas in deep (4,500–6,000 m) and ultra-deep (>6,000 m) layers in China has increased, and important progress has been made successively in the marine deep and ultra-deep layers of the Tarim Basin, Sichuan Basin, and Ordos Basin. According to statistics, the total resource volume of oil and gas in the marine strata of these three basins is 356.63×10^8 t oil equivalent, a considerable part of which is distributed in the deep and ultra-deep layers. For example, the oil and gas resources in the Tarim Basin are mainly distributed in the deep (4,500–6,000 m) and ultra-deep (greater than 6,000 m) layers, where the resources of oil and natural gas in the ultra-deep layer account for 46% and 51% of the total oil and gas resources in the basin, respectively (Yang et al., 2021). The proportion of oil and gas reserves discovered annually in the deep and ultra-deep layers of the Tarim Basin increased from 66% in 2000 to 92% in 2013, and the proportion of ultra-deep wells reached as high as 84% in 2020. Therefore, the exploration potential of deep oil and gas resources in China is relatively large and is an important area for future oil and gas exploration.

7.5.2 Distribution Characteristics of Global Oil and Gas Resources

Currently, approximately 160 oil and gas-bearing basins have been discovered worldwide, and their distribution is generally controlled by the tectonic characteristics of the major plates and the intervening fold belts. Looking at the global distribution outline of oil and gas, the distribution of oil and gas resources on the Earth's crust is quite widespread, with oil and gas fields found in continents, oceans, deserts, and swamps. However, the distribution of oil and gas on the Earth's crust is unbalanced, both in spatial and temporal distribution.

(1) Distribution of Oil and Gas Resources on the Plane

Among the 238 basins worldwide that produce commercial oil, a total of 29,225 oil and gas fields have been discovered, of which the number of onshore and offshore oil and gas fields accounts for 72.8% and 27.2%, respectively. The recoverable reserves of oil and gas in the world's 13 giant (oil equivalent $>68.49 \times 10^8$ t) and 43 large (13.7×10^8 t < oil equivalent < 68.49×10^8 t) oil and gas-bearing basins account for 67.3% and 19.7% of the total reserves, respectively. More than 70% of the world's oil reserves are concentrated in over 300 large oil fields, which are relatively concentrated in oil and gas-bearing basins such as the Persian Gulf, the Gulf of Mexico and the Caribbean Sea area, West Siberia, the Volga-Ural, the Pre-Caucasus-Caspian basins, the North Sea, Canada's Alberta, the Central Stable Platform basins of the United States, the North African basins, Eastern China, and the South China Sea basins (Fig. 7.25).

The remaining recoverable oil and gas reserves in the world are mainly concentrated in 39 basins (with more than 10×10^8 t oil equivalent), and the remaining

Fig. 7.25 Geographic distribution of major oil and gas fields in the world (according to "Foreign Oil News")

recoverable oil and gas reserves in the Arabian, Eastern Venezuela, West Siberian, and Zagros basins account for 63.1% of the global remaining recoverable reserves (Fig. 7.26).

The world's 42 super-large oil and gas fields are only distributed in 10 basins, with 20 in the Persian Gulf Basin and 10 in the Siberian Basin. The oil reserves of

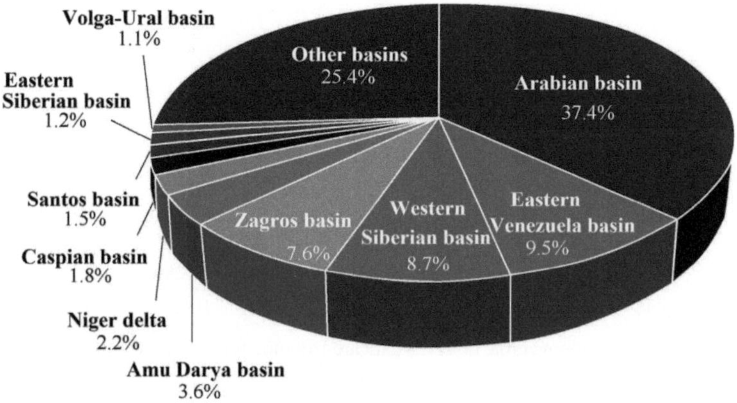

Fig. 7.26 Pie chart of the remaining recoverable reserves in major basins worldwide (Ma et al., 2021)

the 42 super-large oil and gas fields account for half of the world's total reserves. The combined reserves of the Ghawar and Burgan oil fields account for one-fifth of the world's oil reserves. The Urengoy and Yamburg gas fields account for 1/8 of the world's natural gas reserves.

Remaining recoverable reserves represent the potential for future development. According to statistics from BP (Table 7.6), as of the end of 2020, the world's remaining proven recoverable oil reserves were 2444×10^8 t, of which the Middle East accounted for 48.3%, Central and South America for 18.7%, North America for 14.0%, Central Asia-Russia for 8.4%, Africa for 7.2%, and North America and the Asia–Pacific region together accounted for 3.4%. The world's remaining proven recoverable natural gas reserves were 188.1×10^{12} m^3, of which the Middle East accounted for 40.3%, Central Asia-Russia for 30.1%, the Asia–Pacific and North America each accounted for 8.8% and 8.1%, respectively, Africa for 6.9%, and Central and South America and Europe together accounted for 5.9%; the proportions of onshore and offshore were 59.8% and 40.2%, respectively.

Undiscovered recoverable resources mainly represent the potential and direction for future risk exploration. The world's undiscovered recoverable oil reserves and natural gas reserves are $1,309.69 \times 10^8$ t and 193.72×10^{12} m^3, respectively. They are mainly distributed in countries such as Russia, Brazil, and the United States; concentrated in basins such as the Arabian, West Siberian, and Zagros basins; the proportions of onshore and offshore are 50.9% and 49.1%, respectively (Table 7.6).

Table 7.6 Distribution of oil and gas resources in different regions of the world

Region	Oil				Gas			
	RRRR	Proportion (%)	URR	Proportion (%)	RRRR	Proportion (%)	RRR	Proportion (%)
North America	361	14.0	160.08	12.22	15.2	8.1	24.58	12.69
Central and South America	508	18.7	404.57	30.89	7.9	4.2	15.57	8.04
Europe	18	0.8	56.41	4.31	3.2	1.7	9.56	4.93
Central Asia-Russia	199	8.4	199.05	15.20	56.6	30.1	71.02	36.66
Middle East	1,132	48.3	295.55	22.57	75.8	40.3	39.11	20.19
Africa	166	7.2	133.26	10.17	12.9	6.9	21.89	11.30
Asia–Pacific	61	2.6	60.77	4.64	16.6	8.8	11.99	6.19
Total	2,444	100.0	1,309.69	100.0	188.1	100.0	193.72	100.0

Source Remaining proven recoverable reserves data are from the report released by BP in 2021; undiscovered recoverable resources data are from the "Global Oil and Gas Resource Potential and Distribution" written by the Petroleum Exploration and Development Research Institute in 2021, and the statistical data refer to the information released by USGS, IEA, etc.

Note RRRR—Remaining Proven Recoverable Reserves, $\times 10^8$ t(oil), $\times 10^8$ m^3 (gas), URR—Undiscovered Recoverable Resources, $\times 10^8$ t (oil), $\times 10^8$ m^3 (gas)

The world's unconventional oil and gas technically recoverable resources are 6,352.3 × 10^8 t oil equivalent, of which unconventional oil technically recoverable resources are 4,049.3 × 10^8 t, accounting for 63.7%; unconventional natural gas technically recoverable resources are 269.5 × 10^{12} m^3, accounting for 36.3% (according to Ma Xinhua et al., "Global Oil and Gas Resource Potential and Distribution" (2021)). The distribution of technically recoverable resources from large to small is North America, Central and South America, Russia, Africa, Europe, the Middle East, the Asia–Pacific, and Central Asia. With the continuous progress of economically effective development technology, unconventional oil and gas will become a realistic alternative resource in the future, especially shale oil and gas resources 75.34% of the world's unconventional oil recoverable resources are distributed in North America, Central Asia-Russia, and Central and South America (Table 7.7). The top five countries, including the United States, Russia, and Canada, account for more than 66%. The top 20 basins, including Alberta, Eastern Venezuela, and Arabia, are enriched with 70% of the unconventional oil recoverable resources. For different types of unconventional natural gas, the technically recoverable resources of shale oil (including tight oil) are 738 × 10^8 t, accounting for 11.6% of unconventional resources. Heavy oil technically recoverable resources are 1,274.79 × 10^8 t, accounting for 20.1%. Oil sands technically recoverable resources are 631.42 × 10^8 t, accounting for 9.9%. Oil shale technically recoverable resources are 1,405.07 × 10^8 t, accounting for 22.1.

64.25% of the world's unconventional natural gas recoverable resources are concentrated in North America, Central and South America, and Central Asia-Russia (Table 7.8). The top five countries, including the United States, Russia, and Canada, account for over 63%. 26 basins such as Alberta, the Gulf of Mexico, and the Appalachian Basin are rich in 80% of unconventional natural gas resources. For different types of unconventional natural gas, the technically recoverable resources of shale gas are 223.81 × 10^{12} m^3, accounting for 30.1% of unconventional resources: the technically recoverable resources of coalbed methane are 38.68 × 10^{12} m^3, accounting for 5.2% of unconventional resources; the technically recoverable resources of tight gas are 7.0 × 10^{12} m^3, accounting for 0.9% of unconventional resources.

(2) Distribution of Oil and Gas Resources in Geological History

Oil and natural gas are widely distributed in geological history, from the oldest Archean to the newest Quaternary, industrial oil and gas reservoirs have been found, and there are data indicating that primary oil and gas reservoirs have been found from the Mesoproterozoic to the Quaternary. However, the known oil and gas reserves in the world are unevenly distributed in geological history, with the most in the Mesozoic-Cenozoic, where oil accounts for about 80%, and natural gas accounts for about 60%. The oil reserves in the Paleogene-Neogene account for about 26.9% of the world's oil reserves, and the oil reserves in the Cretaceous account for about 28.21% of the world's oil reserves. The oil reserves in the Jurassic are also relatively abundant, accounting for about 14.33% of the world's oil reserves. The distribution of natural gas in the world's large gas fields (reserves $\geq$ 85 billion cubic meters) is

Table 7.7 Distribution of main unconventional oil recoverable resources in different regions of the world

Region	Shale oil		Heave oil		Oil sands		Oil shale		Total	
	Resources ($\times 10^8$ t)	Proportion (%)	Resources ($\times 10^8$ t)	Proportion (%)	Resources ($\times 10^8$ t)	Proportion (%)	Resources ($\times 10^8$ t)	Proportion (%)	Resources ($\times 10^8$ t)	Proportion (%)
North America	313.59	42.49	324.7	25.47	403.39	63.89	544.53	38.75	1,586.21	39.17
Central and South America	89.38	12.11	418.2	32.81	0	0.00	153.23	10.91	660.81	16.32
Europe	23.63	3.20	84.24	6.61	17.93	2.84	200.09	14.24	325.89	8.05
Central Asia-Russia	146.87	19.90	133.53	10.47	185.07	29.31	338.18	24.07	803.65	19.85
Middle East	59	7.99	180.59	14.17	0	0.00	62.65	4.46	302.24	7.46
Africa	63.35	8.58	64.77	5.08	25	3.96	69.68	4.96	222.8	5.50
Asia–Pacific	42.21	5.72	68.76	5.39	0.03	0.00	36.71	2.61	147.71	3.65
Total	738	100	1,274.79	100	631.42	100	1,405.07	100	4,049.31	100

Source The data represents the technically recoverable resources of oil and gas, derived from the "Global Oil and Gas Resource Potential and Distribution" compiled by the Research Institute of Petroleum Exploration and Development of China in 2021. The statistical data refers to materials published by USGS, IEA, and others

Table 7.8 Distribution of main unconventional natural gas recoverable resources in different regions of the world

Region	Shale gas		Coalbed methane		Tight gas		Total	
	Resources ($\times 10^{12}$ m^3)	Proportion (%)	Resources ($\times 10^{12}$ m^3)	Proportion (%)	Resources ($\times 10^{12}$ m^3)	Proportion (%)	Resources ($\times 10^{12}$ m^3)	Proportion (%)
North America	74.32	33.21	16.98	0.08	5.41	77.62	96.71	35.89
Central and South America	40.48	18.09	0.03	5.20	0.11	1.58	40.62	15.07
Europe	16.69	7.46	2.01	33.89	0.73	10.47	19.43	7.21
Central Asia-Russia	22.34	9.98	13.11	0.00	0.34	4.88	35.79	13.28
Middle East	16.08	7.18	0	1.53	0.18	2.58	16.26	6.03
Africa	31.66	14.15	0.59	15.41	0	0.00	32.25	11.97
Asia–Pacific	22.24	9.94	5.96	100.00	0.2	2.87	28.4	10.54
Total	223.81	100	38.68	100	6.97	100	269.46	100

Data Source The data represents the technically recoverable resources of oil and gas, derived from the "Global Oil and Gas Resource Potential and Distribution" compiled by the Research Institute of Petroleum Exploration and Development of China in 2021. The statistical data refers to materials published by USGS, IEA, and others mainly concentrated in the Permian and Cretaceous strata, with the former accounting for about 32.41% of the world's natural gas reserves, and the latter about 29.85%. The natural gas reserves in the Paleogene-Neogene are also relatively abundant, accounting for about 15.12% of the world's natural gas reserves (according to He et al., 2010).

The remaining recoverable oil and gas reserves in the world are mainly distributed in the Cretaceous, Jurassic, Permian, Paleogene, and Neogene, accounting for 27%, 18.9%, 17.5%, 14.4%, and 11.7% respectively, and the remaining recoverable reserves in other strata are relatively small, with a total proportion of only 10.5% (according to Ma Xinhua et al., "Global Oil and Gas Resource Potential and Distribution" (2021)). The remaining recoverable reserves in the Cretaceous are mainly distributed in the Arabian Basin, West Siberian Basin, Zagros Basin, Santos Basin, and Sirte Basin, etc.; the remaining recoverable reserves in the Jurassic are mainly distributed in the Arabian Basin, Amu Darya Basin, West Siberian Basin, Eastern Barents Sea Basin, North Sea Basin, etc.; the remaining recoverable reserves in the Permian are mainly distributed in the Arabian Basin, Zagros Basin, North Slope of Alaska, Volga-Ural Basin, etc.; the remaining recoverable reserves in the Precambrian are mainly distributed in the East Siberian Basin, Oman Basin, and the Precambrian basement reservoirs of some basins.

For unconventional oil and gas, the technically recoverable resources of unconventional oil in the world are mainly distributed in the Cretaceous, Jurassic, Paleogene, Neogene, and Carboniferous, accounting for 32%, 15.8%, 15.7%, 12.5%, and 7% respectively, and the resource potential in other series is relatively small, with a total proportion of 17%; the technically recoverable resources of unconventional natural gas in the world are mainly distributed in the Cretaceous, Jurassic, Devonian, Carboniferous, and Silurian, accounting for 29.9%, 19.8%, 17.3%, 9.8%, and 9.4% respectively, and the resource potential in other series is relatively small, with a total proportion of 13.8% (according to Ma Xinhua et al., "Global Oil and Gas Resource Potential and Distribution" (2021)).

(3) Vertical Distribution of Oil and Gas Resources

Practical exploration of oil and gas fields has proven that oil and gas reservoirs have been found from the surface to deep underground. The maximum depth of proven oil reservoirs has exceeded 8500 m, and the maximum depth of both condensate gas reservoirs and gas reservoirs has exceeded 8700 m. The distribution of oil and gas reserves along the burial depth is not uniform.

Statistical results show that the average burial depth of the world's discovered oil reservoirs is 1465 m, 80% of the oil and gas reserves are distributed between 600–3,000 m in depth, and the peak of reserves is between 800–1,900 m. With the increase of burial depth, oil and gas reservoirs will be replaced by condensate gas reservoirs and dry gas reservoirs. Below 5000 m, it is mainly gas layers, and oil layers only account for 1/5 of the total number of oil and gas layers. At present, the development of deep and ultra-deep oil and gas exploration at home and abroad is very fast, and several deep and ultra-deep oil and gas reservoirs have been found successively, constantly providing more oil and gas resources for human beings.

In terms of the rock types of oil and gas producing layers, sandstone, limestone, and dolomite are the most important, accounting for more than 99% of the world's total oil and gas reserves, almost equally distributed between sandstone and carbonate rocks, with only a very small amount stored in other types of rocks. According to statistics, in the world's remaining recoverable oil and gas resources, clastic rock and carbonate rock reservoirs account for 49.3% and 50.7% respectively. The remaining recoverable reserves in clastic rock reservoirs are 2,101.5 $\times$ 10^8 t oil equivalent, mainly distributed in basins such as Eastern Venezuela, West Siberia, Arabia, and the Niger Delta, with their remaining recoverable reserves all exceeding 100 $\times$ 10^8 t oil equivalent. The remaining recoverable reserves in carbonate rock reservoirs are 2,161.2 $\times$ 10^8 t oil equivalent, mainly distributed in basins such as Arabia, Zagros, Amu Darya, and the Caspian Sea, with their remaining recoverable reserves all exceeding 100 $\times$ 10^8 t oil equivalent (Ma et al., 2021).

In summary, the distribution of the world's oil and gas resources is both universal and obviously uneven in time and space. The distribution of oil and gas resources generally follows certain patterns, which are mainly controlled by large-scale tectonic conditions and lithofacies paleogeographic conditions. In-depth research on the conditions of oil and gas generation, migration, accumulation, and preservation in oil and gas basins can correctly make oil and gas prospect evaluation and effectively guide oil and gas exploration.

7.6 Controlling Factors of Oil and Gas Distribution

The distribution of oil and gas is characterized by its extensive nature and unevenness, being highly imbalanced both spatially and temporally. Similarly, within the same basin or the same oil and gas system, the distribution of oil and gas also exhibits distinct unevenness.

7.6.1 Main Controlling Factors of Oil and Gas Distribution on a Macro Level

On a macro level, the unevenness of oil and gas distribution is primarily caused by differences in tectonic conditions, paleogeographic conditions, and paleoclimatic conditions. Tectonic conditions mainly influence the degree and mode of crustal activity, leading to variations in sedimentary basin types, subsidence rates, and geothermal histories. In contrast, paleogeographic and paleoclimatic conditions affect the proliferation of organisms, the preservation of organic matter, and the degree of its transformation into oil and gas. These differences determine the uneven distribution of oil and gas within large crustal regions and across significant stratigraphic time units.

(1) Tectonic Conditions

Tectonic conditions are the key issue leading to differences in the macro distribution of oil and gas. Generally, basins with long-term continuous tectonic activity and multiple tectonic cycles are conducive to the formation of thick sedimentary deposits and multiple sets of source-reservoir-caprock combinations, which are beneficial for the generation and evolution of oil and gas, thereby facilitating the enrichment of oil and gas in multiple layers.

Oil and gas enrichment areas in the world are often regions dominated by long-term subsidence, characterized by thick sedimentary rock layers and well-developed source-reservoir-caprock layers. Only with continuous subsidence can a basin maintain a relatively stable depositional environment and form thick sedimentary deposits, which form the material basis for the generation of oil and gas. As the basin continues to subside over the long term, the water body deepens, the basin expands, and a closed water body environment is formed, creating a reducing water medium. This environment is conducive to the proliferation of organisms and the accumulation of organic matter, which is favorable for the generation and evolution of oil and gas.

Multiple tectonic cycles give sedimentation a multi-cyclic nature, leading to the development of multiple sets of source-reservoir-caprock combinations, which form the geological basis for the formation of multiple oil and gas-bearing layers. The different tectonic evolution histories of various parts of a basin result in different source-reservoir-caprock rock series ages, leading to diverse distribution characteristics of oil and gas-bearing layers in different regions.

(2) Lithofacies Paleogeographic Conditions

The sedimentary environment conducive to the proliferation of organisms and the preservation of organic matter is the relatively stable shallow sea, deep lake, and semi-deep lake facies. The long-term and large-scale development of such environments can form thick hydrocarbon-generating series, which is a fundamental condition for the formation of large oil and gas fields. On the contrary, areas that do not meet these conditions cannot form oil and gas enrichment zones.

(3) Paleoclimatic Conditions

The amount of organic matter in sedimentary basins depends on the paleoclimatic and paleowater medium conditions during the formation of sediments. Generally, in humid and semi-humid climatic depositional environments, the organic matter content is high. In contrast, in semi-arid and arid climatic environments, ancient lake basins become increasingly saline, leading to a significant decrease in organic matter content.

7.6.2 Main Controlling Factors of Oil and Gas Distribution Within Basins

The basin petroleum theory proposes that the generation, evolution, migration, accumulation, preservation, and destruction of oil and gas all take sedimentary basins as the basic geological structural units. The type, structure, and development of petroliferous basins control the distribution of oil and gas both horizontally and vertically. The uneven distribution of oil and gas within a basin or an oil and gas depression is mainly related to the differences in geological conditions between different structural units within the basin (such as the hydrocarbon supply capacity of source rocks, oil and gas migration, accumulation, and preservation), leading to varying degrees of oil and gas enrichment in different structural parts of the basin.

7.6.2.1 The Development Characteristics of Source Rocks Control the Distribution of Oil and Gas

Source rocks are the material basis for the formation of oil and gas and are the primary controlling factor in the distribution of oil and gas. Extensive oil and gas exploration practices both domestically and internationally have shown that the distribution of oil and gas within a basin is closely related to the development layers, maturity, and hydrocarbon generation and expulsion centers of source rocks.

(1) The hydrocarbon generation and expulsion centers of source rocks control the horizontal distribution of oil and gas

The sedimentary center of the source layer within an oil and gas depression determines the distribution of the oil-generating area (oil source area), and the source layer and effective oil-generating depression essentially control the range of oil and gas field distribution. From the distribution of oil and gas reservoirs, they are mainly distributed around the hydrocarbon generation center, and most oil and gas reservoirs appear within the effective hydrocarbon expulsion range of the source rocks.

On the plane, oil and gas fields are generally distributed within or near the effective oil-generating areas, while areas far from the effective oil-generating areas rarely have oil and gas fields. For example, the central depression of the Songliao Basin is the effective oil-generating area of the basin, and the discovered oil and gas fields are basically within or near the effective oil-generating area, as shown in Fig. 7.8. Similarly, in the Dongying Depression of the Bohai Bay Basin, oil and gas fields are distributed around the center of the depression, which is also the oil-generating center, especially the Liujin Depression with the best oil-generating conditions, which controls the distribution of large oil fields.

Between 1959 and 1962, during the oil and gas exploration work in the Songliao Basin, Chinese scholars summarized the laws that the oil-generating area basically controls the distribution of oil and gas, and proposed the theory that the migration distance of oil and gas is short and the oil source area controls the distribution of oil and gas, namely the "source control theory" (Hu 1982). The basic idea of the "source control theory" is that the distribution area of effective hydrocarbon source rocks essentially controls the approximate distribution range of oil and gas fields. After the hydrocarbons are generated and expelled from the hydrocarbon source rocks, they gather nearby in the favorable oil-generating areas or their vicinity. In the oil and gas exploration of the Bohai Bay Basin from the 1960s to the 1980s, the source control theory was further enriched and developed, and a series of significant exploration achievements were obtained under the guidance of this theory (Hu et al., 1991). In the past 20 years, the source control theory of oil and gas distribution has been further developed, and theories such as the "rich oil depression is full of oil-bearing theory" (Zhao et al., 2004) and the "depression control oil theory" (Zhao et al., 2011) have been proposed, effectively guiding the oil and gas exploration in mature exploration areas.

The "source control theory" is not only applicable to continental sedimentary basins but also to marine sedimentary basins. This is because it objectively reflects the material basis role of hydrocarbon source rocks in the formation of oil and gas reservoirs. In 1971, B. Tissot studied the Jurassic hydrocarbon source rocks in the Paris Basin and found that all oil fields and oil outcrops were in the favorable oil-generating areas. In areas with an oil generation rate of less than 500 g/t, only dry wells were found. The large oil and gas fields in the Persian Gulf Basin, especially the super-large oil and gas fields (referring to oil and gas fields with recoverable reserves exceeding 6.85×10^8 tons of oil equivalent), are mainly distributed in the Persian Gulf and its peripheral areas. These areas have developed several sets of main hydrocarbon source rocks that are spatially superimposed on each other, which have gathered 94.1% of the oil and 96.5% of the natural gas of the total reserves of the basin. The oil and gas have the distribution characteristics of gathering near the hydrocarbon source rocks (Bai, 2007).

The oil-generating range of a sedimentary basin is only limited to the deeper areas of the basin depression, which belongs to the sedimentary strata during the long-term continuous subsidence stage of the basin. Due to the stable sedimentary environment of the deep-water facies, with closed and reducing conditions, a large amount of biological reproduction, and rich organic matter, it is conducive to the accumulation and preservation of organic matter and its transformation into oil and gas. In an oil-bearing basin, the oil-generating favorable area only accounts for a part of the whole basin. The large basin area and small oil-generating range are characteristics of continental sedimentary basins. However, the effective hydrocarbon source rock area in rich oil and gas depressions is relatively large, accounting for more than half of the depression. For example, in the Junggar Basin, the effective hydrocarbon source rock area of the Permian system accounts for 51% of the basin. In the Ordos Basin, the effective hydrocarbon source rock area of the Triassic system accounts for 60% of the basin; and in the Dongying Depression, the effective hydrocarbon source rock area of the Paleogene system accounts for 69% of the depression (Zhao et al., 2004).

(2) The stratigraphic position of source rocks controls the stratigraphic distribution of oil and gas

Source rocks and effective oil-generating depressions also exert control over the distribution of oil and gas fields (reservoirs) in the vertical direction. Vertically, oil and gas-bearing layers are generally located within or adjacent to the oil-generating layers. In other words, the source-reservoir-caprock combinations control the distribution of the main oil and gas reservoirs in the profile.

In basins with frequent tectonic activity and well-developed faults, due to the strong vertical migration of oil and gas, the stratified nature of this oil and gas distribution may change. The main oil and gas-bearing series can be located far above the source rock series. The specific series in the overlying strata where the oil and gas are distributed depends on the conductive capacity of the oil-source faults and the matching degree of the vertical caprock.

Take the Bohai Bay Basin as an example. From the marginal depressions of the basin to the central Bozhong Depression, with the migration of the subsidence and sedimentation center, the main hydrocarbon source rock series of different depressions show a gradually increasing and becoming newer migration trend. The contribution of the upper Sha 1 member and Dongying Formation source rock series gradually increases (Fig. 7.27). The migratory nature of the main hydrocarbon source rock series in different depressions also causes the migratory nature of the oil and gas-bearing series in each depression. From the marginal depressions of the basin to the central Bozhong Depression, the oil and gas-bearing series also gradually become newer, and overall, the series within and adjacent to the oil and gas-generating series are the main oil and gas-bearing series.

(3) The maturity of source rocks controls the phase state of oil and gas

The maturity of source rocks plays a key role in the phase state and distribution of oil and gas in the basin. Different thermal histories and maturities of source rocks lead to different oil and gas phase states. The over-mature marine Paleozoic in the Sichuan Basin and Ordos Basin can only yield gas, while the marine oil in the Tarim Basin is mainly distributed in the centrally located, north–south elongated rectangular block within the basin, with gas being predominant on the eastern and western sides (Zhang et al., 2004).

7.6.2.2 Secondary Structural Zones Control the Accumulation of Oil and Gas

Chinese scholars place great emphasis on the study of "oil and gas accumulation zones." Combining the oil-forming characteristics of China's continental oil and gas basins, they proposed in the 1960s that "secondary structural zones are oil and gas accumulation zones." In the late 1960s and early 1970s, based on a systematic summary of the distribution patterns of oil and gas in the faulted basins of eastern China, the concept of "composite oil and gas accumulation zones" was put forward

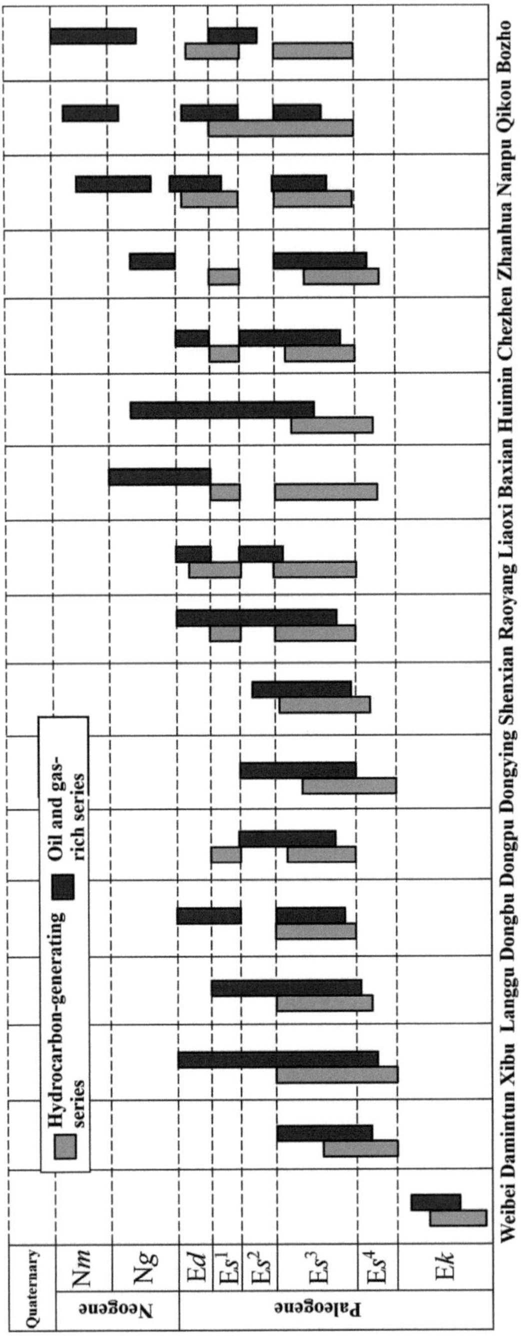

Fig. 7.27 Relationship between the main source rock series and oil-rich series in different sags of the Bohai Bay Basin (Jiang et al., 2021)

(Hu et al., 1991). It was pointed out that under the background of secondary structural zones, oil and gas-bearing zones with different series and different trap types superimposed on each other can be formed.

Secondary structural zones, as the second-order tectonic units within oil and gas depressions, have a significant control effect on the regional migration and accumulation of oil and gas. They are an important foundation and geological background for oil and gas accumulation within the depression. Those long-inherited secondary structures located within or near the hydrocarbon generation and expulsion centers have an important control effect on the development of reservoir layers, the formation of traps, and the migration and accumulation of oil and gas within the basin. They often form structural, structural-lithological, and stratigraphic oil and gas reservoirs (Fig. 7.28). According to statistics, the proven geological reserves of oil accumulated in the secondary structural zones of the Bohai Bay Basin amount to 90.2×10^8 tons, accounting for 48% of the total oil resource quantity (Tian et al., 2002).

During the development of a basin, secondary structural zones may be in a relatively uplifted state during a certain period or have long-term inheritance, exerting important control over sedimentation. When secondary structural zones are still underwater during the sedimentation process, the lithology of the sediments is coarser and better sorted, which is conducive to the formation of good reservoir layers. If the secondary structure is uplifted too high and exposed above the water surface, it is often subject to erosion, which is unfavorable for the preservation of oil and gas fields. For example, in the Songliao Basin, the Yangdachengzi Uplift and Denglouku Structural Zone on the southeast uplift belt, due to the high uplift, the Qingshan Kou Formation mudstone cap layer has been partially or completely eroded, resulting in an incomplete source-reservoir-caprock combination, which is unfavorable for the formation of oil and gas fields.

The relatively uplifted position of secondary structural zones often places them within oil-generating depressions or between different depressions, acting as regional low-energy zones. Buoyancy or overpressure-buoyancy together serve as the main driving forces for the migration of oil and gas towards the secondary structural zones. If the secondary structural zone is surrounded by a depression area (oil-generating area), oil and gas can migrate regionally from all directions towards the secondary structural zone and accumulate into reservoirs. If the secondary structure is located on one side of the depression or on the slope, oil and gas will migrate directionally from the depression side towards the secondary structure. In addition, with the development and formation of secondary structures, regional fractures and unconformities often occur, providing good pathways for the migration of oil and gas towards the secondary structures. From the above analysis, secondary structures control the regional migration and accumulation of oil and gas.

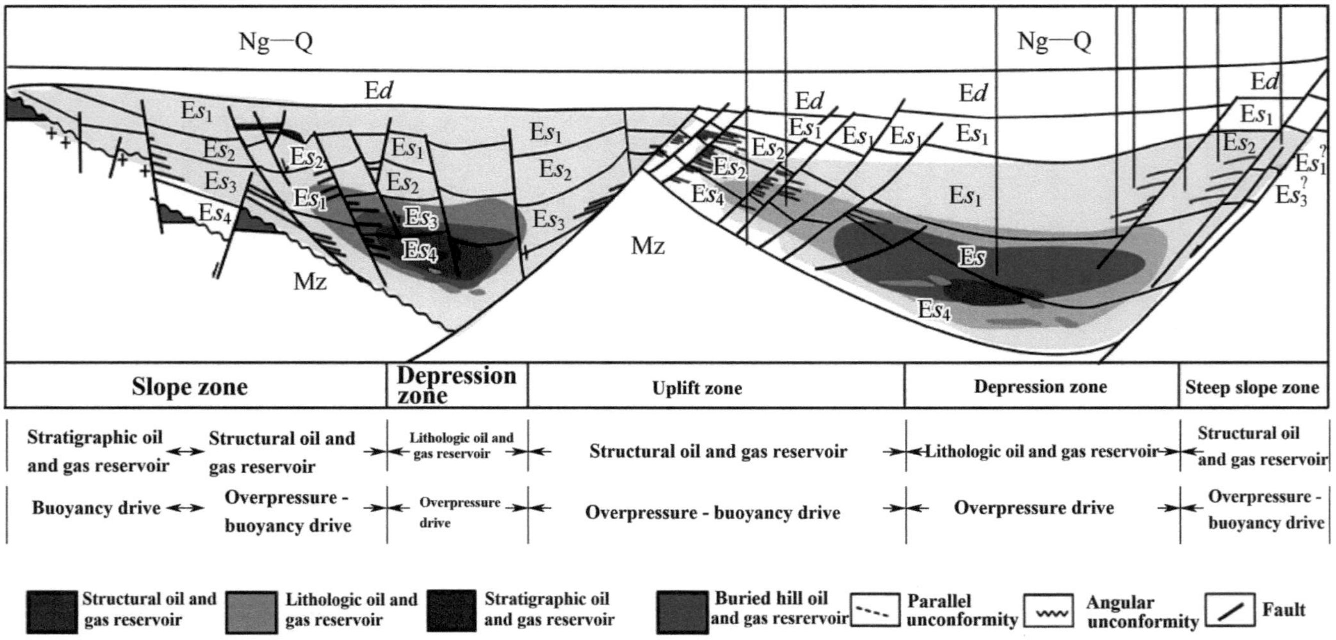

Fig. 7.28 Relationship between secondary structural zones in typical rift basins and the formation and distribution of hydrocarbon reservoirs

7.6.2.3 The Controlling Role of Favorable Sedimentary Facies Zones on the Distribution of Oil and Gas

Sedimentary facies zones not only control the formation of certain types of oil and gas accumulation zones but also control the formation of oil and gas fields. In clastic sedimentary basins, sedimentary facies zones control the development of sandstone bodies, thereby controlling the development of reservoir layers. Basin margin facies, underwater uplift facies, flank slope facies of uplifts, fluvial-deltaic facies, and deep-water turbidite facies are favorable facies zones for oil and gas accumulation and are also favorable areas for finding high-yield oil and gas fields.

Looking at the sedimentary facies zones of various oil and gas basins, there is often a complete facies change from the basin margin to the center. From the marginal clastic facies, moving inward, there are successively fluvial, swamp, deltaic, lacustrine-margin, and lacustrine facies. Circum-lacustrine annular facies zone often forms around the lake area. The central part of the basin generates oil, and the favorable facies zones in the marginal uplift areas are the targets for oil and gas migration, often forming large oil and gas fields, such as the Daqing Placanticline, Karamay Oilfield, Shengtuo Oilfield, etc., indicating that oil and gas migration is strictly controlled by facies zones. Underwater uplifts and the flank slope areas on both sides of the uplifts, during the sedimentation process, due to the crustal movement-controlled basement undulations, result in shallow water bodies, thin sediments, coarse facies, and good physical properties. If in a marine environment, biological reefs and oolitic limestones are formed. If they were once exposed above the water surface, they might have formed caves and fractures due to leaching, which is even more favorable for storing oil and gas. These ancient underwater uplift areas, due to their long-term existence, form a certain pressure difference between the uplift and the depression during the long-term diagenetic process, thus favoring the early migration and accumulation of oil and gas. The deep-water lacustrine turbidites located in the oil-generating deep depressions are conducive to early oil and gas accumulation and are favorable facies zones for finding lithological oil reservoirs.

Understanding the sandstone bodies within sedimentary facies zones is key to grasping the distribution patterns of favorable reservoir layers. The development of sandstone layers and their oil storage properties vary in different parts of the sedimentary facies zones. Generally, in the core and frontal zones of sandstone bodies, the sandstone is thick, has more layers, good connectivity, and favorable oil storage properties, and is also closer to the oil-generating areas, making it the most favorable zone for oil and gas enrichment. When combined with favorable structural conditions, large reserves and high-yield oil and gas fields can often be formed. Although the main part of the sandstone body has well-developed sandstone and good physical properties, it is too far from the oil-generating area. The discontinuous distribution zone, although closest to the oil-generating area, has poorly developed sandstone, thin sandstone layers, and poorer oil storage properties. Generally, only medium-sized reserves and production oil and gas fields can be formed. For example, in the Songliao Basin, the middle part of the source-reservoir-caprock combination is the most important oil and gas-bearing combination in the basin, and its oil and gas fields are concentrated in the core and frontal zones of the sandstone bodies.

There are differences in the oil storage capacity of sandstone bodies in different sedimentary facies zones. Deltaic sandstone bodies, coastal sandstone bodies, lacustrine sandstone bodies, fluvial sandstone bodies, and turbidite sandstone bodies are important reservoir rocks and have significant control over the distribution of oil and gas. Among them, the deltaic sedimentary system spans the alluvial plain and the river–lake or sea entrance. The sand bodies located on the coastal river mouth sandbars, distributary channel sands, and the favorable facies zones at the front are connected to the deep-water mudstone oil-generating areas before and after, forming a favorable source-reservoir-caprock combination, which is a favorable area for the distribution of oil and gas fields. There are many oil and gas fields in the world related to deltaic facies, many of which are large and super-large oil and gas fields. For example, the Burgan oil field in Kuwait and the Bolivar Coastal oil field in the Maracaibo Basin of Venezuela have recoverable reserves of 9.4 billion and 4.2 billion tons, respectively, ranking as the second and third super-large oil fields in the world. The reason why deltas have abundant oil and gas accumulation is the result of the good combination of various favorable oil and gas accumulation conditions. First, the prodelta subfacies and the adjacent deep lake-semi-deep lake mudstone in the deltaic sedimentary system are favorable source rocks, which can provide a rich source of oil and gas. Secondly, the deltaic sedimentary area develops a variety of good sandstone bodies, such as the distributary channel sandstone bodies in the delta plain subfacies, the underwater distributary channel sandstone bodies, the river mouth dam sandstone bodies, the distal sand dam sandstone bodies, and the front sheet sandstone bodies in the delta front subfacies, which are common good reservoir layers. Furthermore, the mudstone located in the delta plain subfacies can serve as a good caprock. Due to the alternating occurrence of progradation and water immersion in the delta area, the regular arrangement of different types of sediments, plus the development of synsedimentary faults, can form various favorable source-reservoir-caprock combinations with good oil and gas conduction capacity, creating favorable conditions for full hydrocarbon expulsion and nearby accumulation. In continental basins, a large delta system often controls a large oil field and several oil fields and enrichment areas. For example, the giant delta system in the northern Songliao Plain, covering nearly ten thousand square kilometers, controls the formation of the Daqing Oilfield; the Yong'an Town Delta in the Dongying Depression controls the formation of the Shengtuo and Yong'an Town oil fields; and the four near-source deltas in the western slope zone of the Liaohe River control the enrichment areas of the Shuguang and Huanxi Oilfields, etc.

7.6.2.4 Faults Play a Significant Controlling Role in the Generation, Migration, and Accumulation of Oil and Gas

Practical oil and gas exploration has proven that the formation and distribution of oil and gas reservoirs in most petroliferous basins (especially faulted basins) are closely related to faults, with their formation and distribution largely controlled by faults. Faults play a role in controlling reservoir development and modifying reservoir properties, controlling the direction and timing of oil and gas migration, controlling the formation of different types of traps, and controlling the distribution of oil and gas reservoirs. Faults not only have an important controlling effect on the formation of hydrocarbon-generating centers in some types of basins but also serve as concentrated distribution zones for oil and gas reservoirs or fields. The formation and evolution of faults are coexistent with the formation and evolution of tectonic units such as basins, depressions, structural zones, and even traps. Therefore, the role of faults of different levels in oil and gas generation, migration, and accumulation varies.

First and second-order major faults are boundary faults that control the development of basins and depressions. They have large displacements, extend over long distances, and have a long duration of activity. The downthrown side of these faults forms deep depressions, which are favorable sites for oil and gas generation and accumulation. As the main channels for vertical oil and gas migration, hydrocarbons can migrate along these large faults and be redistributed, forming secondary large oil fields in the overlying strata. In faulted basins, large fault-controlling depressions are growth faults, which deposit and move simultaneously, forming deep depressions on the downthrown side, where the thickness of the strata significantly increases. For example, on the side of the fault in the Bohai Bay Basin, the Paleogene in the deep depression has the characteristics of rapid subsidence and thick deposition, with the maximum subsidence amplitude reaching 6,000–9,000 m. These deep depression areas are mainly characterized by deep lake-semi-deep lake subfacies, forming thick hydrocarbon source rocks. The thickness of the source rocks in the Dongying, Baxian, Raoyang, and western Liaohe depressions reaches more than 1,500–2,000 m. In the Bohai Bay Basin, the average proven geological reserves per square kilometer in the depressions controlled by growth faults reach 100,000–150,000 tons. Similarly, for the large thrust fault zones in foreland basins, they are also favorable sites for controlling oil and gas generation, migration, and accumulation. In the United States, the first oil field was discovered in the Rocky Mountain thrust fault zone in 1975, and subsequently, more than 20 oil and gas fields were found in the Cambrian and Ordovician of the thrust fault zone between Wyoming and Utah. The K-U thrust fault zone on the northwest margin of the Junggar Basin in China concentrates more than half of the oil and gas reserves of the basin, with both the source layers and oil-bearing layers dipping under the thrust fault zone.

Third-order faults are those that control the structural zones within depressions. They are important channels for oil and gas migration and have a significant controlling effect on vertical oil and gas migration, constraining the vertical distribution series of oil and gas along the faults. Generally, many third-order faults develop in a basin, but not all these faults can become conduits for upward oil and gas migration. Only the faults that connect the source rocks with the overlying reservoirs and are active during the period of intense hydrocarbon generation and expulsion from the source rocks, that is, the oil-source faults, can serve as conduits for oil and gas migration. The oil-source faults among the third-order faults are important factors controlling the enrichment of oil and gas in traps above the source rocks. Generally, the main oil-source faults have a long duration of activity, cut shallow stratigraphic levels vertically, and have significant activity during the accumulation period, which is very favorable for the upward conduction of oil and gas, leading to the enrichment of oil and gas above the source. On the other hand, there are also many faulted depressions with oil and gas enrichment above the source. Multiple oil-source faults develop planarly, with a high density of oil-source fault development, and the oil-source faults are distributed evenly in the plane, throughout various parts of the depression, effectively conducting oil and gas from all parts of the hydrocarbon-generating series. In the profile, the oil-source faults present various combinations such as cabbage-type and compound Y-type, forming a multi-level fault step conduction pattern in space. These fault systems connect the source rocks with the shallow reservoirs vertically, effectively linking the deep source rocks and shallow reservoirs, providing migration channels for the vertical migration of oil and gas, which is an important condition for the enrichment of oil and gas above the source. For example, the Ying 8 fault in the Dongxin area of the Dongying Depression controls the formation and evolution of the Dongxin structural zone, as well as the distribution of oil and gas (Fig. 7.29). The fault zone has strong activity, with the deepest part reaching the hydrocarbon-generating layer series of the lower submember of Sha 3, which is conducive to the upward migration of oil and gas, forming the characteristic of multi-series oil and gas reservoirs distributed along the fault and forming fault zone asphalt along the main oil-source fault (according to Jiang & Liu, 2010).

Fourth-order faults have a controlling effect on the local structures of each structural zone (such as traps). On one hand, they provide the conditions for the formation of traps, and on the other hand, they control the degree of oil and gas enrichment. During the sealing period, fourth-order faults can provide blocking conditions for oil and gas, forming fault block traps, fault-blocking traps, and composite traps such as

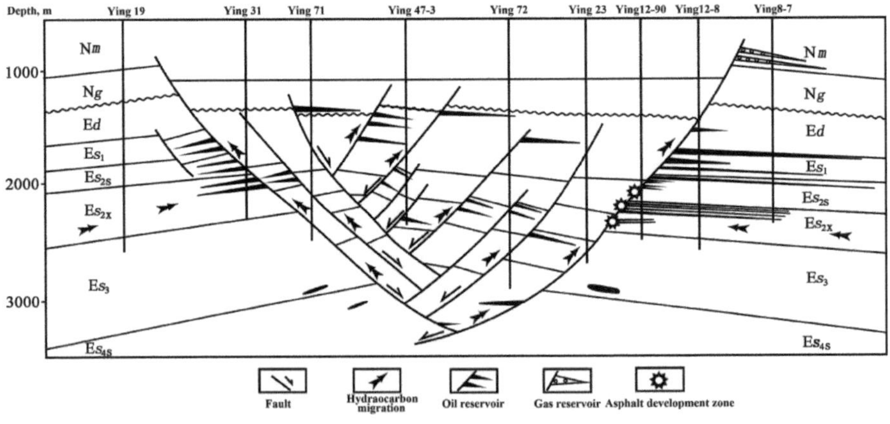

Fig. 7.29 Fault-controlled hydrocarbon accumulation pattern in the Dongxin area of the Dongying Depression (according to Jiang & Liu, 2010)

fault-lithological traps. When the faults are active, they can not only serve as channels for vertical oil and gas migration but also give rise to rollover anticlinal traps and inversion anticlinal traps (Lv et al., 2013). The amount of resources sealed in these fault-type traps depends on the sealing capacity of the fourth-order faults as blocking conditions.

7.6.2.5 Regional Cap Rocks Control the Vertical Distribution of Oil and Gas

Regional cap rocks can seal the entire oil and gas accumulation system from the source rock to the trap, therefore, regional cap rocks play an important controlling role in the enrichment levels of oil and gas. A large amount of practical oil and gas exploration has shown that almost all oil and gas fields are covered by thick dark mudstone, shale, or gypsum salt layers, forming good preservation conditions. These thick mudstones and shales often seal faults, with faults disappearing into the thick mudstone or with reduced displacement, preventing the vertical migration and redistribution of oil and gas. For example, the oil and gas in the Daqing Oilfield terminate below the thick mudstone of the Nenjiang Formation, and in the Sanzhao Depression, the thick hydrocarbon source rock of the Qingshan Kou Formation also serves as a cap rock, with oil and gas mainly migrating downward into the Fuyu oil layer.

Thick, widely distributed, and relatively stable regional cap rocks play an important role in the migration and accumulation of oil and gas in basins or depressions. They combine with hydrocarbons, reservoirs, and faults to jointly control the vertical distribution levels of oil and gas reservoirs. Exploration practice has shown that the reservoirs immediately adjacent to high-quality regional cap rocks are often the most enriched levels of oil and gas. In the Arabian sub-basin of the Persian Gulf Basin, the Upper Permian fractured carbonate reservoirs and the overlying Lower Triassic shale cap rocks containing gypsum, as well as the Upper Jurassic Arab Formation granular carbonate reservoirs and the overlying Hith gypsum salt cap rocks, belong to this type of reservoir-cap rock combination. Among them, the natural gas reserves stored in the Upper Permian account for 78.8% of the total natural gas reserves, and the oil reserves stored in the Upper Jurassic account for 33.6% of the total oil reserves. The Tanan Depression develops two sets of regional cap rocks of argillaceous rocks in the upper part of the Nan 1 member and the Da 1 member, which are thick, with a high mud-to-ground ratio, controlling the enrichment levels of oil and gas (Fig. 7.30). In addition, exploration practice has proven that the accumulation and preservation of oil and gas in tectonically active areas require high-quality gypsum salt cap rocks.

The above controlling factors for oil and gas distribution are proposed based on conventional oil and gas exploration. With the increasing depth of unconventional oil and gas exploration, it has been gradually recognized that the factors affecting the distribution of unconventional oil and gas are different. For example, the abundance of organic matter and the degree of thermal evolution are the main controlling factors for the enrichment of shale oil and gas reservoirs, and the brittleness coefficient of

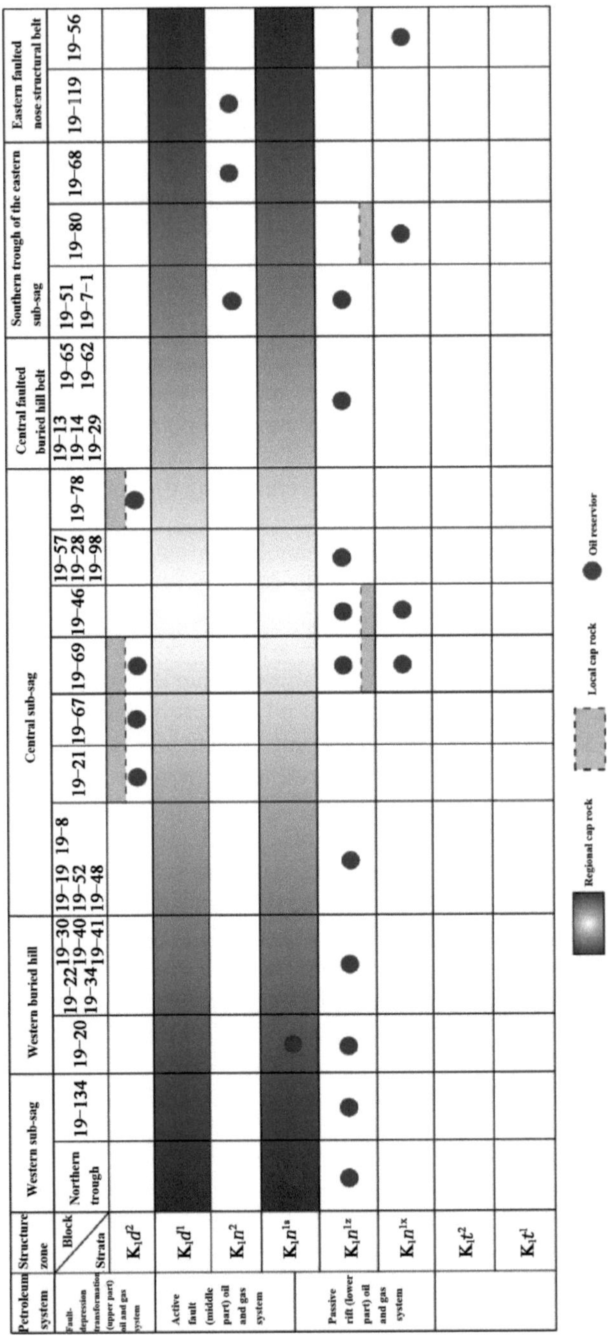

Fig. 7.30 Relationship between cap rocks and oil and gas distribution in the Tanan Depression (according to Fu et al., 2011)

rocks is the key to exploration and development; the widely covered effective mature hydrocarbon source rocks and the source-reservoir integration or close contact are the main controlling factors for the distribution of tight rock oil and gas.

Questions for Discussion

1. The basic concepts of oil and gas fields, oil and gas accumulation zones, oil and gas bearing areas, and oil and gas basins, as well as their interconnections.
2. What is an oil and gas basin? What are the main classification methods?
3. What are the main petroleum geological characteristics of foreland basins, cratonic basins, and rift basins?
4. What is an oil and gas system? What are the main contents of oil and gas system research?
5. What is an oil and gas accumulation system? What is the difference from an oil and gas system?
6. What is a total petroleum system? What is the sequence of unconventional and conventional oil and gas distribution?
7. What are the main characteristics of oil and gas resource distribution in the world and in our country?
8. What are the main factors controlling the macro distribution of oil and gas?
9. What are the main geological factors controlling the distribution of oil and gas within a basin?

References

Bai, G. (2007). Preliminary exploration of the main controlling factors of oil and gas distribution in the Persian Gulf Basin. *Journal of China University of Petroleum (Natural Science Edition)*, (3), 28–32+38.

Beydoun, Z. R., Clarke, M. H., & Stoneley, R. (1992). Petroleum in the Zagros Basin: a late Tertuary foreland basin overprinted onto the outer edge of a vast hydrocarbon-rich Paleozoic passive-margin shelf. In: Macqueen, R. W., Leckie, D. A., (eds.) Foreland basins and fold belts. *AAPG Memoir, 55*, 309–339.

Dickinson, W. R. (1976). Plate tectonic evolution of sedimentary basins. American Association of Petroleum Geologists Continuing Education Course Notes, *1*, 1–56.

Fu, X., Wang, Y., Qu, Y., et al. (2011). Analysis of the distribution patterns and main controlling factors of oil and gas in passive rift basins: Taking the Tanan depression in the Tamuchaig Basin as an example. *Geological Science, 46*(4), 1119–1131.

He, S., Ye, J., Xu, S., et al. (2010). *Petroleum and natural gas geology*. China University of Geosciences Press.

Hu, C. (1982). Source bed controls hydrocarbon habitat in continental basins, east China. *Acta Petrolei Sinica. 2*, 9–13.

Hu, J., Huang, D., et al. (1991). *Theoretical basis of continental petroleum geology in China.* Petroleum Industry Press.

Hu, C., & Liao, X. (1996). The concept of petroleum system and its application in China. *Journal of Petroleum Science, 17*(1), 10–16.

Jia, C., Pang, X., Guo, Q., et al. (2023). Evaluating oil and gas resources based on the genetic method: The theory of the whole oil and gas system and the new-generation basin simulation technology. *Journal of Petroleum Science, 44*(9), 1399–1416.

Jiang, Y., & Liu, H. (2010). Fracture asphalt belt and its geological significance for oil and gas. *Journal of Petroleum Science, 31*(1), 36–41.

Jiang, Y., Su, S., Liu, H., et al. (2021). Differences in oil and gas accumulation periods and dominant controlling factors in the Bohai Bay Basin. *Petroleum and Natural Gas Geology, 42*(6), 1255–1264.

Jiang, Y., Tan, L., Rong, Q., et al. (2002). The concept of petroleum accumulation system and its application in the Boxin area of the Dongying depression. *Journal of Petroleum University (Natural Science Edition), 26*(6), 12–16.

Levorsen, A. I. (1967). *Geology of Petroleum.* San Francisco: W. H. Freeman and Company.

Li, G., & Tang, Y. (2002). *Atlas of oil and gas basins in China* (2nd ed.). Petroleum Industry Press.

Liu, H. (1993). Geodynamic classification of sedimentary basins and analysis of tectonic styles. *Earth Science, 18*(6), 699–724.

Lv, Y., Fu, G., Fu, X., et al. (2013). *The conduction and sealing of oil and gas by faults.* Petroleum Industry Press.

Ma, X., Dou, L., & Shi, B. (2021). *Global oil and gas resource potential and distribution.* Petroleum Industry Press.

Magoon, L. B., & Dow, W. G. (1994). The petroleum system: From source to trap. *AAPG Memoir, 60*, 3–24.

Qi, J., Zhang, Y., Lu, K., et al. (1995). Extensional patterns and dynamic processes of the Cenozoic Rift Basin in the Bohai Bay. *Petroleum Experimental Geology, 4*, 316–323.

Tian, Z., Shi, B., Luo, P., et al. (2002). Fields for further exploration in high-exploration-degree areas of the composite oil and gas accumulation belt in the Bohai Bay Basin. *Journal of Petroleum Science,* (03), 1–5+9.

Yang, X., Tian, J., Wang, Q., et al. (2021). Geological understanding of ultra-deep oil and gas in the Tarim Basin and favorable exploration areas. *China Petroleum Exploration, 26*(4), 17–28.

Zhang H., Fang C., Gao X., et al. (1999). *Petroleum geology* (3rd Ed). Beijing: Petroleum Industry Press.

Zhang, M., Bai, G., Pan, L., et al. (2013). Comparative analysis of oil and gas distribution characteristics in the world's three major oil-rich foreland basins. *Modern Geology, 27*(05), 1233–1243.

Zhang, S., et al. (2004). *The generation of marine oil and gas in the Tarim Basin.* Petroleum Industry Press.

Zhao, W., Zou, C., Wang, Z., et al. (2004). The theory of "full basin oil and gas" in oil-rich depressions: Connotation and significance. *Petroleum Exploration and Development, 31*(2), 130–142.

Zhao, X., Jin, F., Wang, Q., et al. (2011). The theory of oil accumulation in the depressions of continental rift basins and its application: Taking the Jizhong depression in the Bohai Bay Basin and the Erlian Basin as examples. *Journal of Petroleum Science, 32*(1), 18–24.

MIX
Papier aus verantwortungsvollen Quellen
Paper from responsible sources
FSC® C105338

If you have any concerns about our products,
you can contact us on
ProductSafety@springernature.com

In case Publisher is established outside the EU,
the EU authorized representative is:
Springer Nature Customer Service Center GmbH
Europaplatz 3, 69115 Heidelberg, Germany

Printed by Libri Plureos GmbH
in Hamburg, Germany